Green Energy and Technology

Climate change, environmental impact and the limited natural resources urge scientific research and novel technical solutions. The monograph series Green Energy and Technology serves as a publishing platform for scientific and technological approaches to "green"—i.e. environmentally friendly and sustainable—technologies. While a focus lies on energy and power supply, it also covers "green" solutions in industrial engineering and engineering design. Green Energy and Technology addresses researchers, advanced students, technical consultants as well as decision makers in industries and politics. Hence, the level of presentation spans from instructional to highly technical. **Indexed in Scopus**.

More information about this series at http://www.springer.com/series/8059

Ashwani K. Gupta · Ashoke De ·
Suresh K. Aggarwal · Abhijit Kushari ·
Akshai Runchal
Editors

Innovations in Sustainable Energy and Cleaner Environment

Springer

Editors
Ashwani K. Gupta
Department of Mechanical Engineering
University of Maryland
College Park, MD, USA

Ashoke De
Department of Aerospace Engineering
Indian Institute of Technology Kanpur
Kanpur, Uttar Pradesh, India

Suresh K. Aggarwal
Department of Mechanical Engineering
University of Illinois at Chicago
Chicago, IL, USA

Abhijit Kushari
Department of Aerospace Engineering
Indian Institute of Technology Kanpur
Kanpur, Uttar Pradesh, India

Akshai Runchal
Analytic & Computational Research, Inc.
(ACRi), The CFD Innovators
Los Angeles, CA, USA

ISSN 1865-3529 ISSN 1865-3537 (electronic)
Green Energy and Technology
ISBN 978-981-13-9011-1 ISBN 978-981-13-9012-8 (eBook)
https://doi.org/10.1007/978-981-13-9012-8

This Springer imprint is published by the registered company Springer Nature Singapore Pte Ltd.
The registered company address is: 152 Beach Road, #21-01/04 Gateway East, Singapore 189721, Singapore

Local Organization Team and Message from the Director, NIT-Kurukshetra

Local Organization Team

Dr. Satish Kumar, Director NIT-Kurukshetra
Dr. Pankaj Chandna
Dr. Vinod Mittal
Dr. Gulshan Sachdeva

Message from the Director, NIT-Kurukshetra

In my professional career which spans over three and a half decades, I was associated with the aerospace industry for a significant period in the fields of missiles, strategic systems, and hypersonic propulsion. Power and propulsion play a very important role in the fields of aviation, defense, etc., and there is a pressing need to amalgamate these areas with the discipline of sustainable energy to produce technology which is clean and green, which would equally serve the present needs and tend to the future requirements where the conventional sources of energy may be extinct. I say it with extreme pride that the National Institute of Technology (NIT) Kurukshetra and Terminal Ballistic Research Laboratory (TBRL), Chandigarh, are jointly organizing an International Workshop on Sustainable Energy, Power and Propulsion (ISEPP-2018) which is ninth in a series of workshops on the broad theme of energy, power, and propulsion that started in 2004. Over the years, this workshop grew into a world-class forum enquiring into the status of sustainable developments in power and propulsion. This interaction of the workshop is in collaboration with Indian Institute of Technology (IIT) Kanpur; University of Maryland, College Park; University of Illinois at Chicago; Analytic & Computational Research Inc., USA. The focal point of this workshop lies at the general theme of clean energy production and utilization. Over 150 distinguished delegates from all over the globe have been invited to give their contributions in the fields of propulsion, pulse detonation, environmental sustainability, combustion, emissions, etc., to name a few. A paper presentation competition for students is also being organized to bring the young talent into limelight. This attempt at the sustainable developments in the areas of power and propulsion is definitely one of its kinds. I wish everyone associated with this workshop rich research experiences, and I am very confident that at the end of this workshop, we all would move a step closer towards our understanding to have a more cleaner and sustainable environment.

Preface

There is worldwide interest in developing renewable and alternative energy sources in a sustainable manner. This is motivated by our excessive reliance on finite fossil energy sources, concerns about greenhouse gas emissions and climate change, and our growing energy needs especially due to emerging economies and population growth. As an example, the world population is estimated to grow approximately by 83 million people each year, reaching approximately 8.6 billion by 2030 and 9.8 billion by 2050. In particular, the middle class is expected to grow from 3.2 billion in 2016 to 5.2 billion by 2028. Per IEA, the energy consumption, which correlates well with population growth and prosperity, is expected to increase from 549 quadrillion BTU in 2012 to 815 quadrillion BTU by the year 2040. While a favorable shift to renewable energy is making great strides in the total energy mix, it is still highly tilted toward fossil fuels, which are expected to provide about 78% of the world's energy by 2040. This scenario of threatening energy security and dangers posed by climate change highlights many challenges and opportunities for various stakeholders in the energy sector, including governments, industry, and academia.

During the last three decades, there have been major advances in the development of renewable and alternative energy sources, such as solar PV, wind, biomass, algae, and hydro. According to IEA, renewable sources currently contribute over 13% to the total worldwide energy consumption. Moreover, there is a worldwide effort to increase renewable contribution at an accelerating pace in the coming years. We have also seen remarkable improvements in the performance of fossil fuel power and propulsion systems, in terms of both fuel consumption and emission reduction. For instance, NOx and PM emissions from vehicles in the USA have been reduced by approximately 95% during the last two decades. Simultaneously, remarkable progress has been reported in our analytical, computational, and diagnostic capabilities. From its modest beginnings in the 1960s with the advent of the electronic computer, computational sciences have revolutionized our analysis and design capabilities, and computational fluid dynamics (CFD) tools have now become ubiquitous in academia and industry. Concurrently, advanced noninvasive diagnostic techniques continue to be developed, providing spatially and temporally resolved measurements at higher frequency in complex non-reacting and reacting flows. In addition, major advances in sensors and active control technology are enabling us to further improve system efficiency, reduce pollutants emission, prolong equipment life, and mitigate catastrophic failures. In spite of these major advances in our predictive and diagnostic capabilities for designing more efficient and cleaner energy systems, challenges still remain due to growing energy demand, fuel flexibility, and environmental concerns, as the world GDP is expected to double and energy consumption to increase by some 50% over the next 25 years. Hypersonics is another area that has gained significant momentum in recent years. We expect this effort to continue to grow at a faster pace that is expected to bring faster and safer travel over long hauls so that one can travel from New York to Sydney in just two hours. It is through the dedicated research and development efforts and international networking that such vision can be nurtured.

In order to provide a forum to discuss these challenges and promote international collaboration, we initiated a workshop series-International Workshop on Sustainable Energy, Power, and Propulsion (ISEPP)-in 2004. The last ninth workshop in this series was held at the National Institute of Technology (NIT) Kurukshetra, India, during March 18–22, 2018. The Office of Naval Research (ONR) and National Science Foundation in the USA, and Defense Research and Development Organization (DRDO), the Department of Science and Technology (DST), and World Bank-sponsored Technical Education Quality Improvement Programme (TEQIP) from India sponsored the workshop. In addition, the workshop was co-sponsored by several leading research organizations, universities, and industries from the USA and India. Many world-renowned scientists and researchers from the USA, UK, India, Japan, Thailand, France, Italy, Brazil, and Malaysia presented their latest research and development findings and shared their ideas in formal talks and exchange forums. This research monograph brings together the latest research and development efforts from the wealth of knowledge presented by the eminent scientists and engineers at the workshop. A common theme of the monograph is energy, power, and propulsion. The chapters are grouped together into parts that deal with various aspects of this theme. The first part contains chapters dealing with high-speed propulsion including hypersonics. Chapters on renewable fuels and alternative energy sources focus on more efficient and environmental-friendly techniques and systems for power and propulsion. Dynamics and stability of flames in combustion systems is a major concern. Some chapters present novel ideas on several of the topics discussed at the workshop. Sustainable energy technologies provide novel methods of harnessing biofuels as well as their challenges and opportunities. Finally, the gap in our knowledge and understanding of physical processes that underlie computational simulations and analysis are covered in chapters dealing with transport phenomenon, turbulence, advanced cooling technologies, and ignition processes.

College Park, USA — Ashwani K. Gupta
Kanpur, India — Ashoke De
Chicago, USA — Suresh K. Aggarwal
Kanpur, India — Abhijit Kushari
Los Angeles, USA — Akshai Runchal

Contents

About the Editors

Dr. Ashwani K. Gupta is a Distinguished University Professor at the University of Maryland (UMD). He obtained his Ph.D. and higher doctorate (D.Sc.) from the University of Sheffield, was awarded a D.Sc. from Southampton University and has received honorary doctorates from the University of Wisconsin Milwaukee, King Mungkut University of Technology North Bangkok, and University of Derby, UK. He received many honors and awards from AIAA and ASME and the President Kirwan Research Award and College of Engineering Research Award at UMD. Professor Gupta has authored over 750 papers, 3 books, and edited 12 books in the areas of combustion, swirl flows, high temperature air combustion (HiTAC), distributed combustion, waste to energy, acid gas treatment, fuel reforming, and air pollution. He is Honorary Fellow of ASME and Fellow of AIAA, SAE, AAAS and RAeS (UK).

Dr. Ashoke De is an Associate Professor at the Department of Aerospace Engineering, Indian Institute of Technology Kanpur. He has been the recipient of the Humboldt Research fellowship for Experienced Researchers, IEI-Young Engineer's Award-2014, DST Young Scientist award-2015, and P. K. Kelkar Research Fellowship from IIT Kanpur. He is a member of ASME, SIAM, FMFP, ISHMT and Combustion Institute. He received his M.S. from IIT Kanpur, and Ph.D. from Louisiana State University. He has authored over 90 peer reviewed papers. His research interests include combustion modeling, hybrid RANS/LES model development, supersonic flows and Fluid-Structure interactions (FSI) with focus on computational mechanics in combustion and turbulent flows.

Dr. Suresh K. Aggarwal received his Ph.D. from Georgia Institute of Technology. He was a member of research staff at Princeton University and Senior Research Engineer at CMU. He then joined the University of Illinois at Chicago, and was promoted to Professor in 1995. He served as Director of Graduate Studies, and was also Visiting Scientist at ANL, Visiting Professor at Ecole Centrale-Paris, France and Guest Professor at Jiangsu University, China. His research interests include combustion, multiphase reacting flows, emissions, clean energy, and fire suppression. He has authored over 360 publications and is a Fellow of ASME, AAAS and Associate Fellow of AIAA.

Dr. Abhijit Kushari is a Professor at the Department of Aerospace Engineering, Indian Institute of Technology Kanpur. Dr. Kushari received his Ph.D. from Georgia Institute of Technology, in 2000 and joined the IIT Kanpur as an Assistant Professor in 2001. He became a Professor in 2014. His research interests are aerospace propulsion, gas turbine engines, turbo-machinery, liquid atomization flow control and combustion dynamics. He has authored more than 130 technical papers.

Dr. Akshai Runchal is the founder and Director of CFD Virtual Reality Institute and ACRI group of companies. His expertise is in Computational Fluid Dynamics (CFD). He has consulted widely on flow, heat and mass transfer, combustion, environmental impact, hazardous and nuclear waste, ground water and decision analysis to over 200 clients in 20 countries. He is the principal author of PORFLOW®, TIDAL®, ANSWER® and RADM™ simulation models that are used worldwide. He received his Ph.D. from Imperial College (London), and Bachelor's in Engineering from PEC. He has taught in the USA, India and the UK. He has authored 7 books and over 200 technical publications. He has received many honors and awards, and given many invited contributions at conferences.

Introduction

This research monograph provides state-of-the-art advances in areas of energy, power, and propulsion using renewable and non-renewable fuels, their efficiency and performance improvement, and their environmental impact. Special attention is given to renewable and alternative energy resources as well as energy sustainability. This book is written by internationally renowned experts from and around the globe on specific topics of current interest to students, researchers, and engineers from academia, industry and government labs to provide the latest innovations in cleaner energy utilization for a wide range of applications. Novel developments in the areas of renewable energy resources are also presented along with thermal management, emission control, energy efficient buildings and environmental issues. The energy and environment sustainability requires a multipronged approach involving development and utilization of new and renewable fuels, design of fuel-flexible combustion systems, and novel and environmentally friendly technologies for improved fuel use. This monograph will serve as a reference source for practicing engineers, educators, and research professionals for all energy using sectors.

Meeting global energy demand in a sustainable manner continues to remain among the major challenges of the twenty-first century. The word "sustainable" implies providing energy security, addressing climate change concerns caused by greenhouse gas (GHG) emissions, and providing energy in a carbon-neutral manner. The global energy demand continues to rise and is expected to increase by 48%, as the population grows from 7 to 9 billion by the year 2040. Most of this growth is expected to occur in developing nations, with non-OECD countries contributing some 71% of this increase. The world GDP (expressed in purchasing power parity) is a key determinant of growth in energy demand and expected to grow by 3.3% per year during this period. Fossil fuels have been the dominant source of global energy and thus of GHG emissions. They will continue to play a similar role in the foreseeable future and are estimated to provide some 78% of energy demand by 2040. However, challenges posed by climate change and growing energy needs have led to increasing global effort to develop alternative and renewable energy resources.

During the last three decades, we have seen remarkable progress in our analytical and diagnostic capabilities. Major strides have been made in the computational fluid dynamics (CFD) simulations of reacting and non-reacting flows, as well as in non-intrusive diagnostic capabilities. As a consequence, both computational and diagnostic tools are increasingly being used to analyze, test, modify, develop, and improve energy conversion devices, and methodologies, and to design novel concepts and products at both academia and industry. Rapid advances in computational power, numerical algorithms, physical models, turbulence models, and detailed reaction mechanisms for a variety of fuels, as well as advanced diagnostics, are further accelerating the pace of development of more efficient and less polluting energy and propulsion systems. Significant developments have also occurred in computational techniques and simulations of complex flows in realistic geometries. Examples include shock-capturing techniques for supersonic flows, and various methodologies, based on Reynolds-averaged Navier–Stokes (RANS), large eddy simulation (LES), and direct numerical simulations (DNS), for turbulent reacting flows. Concurrently, new noninvasive diagnostic techniques are becoming firmly established with the development of advanced instrumentation that provides spatial and temporally resolved measurements at high frequency for use in complex flows at microscales. Also, in parallel, advances in sensors and active and passive control technology have further helped to improve system efficiency, reduce emissions, and mitigate catastrophic failures. In spite of these major advances on many fronts, challenges still remain due to growing demands on energy and dangers posed by climate change. We anticipate that the research and development efforts will continue to grow nationally and internationally, so that the engineers can materialize even greater efficiency from the existing and new combustion devices. We also anticipate that broadening of the overall energy portfolio by increasing the contributions from all kinds of energy resources will help energy sustainability to foster cleaner efficient use from all fossil and renewable energy sources.

In order to provide a global perspective on the advances and challenges outlined above, we have been organizing international workshops in India on the broad theme of energy, power, and propulsion, since 2004. To date, we have organized nine international workshops, which have been sponsored by various agencies, universities, and industries in the USA and India. The last workshop, "International Workshop on Sustainable Energy, Power, and Propulsion," was held at NIT Kurukshetra, India, during March 18–22, 2018, and was sponsored by NSF, ONR Global, DRDO, DST, and TEQIP organizations in India. It was attended by many world-renowned scientists and researchers from the USA, UK, Europe, India, Thailand, Malaysia, and Japan who presented their latest research findings and shared their ideas in formal talks and exchange forums. This research monograph brings together the latest research presented by the world-renowned scientists and engineers at this workshop. A common theme of the monograph was energy, power, propulsion, hypersonics, and alternative and renewable fuels.

The monograph is divided into six parts dealing with various aspects of this theme. Part I contains chapters on high-speed propulsion systems for improved efficiency and better design using numerical and experimental techniques. Pressure gain combustion devices, cooling techniques for reentry vehicles, ignition characteristics of rocket engines, and drag reduction of projectiles continue to be of major concern. Part II comprises review chapters dealing with the production of renewable fuels from various sources and their utilization in transportation and power generation systems. Chapters in Part III and Part IV provide reviews of current research dealing with various aspects of turbulence, including the modeling of turbulence–chemistry interactions, and simulations of turbulent reacting flows, as well as two-phase flows in flames and reacting systems. Chapters in Part V provides reviews of current research dealing with various aspects of transport processes in energy systems. The last Part VI comprises a review chapter on biofuel energy: past, present, and future as well as chapters that deal with several specific aspects of sustainable energy technologies. Advances in solar power generation using state-of-the-art very-high-efficiency multi-junction solar PV cells are included that have demonstrated the efficiency of over 44%. We conjecture the area of PV to grow even further with cost-effective fuel cells from higher efficiency and use of concentrated solar in PV applications for harnessing greater share of solar energy in the energy mix in power generation. Each chapter is self-contained with a review of the latest research and comprehensive references on each topic. This monograph is expected to be of special interest to researchers and engineers as well as graduate students working in the field of energy, renewable biofuels, alternative energy, combustion, power, propulsion, hypersonics, and environmental issues. It will also serve as supplementary reading material to decision makers from government industry, in addition to providing cutting-edge research and development technologies on the various topics presented herein.

Ashwani K. Gupta
Ashoke De
Suresh K. Aggarwal
Abhijit Kushari
Akshai Runchal

Part I
High-Speed Propulsion

Recent Developments in the Research on Pressure-Gain Combustion Devices

Kazhikathra Kailasanath

Keywords Pressure-gain combustion · Detonations · PDE · RDE

1 Introduction

Conventional propulsion systems in use today are based primarily on some variants of the constant-pressure Brayton cycle. Thermodynamically, energy release under constant-volume conditions is much more efficient than under constant-pressure conditions. There is a pressure increase or pressure gain during the combustion process, in addition to the increase in temperature. Detonations provide a practical way of attaining heat addition under near-constant-volume conditions.

During the past seventy years or so, there have been numerous research efforts at harnessing the potential of detonations for propulsion applications [1]. Since the early 1990s, there has been a renewed interest in intermittent or pulsed detonation engines [2, 3]. The basic theory, design concepts and work in the early 1990s related to pulse detonation engines have been discussed by Bussing and Pappas [4]. Reviews of specific aspects such as performance estimates and nozzles for pulse detonation combustion systems have also been presented [5, 6]. A comprehensive description of the challenges is involved, and the status then was published in 2004 [7]. More recent updates on the status are also available [8, 9]. The primary focus of the studies in the 1990s and early 2000s has been on the "pulsed" detonation engine (PDE) concept. While the PDE concept has matured significantly, there is the potential to increase further the substantial performance gains made by the PDE concept over the traditional constant-pressure cycle-based propulsion engines by moving from

K. Kailasanath (✉)
Lorton, Virginia, USA
e-mail: kkailasanath@gmail.com

A. K. Gupta et al. (eds.), *Innovations in Sustainable Energy and Cleaner Environment*, Green Energy and Technology,
https://doi.org/10.1007/978-981-13-9012-8_1

"pulsed" or "intermittent" detonations to a more "continuous" mode of operation with detonations. The rotating detonation engine (RDE) concept [10–13] is an example of a continuous detonation engine. The objective of this paper is to first present some fundamentals of detonation engines and then provide an update of the previous reviews, focusing on the more recent developments in the research and development of detonation-based pressure-gain combustion systems. The review will be restricted to work openly available in the literature but includes ongoing efforts from around the world. While an attempt is made to cover a broad range of the reported researches, the sheer volume of papers published makes it impractical to be exhaustive.

2 Fundamentals

Before discussing specific devices and their performances, the basic case for using detonations for propulsion and power applications is examined. Then, some of the early attempts are discussed with their potential shortcomings. Finally, a basic laboratory-scale device is described along with experiments and simulations to establish the ground-based performance of this idealized device, considered representative of a pulse detonation engine.

2.1 Why Detonations for Propulsion and Power?

As mentioned before, very rapid material and energy conversion is a key feature of detonations. This rapid "burning" or material conversion rate, typically tens of thousands of times faster than in a flame, can lead to several advantages for propulsion such as more compact and efficient systems. Because of the rapidity of the process, there is not enough time for pressure equilibration and the overall process is thermodynamically closer to a constant-volume process than the constant-pressure process typical of conventional propulsion systems. To illustrate this point, three idealized thermodynamic cycles are compared in Fig. 1.

For purposes of comparison, the only process that is different in the three cycles is the energy conversion or heat addition. For the three cases, heat is added at constant pressure, constant volume or in a detonation. Hence, the three cycles have been referred to as "constant pressure", "constant volume" and "detonation" cycles, respectively. The amount of heat added is also kept the same at 50 kcal/mol (a value typical of hydrocarbon fuels) for the three cycles. In all cases, the fuel–air mixture is initially compressed adiabatically from 1 to 3 atm before heat addition. After heat addition, the products of combustion are expanded adiabatically to 1 atm. Finally, the system is returned to its initial state. The work done during the three cycles is obtained from the area enclosed (Fig. 1). Since all the processes except for heat addition have been maintained the same, the work done or relative thermodynamic efficiency of the three combustion processes can be obtained by comparing the three

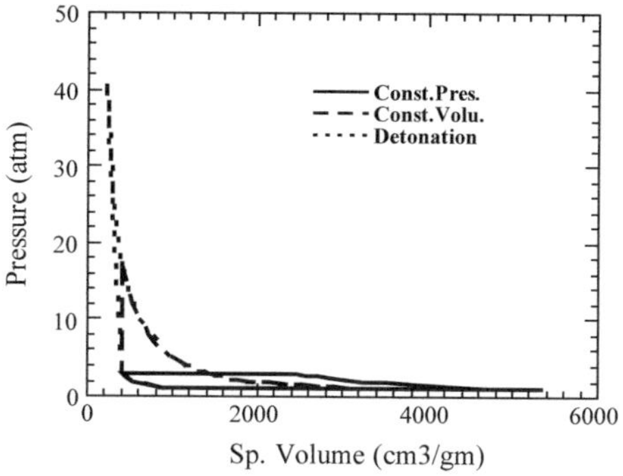

Fig. 1 Comparison of idealized thermodynamic cycles for constant pressure, constant volume and detonation modes of combustion

areas. For the efficiency, the work output is divided by the heat input which was set to be the same for the three cycles. The thermodynamic efficiencies for the three cycles are as follows: 27% for constant pressure, 47% for constant volume and 49% for detonation. From the figure and the above numbers, we see that the thermodynamic efficiency of the detonation cycle is close to that of the constant-volume cycle. The process itself is different with a decrease in a specific volume and a significantly higher pressure being attained during detonations.

2.2 *Early Research*

With the case being made that the key advantage of a detonation cycle over the Brayton cycle is on the basis of thermodynamics, one wonders why it has not been implemented in practice earlier. A historical survey [1] clearly shows that the fundamental advantage has been known for a long time but the lack of successfully operating devices has been due to the various technical challenges that must be overcome to translate the thermodynamic efficiency into propulsive efficiency with acceptable losses. A brief survey of the early research is discussed next.

Specific reference to propulsion applications of detonations appears in the literature as early as the 1940s [14, 15]. Even at this early time, both standing (or stabilized) and unsteady (or intermittent) detonations were explored. In the work of Hoffmann [14], both gaseous (acetylene) and liquid (benzene) hydrocarbon fuels were employed with oxygen. Intermittent detonation appears to have been achieved, but attempts to determine an optimum cycle frequency were less successful. The development of the concept of pulse detonation engines (PDEs) has been traced back to this pioneering work in a number of papers. The proposals of Roy [15] inspired further work in France on the design of systems to stabilize combustion in supersonic flows (e.g., [16]). Soon, the work was also begun in the USA [17–23] and Russia [24, 25]. Bitondo and Bollay [17] conducted an analytical study that indicated that a pulsating detonation engine can be helpful for helicopter propulsion. Gross [18], Gross and Chinitz [19] and Nicholls et al. [20–23] studied means to stabilize detonation waves with a view of applications to hypersonic ramjet propulsion. The possibility of using oblique detonations was also introduced. Theoretical performance comparable to conventional ramjets was reported for appreciably higher flight Mach numbers. However, the laboratory development and demonstrations were not as successful with achieved performance not significantly better than current deflagration-based combustion devices [26]. In hindsight, one of the problems appears to have been initiating detonations in a reliable and efficient manner.

3 Pulse Detonation Engines (PDEs)

3.1 Key Processes in an Idealized Pulse Detonation Device

An idealized PDE consists of a tube closed at one end and open at the other. Although this is often referred to in the literature as an "engine", one must remember it is really an idealized representation of the combustor and additional considerations such as inlets and nozzles must be considered before even a simple representative propulsion device is achieved. With this caveat, let us first look at the key processes that take place in such a combustion device. The various processes in a PDE are illustrated in Fig. 2 using the example of the idealized device. The various colours in the figure correspond to different levels of pressure with red being the highest (30 atm) and the violet (or purple) being the lowest (1 atm). The tube is initially filled with a fuel–air mixture at 1 atm. A shock wave is initiated at the closed end (left end, in the figure). Chemical reactions in the shock-heated mixture generate pressure waves which couple with the shock wave to form a detonation. After a brief transition period, the detonation travels towards the open end of the tube at a nearly constant velocity called the Chapman–Jouguet (CJ) velocity. This CJ velocity and the corresponding pressure behind the detonation wave (CJ pressure) are characteristics

Fig. 2 Process in a pulse detonation engine cycle

of the fuel–air mixture. The detonation initiation process also generates expansion waves that travel towards the closed end (called head end) of the tube.

As the detonation leaves the tube, additional expansion waves form at the open end and travel towards the closed end. These expansion waves are important because they help evacuate the tube. Once the tube is evacuated, fresh fuel and air are injected into the tube. When the injection and mixing are completed, a detonation is initiated again at the closed end and the whole process is repeated. The cyclical nature of the entire process is what gives rise to the term, pulsed detonation engine (PDE). These brief descriptions of the processes that must take place within a PDE also serve to highlight some of the key difficulties. First of all, the fuel and air must be rapidly injected and mixed to achieve a near-uniform detonable mixture before initiation is attempted. The ignition must result in a detonation wave rather than a shock wave, followed by an uncoupled flame at some significant distance behind it. Once the detonation leaves the tube, the burnt gases must be evacuated rapidly enough so that the tube can be refilled with a fresh detonable mixture and the processes repeated. Valving, fuel–air mixing and reliable and repeatable low-energy detonation initiation all appear to be important factors.

3.2 Propulsive Performance of the Idealized Pulse Detonation Device

At first glance, it should be easy to estimate the idealized performance from the operation of such a device. The propulsive thrust and other measures of the performance of such an idealized device can be determined from either the momentum of the gases flowing out of the tube or the pressure at the closed end of the tube. At the turn of the century, a survey of performance estimates presented in the literature [27] showed that even for a tube filled with a premixed stoichiometric hydrogen–air mixture, there was a remarkable variation in the computed performance ranging from 3000 to 8000 s. Detailed numerical simulations [28] conducted to resolve this discrepancy have shown that most of this variation can be explained on the basis of the initial conditions and boundary conditions used in various computational studies. The performance predicted by these detailed simulations for the stoichiometric hydrogen–air mixture at 1 atmosphere static condition was about 4100 s. It is noteworthy to compare this number to that achieved in the early laboratory studies of about 2000 s [20].

Validation of performance estimates. The thrust from the idealized PDE can be obtained experimentally either directly using a thrust balance or from integrating the pressure history at the head end of the tube. Using the pressure history provides a direct means of comparing the experimental data to the results of numerical predictions. Thrust balance measurements provide an additional check on the calculated thrust or impulse. The equivalence of the two methods has been shown experimentally by Cooper et al. [29]. Their study also showed that when DDT enhancement devices such as obstacles were used, the head-end pressure was no longer a reliable measure of the performance and direct measurements were needed.

In performing direct comparisons between the predictions of numerical simulations and experimental data, the effects of factors such as the initiators used and the boundary conditions at the open end of the tube must be considered. The difficulties in isolating the effects of detonation initiators or transition devices can be reduced by focusing on easily detonable mixtures such as acetylene–oxygen or ethylene–oxygen. The ambiguity with specifying the open boundary conditions can be removed by conducting multidimensional simulations in which the exit plane of the thrust tube is an interior point. These simulations are expensive because the regions outside the tube (such as a dump tank) also need to be included in the computational domain. In addition to providing information for direct comparison with experimental data, the results from these simulations can be used to determine better approximations for open boundary conditions for use in one-dimensional simulations. Such two-dimensional and one-dimensional simulations have been carried out [28, 30] and their results compared to the measured head-end pressure history [31] from experiments at Stanford University for a stoichiometric C_2H_4–O_2 mixture. The very good agreement in the pressure histories also results in very good agreements in the impulses and specific impulses estimated from the pressure histories. The impulse calculated from the pressure history for the 1.35 m long tube is 2807 and 2820 N s/m^2 for the two- and one-dimensional simulations, respectively. These impulses correspond to 2079–2089 N s/m^3 for the impulse per unit volume and that is in excellent agreement with the estimates of about 2100 N s/m^3 from a different set of experiments at Cal Tech [29]. In summary, good estimates for the propulsive performance measures of a stoichiometric ethylene–oxygen mixture initially at 1 atmosphere are about 2100 N s/m^3 for the impulse (independent of length), about 163–168 s for the mixture-based Isp and 725–745 s for the fuel-based Isp. With further advances in experimental methodology, measurements of the performance using hydrogen–air mixtures have also confirmed the computational prediction of 4100 s by different laboratories around the world [32].

3.3 Comparison of Key Physical Parameters

The very good agreement between the computed and measured performance suggests that the underlying models used in the detailed simulations have achieved a sufficient level of maturity that they can be relied upon when sufficient reliable experimental data is not available. This has been further verified by directly comparing various physical parameters such as pressure, temperature and velocity between experiments and simulations. Experimental data [33] on the time variation of the pressure and fluid velocity at 0.9 L from the head end of a 160 cm (L) long tube filled initially with a stoichiometric ethylene–oxygen mixture is shown in Fig. 3, along with the computed values. The peak values as well as the variations in the overall shape of the pressure and velocity are captured very well by the numerical simulation. Other

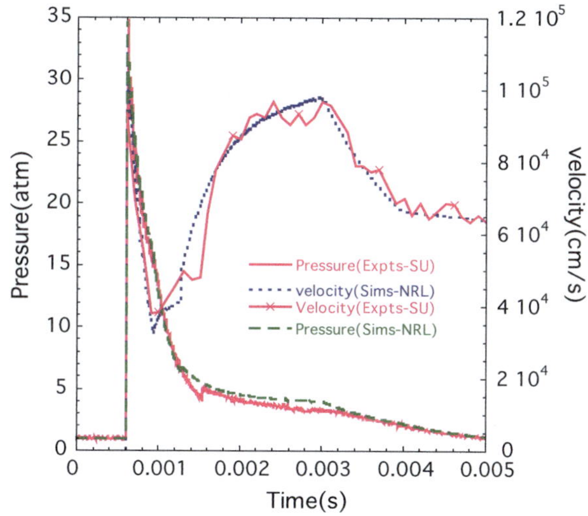

Fig. 3 Comparisons of predicted detonated gas velocities and pressures from NRL simulations to measured velocities and pressures from experiments at Stanford University (from Ref. [33])

general considerations on the verification and validation of such simulations have been addressed in Ref. [34].

3.4 Other Basic Research Issues

The key issues that need to be resolved have been addressed in a number of papers (e.g., [35, 36]). These include fuel–air injection and mixing, low-energy detonation initiation, detonation of liquid fuels and system-level performance. As discussed above, the performance of idealized laboratory pulse detonation devices is one of the topics that have been resolved. However, the performance of these devices under flight conditions is a controversial topic. This is partly because of the uncertainty in the "proper" design of nozzles and inlets for these engines. Nozzles for the PDE present challenging design and integration issues because of the inherently unsteady nature of the pulse detonation process. For a high frequency, multi-tube system, where the overall system may appear to be nearly steady, there are issues in designing a common flow path for the exhaust from the individual thrust tubes.

Reliable and repeated low-energy initiation of detonations in the high-speed flow in pulse detonation engines operating on practical mixtures is one of the most challenging problems in the development of these engines. Significant research efforts and accomplishments continue to be made on this topic worldwide. Issues with multiphase detonations in tubes are extensive and have been discussed elsewhere and not repeated here for brevity [37]. With the successes in the operation of laboratory-scale devices, attention shifted to the development of flight-scale multi-tube systems.

Fig. 4 Pulse detonation rocket, "Todoroki" (from Ref. [40])

3.5 Flight Demonstrations

There have been efforts around the world in developing a flight-capable pulse detonation engine. Two of the more noteworthy efforts are highlighted here.

In Japan, there has been a sustained developmental work (e.g., [38–40]) on pulse detonation-based rocket engines (PDREs). An engine sliding on rails (see Fig. 4, taken from Ref. [40]) was developed and demonstrated. Interestingly, the thrust obtained was within 4% of the calculated value.

In the USA, a major accomplishment was the successful flight of the manned Long E-Z aircraft, powered by a pulse detonation engine [41]. The engine produced more than 200 lb of thrust. The goals of this demonstration included showing that an aircraft and pilot could withstand the acoustic pressures generated by the detonations. With the successful flight demonstration of the pulse detonation engine, attention shifted to the development of continuous detonation engines and hybrid engines combining the benefits of gas turbines and detonation engines.

4 Rotating Detonation Engines (RDEs)

There is the potential to increase further the substantial performance gains made by the pulsed detonation engine concept over the traditional constant-pressure cycle-based propulsion engines by moving from "pulsed" or "periodic" detonations to a more "continuous" mode of operation with detonations [11]. Again, this is not a recent idea. The feasibility of a continuous detonation-wave engine has been investigated at least from the early 1960s [25, 42–44]. The early studies have been followed up with more recent studies in Russia, Poland, France, Japan, China, Singapore and the USA [11, 45]. The overall flow field and key parameters and issues facing further development of this engine are discussed next.

The schematic of a basic RDE configuration is shown in Fig. 5. Here, the combustion or detonation chamber is an annular ring, where a premixed fuel–air mixture is injected axially at the bottom end of the combustion chamber, and once initiated, a detonation propagates circumferentially around the annular ring near the injection plane. The detonated products are expanded and exhausted out the top end of the combustion chamber, which could also accommodate a nozzle to further harness the energy of the exhaust products. Unlike the PDE, the RDE can provide a nearly steady source of thrust, without having to initiate a detonation repeatedly at very high frequencies. Of course, the RDE has its own challenges. For example, because the detonable mixture is injected axially and the detonation wave runs circumferentially around the combustion chamber, the flow field within an RDE has both very strong axial and azimuthal components, which makes the analysis of the engine more complex and an efficient design potentially more difficult to develop. In addition, because a detonation wave is present continuously in the "combustion chamber", heat transfer to the walls of the chamber and its control could be a major challenge.

Fig. 5 Schematic of a generic rotating detonation-wave engine (RDE)

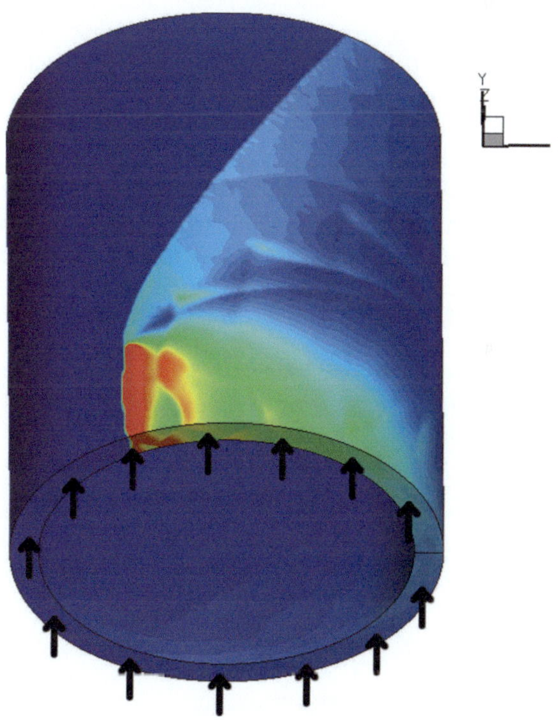

4.1 Basic Flow Field and Wave Structure Within an RDE

The typical wave structure within an RDE is shown in Fig. 6. Because the radial dimension of the generic RDE shown in Fig. 5 is typically small compared to the azimuthal and axial dimensions, there is generally little variation radially within the flow. Because of this, the RDE is usually "unrolled" into two dimensions as shown in Fig. 6, and two-dimensional computations are carried out. Full three-dimensional simulations are also carried out in order to assess the role of three-dimensionality in specific situations [46, 47].

Figure 6 shows the main features of a rotating detonation engine. As before, the premixed fuel–air mixture is injected from the bottom, the detonation is propagating in the azimuthal direction near the injection end, and the detonated products are expanding azimuthally and axially to the exit plane. The secondary shock wave (D) varies considerably depending on the inlet stagnation pressure and the back pressure. The pressure just behind the detonation wave is high enough that the micro-nozzles are completely blocked (F). Experimentally, this can be a problem because of the potential for backflow into the premixture plenum. Further behind the detonation front, the fuel–air mixture can begin to penetrate into the chamber (G). For most of this region, the flow through the micro-nozzles is choked, which is why the premixture region expands almost linearly. Also of interest is the region where the premixture and the reacted gases meet (E). Here, the RDE experiences some non-detonative burning, which will result in a loss of performance (compared to the ideal detonation cycle).

Basic Thermodynamic Cycle within an RDE. Since the overall operation of an RDE is continuous, it is possible to represent the processes within a conventional cycle diagram. One approach to ascertain the effective thermodynamic cycle in RDEs

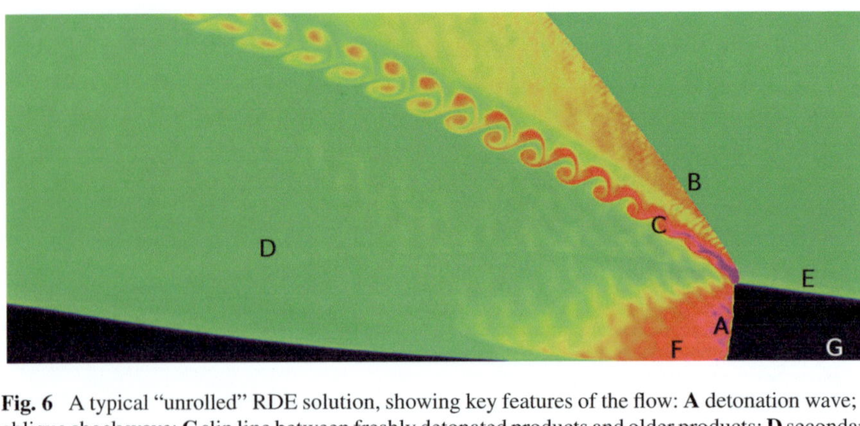

Fig. 6 A typical "unrolled" RDE solution, showing key features of the flow: **A** detonation wave; **B** oblique shock wave; **C** slip line between freshly detonated products and older products; **D** secondary shock wave; **E** mixing region between fresh premixed fuel–air gases and detonated gases; **F** region with blocked injection micro-nozzles and **G** unreacted premixed fuel–air mixture injected from the micro-nozzles

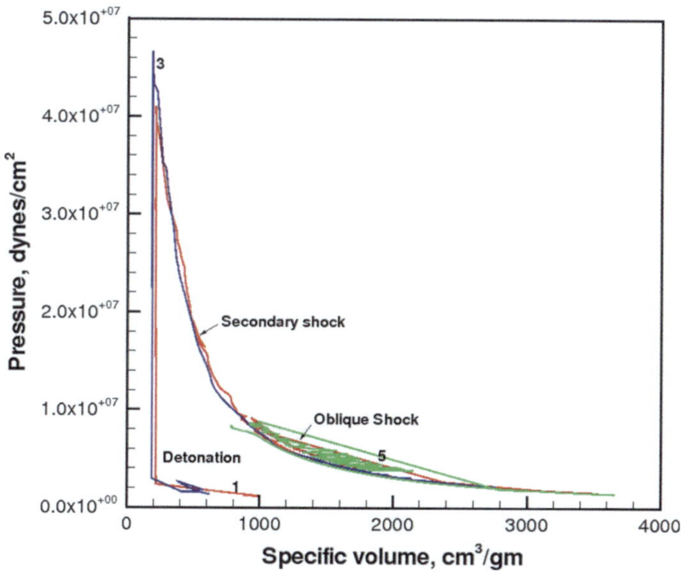

Fig. 7 *p-v* diagram for a typical solution in the RDE shown in Fig. 5

is to take information from the detailed simulation of the flow field and plot the results in a conventional *p-v* diagram framework as shown in Fig. 7. The locus of various points shows that indeed a near "steady" detonation cycle is followed by the combustible material in the RDE. Hence, one can expect better performance than a PDE or "intermittent" detonation engines, where all fluid elements have not been shown to lie on the "detonation cycle" in a *p-v* diagram framework. This approach has been developed further by Nordeen et al. [48].

4.2 A Key Parameter and Overall Performance

The pressure change is one of the important factors for determining the performance of detonation-based engines since these are examples of "pressure-gain" combustion systems. There are two key pressures to be considered in the generic RDE discussed above. The first is the stagnation pressure for the inlet micro-nozzles, and the second is the back pressure at the exit of the detonation/combustion chamber. To highlight the impact of these pressures, the same geometry as in the generic configuration in Fig. 5 has been considered in a parametric study [49] with a stoichiometric hydrogen–air mixture. All of the simulations were run to 4 ms and checked to ensure they were essentially "steady". In this parametric study, the pressure ratio po/pb was varied in two ways. First, the stagnation pressure, po, was set for the micro-nozzles at 10 atm, and the back pressure was varied from 0.5 to 4 atm. Second, the back pressure was

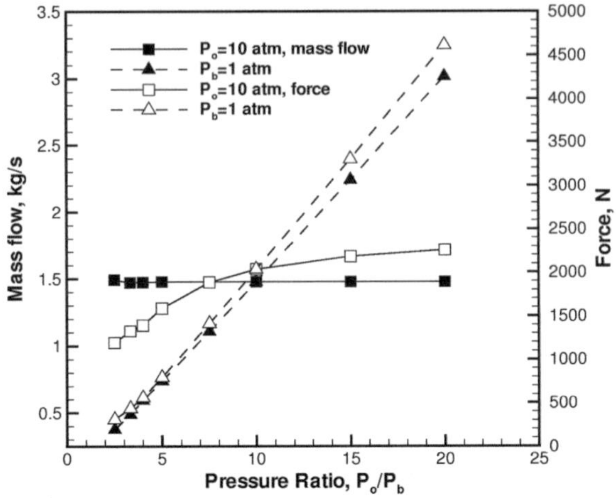

Fig. 8 Impact of stagnation pressure and back pressure on mass flow rate and computed thrust

held constant at 1 atm, and the stagnation pressure was varied from 2.5 to 20 atm. In both cases, the pressure ratio varied from 2.5 to 20. For the various cases, the detonation velocity varied from 1875 to 1920 m/s, compared to the CJ detonation velocity of around 1969 m/s.

The overall performance for the various cases is shown in Fig. 8. The two global quantities that are of particular interest are the mass flow rate of the premixed gases into the chamber and the propulsive force at the exit of the combustion chamber. As expected, the mass flow into the RDE is independent of the back pressure and almost completely depends on the stagnation pressure at the inflow micro-nozzles. The force felt at the exit, however, is both a function of the stagnation and back pressures. This is most clearly seen by holding the stagnation pressure constant, indicated by the square symbols in Fig. 8. Interestingly, as shown in Fig. 9, the specific impulse is dependent solely on the pressure ratio, rather than specific values of either pressure.

It is informative to put the computed specific impulse shown above in the context of other detonation-wave engines. The comparable specific impulse for a multi-tube PDE operating on a stoichiometric hydrogen–air mixture at sea-level static conditions is around 4100 s (see, e.g., Ref. [8]). Thus, we currently estimate a further increase in the performance for the RDE over the PDE of about 33%.

4.3 Effect of Inlet Area Ratio

One of the main guiding parameters for the efficiency of the RDE is the area ratio between the injector throat area and injection face area [50]. As this area ratio is increased, we see a large gain in specific impulse, especially at lower pressure ratios as shown in Fig. 10. These low pressure ratios are the desired operating conditions for an air-breathing RDE. When we conducted simulations including more realistic

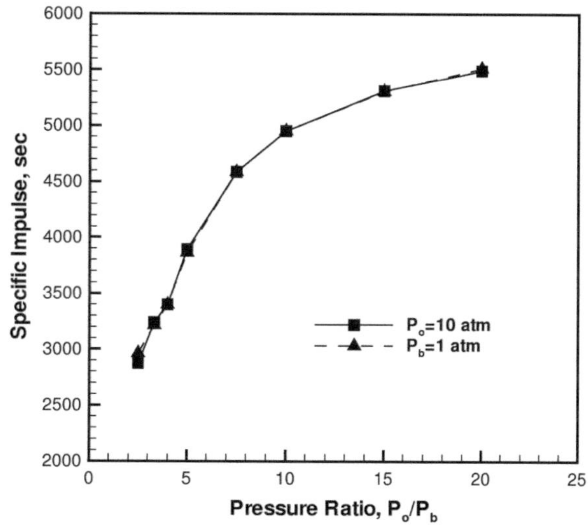

Fig. 9 Dependence of the specific impulse on the stagnation and back pressures

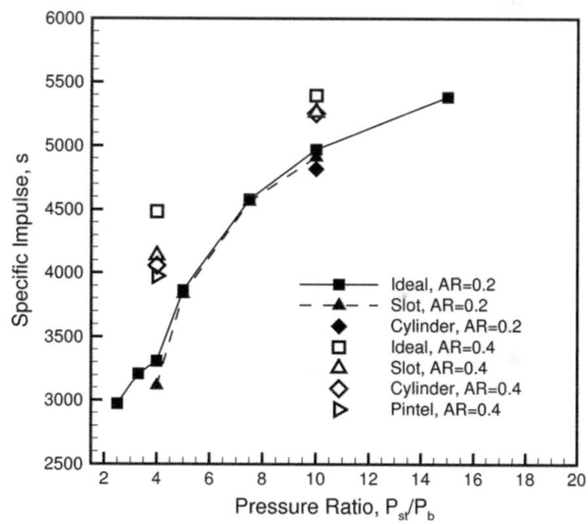

Fig. 10 Specific impulse as a function of pressure ratio and various injectors

injectors, we saw two important effects: there is a decrease in the performance which is especially pronounced at lower pressure ratios; we also see a much greater range of pressure unsteadiness in the mixture plenum compared with a smaller area ratio. Both of these aspects are problematic for an RDE in a practical device. In addition to this, in more practical, non-premixed fuel–air injector designs, the injector plays a critical role in mixing the fuel and oxidizer before the detonation wave reaches this mixture. Effective designs for the injectors that maximize performance while minimizing upstream feedback are still an area of active research.

4.4 Current Global Efforts

There are ongoing research and development efforts on RDEs all over the world. Some of the research efforts that are often presented at international conferences and the open literature are highlighted here. A good source is the series of annual conferences titled, "International Workshop on Detonations for Propulsion (IWDP)".

During the last two decades, there has been a concerted effort in France on developing operational detonation-wave engines for high-speed applications [51–55]. The company MBDA in France has developed a full-scale engine and conducted ground tests. The proposed future application is for a supersonic (Mach 4+) multi-role strike weapon system by the year 2030. They have compared the proposed new missile to the BrahMos [56] missile and estimated that the new "Perseus" supersonic missile would reduce the launch mass from 3000 to 800 kg as well as reduce the length from 8.4 to 5 m. MBDA has also been investigating RDEs for potential replacement of (a) turbojet engine, (b) ramjet engine and (c) liquid rocket engine. According to them, the key advantages for each of these detonation engine systems over their conventional counterparts are as follows: turbojet for improved performance and a simplified system by reducing the compression stage; a detonation Ramjet for improved performance with shorter ram combustor and operation from a lower Mach number; a liquid rocket detonation engine for improved performance, compactness, lower feed pressure and thrust vectoring. Their overall effort is very advanced, and full-scale testing is also underway. In addition to developing and testing an RDE-based missile system, France has also extensively invested in related basic research through the CNRS (Centre National de la Recherche Scientifique) laboratories, most notably the famous Laboratory of Combustion and Detonation (LCD) at Poitiers [57].

Another European country with a long history of basic and applied studies in detonation is Poland, most notably the Warsaw University of Technology and the Warsaw Institute of Aviation [58]. Details of their research efforts can be found in the extensive publications of Wolanski et al. [59–64], with an excellent summary in Ref. [63]. A recent focus of their research efforts has been in developing a rocket engine with a rotating detonation-based combustion chamber. In addition to the RDE-based rocket engine, they have made significant progress in integrating an RDE with a gas-turbine engine.

While the R&D efforts in Asia are dominated by China and Japan, there has also been significant detonation engine-related research conducted in Hong Kong, Malaysia, Singapore, South Korea and Taiwan. Some of the leading institutions in Mainland China conducting detonation engine research and development include the Peking University, the Beijing Institute of Technology and the Northwestern Polytechnic University in Xian. Professor J. P. Wang has delivered several lectures on the progress in RDE research at the Peking University [65–68] at various international conferences and published a review article on this technology recently in the Chinese Journal of Aeronautics [13].

In Japan, there continues to be extensive work going on in both PDEs and RDEs [38–40, 69–77]. Key organizations that have been making open presentations at international conferences include Aoyama Gakuin University, Hiroshima University, Hokkaido University, Kyushu Institute of Technology, Keio University, Nagoya University, University of Tsukuba, as well as JAXA/ISAS. The Temasek Laboratories at the National University of Singapore is leading the efforts in Singapore and is conducting both PDE- and RDE-related research [78]. Both computational and experimental studies [79] of PDEs and RDEs are going on at Pusan National University (in South Korea) under the guidance of Prof. J. Y. Choi. He has been active on this topic for the past decade and is also building a high-performance computing capability at Pusan.

5 Concluding Remarks

Significant progress is noted in the research towards the development and application of detonation-based, pressure-gain combustion systems for air-breathing and rocket propulsion applications. Extensive literature is now available on the basic research of a representative straight-tube pulse detonation device. There is a consensus now on the performance of these devices, both theoretically and experimentally. The maturity of the research effort has led to a flight demonstration of a PDE-powered aircraft as well as a rocket demonstrator. The attention of the research community has largely shifted to rotating detonation-wave engines (RDE).

Rotating detonation-wave engines, a form of the continuous detonation-wave engine, are shown to have the potential to further increase the performance of air-breathing propulsion devices than pulsed or intermittent detonation-wave engines. Promising experimental, theoretical and computational studies have been briefly discussed. Numerical simulations of a representative geometry are used to highlight the complex flow field within the RDE. The results from the simulations are used to show that the flow follows the thermal detonation cycle quite closely. Parametric studies have been discussed varying the pressure conditions both upstream through the inlet micro-nozzles and downstream through the back pressure. The specific impulse is found to depend solely on the ratio of the inlet stagnation pressure to the chamber back pressure. The computed specific impulse is shown to be higher than that of a comparable PDE but slightly lower than the theoretical maximum for an "arbitrary" engine based on a steady, detonation cycle. This suggests that optimizing the geometry and flow field could lead to further potential gains in performance.

References

1. Kailasanath K (2000) Review of propulsion applications of detonation waves. AIAA J 38(9):1698–1708
2. Eidelman S, Grossmann W, Lottati I (1991) Review of propulsion applications and numerical simulations of the pulse detonation engine concept. J Prop Power 7(6):857–865

3. Eidelman S, Grossmann W (1992) Pulsed detonation engine: experimental and theoretical review. AIAA paper 92-3168, AIAA, Reston, VA
4. Bussing T, Pappas G (1996) Pulse detonation engine theory and concepts. In: Murthy SNB, Curran ET (eds) Developments in high-speed-vehicle propulsion systems, Prog Astro Aero, vol. 165. AIAA, Reston, VA, pp 421–472
5. Kailasanath K (2001) A review of PDE research—performance estimates. AIAA paper 2001-0474, AIAA, Reston, VA
6. Kailasanath K (2001) A review of research on pulse detonation engine nozzles. AIAA paper 2001-3932, AIAA, Reston, VA
7. Roy GD, Frolov SM, Borisov AA, Netzer DW (2004) Pulse detonation propulsion: challenges, current status, and future perspective. Prog Energy Combust Sci 30(6):545–672
8. Kailasanath K (2003) Recent developments in the research on pulse detonation engines. AIAA J 41(2):145–159
9. Kailasanath K (2009) Research on pulse detonation combustion systems—a status report. AIAA paper 2009-0631, AIAA, Reston, VA
10. Hishida M, Fujiwara T, Wolanski P (2009) Fundamentals of rotating detonations. Shock Waves 19(1):1–10
11. Kailasanath K (2011) The rotating-detonation-wave engine concept: a brief status report. AIAA paper 2011-0580, AIAA, Reston, VA
12. Wolański P (2013) Detonative propulsion. In: Proceedings of the combustion institute, vol 34. The Combustion Institute, Pittsburgh, PA, pp 125–158
13. Rui Z, Dan W, Wang J (2016) Progress of continuously rotating detonation engines. Chin J Aeronaut 29(1):15–29
14. Hoffmann N (1940) Reaction propulsion by intermittent detonative combustion. Ministry of Supply, Volkenrode Translation
15. Roy M (1946) Propulsion par Statoreacteur a Detonation, (Detonation Ramjet Propulsion), Comptes-Rendus de l'Acad. des Sciences, Paris, vol 222, pp 31–32
16. Reingold L (1950) Recherches sur les combustions permanentes apportees aux foyers a circulation interne supersonique (Investigations of steady combustion attained in internal supersonic flow combustors). ONERA TN2, Department of Energy and Propulsion, Study 728-E, March 1950
17. Bitondo D, Bollay W (1952) Preliminary performance analysis of the pulse-detonation-jet engine system. Aerophysics Development Corp., Report ADC-102-1, April 1952
18. Gross RA (1959) Exploratory study of combustion in supersonic flow. Air force office of scientific research, AFOSR-TN 59-587, June 1959
19. Gross RA, Chinitz W (1960) A study of supersonic combustion. J Aerosp Sci 27(7):517–525
20. Nicholls JA, Wilkinson HR, Morrison RB (1957) Intermittent detonation as a thrust-producing mechanism. Jet Propul 27(5):534–541
21. Dunlap R, Brehm RL, Nicholls JA (1958) A preliminary study of the application of steady-state detonative combustion to a reaction engine. Jet Propul 28(7):451–456
22. Nicholls JA, Dabora EK, Gealer RL (1959) Studies in connection with stabilized gaseous detonation waves. In: Proceedings of the combustion institute, vol 7. The Combustion Institute, Pittsburgh, PA, pp 766–772
23. Nicholls JA, Dabora EK (1962) Recent results on standing detonation waves. In: Proceedings of the combustion institute, vol 8. The Combustion Institute, Pittsburgh, PA, pp 644–655
24. Voitsekhovskii BV (1959) Stationary detonation. Doklady Akad Nauk 129(6):1254–1256
25. Voitsekhovskii BV (1960) Stationary spin detonation. Sov J Appl Mech Tech Phys 3(1):157–164
26. Krzycki LJ (1962) Performance characteristics of an intermittent detonation device. U.S. Naval Ordnance Test Station, NavWeps Rept. 7655, ASTIA 284–312, China Lake, CA
27. Kailasanath K, Patnaik G, Li C (1999) Computational studies of pulse detonation engines: a status report. AIAA paper 99-2634, AIAA, Reston, VA
28. Kailasanath K, Patnaik G (2000) Performance estimates of pulsed detonation engines. In: Proceedings of the combustion institute, vol 28. The Combustion Institute, Pittsburgh, PA, pp 595–601

29. Cooper M, Jackson S, Austin JM, Wintenberger E, Shepherd JE (2001) Direct experimental impulse measurements for detonations and deflagrations. AIAA 2001-3812, AIAA, Reston, VA
30. Li C, Kailasanath K, Patnaik G (2000) A numerical study of flow field evolution in a pulse detonation engine. AIAA paper 2000-0314, AIAA, Reston, VA
31. Jenkins TP, Sanders ST, Kailasanath K, Li C, Hanson RK (2000) Diode laser-based measurements for model validation in pulse detonation flows. In: Proceedings of the 25th JANNAF airbreathing propulsion meeting, Monterey, CA
32. Kailasanath K, Patnaik G, Li C (2001) On factors controlling the performance of pulsed detonation engines. In: Roy GD, Frolov SM, Netzer DW, Borisov AA (eds) High-speed deflagration and detonation: fundamentals and control. ENAS Publishers, Moscow, pp 193–206
33. Sanders ST, Mattison DW, Muruganandam TM, Hanson RK (2001) Multiplexed diode-laser absorption sensors for aeropropulsion flows. AIAA paper 2001-0412, AIAA, Reston, VA
34. Kailasanath K (2001) Verification and validation of pulsed detonation engine computations. In: Proceedings of the second joint meeting of the US sections of the combustion institute, Paper No. 142
35. Eidelman S, Yang X, Lottati I (1995) Pulsed detonation engine: key issues. AIAA paper 95-2754, AIAA, Reston, VA
36. Desbordes D, Daniau E, Zitoun R (2001) Pulsed detonation propulsion: key issues. In: Roy GD, Frolov SM, Netzer DW, Borisov AA (eds) High-speed deflagration and detonation: fundamentals and control. ENAS Publishers, Moscow, pp 177–192
37. Kailasanath K (2006) Liquid-fuelled detonations in tubes. J Propul Power 22(6):1261–1268
38. Kasahara J, Tanahashi Y, Hirano M, Numata T, Matsuo A, Endo T (2004) Experimental investigation of momentum and heat transfer in pulse detonation rockets. AIAA paper 2004-0869, AIAA, Reston, VA
39. Kasahara J, Hirano M, Matsuo A, Sato S, Endo T, Satori S (2004) Flight experiments regarding ethlene-oxygen single-tube pulse detonation rockets. AIAA paper 2004-3918, AIAA, Reston, VA
40. Kasahara J, Hasegawa A, Nemoto T, Yamaguchi H, Yajima T, Kojima T (2007) Thrust demonstration of a pulse detonation rocket "TODOROKI". AIAA paper 2007-5007, AIAA, Reston, VA
41. Barr L (2008) Pulse detonation engine flies into history. Press release, Air force material command. http://www.afmc.af.mil/news/story.asp?id=123098900, May 2008
42. Voitsekhovskii BV, Mitrofanov VV, Topchiyan ME (1963) Structure of a detonation front in gases. Siberian Branch of the USSR Academy of Sciences, Novosibirsk
43. Mikhailov VV, Topchiyan ME (1965) Studies of continuous detonation in an annular channel. Fiz Goreniya Vzryva 2(4):20–26
44. Nicholls JA, Gullen RE, Ragland KW (1966) Feasibility studies of a rotating detonation wave rocket motor. J Spacecraft Rockets 3:893–898
45. Kailasanath K (2017) Recent developments in the research on rotating detonation-wave engines. AIAA paper 2017-0784, AIAA, Reston, VA
46. Schwer DA, Kailasanath K (2011) Numerical study of engine size effects on rotating detonation engines. AIAA paper 2011-0581., AIAA, Reston, VA
47. Schwer DA, Kailasanath K (2014) Effect of low pressure ratio on exhaust plumes of rotating detonation engines. AIAA paper 2014-3901, AIAA, Reston, VA
48. Nordeen C, Schwer DA, Schauer F, Hoke J, Cetegen B, Barber T (2013) Thermodynamic modeling of a rotating detonation engine. AIAA paper 2013-1175, AIAA, Reston, VA
49. Schwer DA, Kailasanath K (2010) Numerical investigation of rotating detonation engines. AIAA 2010-6880, AIAA, Reston, VA
50. Schwer DA, Kailasanath K (2011) Numerical investigation of the physics of rotating-detonation-engines. In: Proceedings of the combustion institution, vol 33. The Combustion Institute, Pittsburgh, PA, pp 2195–2202
51. Daniau E, Falempin F, Bykovxkii FA, Zhdan S (2005) Pulsed and rotating detonation propulsion system: first step toward operational engines. AIAA 2005-3233, AIAA, Reston, VA

52. Falempin F, Daniau E, Getin N, Bykovskii FA, Zhdan S (2006) Toward a continuous detonation wave rocket engine demonstrator. AIAA-2006-7956, AIAA, Reston, VA
53. Falempin F, Daniau E (2008) A contribution to the development of actual continuous detonation wave engine. AIAA 2008-2679, AIAA, Reston, VA
54. Falempin F (2008) Continuous detonation wave engine. In: Advances in propulsion technology for high-speed aircraft. RTO-EN-AVT-150, Paper 8, NATO
55. Davidenko D, Gökalp I, Kudryavtsev A (2008) Numerical study of the continuous detonation wave rocket engine. AIAA 2008-2680, AIAA, Reston, VA
56. BrahMos: https://en.wikipedia.org/wiki/BrahMos
57. Lentsch A, Bec R, Serre L, Falempin F, Daniau E, Piton D, Prigent E, Canteins G, Zitoun R, Desbordes D, Jouot F, Gokalp I (2005) Overview of current French activities on PDRE and continuous detonation wave rocket engines. AIAA 2005-3232, AIAA, Reston, VA
58. Wolanski P, Kindracki J, Fujiwara T (2006) An experimental study of small rotating detonation engine. In: Roy G, Frolov S, Sinibaldi J (eds) Pulsed and continuous detonations. Torus Press, Moscow, pp 332–338
59. Yi T-H, Turangan C, Lou J, Wolanski P, Kindracki J (2009) A three-dimensional numerical study of rotational detonation in an annular chamber. AIAA paper 2009-634, AIAA, Reston, VA
60. Tobita A, Fujiwara T, Wolanski P (2010) Detonation engine and flying object. United States patent US 7784267
61. Kindracki J, Wolanski P, Gut Z (2011) Experimental research on the rotating detonation in gaseous fuels–oxygen mixtures. Shock Waves 21(2):75–84
62. Wolanski P (2011) Rotating detonation wave stability. In: 23rd international colloquium on the dynamics of explosions and reactive systems. IDERS, Irvine, CA, pp 1–10
63. Wolanski P (2013) Detonation propulsion. In: Proceedings of the Combustion Institution, vol 34. The Combustion Institute, Pittsburgh, PA, pp 125–158
64. Wolanski P (2014) Detonation engine research in Poland. In: 2014 international workshop on detonation for propulsion. IWDP, Warsaw, Poland
65. Wang JP, Shi TY, Wang YH, Liu YS, Li YS (2010) Experimental research on the rotating detonation engine. In: 14th shock wave and shock tube conference. ISSW, Huang Shan, China
66. Wang YH, Wang JP, Shi TY, Liu YS (2012) Experimental research on transition regions in continuously rotating detonation waves. AIAA 2012-3946, AIAA, Reston, VA
67. Wang YH, Wang JP, Liu YS, Li Y (2013) Experimental investigation on continuously rotating detonation engines. In: 2013 Asia-Pacific international symposium on aerospace technology
68. Wang YH, Wang JP, Shi TY (2013) Discovery of breathing phenomena in continuously rotating detonation. Proc Eng 67:188–196
69. Kasahara J, Matsuoka K, Sakamoto R, Ikeguchi KB, Sakumi T, Morozumi T, Matsuo A, Funaki I (2012) Multitube-rotary-valved pulse detonation rocket engine for flight test. In: 2012 international workshop on detonation for propulsion. IWDP, Tsukuba, Japan
70. Kasahara J, Kato Y, Ishihara K, Matsuoka K, Matsuo A, Funaki I, Nakata D, Higashino K, Tanatsugu K (2016) Research and development of rotating detonation engine for upper-stage kick motor system. In: 2016 international workshop on detonation for propulsion. IWDP, Singapore
71. Uemura Y, Hayashi AK, Asahara M, Tsuboi N, Yamada E (2013) Transverse wave generation mechanism in rotating detonation. Proc Combust Inst 34(2):1981–1989
72. Tsuboi N, Watanabe Y, Kojima T, Hayashi AK (2013) Numerical estimation of the thrust performance on a rotating detonation engine for a hydrogen–oxygen mixture. In: Proceedings of the combustion institution, vol 35. The Combustion Institute, Pittsburgh, PA, pp 2005–2013
73. Tsuboi N, Hayashi AK, Kojima T (2013) Numerical study on a rotating detonation engine at KIT. In: 2013 international workshop on detonation for propulsion. IWDP, Tainan, Taiwan
74. Hayashi AK (2013) Two-phase detonation and its application to PDE and recent results of RDE. In: 2013 international workshop on detonation for propulsion. IWDP, Tainan, Taiwan
75. Hayashi AK, Kimura Y, Yamada T, Yamada E (2009) Sensitivity analysis of rotating detonation engine with a detailed reaction model. AIAA 2009-0633, AIAA, Reston, VA

76. Yamada T, Hayashi AK, Yamada E, Tsuboi N, Tangirala VE, Fujiwara T (2010) Numerical analysis of threshold of limit detonation in rotating detonation engine. AIAA 2010-0153, AIAA, Reston, VA
77. Nakagami S, Matsuoka K, Kasahara J, Kumazawa Y, Fujii J, Matsuo A, Funaki I (2015) Experimental visualization of the structure of rotating detonation waves in a disk-shaped combustor. AIAA paper no. 2015-4102, AIAA, Reston, VA
78. Teo CJ, Li JM, Li L, Nguyen VB (2016) Detonation engine research at the Temasek laboratories, Singapore. In: 2016 international workshop on detonation for propulsion. IWDP, Singapore
79. Choi JY (2016) Research progress of detonation studies for propulsion in PNU. In: 2016 international workshop on detonation for propulsion. IWDP, Singapore

Drag Reduction with Optimum Designing of a Base Bleed Projectile Using Computational Analysis

Ashoke De and Piyush Chettri

Keywords Base drag · Base bleed · Artillery shell · Boat tail

Nomenclature

$\dot{m}$	Bleed mass flow rate
ρ_∞	Free stream density
V_∞	Free stream velocity
I	Injection parameter
A_b	Projectile base area
C_D	Coefficient of drag
K	Turbulence kinetic energy
T	Temperature
ω	Specific dissipation rate

Abbreviations

CFD	Computational fluid dynamics
RANS	Reynolds-averaged Navier-Stokes
SST	Shear stress transport

A. De (✉) · P. Chettri
Department of Aerospace Engineering, Indian Institute of Technology Kanpur, Kanpur 208016, India
e-mail: ashoke@iitk.ac.in

A. K. Gupta et al. (eds.), *Innovations in Sustainable Energy and Cleaner Environment*, Green Energy and Technology,
https://doi.org/10.1007/978-981-13-9012-8_2

BRL Ballistic research laboratory
2D Two dimensional

1 Introduction

An artillery shell is primarily used to soften the targets in range of 10–35 km of Tactical Battle Area (TBA). Once launched from the gun, it is generally corrected at the terminal end for the next firing round after it hits the target. When the projectile leaves the gun barrel leaving the effects of immediate surroundings, it enters the realm of the exterior ballistics. Here, the projectile is exposed to different kind of forces such as (a) pressure force due to the atmosphere that it is flying through, (b) spinning force (induced due to its spin), and (c) gravity force. Thus the projectile in flight is no longer constrained in lateral motions by the walls of the gun and as a free body can develop motions that are complex and detrimental for the user and its designer. Keeping in mind the deployment of artillery, the projectile is subjected to varying Mach numbers in the supersonic and subsonic region [1–19].

When ammunition is designed, it is done for a maximum Mach number which may be higher than that used in operations, especially when guns fire at higher angles akin to hilly terrain. Over a time period, there have been rigorous efforts to increase the range and accuracy of these field guns. Gun with high muzzle velocity is generally fired with flat trajectory weapons, and they fail to clear the crest and engage the target area (Fig. 1). Thus guns with lower muzzle velocity at a high angle have to be utilized to clear the crest. The increment in range can be either achieved by changes in the design of the gun, propellant material (charge), and design of the ammunition. The third option is much more feasible than the former ones due to cost and safety factors. A significant area of concern in shell design is the total aerodynamic drag.

In order to ensure that the projectile fires from the same gun, the alterations possible for the ammunition are changed in boat-tail angle as the maximum caliber

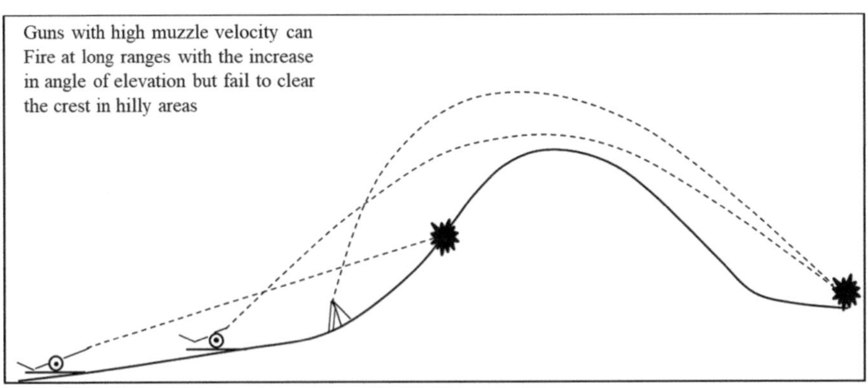

Fig. 1 Variation in a firing range with different angles

of the gun cannot be altered. The steady data achieved in experiments is captured at the beginning of the launch, appreciably away from the muzzle end of the gun, not to have any interior ballistic effect in the experimental reading especially due to the propelling charge.

1.1 Forces on Projectiles

Broadly, the forces acting on the projectile can be divided into lift and drag. The exterior ballistics also includes the effects subjected to the projectile due to atmospheric conditions (Fig. 2).

Drag and Lift:

Aerodynamic drag depends on the size and shape of the projectile, its inclination, velocity, mass, compressibility, and viscosity of the atmosphere. Aerodynamic forces can broadly be divided into two categories, i.e., lift and drag or (lift force or drag force). The designer is always keen to increase the range and terminal velocity of projectiles by reducing the aerodynamic drag. There are in general two types of drag that can act on a body: one is pressure drag (along the surface area) and other one is skin friction drag (normal to the surface). Further due to shock wave, the wave drag (form of pressure drag) can be formed when the local velocity along the surface of the projectile becomes supersonic. The drag force can be recast as:

$$F_{\mathrm{D}} = \frac{1}{2}\rho_{\infty} S C_{\mathrm{D}} V_{\infty}^{2} \tag{1}$$

C_{D} varies as a function of free stream velocity, flow direction, object position, object size, fluid density, and viscosity. For a certain body shape, C_{D} depends on the Reynolds number (Re), Mach number (M), Area S representing the reference area based on the projectile diameter and the direction of flow.

Lift is defined as the aerodynamic force which acts orthogonally to the velocity vector. Aerodynamic Lift is defined as:

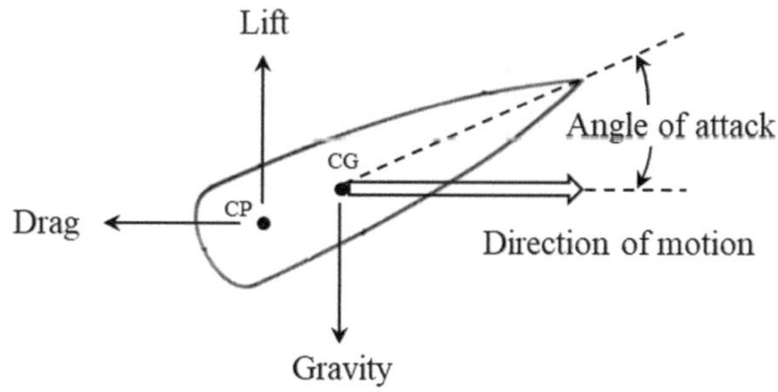

Fig. 2 Drag on projectiles

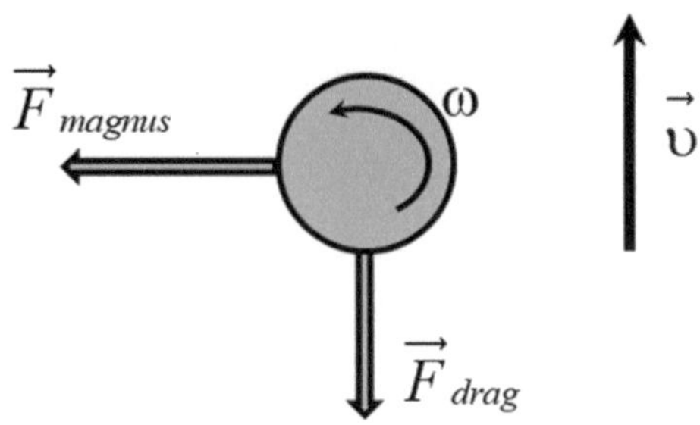

Fig. 3 Magnus effect on projectiles

$$L = \frac{1}{2}\rho_\infty S C_L V_\infty^2 \tag{2}$$

The components of lift and drag are evaluated for balancing the forces, however, in the case of the symmetric projectile (as in this case), there will be no lift at zero angles of attack.

Magnus force:

Magnus force is developed on an immersed body of circular cross section when it spins about its axis. The Magnus moment contributes significantly to the stability of the projectile as shown in Fig. 3. The Magnus force will depend on the crosswind flow which will contribute toward the yaw of the projectile which would be detrimental in evaluating aerodynamic drag. This is subject to the atmospheric conditions which will assist or resist in the direction of rotation of the projectile, thus affecting the drag and stability of the projectile.

Base Bleed:

The concept of base bleed is common in most of the modern projectile. The base drag is due to the wake region behind the base (Fig. 4). The sharp turning angle as the flow goes beyond the boat tail and behind the base causes separation and formation of reverse flow due to adverse pressure gradient is known as the recirculation region or the separation bubble. In the wake region, the point along the axis of symmetry, where the velocity diminishes, is called a shear layer reattachment point. As the shear layer reattaches, the flow is forced to move along with the symmetry, causing the formation of a reattachment. A base bleed reduces the base drag by injecting gas, generated by burning propellant at the base area. To achieve base bleed, a gas generator or base bleed unit is attached to the projectile which is in addition to the main propelling charge (providing thrust). The base bleed unit provides a low-velocity flow behind the projectile and increases the pressure in the wake region behind the base. As the base bleed propellant burns, there is an injection flow which increases initially and then decreases at the end. The burning of the base bleed unit generally lasts from 30 to 38 s depending on the propellant grain. The propellant (base bleed) is kept in a base bleed unit, or the mass injection occurs through a hole in propellant chamber

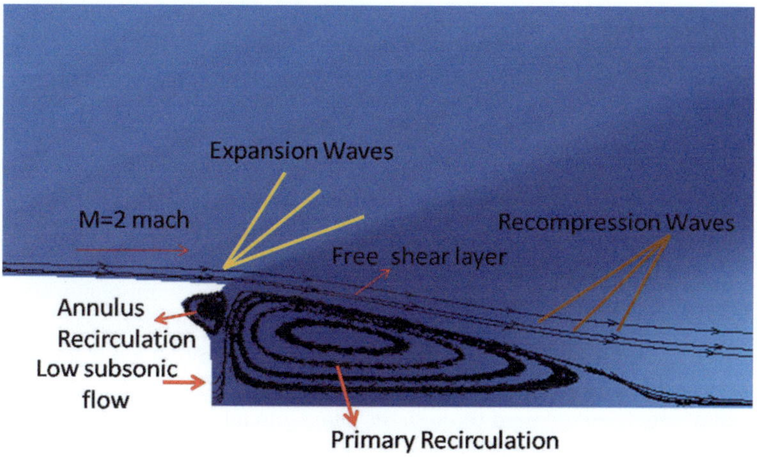

Fig. 4 Base bleed flow for $M = 2$ ($I = 0.006$)

that is located at the base of the projectile. The hole is not a nozzle, such as that found in a rocket-assisted projectile (RAP), so that the thrust resulting from the burning propellant is small.

The ejected gaseous used to destroy the strong vortex adjacent to the base into more than one weak and small vortices. The mass flow from the center of the projectile is estimated to last only the first few kilometers of the flight. However, the first few kilometers are the most important from extended range. Based on the injection parameter, the mass flow injection rate is calculated for numerical analysis. It is defined as the bleed mass flow rate normalized by the product of the base area and the free stream mass flux. The injection parameter (I) is a non-dimensional number given by the following equation:

$$I = \frac{\dot{m}}{\rho_\infty V_\infty A_b} \tag{3}$$

Boat Tailing:

Projectiles and missiles are generally tapered at the end to reduce the wake region at the base. Due to this, the diameter at the base is less than the maximum diameter of the projectile. However, this boat tailing is limited by the operating Mach numbers. The increase in boat-tail angle will lead to an increase in the overall perimeter length which increases the skin friction drag. Considering all the factors and deciding on a maximum operational Mach number, the boat tailing can be maximized to achieve an optimum design. Another factor while considering the boat-tail body is the spinning projectile and yawing which might get affected with the change in boat-tail angle.

2 Motivation

Artillery guns have a specific target area which is limited by the type of ammunition, gun, and terrain. During offensive operations, there is an imperative need to change the target area with the advancement of ground forces. When ammunition is designed, it is done for a maximum Mach number which may be higher than that used in operations. Thus the ammunition may not be fired at such high muzzle velocity especially when firing at high angles. This gives the scope to optimize the range of the gun with the given ammunition by altering its design without any significant changes in its geometric profile. This would allow the artillery forces to engage the enemy from a far distance and minimize the gun movement which is a tedious and hectic task. The question arises if there can be some alterations to the design of ammunition which can be fired through the same gun and have equal effectiveness.

The above requirement translates into further drag reduction of the projectile. Since the pressure (excluding the base) and viscous components of drag generally cannot be reduced significantly without adversely affecting the stability of the shell. Therefore, attempts to reduce the total drag have been directed toward reducing the base drag. Base drag reduction can be made with base bleed, base cavity, and boat tailing. Base drag is influenced by a variety of flow and geometrical parameters. At high Reynolds number, there is a turbulent flow ahead of the nose. The factors to be considered for base drag are (i) freestream Mach number (ii) boundary layer momentum thickness ahead of the base (iii) base diameter (iv) angle of attack (v) boat-tail angle or after body shape (vi) characteristics of the base bleed unit. With the restrictions on the projectile geometry and artillery gun, boat tailing becomes an easy option to optimize the drag reduction. Boat tailing the projectile to its maximum angle along with base injection can lead to increase in range at the same firing angle. The trajectories of bullets and shells could not be estimated without some knowledge of the pressure acting on their blunt bases, and it was recognized that the base pressure made an essential contribution to the total drag of the body.

The testing for increasing the range and continuously altering the design will take a lot of resources and coordination between the testing agency and the users. Therefore, to narrow down that requirement it is necessary that to computationally come to design and reduce the substantial logistic burden for optimizing the range of given ammunition in commensuration with specific military requirements.

3 Literature

3.1 M864 155 mm Artillery Projectile

The M864 is an Army shell that burns a solid propellant in the base region (Fig. 5). It is dual purpose ammunition which is designed for extended range engagements. Generally, the testing of ammunition has been carried out on a similar model but

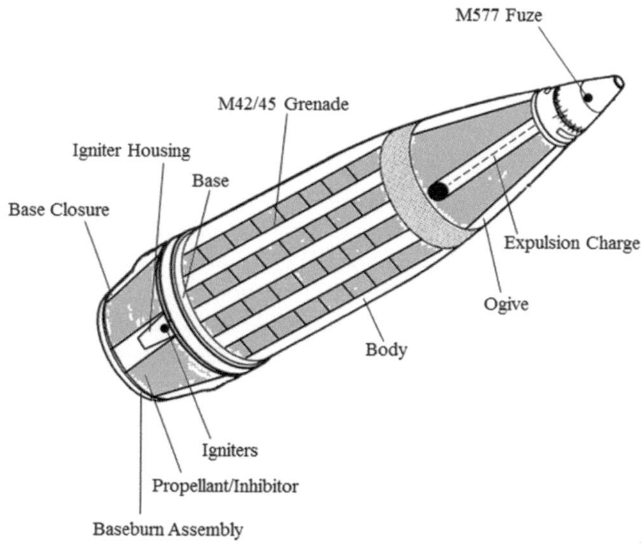

Fig. 5 Schematic diagram of an M864 projectile

keeping in mind the lethality of the ammunition, it may vary with its arsenal and explosives. The most common utilization of this kind of projectile is the one which is capable of delivering several grenades on its target. By dual purpose, we mean that it can be used in anti-armor or anti-personnel. It is considered as the US Army's long-range cargo around. The improved conventional ammunition incorporates base burn technology to increase its range. The projectile is utilized both in plains and mountains and has a maximum range of up to 30 km. The M864 is a streamlined forged steel ogive, a low-drag steel body of generally alloy profile and a steel base closure. A gliding metal drive band encircles the rear of the body. It is a base-ejection projectile that carries 72 DP grenades (48 M42 grenades and 24 M46 grenades). It is the grenades which provide the projectiles with dual destructive capability. However, a third effect can be achieved by replacing the original expelling charge with a spotting charge designed to detonate the entire projectile as if it were a bulk-loaded HE (high explosive). Once fired, the propellant grain present in the base bleed unit ignites the base burner unit, which expels hot gas and increases the projectile's range. At the predetermined time in flight, the grenades are expelled, and they arm themselves while falling. They function upon impact. The base burner unit is inbuilt in the rear of the projectile which has to consider the offset of the weight of the propellant it is carrying versus the base burn that would be achieved in materializing in range extension. The propellant utilized to fire the projectile is also called the charge of a projectile which is different from the propellant used in the base bleed unit which does not work as a rocket-assisted projectile but rather a low-velocity flow to diminish the wake behind the projectile. The release of the grenades from the projectile has to be in time, keeping in account the flight of the projectile and the target area.

3.2 Ballistic Research

In more recent years, ballistic research has given a new drive to the study of base flow. An essential technique to reduce the base drag (increasing the base pressure) is based on the injection of combustible gases from the base. The recent development of computational fluid dynamics (CFD) techniques has enabled us to analyze complicated flow problems with improved numerical analysis of both fore body and base flow problems. The critical aerodynamic behavior of projectiles, indicated by rapid changes in the aerodynamic coefficients, occurs in the transonic speed regime and can be attributed in part to the complex shock structure existing on projectiles at transonic speeds. Therefore, in order to predict the total drag for projectiles, computation of the full flow field (including the base flow) must be made. Initially, the computational analysis is carried out on the flat base projectile and followed by a dome base cavity projectile. The correlation of measurements made in wind tunnels and firing trials has been found to be difficult, and it soon became clear that body shape, Reynolds number, and compressibility effects all played their part in determining drag coefficient.

Nietubicz, Sahu, Danberg, and few others started computation in US Army Ballistic Research Laboratory who had utilized inviscid/boundary-layer coupled techniques and implicit Navier–Stokes codes to study the flow fields for many different projectile configurations [1–9]. However, the results obtained from these Navier–Stokes equation-based codes have shown around 10–20% error which may be considered in good agreement for approximate range calculation, but these have a significant impact during the design of ammunition. Nietubicz et al. [4, 6–8] had used combustion of the base bleed materials into account, however, in this study temperature of 1500 K had been assumed for hot combustive gases, and the achieved numerical results were in good agreement. The average mass injection parameter was kept around 0.006. The data available is for I-.0022, which is supposed to be the initial burning phase of the base bleed unit. Each propellant used in the base bleed unit for base injection flow will have different characteristics which require the knowledge of its composition. However, using an injection parameter as small 0.0022 helps us in approximating it as a hot flow. After validating the flow field for the base bleed, the geometry of the projectile is modified numerically by increasing the boat-tailing angle and further reducing the area for the wake. The calculation has been done for 1.5 and 0.7 M, the maximum angle at which the drag can be further reduced is numerically calculated for an optimum design.

Experimental analysis:

US Army Ballistic Research Laboratory has conducted flight tests of production ammunition over different conditions. Some data is also available by wind tunnel experiments, but due to various limitations, the aerodynamic data is limited.

Kayser et al. [10] launched an instrumented M864 in September 1988. The flight experiments employed Doppler radar to track the trajectory. M864 projectile had been fired in the transonic aerodynamic range, and ranges were in access to the US

Army Ballistic Research Laboratory. The data had been obtained through a smear camera, a HAWK Doppler velocimetry and a Weibel radar system. Various pressure transducer and instruments were attached, keeping in mind there was no effect on flow over the projectile body. The used pressure transducers were completely equipped with mechanical stops to have overload protection up to 40 times the rated pressure. The important factor was transducer sensitivity to acceleration which was very low. As reported, the typical value of sensitivity was found to be 0.0005% of full scale per "g" perpendicular to the diaphragm and 0.0001% transverse to the diaphragm. Also, the maximum projectile acceleration was recorded within the gun tube on the order of 6000 g, while the same was found to be less than 3 g during flight and data acquisition. The temperature inside the propellant chamber was measured using a tungsten-rhenium thermocouple. The thermocouple was inserted into the propellant chamber through the hole which was drilled through the transducer fixture and the front wall of the base assembly. It was difficult to achieve the results initially especially in the transonic region, due to which noise reduction was undertaken using additional instruments. A pressure transducer was placed on the ogive of the projectile and was used to measure the pressure on the forebody of the projectile. Another important motion, i.e., yawing motion, was recorded using a sensing device of a Yawsonde which was mounted in the ogive flush to the exterior of the projectile as solar sensors. The voltage signals obtained through these sensors, i.e., Yawsonde and pressure transducer, were converted to modulated frequencies, which was later combined with the signals received from the base of the projectile. All of the mixed frequency signals were then used to modulate a transmitter carrier frequency of 250 MHz. The signal broadcast from the projectile was received by antennas on the ground near the launch site and recorded on magnetic tape.

The firing trials conducted by the US Army Ballistic Research Laboratory had been conducted several times for different types of experiments. The mode of testing had been almost similar as discussed above, barring some additional instruments and changes in launch conditions. Some of the tests have also been done on ground fixture especially for a cold injection case, where mass flow of air was used in wind tunnel. The transition of testing the inert projectile to base bleed projectile included many changes which mainly included protection of sensors and other instruments required for measuring data.

Numerical Analysis:

CFD offers analysts the ability to test virtually any aerodynamic problem without the limitation of the available experimental apparatus. However, CFD is limited in the accuracy with which it predicts the flow physics and the time which it takes to recreate and solve the problem virtually. In 1985, Sahu and Nietubicz initially started their computational analysis of projectile with and without base bleed [6]. At that time, the testing projectiles were on flat base, and most of the experiments were referred from wind tunnel experiments to validate their CFD results. The most restricting factor was the processing technology of the computers due to which computational grid was limited, and the solution was time-consuming.

In 1993, Charles J. Nietubicz from US Army Ballistic Research Laboratory and Howard J. Gibeling from Scientific Research Associates undertook the numerical analysis of a dome base cavity M864 projectile [4]. In addition to the above work, there was a chemistry model wherein composition of propellant grain was used and combustion was taken into account. The numerical data was validated with the previous flight test data and some wind tunnel experiments. The non-dimensional injection parameter was taken to be 0.0022. Given the lack of detailed experimental data for the M864, comparisons had been made for the zero yaw drag values, which can be obtained from the computations and experimental flight tests. In recent times, some other literature also reported on drag reduction of projectiles using numerical methodology [5, 11, 15–17].

In this study, we have chosen an M864 155 mm artillery ammunition which has a base cavity and base bleed unit incorporated within it with a boat-tail angle of 3° [4, 7]. The M864 uses the base burn theory of reducing base drag by injecting gas, generated by burning propellant, into the base area. In order to numerically finalize a final design utilizing computational approach, it is indispensable to validate the results with experimental analysis. Therefore, the validation is carried out against the aerodynamic data obtained from the firing carried out in the ranges of US Army BRL and flight test [3, 4].

4 Numerical Details

In the present work, a spinning M864 artillery cell has been studied using a numerical approach. Because of the symmetry of the geometry, the present calculations are completed utilizing a two-dimensional axisymmetric configuration. The mass, momentum, energy, and turbulence equations are solved using the Favre-averaged governing equations. The turbulence field is modeled using a two-equation SST $k-\omega$ model. The density-based coupled solver is invoked due to high Mach number regime while the second order upwind scheme is used to discretize all the convective fluxes. The fluid material is taken as air and treated as an ideal gas, while the thermal conductivity is calculated based on kinetic theory and viscosity according to Sutherland law. All computations carried out in the present work are performed using ANSYS FLUENT [20].

Geometry and boundary conditions:

A contour sketch of 155 mm M864 base bleed projectile is shown in Fig. 6. The approximate geometry obtained from this model is made in solidworks and later exported to ICEM for grid generation [20]. Since the projectile is axisymmetric, a 2D geometry is created. Length of the projectile is 0.896 m mm, and its max diameter is 155 mm. The cylindrical section is slightly undercut, and the boat-tail length is 0.505 calibers with the approximately 3° angle. The model nose is a combination of ogive and conical sections. Figure 6 depicts the geometry both for flat base and dome cavity. The geometry is tested first for a flat base and then a dome cavity. Though we do not have experimental data for the flat projectile, it has been compared with data from Sahu [6] as a road map for further validation.

A 2D domain of 16 m in X-axis and 6 m in Y-axis is chosen to ensure proper development of the flow field in the domain (Fig. 7). The geometry has been broadly divided into four parts such as front, projectile body, rear, and hole. The vertical domain is almost 50 times the caliber of the projectile while axis from the base almost 100 times the caliber. This large extension of the domain ensures that appropriate effect of the wake region is taken into account for the base drag. This adaptation becomes more critical as base bleed is activated with the mass flow inlet from the hole. For calculating the optimum design, the boat-tail angle is altered with the same configuration of meshing.

No slip boundary condition is imposed over the projectile body with rotating wall conditions to give the projectile the spinning effect. The top domain is kept at pressure far-field condition as shown in Fig. 7. The ambient condition is maintained as the Mach number at which it is tested. Velocity inlet conditions are maintained at leftmost domain boundary while pressure outlet is imposed at the rightmost boundary of the domain, so that specifically ambient static pressure is used only when the flow is subsonic and all the flow quantities are extrapolated from the interior, should the flow become supersonic. Considering high Reynolds number at the inlet, the turbulent viscosity ratio is kept at 100 while the free stream turbulent intensity is maintained at 5%. For low subsonic flow at the hole in a base bleed case, the turbulent viscosity

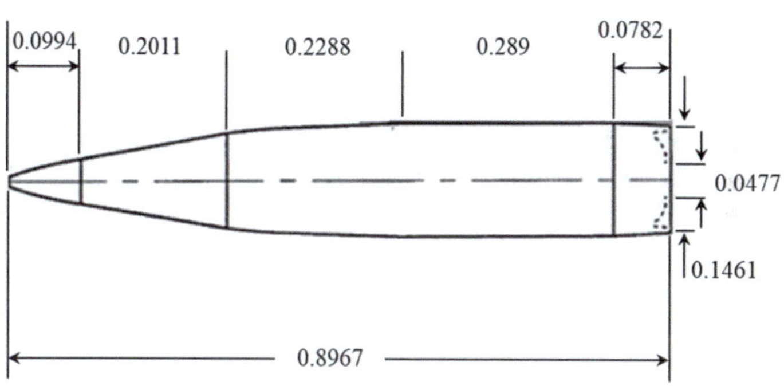

Fig. 6 Geometry of M864 projectile with dimensions in m [4]

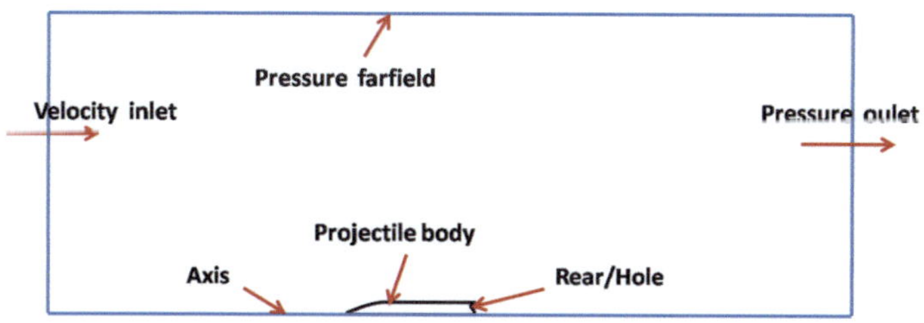

Fig. 7 Computational domain of a projectile

Table 1 Boundary conditions

Inlet	Velocity inlet
Domain wall	Pressure far field
Axis	Axis
Outlet	Pressure outlet
Projectile body	Rotating wall
Front (nose)	Rotating wall
Rear	Rotating wall
Hole	Rotating wall (no injection)/mass flow inlet (injection)

ratio of 1 while intensity at 1% is maintained. The flow is developed at a fixed static pressure (atmospheric pressure), and the surrounding temperature is taken to be 294 K. Details of boundary conditions can be perceived through Table 1. For no injection and cold injection, the temperature is 294 K, however, for hot injection, the temperature at the hole is taken to be 1500 K which is assumed as the combustion temperature of the hot gases generated by the propellant in the gas generator attached to the base. The rotating wall is set to 898 rad per sec (rps) as per the test condition. After carrying out the grid independence test, the best grid of ~0.87 M is chosen for detailed analysis.

Boat tailing of projectile:

The boat tailing of the projectile has been undertaken after the base bleed validation. The boat tail of the M864 projectile geometry has been numerically analyzed gradually on varying degrees 1°, 3°(original model), 5°, 6°, 7°, 8°, and 9° (Fig. 8). In order to evaluate an optimum design, 1.5 Mach in supersonic and 0.7 Mach in the subsonic regime are selected for computational analysis. It is ensured that the aspect ratio and quality are maintained so that the computational analysis is done at the same reference. It is observed that the value of C_D declines continuously as the angle increases to a certain point, where C_D starts increasing with the angle. This will be further elaborated in results and discussion. The computationally designed boat-tail angles (including grid) for M864 model are depicted in Fig. 8.

5 Results and Discussion

The study has been broadly conducted in two phases. The first phase includes the validation of base bleed projectile while the second phase involves boat tailing of a given projectile at different angles to finalize a conclusive design for optimum drag reduction. The validation is carried out with dome cavity without injection and has been compared against the experimental data of US BRL obtained from their transonic aerodynamic range and flight test. The second phase involves base injection wherein the base region is divided into rear and hole. The base injection is

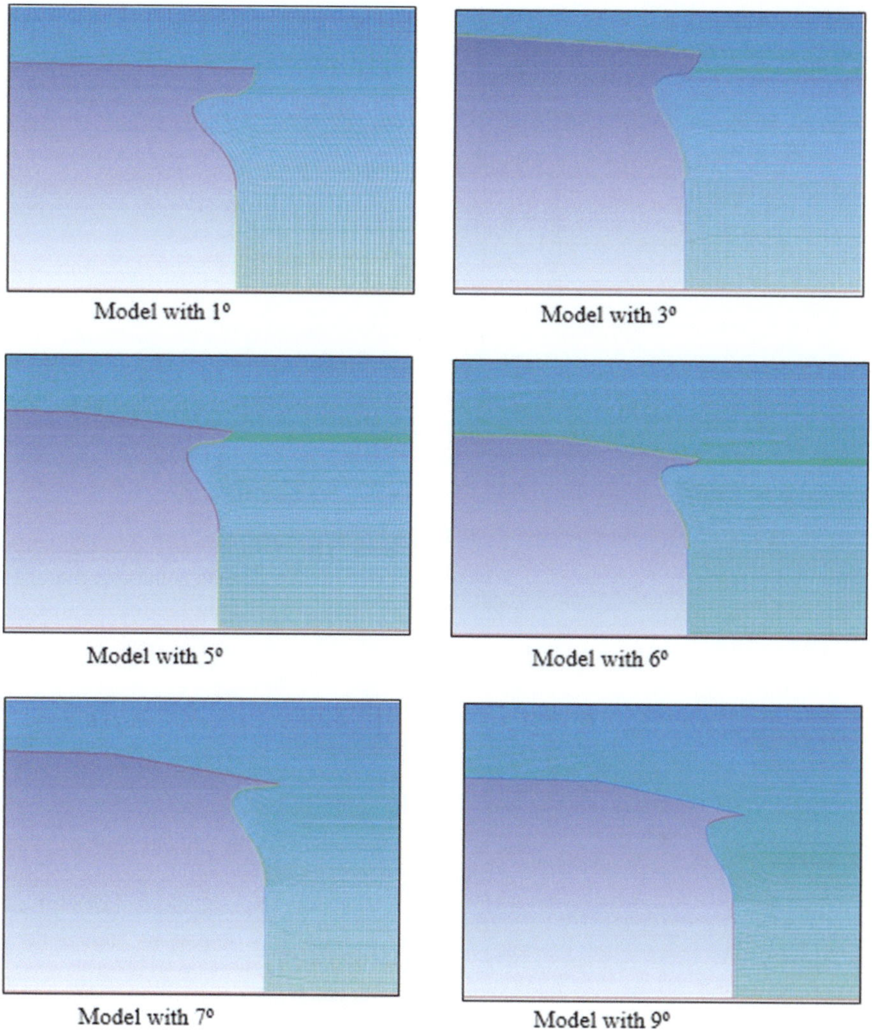

Fig. 8 Different geometry with varying boat-tail angle

done with cold injection at 300 K and hot injection at 1500 K flow from the base of the projectile.

The experimental results are taken from US Army BRL [10] and various flight tests [4]. It is observed that the coefficient of drag (C_D) increases with Mach number in the subsonic regime and then decreases in supersonic flow. Moreover, a drastic change in C_D with Mach number is observed in the transonic range along with the inclusion of drag divergence. Considering that the variation of C_D with Mach number is very gradual, the computation results are in good agreement with not more than 3% error except in the transonic region where it ranges up to 7%.

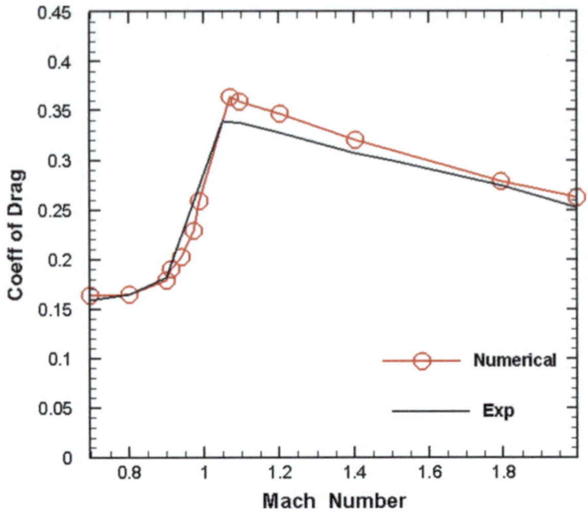

Fig. 9 Comparison with transonic range at US BRL [10]

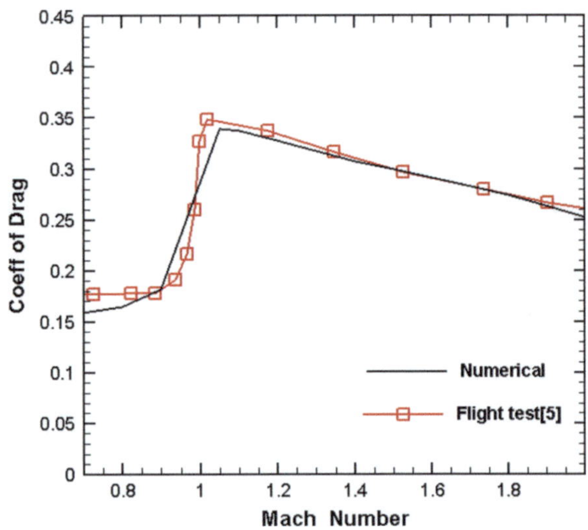

Fig. 10 Comparison with flight test data (inert)

5.1 Inert Base Validation

The initial validation has been carried out without base bleed where the entire base portion has been treated as a wall. The base cavity also assists in reducing some drag as compared to the flat base. Studies have shown that variation in the base cavity can also reduce the drag. The validation without injection is represented in Figs. 9 and 10.

Pressure coefficients (C_p) for subsonic and supersonic are shown in Fig. 11. It is observed from the surface pressure distribution curve that as the flow comes across

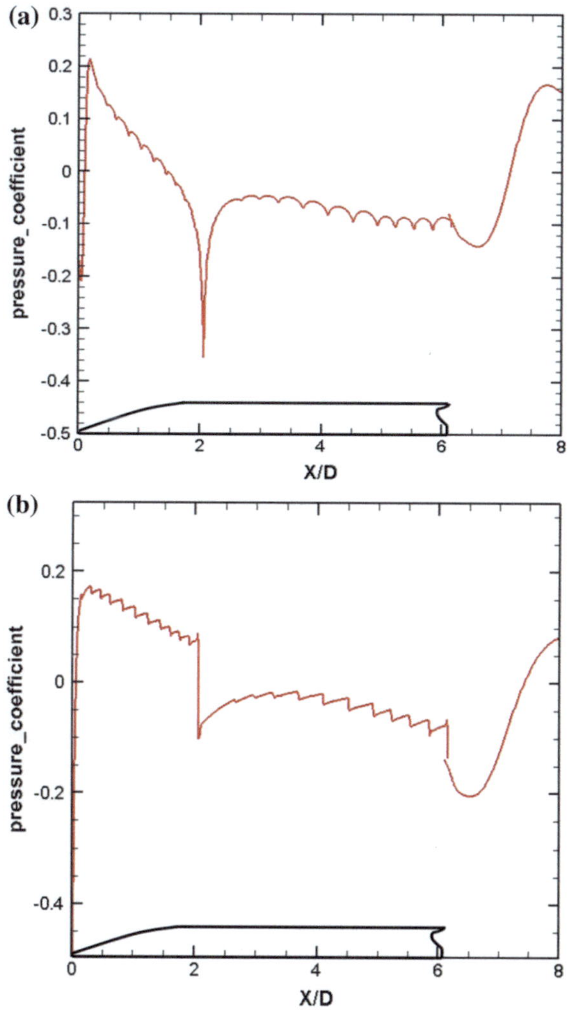

Fig. 11 Surface pressure distribution for $I = 0$: **a** Mach = 0.7 and **b** Mach = 1.5

the projectile body, there is an increase in C_p and it further decreases as the flow moves over the conical portion. There is a steep fall in pressure due to the expansion fan, and then it increases gradually along the projectile surface. Toward the base of the projectile, we can see a drop in pressure due to the wake region. As the flow moves ahead of the recirculation region, the pressure is recovered, and the flow reattaches with main flow.

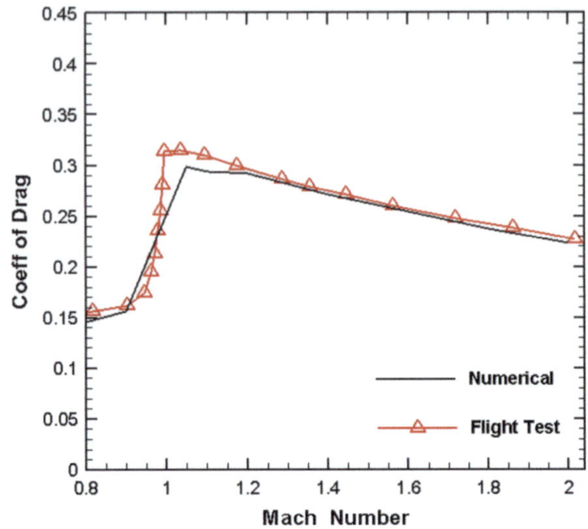

Fig. 12 Comparison with flight test data ($I = 0.0022$) [4]

5.2 *Base Bleed Validation*

The base bleed validation is conducted with $I = 0.0022$ [5] and temperature of 1500 K at the hole portion of the projectile. The computation has been done in the same configuration as that of no injection as depicted in Fig. 12. A base bleed flow is activated by changing the boundary conditions from hole to mass inlet flow. The mass flow rate is calculated by Eq. 3, based on the injection parameter. The injection parameter is at the launching stage of the projectile excluding the effect of the interior ballistics of the gun barrel [5].

The computation for base bleed has been undertaken for cold and hot injection. There is a slight reduction in C_D for hot injection flow as compared to the cold injection case, due to hot temperature flow at the base which increases the pressure in the wake region. The base pressure would further increase as the mass flow rate increases during the flight [4]. An appreciable reduction in C_D can be observed due to base bleed as compared to the inert validation. It is specially brought to notice that the injection parameter which has been set for testing is $I = 0.0022$, which is very low and depicts the launch condition when the mass flow rate is not much as compared to the average mass flow rate which would be around 0.006–0.01.

Figure 13 exhibits the comparison of surface pressure distribution and base pressure distribution. It is observed that the pressure distribution curves over the projectile profile are identical except at the base where an increase in pressure can be seen with the increase in injection parameter. The commensuration with base bleeds is shown in Fig. 3, the flow moves toward the annulus region where appreciable pressure difference can be seen as depicted in Fig. 13b. The predictions are tabulated in Table 2.

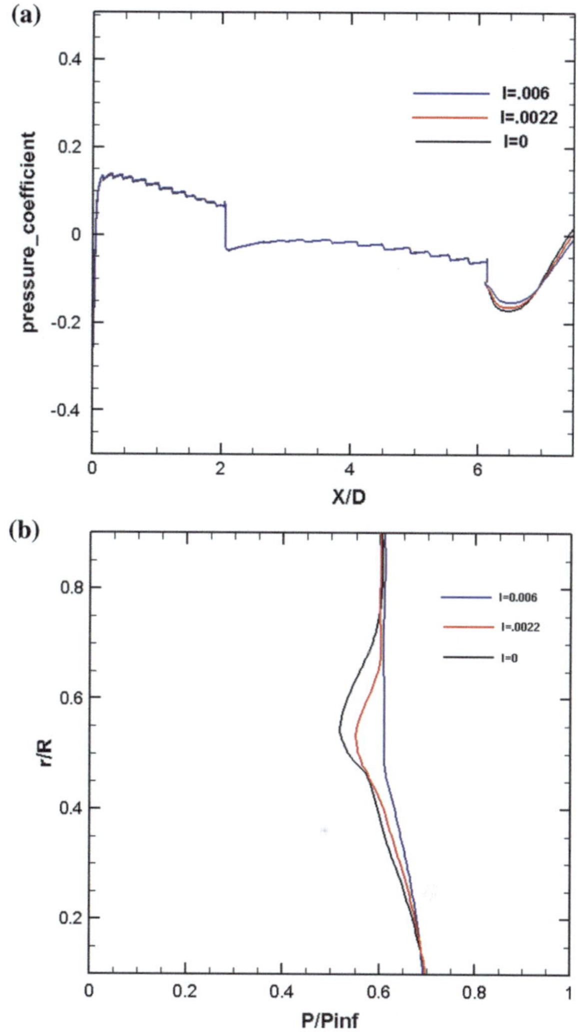

Fig. 13 Flow at Mach 2 (I = 0, 0.0022, 0.006): **a** surface pressure distribution and **b** base pressure distribution

5.3 *Design Optimization with Boat-Tail Angle*

After validating the base bleed projectile, the second phase is to alter the design without changing the profile of the projectile as this would require an entirely different grid system, and the validation would not be of any consequence. Moreover, as an essential design requirement, the ammunition has to be fired from the same gun. Thus the drag optimization is undertaken through boat tailing. The increase in the boat-tail angle reduces the cross-section area of the base by reducing the base diameter; however, as the angle increases it reaches a minimum point for the CD, at that angle

Table 2 Comparative summary: M864 base bleed projectile

Mach No.	C_D (no injection)		C_D (cold injection)	C_D (hot injection)	
	Exp	Num	Num	Exp	Num
0.8	0.177	0.165	0.151	0.156	0.146
0.9	0.183	0.182	0.165	0.162	0.156
1.05	0.350	0.338	0.300	0.314	0.299
1.1	0.345	0.337	0.299	0.311	0.294
1.2	0.335	0.328	0.298	0.297	0.292
1.4	0.311	0.304	0.278	0.273	0.271
1.5	0.301	0.297	0.267	0.265	0.263
1.8	0.275	0.274	0.248	0.242	0.237
2.0	0.258	0.252	0.227	0.227	0.223

Table 3 Variation of bubble length

Mach number	Angle (Degree)	Bubble length (m)
0.7	1	0.23
0.7	3	0.22
0.7	9	0.16
1.5	1	0.235
1.5	3	0.225
1.5	9	0.19

C_D starts increasing. The maximum boat-tail angle that can be achieved depends upon the geometric profile and Mach number at which the projectile is traveling [2].

It is evident from streamlines seen in Figs. 14 and 15 that with the increase in boat-tail angle, the low wake pressure region is decreasing thus resulting in drag reduction. The flow in the wake region is similar to a back-step flow. Thus the bubble length will vary by the Mach number and boat-tail angle. Bubble length at various boat-tail angles is shown in Table 3. It can be observed from the above figures and Table 3 that the measured bubble length of the wake region is reducing with an increase in boat-tail angle. The increase in boat-tail angle is reducing the base diameter also. Thus, overall wake region is decreasing. It is important to mention that as the boat-tail angle is varied, its perimeter length is increasing which leads to an increase in skin friction. However, there is much more reduction in base drag with an increase in boat-tail angle which leads to an overall reduction in drag up to an optimum value for a particular Mach number.

The variation of angle with C_D is shown in Fig. 16 for Mach 1.5 and 0.7. The results reveal that at 8°, drag for Mach = 1.5 again starts increasing marginally while it continues to decrease for Mach = 0.7. Thus, the optimum boat-tail angle for this regime is around 7–8°. It can be observed from Fig. 16 that almost 5% more reduction is achieved at Mach = 1.5 and around 20% more drag reduction at Mach

Fig. 14 Pressure contours at Mach 0.7: **a** 1° and **b** 9°

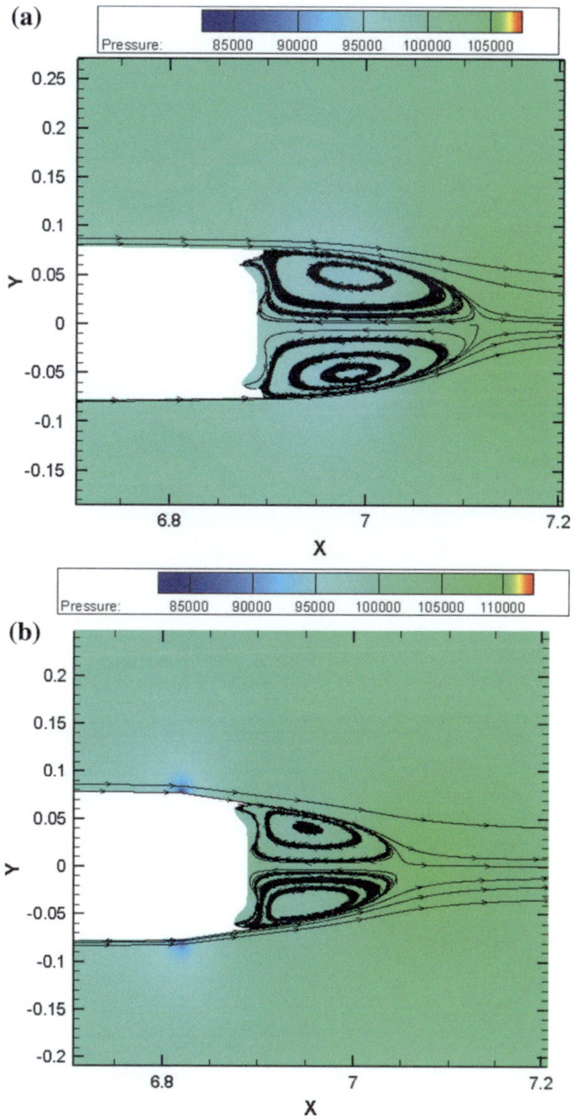

= 0.7. Considering the overall flight of a projectile launched at Mach = 1.5, the average flight velocity may be much lower. Thus optimum boat-tail angle will result in reasonable drag reduction. This can be substantiated with the observation made in base pressure distribution curve (Fig. 17) that there is a rise in base pressure in the wake region with the increase in boat-tail angle.

Fig. 15 Pressure contours at Mach 1.5: **a** 1° and **b** 9°

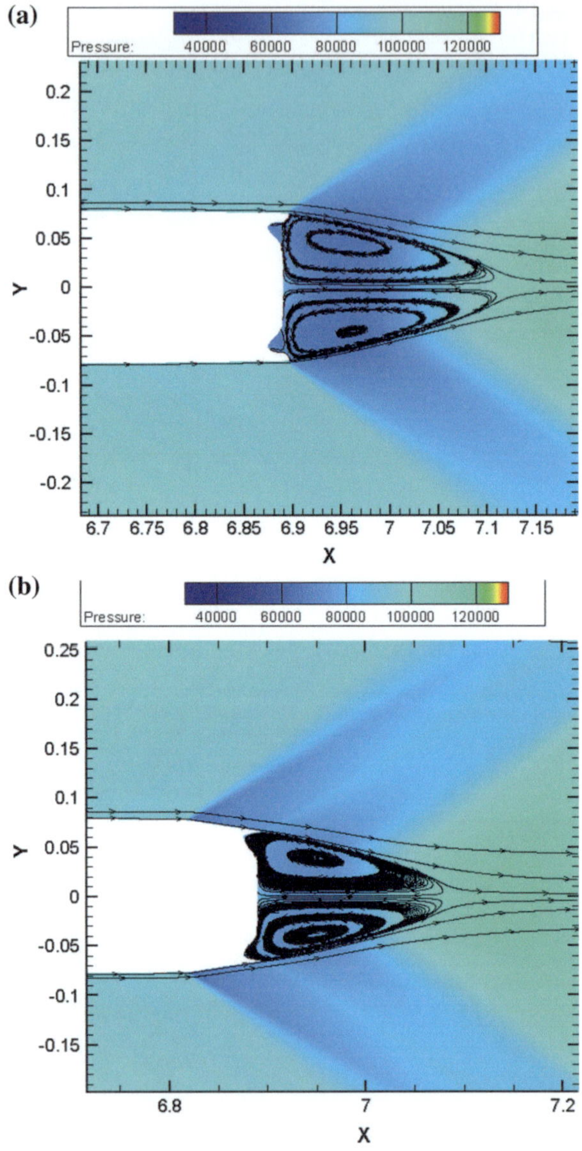

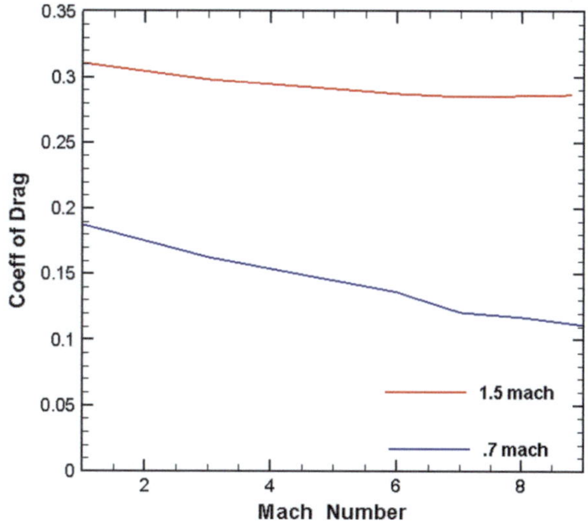

Fig. 16 Variation of C_D with boat-tail angle

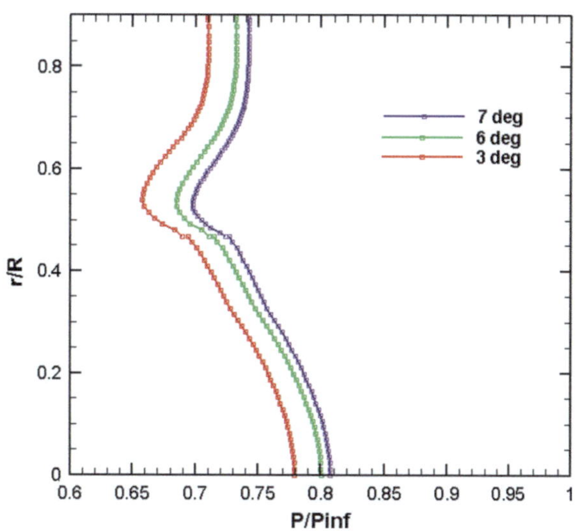

Fig. 17 Base pressure distribution at Mach = 1.5 (3°, 6°, 7°)

6 Conclusion

The study focuses on drag reduction of a given projectile to enhance its range with a combination of base bleed and boat tailing. Main conclusions of this work can be summarized in two parts as follows:

1. Predictions of drag coefficients through validation of the M864 projectile with and without base bleed.
2. Alteration of the boat-tail design to achieve an optimum drag reduction.

The flow domain over the projectile has been studied with an all-out effort to reduce the drag. The external flow is simulated using a 2D axisymmetric spinning projectile. The assumption used for modeling the hot flow as a substitute to combustion is found to be very useful to near launch conditions, however, for intermediate flow, this may show variance and depend, to a certain extent, on the composition of propellant grain. The present predictions on the M864 projectile have been observed to be more accurate as compared to the numerical results achieved in US Army BRL. The earlier studies have issues with the transonic zone, and appreciable under prediction could be observed for coefficient of drag (C_D) in the subsonic region. After the validation, boat tailing has been done on the same geometry, and the results show that there is a scope to further increase the boat-tail angle in the original geometry for further drag reduction. By numerically evaluating the optimum geometry, it would assist the manufacturer in evaluating a final design with an appreciable reduction on logistics and time, thus reducing the efforts required in testing the ammunition experimentally.

Acknowledgements The authors would like to acknowledge the IITK computer center (www.iitk.ac.in/cc) for providing support to perform the computation work, data analysis, and article preparation.

References

1. Sahu J (1986) Drag predictions for projectiles at transonic and supersonic speeds. US Army Ballistic Research Laboratory Aberdeen Proving ground, Maryland. Memorandum Report BRL-MR-3523, June 1986
2. Suliman MA, Mahmoud OK, Al-Sanabawy MA, Abdel-Hamid OE (2009) Computational investigation of base drag reduction for a projectile at different flight regimes. In: 13th international conference on aerospace sciences & aviation technology (ASAT-13), May 26–28 2009, Egypt, ASAT-13-FM-05
3. Lieske RF (1989) Determination of Aerodynamic drag and exterior ballistic trajectory simulation for the 155 mm. DPICM, M864 base-burn projectile. US Army Ballistic Research Laboratory Aberdeen Proving ground, Maryland. Memorandum Report BRL-MR-3768, June 1989
4. Nietubicz N, Gibeling C (1993) Navier-Stokes computations for a reacting, M864 base bleed projectile. In: 31st aerospace sciences meeting & exhibit, 11–13 Jan, Reno, AIAA 93-0504
5. Belaidouni H, Zivkovic S, Samardzic M (2016) Numerical simulations in obtaining drag reduction for projectile with base bleed. Sci Tech Rev 66:36–42
6. Sahu J, Nietubicz CJ, Stegerf JL (1985) Navier-Stokes computations of projectile base flow with and without mass injection. AIAA J 23:1348–1355
7. Nietubicz CJ, Sahu J, Heavey KR (1988) Supercomputer applications in projectile aerodynamics. In: Army science conference proceedings (AD-A203102), vol II, pp 368–383
8. Danberg JE, Nietubicz C (1992) Predicted flight performance of base bleed projectiles. J Spacecraft Rockets 29:366–372
9. Balon R, Komenda J (2006) Analysis of the 155 mm ERFP/BB projectile trajectory. Adv Mil Technol 1:91–144
10. Kayser LD, Kuzan JD, Vazquez DN (1990) Flight testing for a 155 MM Base Burn projectile. Ballistic Research Laboratory, Apr 1990, AD-A222562
11. Kubberud N, Jarle Oye I, Prytz AK, Raufoss ASN (2011) Extended range of 155 mm projectile using an improved Base Bleed unit—simulations and evaluation. In: 26th International symposium on ballistics, Miami, Florida, USA, 12–16 Sept 2011

12. Gibeling R, Buggeln RC (1992) Projectile base bleed technology, part I: analysis and results. Army Research Laboratory November 1992, AD-A258-459
13. Viswanath PR (1996) Flow management techniques for base and afterbody drag reduction. Prog Aerosp Sci 32:79–129
14. Lieske RF, Reiter ML (1966) Equations of motion for a modified point mass trajectory. Ballistic Research Laboratory, March 1966 (AD 485869), Report No. 1314
15. Das P, De A (2017) Numerical study of supersonic flow past a cylindrical afterbody. J Aerosp Sci Technol 69(1):25–35
16. Das P, De A (2016) Numerical study of flow physics in supersonic base-flow with mass bleed. Aerosp Sci Technol 58:1–17
17. Das P, De A (2015) Numerical investigation of flow structures around a cylindrical afterbody under supersonic condition. Aerosp Sci Technol 47:195–209
18. Nietubicz CJ, Sahu J (1991) Navier-Stokes computations of base bleed projectiles. Int J Energ Mater Chem Propul 1:93–105
19. Sturek WB, Nietubicz CJ, Sahu J, Weinacht P (1994) Applications of computational fluid dynamics to the aerodynamics of army projectiles. J Spacecraft Rockets 31:186–199
20. ANSYS Fluent 18.0 User's guide, Canonsburg, PA, USA

An Experimental Investigation of Mass Transfer Cooling Techniques for Atmospheric Entry Vehicles

S. Mohammed Ibrahim and K. P. J. Reddy

Keywords Hypersonic Aerothermodynamics · Mass transfer cooling

1 Introduction

A hypersonic flow is best defined as a flow whose kinetic energy is equivalent to the dissociation energy of the test gas molecules in the flow [1]. Such flows are encountered during planetary entry of spacecraft and other foreign bodies, atmospheric scramjet flight test, ballistic phase of ICBM, etc. An important aspect of hypersonic flows is the associated aerodynamic heating, and it plays a crucial role in the design of these vehicles. When a vehicle travels at hypersonic speed, it encounters a strong shock wave ahead of it. The region between the shock wave and the surface of the body is defined as shock layer. The large kinetic energy of the freestream, associated with hypersonic flight, gets converted into internal energy of the gas across the shock wave [2], creating a very high temperature in the shock layer, temperatures

S. Mohammed Ibrahim (✉)
Department of Aerospace Engineering, Indian Institute of Technology—Kanpur, Kanpur 208016, India
e-mail: ibrahim@iitk.ac.in

K. P. J. Reddy
Laboratory for Hypersonic and Shock Wave Research (LHSR), Department of Aerospace Engineering, Indian Institute of Science, Bangalore 560012, India

A. K. Gupta et al. (eds.), *Innovations in Sustainable Energy and Cleaner Environment*, Green Energy and Technology,
https://doi.org/10.1007/978-981-13-9012-8_3

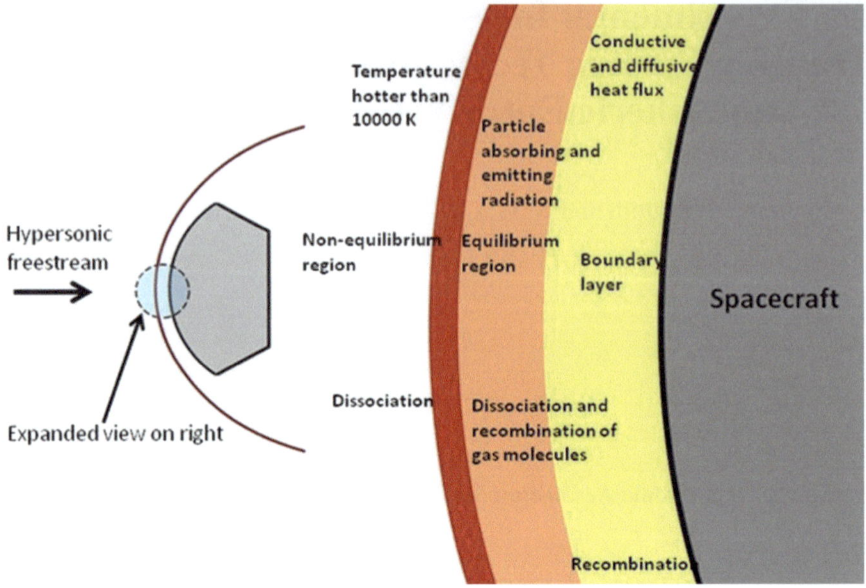

Fig. 1 Schematic of typical shock layer in a hypersonic flight [2, 3]

high enough to cause vibrational excitation, dissociation and recombination of gas molecules and ionization, so-called high temperature effects. The flowfield existing behind the shock wave over a blunt body, in a hypersonic flight, is shown in Fig. 1.

The high temperature existing in the shock layer will be transferred to the spacecraft as heat flux. The two main modes of heat transfer to the spacecraft from the shock layer are convective and radiative transfer. Convective heat transfer is due to the flow of hot gas over the spacecraft, while radiative heat transfer is due to the thermal radiation emitted by the molecules at high temperature existing in the shock layer. Both convective and radiative heat fluxes increase with increasing freestream velocity, with radiative heat transfer dominating at higher velocities [3] as shown in Fig. 2.

Generally, large angle blunt cones are used as forebodies for the spacecraft to minimize the aerodynamic heating [4–6]. These configurations also enhance aerodynamic drag, assisting in reducing the speed of the spacecraft during entry. However, the spacecraft requires an additional thermal protection system, acting as a barrier, to withstand the high temperature existing in the shock layer, for safe landing on the surface. The spacecraft might disintegrate during entry if TPS fails or not designed to the required standards.

The design of an effective TPS depends on the knowledge of total heat transfer to the spacecraft at all points on its trajectory. This requires extensive theoretical, experimental and computational studies and flight testing to get preliminary measures of heat transfer rates. Flight test data are more accurate; however, they are very expensive and can be carried out only after design validation of the vehicle. Though

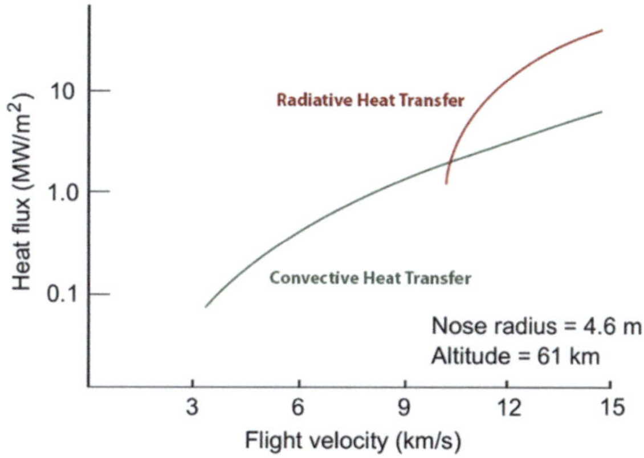

Fig. 2 Variation of convective and radiative heating with flight velocity [2, 3]

computational tools are fast emerging, they still cannot be relied upon, especially in hypersonics, as it requires validation against experimental and flight test data. So we are left with experimental techniques to get information about the high temperature existing in the shock layer and the associated heat transfer rates in a hypersonic flow.

The TPS commonly used for planetary entry of spacecraft is ablative heat shield. This TPS has been used for Apollo missions, several Mars exploration missions, Stardust, Hayabusa missions, etc. Though these systems have been successfully used, they have certain drawbacks which are listed below:

- During ablation, the complex hydrocarbon products formed give rise to chemical reaction in the boundary layer of the spacecraft, thereby making a chemically reacting boundary layer, which can have significant influence on aerodynamic forces and moments [2].
- The shape change occurring during ablation can affect the aerodynamics of the spacecraft resulting in change in flight path.
- If the spacecraft experiences insufficient heat flux to cause pyrolysis, then the TPS material will start conducting heat from the outer surface into the spacecraft, leading to TPS failure. This happened in the case of Viking 1, where the peak heat flux experienced was 21 W/cm^2 [7], not enough to cause pyrolysis, and the TPS acted as thermal insulator than an ablator.

These drawbacks drove researchers around the globe to look for the possibility of using an alternate TPS.

In the presented research work, an experimental campaign was carried out to investigate the feasibility of using mass transfer cooling, film cooling and transpiration cooling techniques. Of particular interest to us was to explore for the case of Martian atmospheric entry, so that the results can support the missions planned for future. The experiments were performed for two different flight velocities, using hypersonic shock tunnel test facility, HST2 and HST3, respectively, at Laboratory for Hypersonic and Shock Wave Research (LHSR), Department of Aerospace Engineering, Indian Institute of Science, Bangalore. The test model used is a large angle blunt sphere-cone configuration, typically used as forebodies for planetary entry spacecraft. Further details about experimentation and obtained results are presented below.

2 Experimental Test Facility

The experimental investigations reported in this work are carried out in two different hypersonic shock tunnel facilities HST2 and HST3. The test facility HST2 is a conventional shock tunnel facility and is capable of simulating flight velocity up to 2 km/s. The HST3 test facility is a free piston driven shock tunnel facility capable of generating flight velocities ranging from 2 to 8 k/s, where high temperature effects can be simulated. The two test facilities are explained in detail below.

2.1 Hypersonic Shock Tunnel HST2

The facility consists of a 50 mm diameter shock tube with 2 m long driver section and 5.12 m long driven section separated by a metallic diaphragm. The driver section is pressurized using helium gas from a high pressure cylinder which ruptures the diaphragm and generates shock wave traveling into the test gas, carbon dioxide (Martian atmospheric gas), in the driven section. The shock wave upon its reflection at the end wall stagnates the test gas, thereby increasing its pressure and temperature, which serves as the reservoir condition for the nozzle. The hypersonic nozzle is separated from the driven side by a thin paper diaphragm, opening into the dump tank containing the 450 mm long test section of size 300 mm $\times$ 300 mm, and is initially maintained at vacuum level of 10^{-5} mbar. PCB pressure sensors are flush mounted on the inner surface of the driven side to measure the shock speed and stagnation pressure, from which the reservoir conditions are calculated. A schematic of the test facility and photographic picture is shown in Figs. 3 and 4.

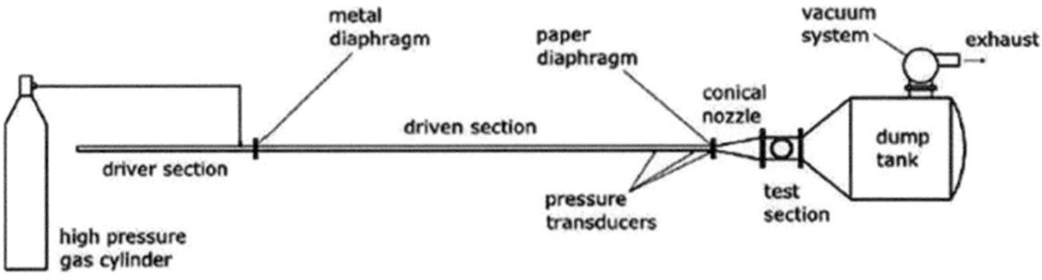

Fig. 3 Schematic of HST2 test facility

Fig. 4 Photograph of HST2

Fig. 5 Diaphragm less mode of operation using ISTA–KB-40 valve

The test facility is also operated in the diaphragm less mode by using a quick acting valve ISTA-KB-40, at the diaphragm station as shown in Fig. 5. The main reasons for using the diaphragm less valve are as follows:

- To get repeatable signals.
- To get a clean flow, flow free from metal diaphragm particles which affect the model surface and sensors mounted on it.

2.2 *Hypersonic Shock Tunnel HST3*

The test facility consists of a 10 m long 165 mm internal diameter compression tube in which a 20 kg piston compresses the helium gas adiabatically with a corresponding increase in temperature which will rupture a metallic diaphragm. The diaphragm rupture generates a strong incident shock wave with a typical shock Mach number of 8.6, which travels into the 5 m long shock tube of 50 mm diameter containing carbon dioxide at sub-atmospheric pressure. The test gas behind the reflected shock wave at the end of the shock tube expands through the convergent-divergent nozzle to

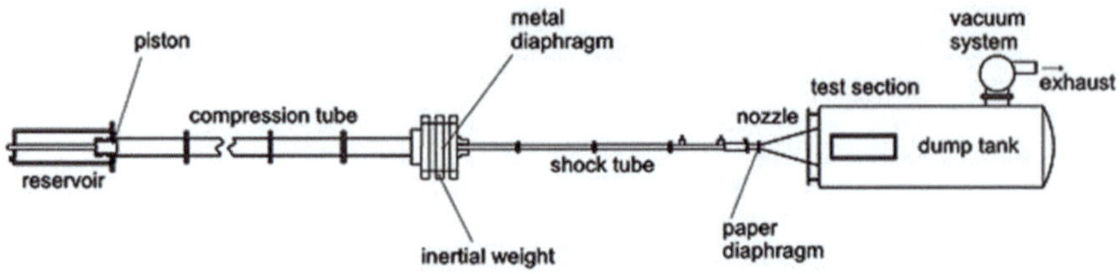

Fig. 6 Schematic of HST3

Fig. 7 Photograph of HST3

hypersonic Mach number in the test section. The test model is mounted on the floor of the dump tank along with a pitot tube. The nozzle and the dump tank assembly separated by thin paper diaphragm from the shock tube are maintained at 10^{-5} mbar vacuum using rotary-diffusion pump assembly. A schematic of the test facility and photographic picture is shown in Figs. 6 and 7.

3 Film Cooling Experiments

Film cooling is a technique where a coolant gas is injected through discrete slots from the body surface into the shock layer, formed by hypersonic flow. The injected coolant gas spreads over the body surface, forms a protective layer and takes away the heat by convection, existing in the high temperature shock layer. The concept of film cooling can be applied locally to zones of high heat transfer rate. Especially in the case of reentry bodies where high heating rates are observed in the nose portion (ballistic missiles and space shuttle) or in the conical region (stagnation zone) of a large angle blunt cone forebody entering at angle of attack. Though the injected coolant disturbs the dynamics of the spacecraft to certain extent, it serves its major purpose, reducing heat transfer substantially. A schematic of the flowfield due to

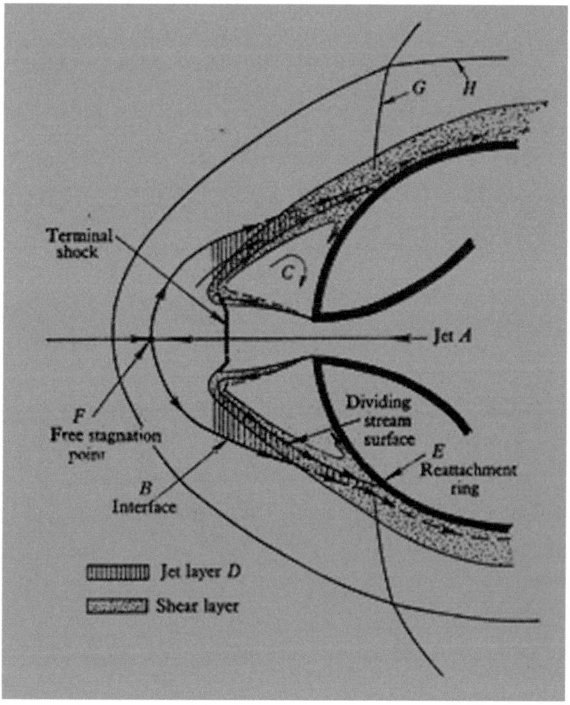

Fig. 8 Flow field during jet injection from nose of a blunt cone [8]

jet injection from the nose of a spacecraft forebody configuration is shown below (Fig. 8).

The effectiveness of film cooling over large angle blunt cones primarily depends on the jet total pressure to freestream pitot pressure ratio, $P = P_{oj}/P_{of}$. For a jet to emerge out of the body, its pressure P_{oj} should be greater than P_{of} [9]. As the value of the pressure ratio P increases from one, the body shock wave is pushed away from the body until the ratio reaches a critical value $P_{critical}$, beyond which injection of gas best serves as a drag reduction device. Hence for heat protection device, the value of P should be $1 < P < P_{critical}$. This critical value of P depends on the jet size and nose radius of the configuration [8].

Several works investigating the effect of film cooling on heat transfer rate distributions, in a hypersonic flow, have been carried out [9–16]. However, these works were limited to low-enthalpy flows, where the high temperature effects will not be significant. In this backdrop, the experiments were planned to investigate the effect of film cooling on high temperature flowfield existing during Martian atmospheric entry. The experiments were carried out in HST2 (low enthalpy) and HST3 (high enthalpy) test facilities and the flow conditions for the present case are tabulated below (Table 1).

Table 1 Flow properties in reservoir and free stream regions for HST2 and HST3

Properties	1.8 MJ/kg	3.8 MJ/kg
P_o (kPa)	2992.5 ± 115.8	10,543 ± 341.74
T_o (K)	1642.7 ± 11.7	3072.5 ± 33.7
ρ_o (kg/m^3)	9.64 ± 0.35	16.38 ± 0.35
M_∞	5.16 ± 0.03	5.97 ± 0.01
P_∞ (kPa)	0.79 ± 0.03	0.59 ± 0.01
T_∞ (K)	519.89 ± 25.93	769.99 ± 7.39
ρ_∞ (kg/m^3)	0.008 ± 0.0002	0.004 ± 0.0001
V_∞ (m/s)	1742.9 ± 6.7	2497.2 ± 14.5
P_{pitot} (kPa)	21.06 ± 0.62	13.55 ± 0.27

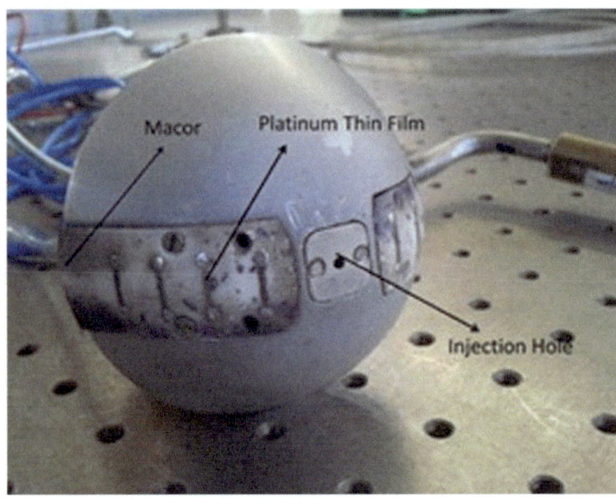

Fig. 9 Photographic picture of test model

3.1 Test Model Configuration

Experiments are carried out over a 60° apex angle blunt cone with a base diameter of 80 mm and nose radius of 35 mm. The nose portion of the model has a square slot of 13 mm × 13 mm area and thickness of 2 mm where the plate containing the 2 mm orifice, through which the coolant gas is injected into the hypersonic mainstream, is housed. A photograph of the model is shown in Fig. 9. The model has a stagnation chamber, and it gets the supply from an external high pressure gas cylinder maintained at atmospheric condition. The supply pressure from the gas cylinder is monitored using a pressure regulator. The total pressure of the jet P_{oj} is measured using a Kulite pressure sensor flush mounted on the wall of the stagnation chamber. Platinum thin film sensors deposited on Macor surface as shown in Fig. 9 are flush mounted on the surface of the model to measure the convective heat transfer rates. A total of eight thin film sensors are used in the present investigation.

3.2 Results and Discussions

The gas dynamic phenomena involved in coolant injection have been well explained in the literature [8]. All experiments in the present investigation have been carried out at 0° angle of attack. Because the injection is from the nose, which is the stagnation point at zero incidence, it is possible to generalize the qualitative trend for other angles of incidence with injection from a corresponding stagnation point (rather than the nose), insofar as the purpose of injection is not to alter the aerodynamic forces drastically by pushing the shock. The jet injection is at a low pressure ratio of $P = 1.2$, which is held the same for all freestream conditions. Because the total pressure of the jet is higher than the freestream pitot pressure, the jet exiting the orifice (minimum cross-sectional area) is considered to be sonic ($M = 1$). The coolant gas investigated is nitrogen for all test cases.

3.2.1 Experiments at 1.8 MJ/Kg

The temperature signals obtained from the thin film gauges are integrated numerically using the algorithm given by Cook and Felderman to obtain the convective heat transfer rates [17]. The heat transfer values, both in the presence and absence of coolant gas injection, are shown in Fig. 10. A total of five experiments were carried out for each set and the mean values were plotted. The run-to-run variation of the heat transfer rate was 4 W/cm^2. The measured heat transfer rate was higher near the nose of the model, which is a typical feature of blunt bodies in hypersonic flows [2], and it reduces gradually as we move downstream toward the conical portion of the model. With the injection of the coolant into the mainstream flow, a reduction in heat transfer rate was observed. The heat transfer rate signal obtained from the gauge located in the nose region of the model, both in the presence and absence of coolant injection, is shown in Fig. 11.

The injection pressure ratio being small, the coolant gas flows into the boundary layer developed over the body and forms a film around it. This takes the heat from the mainstream and modifies its characteristics such that there is a reduction in heat

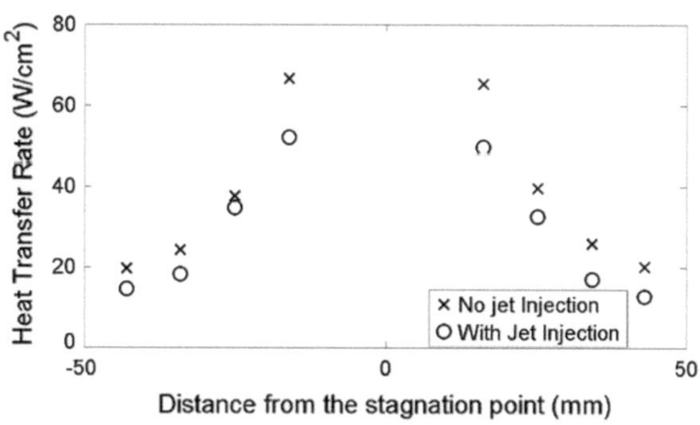

Fig. 10 Heat transfer distribution on the model surface for 1.8 MJ/kg

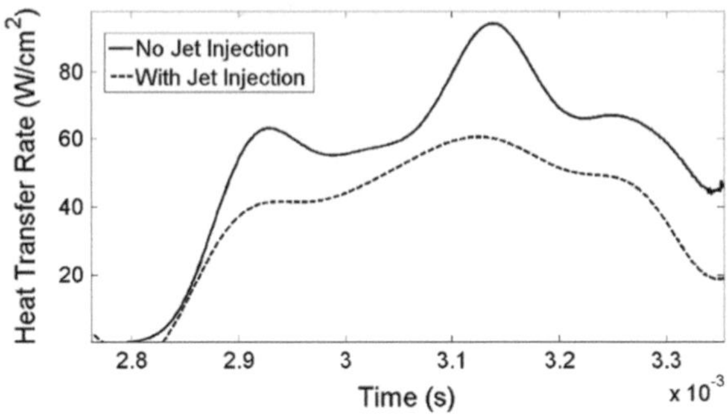

Fig. 11 Heat transfer rate signal for gauge located in the nose region of the test model, $s/R_b = 0.4$

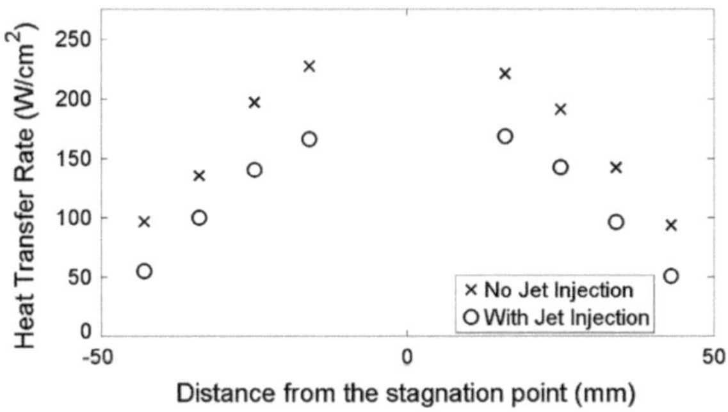

Fig. 12 Heat transfer distribution on the model surface for 3.8 MJ/kg

transfer to the body. The gauge located in the nose region, close to the stagnation point, had higher reduction in heat transfer rate than those in the conical segment of the model. This is because the coolant gas injected from the geometric stagnation point of the model diffuses faster near the nose portion than in the conical region of the model [9].

3.2.2 Experiments at 3.8 MJ/Kg

The measured heat transfer rates, both in the presence and absence of coolant gas injection, are plotted in Fig. 12. A total of four runs were carried out for each case and the run-to-run variation in heat transfer rates was less than 8 W/cm^2. A trend similar to the 1.8 MJ/kg case was observed. However, as expected, the heat transfer rates were significantly higher than the 1.8 MJ/kg case.

This is due to the very high energy content of the freestream, the kinetic energy of which gets converted into thermal energy behind the body shock [2]. The temperature in the shock layer could be sufficient to initiate vibrational excitation, dissociation and recombination of carbon dioxide molecules, giving rise to the so-called high temperature or real gas effects. These effects will not be significant at the lower enthalpy of 1.8 MJ/kg. However, at higher enthalpies, this may be significant, and so, CHEMKIN, a chemical kinetics software, was used to predict the dissociation

Table 2 Species and their respective mole fraction predicted using CHEMKIN

Species	Mole fraction
CO_2	0.894
CO	0.070
O_2	0.035
O	0.001
C	1.0 e^{-13}

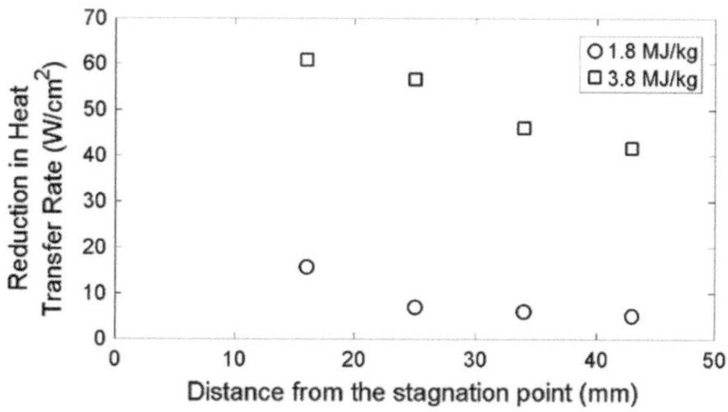

Fig. 13 Comparison of reduction in heat transfer due to film cooling at 1.8 and 3.8 MJ/kg

of carbon dioxide molecules at 3.8 MJ/kg, corresponding to a total temperature of 3072 K [18]. The mole fractions of the resulting species are listed in Table 2. Dissociation of carbon dioxide molecules was not predicted by CHEMKIN at 1.8 MJ/kg.

The reduction in heat transfer rate due to coolant injection at 3.8 MJ/kg was higher than the 1.8 MJ/kg case, and they are compared in Fig. 13. Owing to the symmetry in heat transfer rate distribution, the values are compared for one side of the body. For 3.8 MJ/kg, the reduction in heat transfer was four times higher than the 1.8 MJ/kg in the nose region of the model. Also, along the conical portion, the reduction was an order of magnitude higher than the corresponding low-enthalpy value.

The trend may be understood better by comparing normalized reduction in heat transfer rates for the two cases, given in terms of percentage reduction in heat transfer rate with respect to the heat transfer rate without coolant injection for the specific case (at given location). A comparison of percentage reduction in heat transfer rates between the two different enthalpies along the surface of the blunt cone is shown in Fig. 14.

Everywhere on the surface, the percentage reduction in heat transfer rate is significantly higher for the 3.8 MJ/kg case. The higher reduction in heat transfer rate at 3.8 MJ/kg occurs despite the fact that, by maintaining the same pressure ratio, the mass flow rate is lower at 3.8 MJ/kg than at 1.8 MJ/kg. This is due to the lower pitot (as well as static) pressure at 3.8 MJ/kg. Whereas the mass flow rate is $2e^{-4}$ kg/s at 1.8 MJ/kg, it is only $1.4e^{-4}$ kg/s at 3.8 MJ/kg. The present experimental results indicate concretely that the cooling is more effective at higher velocity (or equivalently higher enthalpy) for the same injection pressure ratio ($P = 1.2$). Though it must be admitted that the maximum freestream velocity in the present experiments

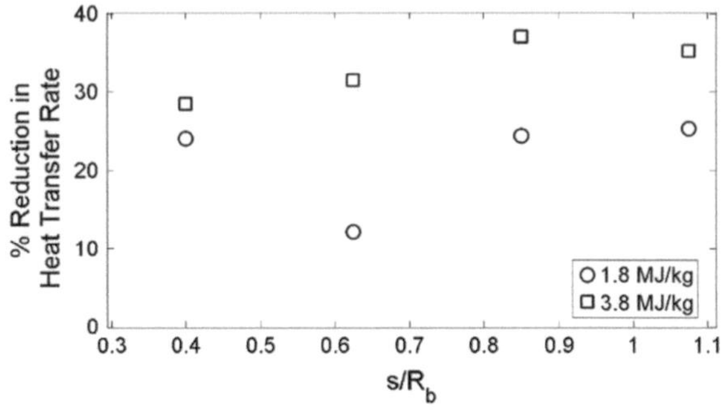

Fig. 14 Comparison of percentage reduction in heat transfer due to film cooling at 1.8 and 3.8 MJ/kg

is around 2500 m/s (corresponding to 3.8 MJ/kg), the predicted increase in cooling effectiveness with increase in freestream velocity could be useful and encouraging for extending the conception of film cooling for realistic Mars reentry velocities.

The higher percentage reduction in heat transfer rate at higher enthalpy may be attributed to the excitation of vibrational modes of the coolant gas nitrogen. At higher temperatures, the heat from the mainstream is absorbed considerably in exciting the vibrational modes of nitrogen, due to which the temperature (due to translational mode) of the coolant gas is not increased correspondingly. Because these phenomenons are more effective at higher temperatures, the reduction in heat transfer rate is higher downstream of the stagnation zone, because the coolant gas starts experiencing a rise in temperature only in the stagnation zone where it comes in contact with the main flow for the first time. Such real gas effects are absent at lower enthalpies.

4 Transpiration Cooling Experiments

In transpiration cooling, the coolant gas from the plenum chamber is forced into the boundary layer through a porous wall. As a result of which the coolant gas comes out as a continuous mass and forms a layer of it over the model surface as shown in Fig. 15. The mechanism of transpiration cooling involves two steps. First, the coolant gas on passing through the porous wall absorbs the heat flux conducted into the material of the wall from the shock layer. Second, having passed through the porous wall, the coolant gas forms a film on the model surface, absorbs the heat flux partially through convection and thus reduces the heat flux conducted into the wall from the high temperature gas existing in the shock layer. The heated coolant gas is flushed downstream by the continuous supply of gas from the plenum chamber. In this way, the heat transferred to a forebody traveling at hypervelocities can be greatly reduced. With the development of ceramic matrix composites (CMC) like C/C which can withstand temperature exceeding 2500 K and having natural porosity, transpiration

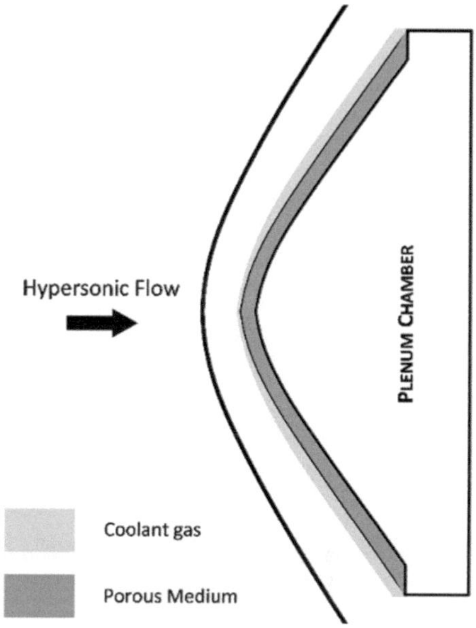

Fig. 15 Schematic of transpiration cooling technique

cooling seems to be a promising technique which can be applied to vehicles traveling at hypersonic speeds [19].

Several research works have been carried out studying the effect of transpiration cooling on heat transfer distribution [20–26]. However, most of the transpiration cooling studies were carried out on simple geometries like at plate, cone and cylinder. Data for realistic planetary entry configuration like large angle blunt cone forebodies were not available in literature to the best of authors knowledge. Also, the two important parameters that influence transpiration cooling are coolant gas mass flow rate and coolant gas specific heat capacity. Hence it was decided to carry out transpiration cooling studies over a large angle blunt cone planetary entry forebody configuration at two different flow enthalpies using nitrogen and helium as coolants (two different types of gas having different thermophysical properties) in order to study its effectiveness at different flow enthalpies.

4.1 Test Model and Instrumentation

The test model used is a 60° apex angle blunt cone forebody with nose radius of 35 mm and base radius of 40 mm, similar to the one used in film cooling experiments. The forebody wall was made of plaster of Paris (POP), scientifically known as calcium sulfate hemihydrate, with 10 mm wall thickness and 40% porosity. POP was chosen as wall material because of its ease of preparation and cost-effectiveness compared to sintered porous materials. A schematic view and photographic picture of the forebody

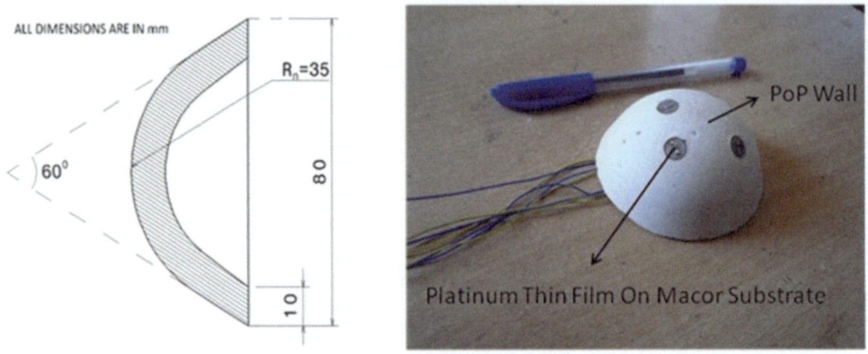

Fig. 16 Schematic and picture of PoP forebody

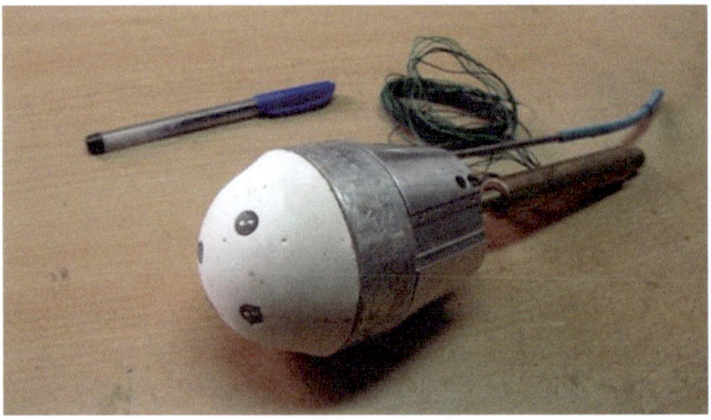

Fig. 17 Photographic picture of test model

and the model are shown in Figs. 16 and 17, respectively. The heat transfer rates on the model surface are measured using platinum thin film sensors deposited on a Macor surface and flush mounted with the model. Four sensors were used, one in the nose region, one in the sphere–cone junction and the remaining two in the conical region.

The model gets the supply of coolant gas from an external high pressure gas cylinder. The total supply pressure of the coolant gas was held at 6 ± 0.2 bar and was monitored using the pressure regulator. A pipeline from the supply cylinder connects to the flow meter, to regulate the coolant gas mass flow rate, with an accuracy of $\pm 5\%$ of total range (0–500 l/min). The output of the flow meter is connected to the test model mounted inside the test section.

Table 3 Flow properties in reservoir and free stream regions for HST2 and HST3

Properties	1.8 MJ/kg	3.8 MJ/kg
P_o (kPa)	2797.08 ± 89.93	10233.5 ± 643
T_o (K)	1246 ± 5.34	2877 ± 82
ρ_o (kg/m^3)	12.74 ± 0.42	14.12 ± 0.24
M_∞	5.07 ± 0.02	5.87 ± 0.1
P_∞ (kPa)	1.06 ± 0.04	0.572 ± 0.055
T_∞ (K)	386.88 ± 2.87	739.85 ± 9
ρ_∞ (kg/m^3)	0.0145 ± 0.0004	0.0041 ± 0.0004
V_∞ (m/s)	1487.4 ± 1.65	2364.6 ± 12.35

4.2 Experimental Results and Discussions

The effectiveness of transpiration cooling over a 60° blunt cone forebody was investigated at two flow enthalpies, 1.4 MJ/kg (low enthalpy) and 3.6 MJ/kg (high enthalpy), in HST-2 and HST-3, respectively. The nozzle reservoir and freestream conditions generated for the two enthalpy cases are tabulated in Table 3. The model was tested at 0° angle of attack in the presence and absence of transpiration cooling using nitrogen and helium gas as coolant.

4.2.1 Experiments at 1.4 MJ/Kg

A total of four experiments were carried out for each case, and the mean values in heat transfer rates are taken and plotted. The shot-to-shot deviation in measured heat transfer rates was less than ±4 W/cm^2. The effect of coolant gas thermophysical properties and coolant gas mass flow rates on heat transfer distribution is explained next.

Influence of Coolant Gas

The effect of coolant gases, helium and nitrogen, on heat transfer rate distribution is plotted in Fig. 18. The supply mass flow rate was maintained the same, 2.6e^{-4} kg/s, for both coolants. In the absence of cooling, the heat transfer was higher in the nose region (stagnation zone) of the model and gradually decreased as we move downstream toward the conical region. With transpiration cooling, a reduction in heat transfer rate was observed on the model surface, and the trend in heat transfer distribution was preserved.

The percentage reduction in heat transfer due to transpiration cooling of helium and nitrogen gas is plotted in Fig. 19. In the nose region, the reduction in heat transfer was 25% and gradually increased to 41% in the conical region for helium, whereas for nitrogen coolant, it was 15% in the nose region and 22% in the conical region.

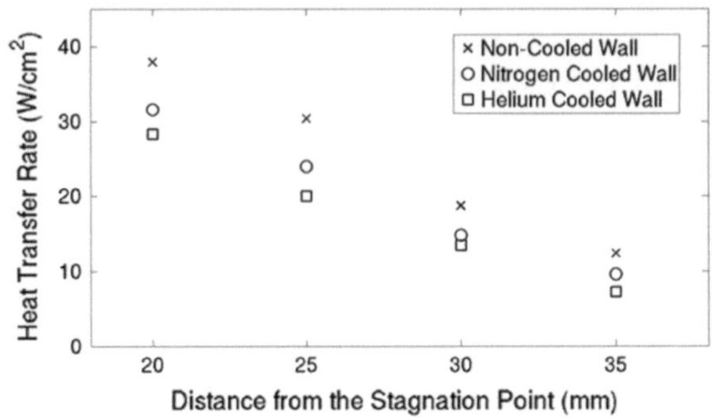

Fig. 18 Heat transfer rate distribution on the model surface in the absence and presence of transpiration cooling at 1.4 MJ/kg

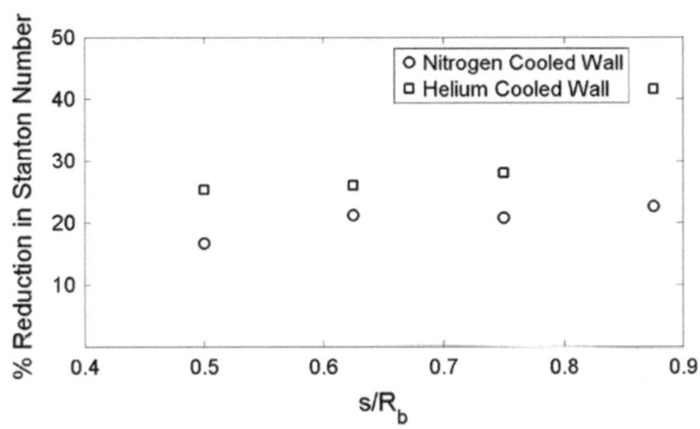

Fig. 19 Percentage reduction in heat transfer for nitrogen and helium cooled wall at 1.4 MJ/kg

This clearly shows that helium gas serves as a better coolant compared to nitrogen gas. This is due to the fact that the specific heat capacity of helium is higher (~5 times) compared to nitrogen [20]; helium gas takes more heat from the shock layer compared to nitrogen. Also, for the same coolant gas mass flow rates, the volume flow rate and hence the velocity of helium gas diffusing out from the model surface are ~7 times higher than nitrogen. This leads to a higher displacement of the boundary layer, hence an increase in its thickness and a lower temperature gradient across it, resulting in a larger reduction in heat transfer compared to nitrogen coolant gas.

An increasing trend in percentage reduction in heat transfer was observed from the nose to the conical region of the test model. The surface pressure distribution imposed by the freestream flow will be higher in the nose region, being maximum at the stagnation point and gradually reducing toward the conical region of the model. This results in a higher blowing of coolant gas and hence higher percentage reduction in heat transfer rate with increasing distance from the stagnation point along the model surface.

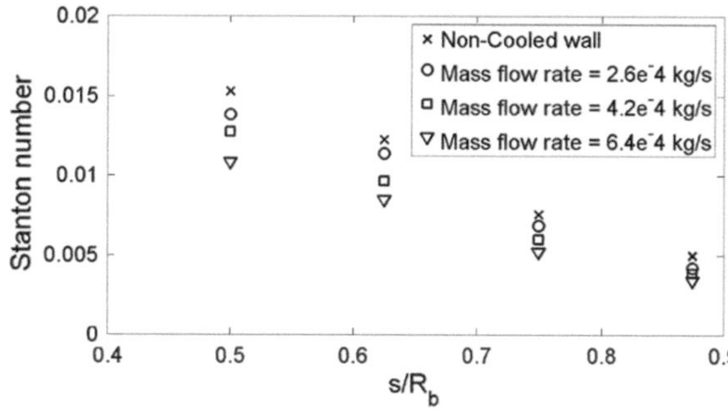

Fig. 20 Heat transfer rate distribution for varying nitrogen coolant mass flow rate

Influence of Mass Flow Rate

The effects of coolant gas mass flow rate on heat transfer rate distribution on the model surface are plotted in terms of Stanton number in Fig. 20. Experiments were carried out using nitrogen gas as coolant at mass flow rates of $2.6e^{-4}$, $4.2e^{-4}$ and $6.4e^{-4}$ kg/s. A reduction in heat transfer was observed with an increase in coolant mass flow rates. This reduction with increasing coolant mass flow was very significant in the nose region. However, as we move downstream toward the conical region, this was not the case. This is because, in the nose region (stagnation zone), the temperatures are higher, and more mass of coolant will take away more heat in the shock layer, whereas in the conical region, the temperatures are lower, and a little quantity of coolant gas will be sufficient to reduce the heat transfer.

4.2.2 Experiments at 3.6 MJ/Kg

Experiments are carried out at high enthalpy to investigate the effectiveness of coolant gases used for transpiration cooling. The coolant gas mass flow rate was held the same $2.6e^{-4}$ kg/s for both nitrogen and helium coolants. A total of three experiments were carried out for each case and the deviation in heat transfer measurements from shot to shot was ± 7 W/cm^2. Heat transfer rate over the test model, in the absence and presence of transpiration cooling, is plotted in Fig. 21. The trend in heat transfer distribution, in the absence of transpiration cooling, was similar to 1.4 MJ/kg case. However, as expected, the heat transfer rates were higher. This is due to the real gas effects (high temperature effects) as mentioned previously. Reduction in heat transfer was observed in the presence of transpiration cooling for both nitrogen and helium coolant gas. The percentage reduction in heat transfer due to transpiration cooling for nitrogen and helium coolant is plotted in Fig. 22.

Similar to the experiments at low enthalpy, an increasing trend in percentage reduction in heat transfer along the model surface was observed. A higher reduction in heat transfer was observed when using nitrogen coolant.

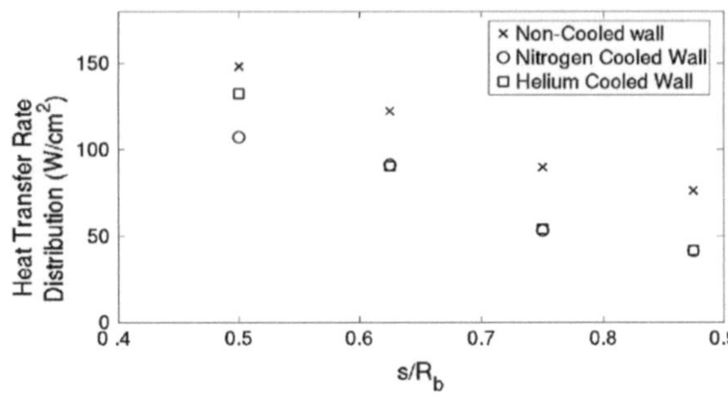

Fig. 21 Heat transfer rate distribution on the model surface in the absence and presence of transpiration cooling at 3.6 MJ/kg

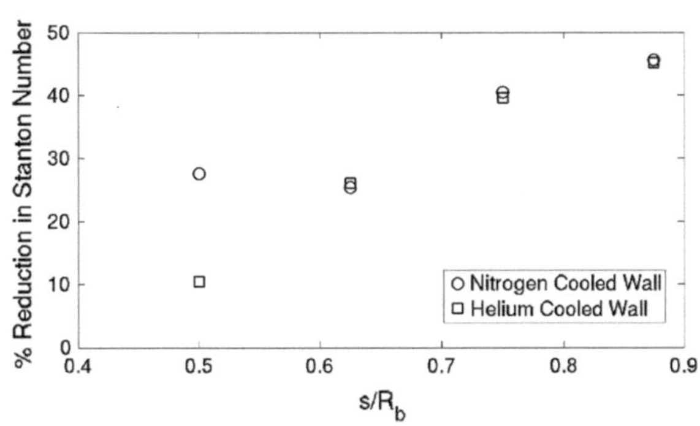

Fig. 22 Percentage reduction in heat transfer for nitrogen and helium cooled wall at 3.6 MJ/kg

In the nose region of the model, at $s/R_b = 0.5$, the percentage reduction in heat transfer was 28% for nitrogen coolant, whereas for helium, it was 10%. Also, along the conical portion of the test model, the percentage reduction for both coolants was almost the same. A completely different trend in heat transfer reduction was observed compared to the low enthalpy case. Nitrogen coolant gas performed more effectively despite the fact that its heat capacity is lower compared to helium gas.

The better performance of nitrogen at high enthalpy case may be attributed to the diatomic nature of the gas, having translational, rotational and vibrational degrees of freedom. The temperature in the shock layer at 3.55 MJ/kg case is high enough to cause vibrational excitation of nitrogen molecules coming out from the model surface. As a result, considerable amount of heat (energy) from the shock layer is absorbed by vibrational mode and the remaining is used to increase the temperature (translational mode) of coolant which is felt by the body as heat. In the case of helium coolant (monoatomic), having only translational energy mode, the temperature of the injected coolant, after energy transfer from shock layer, will be higher compared to nitrogen gas. Hence nitrogen gas performs as a better coolant than helium gas at high enthalpy case.

5 Conclusion

The use of film and transpiration cooling techniques, as a possible thermal protection system for atmospheric entry vehicles, was investigated experimentally. The methods were explored for two different flight velocities (low and high enthalpy), where the effect of these cooling techniques on heat transfer rate distribution was studied for a planetary entry forebody configuration. The investigations showed that both these techniques resulted in reduction of heat transfer rate to model surface and their effect was significant at higher flight velocities. It was also observed that the type of gas (monoatomic or diatomic) to be used as coolant also has a significant role to play in reducing heat transfer rate. The use of diatomic gases having additional energy modes compared to monoatomic gases resulted in a higher reduction in heat transfer at higher flight velocities (high enthalpy). Though the results obtained were encouraging, a detailed experimental campaign covering the entire flight trajectory, starting from atmospheric entry, say 12 km/s, to low hypersonic velocity, 1 km/s, has to be carried out, before possible implementation of these techniques for realistic vehicles.

Acknowledgements We would like to thank Defence Research and Development Organization (DRDO) for the financial support provided for carrying out these investigations. We also thank American Institute of Aeronautics and Astronautics, Inc. for providing us the copyright permission.

References

1. Anderson JD (1990) Modern compressible flow. MacGraw Hill Book Company, ISBN 0-07-124136-1
2. Anderson JD (1989) Hypersonics and high temperature gas dynamics. McGraw Hill Publications
3. Sheikh UA (2014) Re-entry radiation aerothermodynamics in the vacuum ultraviolet. Ph.D. Dissertation, University of Queensland
4. Peter FI, Donn BK (1987) High-speed aerodynamics of several blunt cone configurations. J Spacecraft Rockets 24(127)
5. Stewart DA, Chen YK (1994) Hypersonic convective heat transfer over 140 blunt cones in different gases. J Spacecraft Rockets 31(735)
6. Hollis BR, John NP (1997) High enthalpy aero thermodynamics of a mars entry vehicle part 1: experimental results. J Spacecraft Rocket 34(449)
7. en.wikipedis.org/wiki/Atmospheric_entry#Thermal_protection_ systems
8. Finley PJ (1966) The flow of a jet from a body opposing a supersonic free stream. J Fluid Mech 26(2):337–368
9. Sahoo N, Kulkarni V, Saravanan S, Jagadeesh G, Reddy KPJ (2005) Film cooling effectiveness on a large angle blunt cone flying at hypersonic speed. Phys Fluids 17:036102
10. Kulkarni V, Jagadeesh G, Reddy KPJ (2008) Enhancement in counterflow drag reduction by supersonic jet in high enthalpy flows. Phys Fluids 20:016103
11. Warren CHE (1959) An experimental investigation of the Effect of ejecting a coolant gas at the nose of a bluff body. J Fluid Mech 8(3):400–416
12. Libby PA, Cresci RJ (1961) Experimental investigation of downstream influence of stagnation point mass transfer. J Aerosp Sci 26(6):51–63

13. Baron JR, Alzner E (1962) An experimental investigation of a two-layer invicid shock cap due to blunt-body nose injection. J Fluid Mech 15(3):442–448
14. Romeo DJ, Sterrett JR (1963) Exploratory investigation of the effect of a forward-facing jet on the bow shock of a blunt body in a mach number 6 free stream. NASA TN D-1605
15. Edward R, Barber Jr (1965) An experimental investigation of stagnation point injection. J Spacecraft Rockets 2(5):770–774
16. Sriram R, Jagadeesh G (2009) Film cooling at hypersonic mach numbers using forward facing array of micro-jets. Int J Heat Mass Transf 54:3654–3664
17. Cook WJ, Felderman EJ (1966) Reduction of data from thin-film heat transfer gages: a concise numerical technique. AIAA J 4–3:561–562
18. CHEMKIN (2012) Chemical Kinetic Software, Ver. 10113. http://www.reationdesign.com/lobby/open/index.html
19. Greuel D, Herbertz A, Haidn OJ, Ortelt M, Hald H (2004) Transpiration cooling applied to C/C liners of cryogenic liquid rocket engines. AIAA Paper 2004–3682
20. Leo TC, Howard SC (1955) Exploratory test of transpiration cooling on a porous 8 cone at M = 2.05 using nitrogen gas, helium gas and water as the coolants. NACA RM L55C29
21. Bernard R (1957) Exploratory investigation of transpiration cooling of a 40 double wedge using nitrogen and helium as coolants at stagnation temperatures of 1295 to 2910 F. NACA RM L57F11
22. Hirotaka O, Fujita K, Ito T (2007) Application of the transpiration cooling method for reentry vehicles. AIAA Paper 2007-12029
23. Yuan QL, Pei XJ, Shao SJ, JI GS (2010) Transpiration cooling of a nose cone by various foreign gases. Int J Heat Mass Transfer 53:5364–5372
24. Sreekanth M, Reddy NM (1994) Transpiration cooling analysis at hypersonic mach numbers using Navier-Stokes equations. AIAA Paper 94-2075
25. Sreekanth M, Reddy NM (1995) Numerical simulation of transpiration cooling over blunt bodies at hypersonic mach number. AIAA Paper 95-2082
26. Kulkarni VN (2003) Numerical and experimental investigation of transpiration cooling in carbon dioxide atmosphere at hypersonic mach numbers. M.Sc. Dissertation, Department of Aerospace Engineering, Indian Institute of Science, Bangalore

Numerical Investigation of Hypergolic Combustion Characteristics in Rocket Engines

Ashoke De, Malay K. Das, Rupesh K. Sinha and Sindhuja Priyadarshini

Keywords Hypergolic combustion · MMH · RFNA · DPM

1 Introduction

Rocket is a propulsive device whose function is to propel the object of a given mass to a well-designated position. Rocket combustion technology is the most challenging field of research as it requires the highest level of accuracy with the utmost care to develop a full-fledged rocket system. This work is just a small effort toward the understanding of combustion characteristics in one of the most demanding type of rocket combustion systems such as hypergolic combustion. In this work, the focus is to have a better understanding of underlying physics related to the combustion of such system and in that way to find a better way to control, improve, modify, and enhance the rocket performance to a possible degree.

Hypergolic propellants are pronounced as a combination of propellants having the ability to ignite spontaneously when coming in contact. These are two propellant components usually consist of a fuel and an oxidizer. Hypergolic propellants are challenging to handle because of their extreme toxicity and corrosiveness. These propellants can be stored as liquids at room temperature while having the ability for a repeatable start, smooth combustion, and easy storage are among some other char-

A. De (✉) · S. Priyadarshini
Department of Aerospace Engineering, Indian Institute of Technology Kanpur, Kanpur 208016, India
e-mail: ashoke@iitk.ac.in

M. K. Das · R. K. Sinha
Department of Mechanical Engineering, Indian Institute of Technology Kanpur, Kanpur 208016, India

A. K. Gupta et al. (eds.), *Innovations in Sustainable Energy and Cleaner Environment*, Green Energy and Technology,
https://doi.org/10.1007/978-981-13-9012-8_4

acteristics. These properties make hypergolic propellant as one of the most desired class of propellants for the rocket propulsion systems. They have advantages over solid propellant in higher specific impulse and throttling ability which felicitates the controlled operation of the propulsive device. Due to such characteristics, many current launch vehicles use such kind of propellant. Future launch vehicle uses a gelled propellant version of hypergolic propellants. The gelled hypergolic propellant has the advantages of less spillage since these propellants are highly toxic and corrosive, they provide us with a smooth transport and handling option. Many modern-day rockets, e.g., PSLV, GSLV, TITAN, PROTRON, ARIANE, use hypergolic combustion.

Due to advancement in the rocket propulsion system, there build up a need to understand the combustion characteristics of such systems. Lack of reaction mechanism has halted the research in the area of hypergolic combustion for so long. Hypergolic propellants have ignition capability at a relatively low temperature, and it has self-igniting nature. In addition to this, hypergolic propellant cannot be easily handled and experimenting will not be good for the environment as well as it will cause extra cost in the development program of such rocket engines. Here only the numerical approach comes in handy, and all the numerical simulations can be performed before the actual test that saves money, energy, time and effort. Carrying out a numerical simulation of such complex system is challenging and at the same time very interesting as it helps to understand such complex mechanism through the border eye of computational technology, which gives us more insight with wonderful details.

Hypergolic propellants consist of multiple species, and detailed reaction chemistry is not available for these species. This is the reason numerical research has been halted in this domain. Recently, some chemical mechanism is developed by Anderson et al. in 2010 [1]. The mechanism is developed by Anderson consists of 81 species and 513 reactions, and employing such a detailed mechanism for computing reacting flow presents an expensive prospect for numerical analysis. A reduced mechanism Rchem1, for MMH–RFNA propellant combination, was proposed by Labbe et al. [2], which consists of 27 species and 99 reactions and presents a good prospect for numerical simulation. The work done by Sardeshmukh et al. [3] affirms the use of the reduced chemistry as they compare well with full reaction chemistry. Minimal numerical work is done in the domain which includes the work done by Sardeshmukh et al. [3] in the 2D domain and Park et al. [4] in a 3D domain with gas phase injection only. Tani et al. [5, 6] carried out some work with hydrazine–NTO combination with gas phase injection in 2D and 3D domain and Daimon et al. [7] have done simulation in a 2D domain with liquid and gas phase injection.

Present work begins with the validation of Park et al. [4] work in the 2D domain with equilibrium chemistry using two mixture fraction model and extends the simulations in 2D and 3D domain with the Rchem 1 mechanism (25 species and 98 reactions). The comparison is carried out with the same impingement zone geometry and injection condition between 2D and 3D domain to ascertain the use of the appropriate computational domain. Later liquid–gas phase injection simulation is carried out in this 3D domain. Effect of different droplet distribution type, different droplet diameter, and different oxidizer temperature is studied in the liquid–gas phase injection system with the temperature contour and vortices structure.

2 Background

Hypergolic combustion was first experimented by the Soviet researcher as early as 1931. These propellants are highly toxic, corrosive, explosive, and not readily available; hence, very few experiment work and numerical result are available to the present date. Here, we report some of the important and useful works as available in the literature.

2.1 Experimental Work

Drop test

Wang et al. [8] investigated the evaporation and combustion mechanism for hypergolic droplet particle in details. They have selected monomethylhydrazine as a fuel and nitric acid as an oxidizer. Fuel(MMH) droplet was dropped into the oxidiser (HNO_3) pool, and the mechanism that led to the combustion was studied using a high-speed camera. They have found out that MMH droplet combustion mainly consists of three stages: Stage 1 being liquid phase reactions during which ionic compound monomethylhydrazine nitrate formed along with oxidation products methyl nitrate, methyl azide, N_2O, H_2O, and N_2. The rapid rise in temperature was noticed during this stage from the ambient level to the boiling point of the droplet. Stage 2 did signify the formation of an aerosol cloud which was mainly composed of monomethylhydrazinium nitrate. Relatively small temperature rise rate was noticed in the second stage. The final stage was the thermal decomposition of MMH droplet into smaller species, e.g., H_2O, HONO, CH_3ONO_2, CH_3ONO, CH_3N_3, CH_3OH, CH_3NH_2, CH_4, N_2O, NO, N_2, where very rapid temperature rise rate was noticed during this time.

Impingement test

Dennis et al. [9] carried out an experimental test based on realistic combustion condition. They had selected an impinging jet apparatus with the hypergolic combination gelled monomethylhydrazine and red fuming nitric acid. In this work, they had varied the injection velocity of both MMH and RFNA to check the effect of Reynolds number and contamination (with the gel condition and immaculate condition) on the combustion characteristic of the hypergolic mixtures. They had carried out a series of test, and their work is significant for the numerical study as they have used a simple geometry with very few varying parameters. Also, they reported ignition delay for the various combination of the inlet velocity. These works were later used for various numerical work in 2D and 3D domain.

2.2 Reaction Mechanism Development

MMH–RFNA mechanism

To numerically model the ignition and combustion characteristics of monomethylhydrazine (MMH) and red fuming nitric acid (RFNA), the chemical kinetics becomes important. The detailed finite-rate-based kinetics for these reactants was developed by Anderson et al. [1], which is commonly known as ARL mechanism that consists of 81 species and 513 reactions. This particular mechanism was a collection of different rate expressions and thermochemical data, taken either individually or in sets from other kinetics, which was validated and established for other reacting systems. In general, this mechanism is reasonably well validated and thus quite useful to be considered as a starting point for achieving more refined and detailed kinetic mechanism for MMH–RFNA (or MMH–IRFNA) reaction.

Later on, Labbe et al. [2] developed two reduced sets of the mechanisms from the ARL mechanism. One is Rchem 1 that consists of 25 species and 98 reactions and while the other one is Rchem 2 which consists of 41 species and 200 reactions. In this work, we use Rchem 1 to reduce the computation cost as it has less number of species and reaction steps. Usually, it is quite well known that this reduced mechanism, i.e., Rchem 1, is capable of predicting ignition delays close to the original mechanism. In this mechanism, the reaction begins with the successive abstraction of the hydrogen atom from the nitrogen atoms followed by abstraction of methyl radical and finally the oxidation of carbon and nitrogen atom to form the final products, e.g., nitrogen, water, and carbon dioxide. This reduced mechanism, i.e., Rchem 1, has some limitations such as s single path is responsible for hydrogen abstraction from MMH and it does not account for hydrogen abstraction from a methyl radical. Despite this inherent limitation, Rchem1 predicts reasonably well ignition delay which is quite comparable with the predictions obtained from some of the novel mechanisms that consider alternate paths [2] of hydrogen abstraction.

Hydrazine–NTO mechanism

Daimon et al. [10] developed a chemical kinetic mechanism for N_2H_4(Hydrazine) and N_2O_4–NO_2 which is a hypergolic combination. This mechanism is followed as the successive abstraction hydrogen atom from the hydrazine. Daimon et al. [10] have found out that abstraction of a first hydrogen atom is an endothermic but successive abstraction of 3, 2, and 1 hydrogen atoms from the hydrazine species are exothermic. Rate coefficients are evaluated using different theories such as unimolecular rate theory, transition state theory, and master equation analysis with quantum chemical calculations of potential energy curves. Reactions of N_2H_4 with N_2O_4 isomers were also considered. Lower temperature ranges of a kinetic mechanism for gas phase hypergolic ignition of N_2H_4/NTO mixtures at temperatures down to 200 K. This mechanism consists of 39 species and 261 reactions.

2.3 Numerical Work

2D combustion with gas phase injection

Sardeshmukh et al. [3] investigated numerical modeling of MMH–RFNA combination using mechanism proposed by Labbe et al. [2]. They have used an in-house computational fluid dynamics (CFD) solver, i.e., General Equations and Mesh Solver with multiple approaches (GEMS) for their work. The solver is unsteady in nature and capable of handling reacting flows, where the governing equations are solved in coupled fashion. It is second-order accurate in time and space with dual time stepping. The turbulence was modeled using hybrid LES-RANS approach where Wilcox's k-ω model was invoked for RANS. In this work, numerical study was investigated on zero-dimensional, one-dimensional, and two-dimensional domains. The numerical simulations were carried out to study the combustion characteristics of gas phase hypergolic propellants where the fuel was mono-methylhydrazine (MMH) and the oxidizer was red fuming nitric acid (RFNA). The overall objective was to characterize the involved fundamental processes for the considered experimental setup of impinging injectors. The study involved variable operating pressure starting from 0.1 to 10 atm. The heat and mass diffusion of different composition was an important factor which was studied by invoking three different inert gases such as helium, neon, and argon. Further, the gas phase ignition and flame spreading were studied using recently developed reduced kinetic mechanisms Rchem1 and Rchem2 given by Labbe et al. [2]. The gas phase ignition delay was in a good match with experimental measurements at 1 atm and varied inversely to the pressure. Notably, at the upstream of the ignition location, the pressure effect on flame spread was directly proportional and there observed to have significant influence of the type of inert gas used.

Tani et al. [6] carried out a numerical simulation of N_2H_4/NTO combination. In their work, they investigated the combustion mechanism for hypergolic ignition processes of N_2H_4/NTO reacting systems. The considered configuration was a co-flowing planner jet with N_2H_4/NTO mixtures operated under low-pressure and low-temperature conditions. In addition to that, the distinct chemical reaction, e.g., hydrogen abstraction by NO_2 and thermal decomposition of N_2H_4, was also investigated using the direct numerical simulation methodology. The chemical reaction mechanism used consists of 39 species and 262 reactions.

3D combustion with gas phase combustion

Park et al. [4] extended the work of Sardeshmukh et al. [3] form the 2D domain to the 3D domain. In their work, they studied the combustion behavior of gas–gas hypergolic propellants (MMH and RFNA) using a detailed chemical reaction mechanism. Initially, 2D domain ignition characteristic was studied with the different inlet kinetic energy level to estimate the effect on the ignition delay. Their study involved the analysis of ignition characteristics due to the inclined wall, especially contact time/location and ignition delay/location had been investigated. Additionally, the wall effect was assessed concerning mixing and flame spread. The baseline three-dimensional simulation results compared with and without the inclined wall, under the same inlet flow conditions. Boundary condition and the computational domain were selected based on the experimental investigation of Dennis et al. [9].

Daimon et al. [7] used 3D gas–gas combustion of N_2H_4/NTO and investigated the hypergolic ignition processes in a N_2H_4/N_2O_4 bipropellant thruster. They reported the effect of 3D flame structures on ignition process and flame holding mechanism for two different inlet gas temperatures of 400 and 600 K. Additionally, the effect of chemical reaction (induction time) on the 3D flame characteristics was also considered.

2D combustion with liquid–gas phase injection

Daimon et al. [7] studied the injection of liquid hydrazine into the gaseous NTO. In this work, different oxidizer (NTO) temperature was used for the simulation to study the effect of gaseous temperature on the vaporization rate and flame propagation. The influence of the evaporation of N_2H_4 droplets upon the hypergolic ignition characteristics including flame structures was investigated. N_2H_4 spray droplet diameter was varied from 10 to 30 μ. In addition to that different temperature of the inlet, oxidizer was also used. Before the auto-ignition, the preheating of N_2H_4 vapor and the ambient NTO gas mixtures was carried out, whereas the temperature of the gas mixtures was reduced due to the evaporation of the N_2H_4 droplet. However, due to sufficiently high temperature in NTO flow, the auto-ignition took place near the leading edge of the N_2H_4 spray. In this case also, double flame structure (outside diffusion flame and inside decomposition flame) was observed like the gas–gas N_2H_4/NTO co-flowing jets. However, liquid N_2H_4 with gaseous NTO mixture showed a sinusoidal behavior of the inner decomposition flame. Further, they explained instantaneous flame structure along with different flame composition.

3 Numerical Details

In the present study, ANSYS FLUENT [11] is used to carry out the numerical simulations while the grid generation is carried out in ICEM-CFD [11]. The representative governing equations are Favre averaged which include mass, momentum, energy conservation including turbulence which has the following general form as:

$$\frac{D(\rho\tilde{\phi})}{Dt} = \nabla^2(\Gamma\,\rho\tilde{\phi}) + \langle S_\phi \rangle \tag{1}$$

where $\tilde{u}_j$ is the Favre averaged velocity, ρ is the mean density, $\tilde{\phi}$ is Favre averaged scalar in turbulent flow field, $\langle S_\phi \rangle$ is the mean source, and Γ is the coefficient of scalar diffusion. To close the turbulent eddy viscosity, various Reynolds average Navier stroke (RANS) models, i.e., Standard K-ω , SST K-ε, and detached eddy simulation (DES)-based approach, are invoked [11].

3.1 Non-premixed Combustion Model: Steady Laminar Flamelet Approach

In this particular approach, we solve the transport equation of conserved scalar (mixture fraction: Z) and its variance (Z'^2) as recast as:

$$\frac{D}{Dt}(\rho\overline{Z}) = \frac{\partial}{\partial x}\left(\frac{\mu_{\text{eff}}}{\sigma_t}\left(\frac{\partial\overline{Z}}{\partial x}\right)\right) \tag{2}$$

$$\frac{D}{Dt}\left(\rho\overline{Z'^2}\right) = \frac{\partial}{\partial x}\left(\frac{\mu_{\text{eff}}}{\sigma_t}\left(\frac{\partial\overline{Z'^2}}{\partial x}\right)\right) + C_g\mu_t\left(\frac{\partial\overline{Z}}{\partial x}\right)^2 - C_d\rho\frac{\varepsilon}{k}\overline{Z'^2} \tag{3}$$

where $Z' = Z - \overline{Z}$ and values of the constants σ_t, C_g and C_d are 0.85, 2.86 and 2.0, respectively [12].

In the turbulent flow field, the flame surface is considered to be as an iso-surface of the mixture fraction. Hence, a laminar diffusion flame with counter-flow configuration is invoked to represent the thin reactive-diffusive layers in the turbulent flow field. Therefore, the flame equations form physical space gets transformed into mixture fraction space and is given by [12]:

$$\frac{\partial Y_i}{\partial t} = \frac{1}{2}\chi\rho\frac{\partial^2 Y_i}{\partial f^2} + S_i \tag{4}$$

$$\rho\frac{\partial T}{\partial t} = \frac{1}{2}\rho\chi\frac{\partial^2 T}{\partial f^2} - \frac{1}{C_{\text{p}}}\sum_i H_i S_i + \frac{1}{2C_{\text{p}}}\rho\chi\left[\frac{\partial C_{\text{p}}}{\partial f} + \sum_i C_{p,i}\frac{\partial Y_i}{\partial f}\right]\frac{\partial T}{\partial f} \tag{5}$$

where Y_i and T represent species mass fraction and temperature, respectively.

Now, the species mass fraction and mean temperature are mapped back to physical space using the scalar dissipation rate and mixture. Scalar dissipation rate is written as:

$$\chi = 2D\left(\frac{\partial\overline{Z}}{\partial y}\right)^2 \tag{6}$$

where D is the diffusion coefficient and χ is the function of the mixture fraction. The turbulent flame is considered to be a collection of discrete diffusion flamelets, which is usually parameterized by $\overline{Z}$ and χ_{st} for adiabatic systems, while using β-PDF of $\overline{Z}$ and χ_{st} one can obtain the species mass fraction and mean temperature as:

$$\overline{\phi} = \iint \phi(\overline{Z}, \chi_{\text{st}})p(\overline{Z}, \chi_{\text{st}})d\overline{Z}d\chi_{\text{st}} \tag{7}$$

where φ represents species mass fractions and temperature. More details can be found in the literature [12].

3.2 Discrete Phase Model

In the dispersed phase, a large number of particles are tracked in the flow field; while this phase also exchanges mass, momentum, and energy with the continuous phase. One of the fundamental assumptions in this approach is that second phase (dispersed) has low volume fraction even in the condition with high mass loading $\left(\dot{m}_{\text{particle}} \geq \dot{m}_{\text{fluid}}\right)$ [11]. Individual particle trajectories are tracked at given intervals during calculation. Typically, the trajectory of a discrete phase particle is computed through the force balance on the particles, and it is recast in a Lagrangian reference framework as:

$$\frac{\mathrm{d}u_{\mathrm{p}}}{\mathrm{d}t} = F_D\left(u - u_{\mathrm{p}}\right) + \frac{g_x\left(\rho_{\mathrm{p}} - \rho\right)}{\rho_{\mathrm{p}}} \tag{8}$$

where F_D is the drag force per unit particle mass.

$$F_D = \frac{18\mu}{\rho_{\mathrm{p}} d_{\mathrm{p}}^2} \frac{C_d Re}{24} \tag{9}$$

Here, u is the fluid phase velocity, u_{p} is the particle velocity, μ is the molecular viscosity of the fluid, ρ is the fluid density, ρ_{p} is the density of the particle, and d_{p} is the particle diameter. *Re* is the relative Reynolds number, which is defined as:

$$Re = \frac{\rho d_{\mathrm{p}}\left|u_{\mathrm{p}} - u\right|}{\mu} \tag{10}$$

3.3 Real Gas Equation of State

It is well known that ideal gas assumption fails at very high pressure or very low-temperature conditions, and thus, the flow field cannot be accurately modeled using this assumption. The real gas model solves accurately for the fluid flow and heat transfer problems where the working fluid behavior deviates from the ideal gas assumption. The most straightforward equation of state is the ideal gas law, which is approximately valid for the low-pressure gas region of the *P-T* and *P-V* diagrams. Ideal gas behavior can be expected when

$$P/P_c << 1 \quad \text{or} \quad T/T_c > 2 \text{ and } P/P_c < 1 \tag{11}$$

Beyond this region original equation of state is used, and in the present work *Soave–Redlich–Kwong* real equation of state is invoked and recast as:

$$p = \frac{RT}{V_m - b} - \frac{a\alpha}{V_m(V_m + b)} \tag{12}$$

where $a = \frac{0.42747 R^2 T_c^2}{P_c}$, $b = \frac{0.08664}{P_c} T_c$, $\alpha = (1 + (0.48508 + 1.55171\omega - 0.1561\omega^2)(1 - T_r^{0.5}))^2$ and $T_r = \frac{T}{T_c}$. Here, ω is an acentric factor.

4 Results and Discussion

4.1 2D Equilibrium Model with Gaseous Phase Inlets

Numerical investigation of the ignition and combustion characteristics of the hypergolic combustion carried out in a 2D domain with gaseous fuel and oxidizer inlet, for combustion modeling equilibrium PDF model used with secondary mixture fraction model to incorporate three fluid systems. The 2D computational domain is depicted in Fig. 1. The 2D domain has a streamwise length of 14.91 cm and a transverse length of 10.7 cm. The wall angle is 60 degrees away from the flat top surface. Zoomed view of the computational domain is given, which shows the relative spacing of MMH and RFNA inlet as well as the point of impingement and height of the top wall from the point of impingement (Fig. 1b). The total length of these propellants feeder tubes is 4.2 cm which includes 2.4 cm of inclined length and 1.8 cm in straight length. MMH feeder tube has a width of 0.20 cm, and RFNA has a width of 0.24 cm. The 2D mesh is constructed using 0.1 million cells, and the height of the first cell in the boundary layer is chosen to be 2 μm to satisfy the condition of $y^+ < 1$ for the turbulent boundary layer. The cell size near the impingement point is 0.1 mm. The cell aspect ratio is limited to 7, and the skewness is limited to 35°.

Figure 2 shows a schematic configuration of the 2D computational domain with imposed boundary conditions. Flow is initialized using the fixed velocities as per given in Table 1 at both the inlets. Adiabatic no-slip wall boundary condition is set for inlet tubes wall. An ambient inlet is provided next to the injector wall. At this ambient inlet, the inert gas argon is chosen with the fixed static pressure of 101,325 Pa. All other sides of the domain are set as a pressure outlet boundary condition, and the operating condition is set as an atmospheric condition [4]. To study the effect of different inlet energy level on the ignition delay, the following cases are analyzed.

Initially, the computations are performed to decide on the tube length, and after a few iterations, the tube length of 7.5D is selected for the present work (D stands for the diameter of RFNA inlet tube). To compare the tabulated cases, the following parameters are used and defined.

Contact time: Time required for propellants to meet and contain 10% MMH and 10% HNO_3 at a location.
Contact distance: The distance from the top wall to a location where 10% MMH and 10% HNO_3 exist.

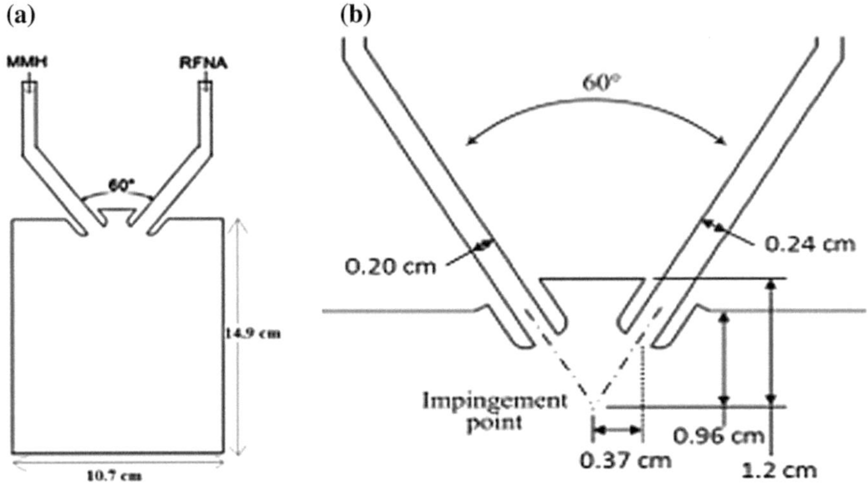

Fig. 1 **a** 2D computational domain, **b** Near impingement zone zoomed view [4]

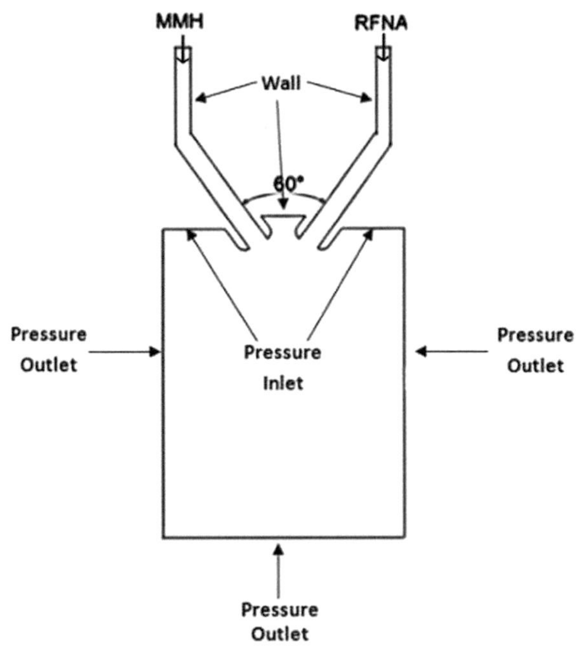

Fig. 2 2D domain with imposed boundary conditions

Table 1 Inlet boundary conditions for 2D domain

Case No.	Inlet gas	Velocity or pressure	Temperature (K)	Turbulent kinetic energy (m^2/s^2)	Turbulent dissipation (1/s)
Case 2D-1	MMH	5.4 m/s	800	0.158	5857
	RFNA	5.2 m/s	800	0.15	5050
	Argon	101,325 Pa	800	0.1	1000
Case 2D-2	MMH	10.7 m/s	800	0.538	10,808
	RFNA	10.5 m/s	800	0.5	9221
	Argon	101,325 Pa	800	0.1	1000
Case 2D-3	MMH	21.5 m/s	800	1.809	19,819
	RFNA	21.0 m/s	800	1.666	16,832
	Argon	101,325 Pa	800	0.1	1000

Ignition delay: Time required for the temperature to exceed 1500 K at a location from the contact time.
Ignition location: The distance from the top wall to a location where the temperature is higher than 1500 K.

Contact time obtained for the case 2D-1 is 8.3 ms which is the same as obtained by Park et al. [4]. The relative position of the 10% MMH and 10% RFNA at the contact time is depicted in Fig. 3 where the contact location obtained here is 0.88 cm from the top surface. Here, lines shown in the impingement zone are iso-surface mixture fraction lines for 10% MMH and 10% RFNA and point tool is at the contact point of these lines. Similarly, the ignition time obtained for the case 2D-1 is 9 ms as compared to 10 ms obtained by Park et al. [4]. Ignition location for the same case 2D-1 is found to be 1.07 cm from the top. In the present case, the ignition delay is 0.7 ms as compared to 1.7 ms obtained by Park et al. [4]. The difference between these two results is mainly attributed to the use of different solver and combustion models as well.

Similarly, the contact time for the case 2D-2 is 4.0 ms which is the same as obtained by Park et al. [4]. The contact location is found to be 0.81 cm from the top surface, while the ignition location for this case is 1.07 cm from the top. Ignition time obtained for the case 2D-2 is 4.4 ms whereas the ignition time obtained by Park et al. [4] is 5.5 ms. The simulations are repeated for the other case, and a similar observation is made.

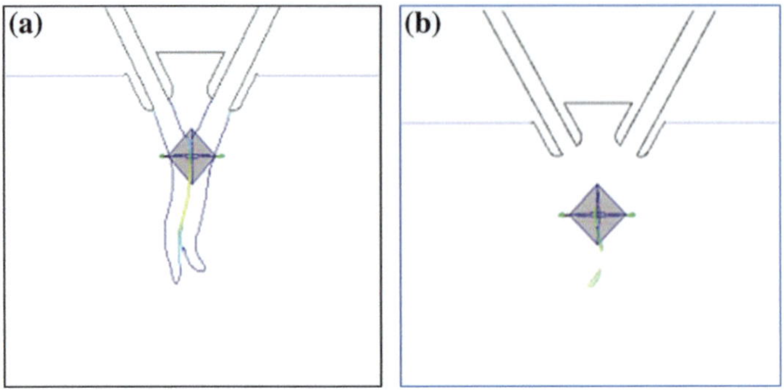

Fig. 3 2D Case: Schematic of **a** Contact location, **b** Ignition location

4.2 3D Domain with Steady Flamelet Model

In the previous section, the results with 2D domain show a significant deviation from the published results. To minimize the differences, the present section reports the results with Rchem 1 chemical mechanism (27 species and 98 reactions) where the comparison between 2D and the 3D domain is also carried out, and results are compared with published results. In addition, the flame spreading behavior is also studied.

Computational Domain and Grid Generation

3D domain geometry (Fig. 4) has been designed with the view of 2D domain geometry. To consider that, near impingement zone geometry for 3D domain kept same as 2D domain, i.e., depth, the inclination of the wall, etc. Length and diameter of the feeder tube are also kept same. All other dimensions are kept such that it reduces the size of the overall computational domain by focusing on the near-injector area. The length of this domain is 7.4 cm, width is 4.2 cm, and depth is kept 12.5 cm. These are same as used in the Park et al. [4] for 3D domain case.

Since zone of interest is near impingement zone where gaseous MMH and Gaseous RFNA come in contact with each other, and due to unsymmetrical nature of the geometry, tetrahedral mesh generation is used in the central core zone. Hybrid type mesh generation strategy is used here which has 3.0 million elements (Fig. 5) and consists of hexahedral meshing in the remaining part of the 3D domain.

Boundary conditions used for the 3D computational domain are same as used for the 2D computational domain. The only difference is due to 3D geometry there will be so many surfaces, out of which only back surface is set as an adiabatic no-slip wall boundary condition. Both the computational domains are subjected to same inlet boundary condition (Table 2). A grid convergence study has been carried out for both the 2D and 3D domains using the flamelet model. To have a comparison between the different grid sizes, two parameters are defined here.

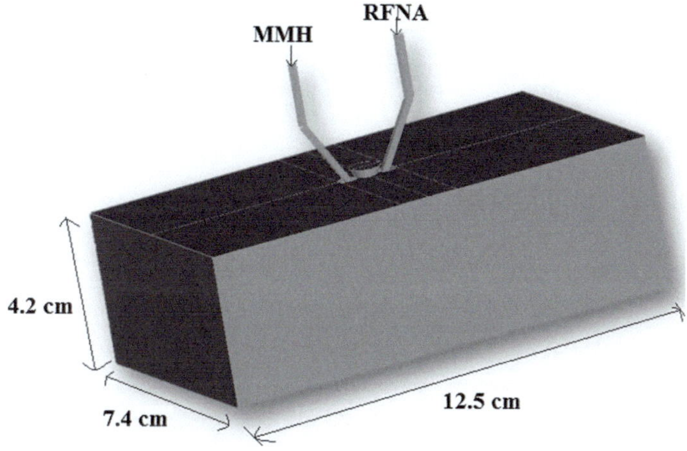

Fig. 4 3D computational domain

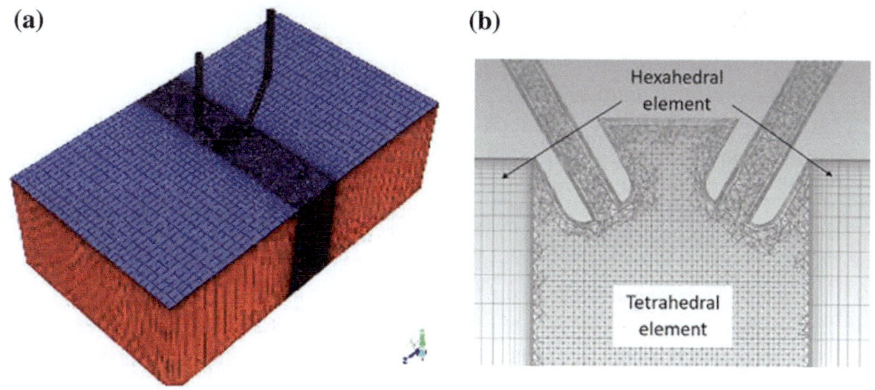

Fig. 5 Grid for **a** 3D computational domain, **b** Zoomed view near impingement zone

Table 2 Inlet boundary conditions for a 2D and 3D domain with flamelet model

Inlet gas	Velocity or pressure	Temperature (K)	Turbulent kinetic energy (m^2/s^2)	Turbulent dissipation (1/s)
MMH	10.7 m/s	800	0.538	10,808
RFNA	10.5 m/s	800	0.5	9221
Argon	101,325 Pa	800	0.1	1000

Table 3 2D grid independence test results using the flamelet model

Parameter	Mesh 1	Mesh 2	Mesh 3
Mesh size (Millions)	0.1	0.16	0.4
Contact time (ms)	3.9	3.8	3.8
Ignition time (ms)	4.6	4.5	4.5
Ignition delay (ms)	0.7	0.7	0.7

Table 4 3D grid independence test results using the flamelet model

Parameter	Mesh 1	Mesh 2	Mesh 3
Mesh size (Millions)	0.6	1.6	3.0
Contact time (ms)	4.2	3.6	3.5
Ignition time (ms)	5.1	4.6	4.5
Ignition delay (ms)	0.9	1.0	1.0

(a) **Contact time**: It is defined as the instant, the temperature near the impingement region starts rising from the initial temperature of 800 K.
(b) **Ignition time:** It is defined as the instant when the maximum temperature in the impingement zone is reached 1500 K.

2D Domain

A grid independence study for the 2D computational domain has been carried out for 3 grid of sizes 0.1 million, 0.16 million, and 0.4 million. A constant time step of 10^{-6} s is used. The obtained results depict that the ignition delay for coarse, medium, and fine grids are in close agreement. For medium and fine grids, contact time and ignition time are very much same; hence, mesh 2 is used for further analysis (Table 3).

3D domain

A grid convergence study for the 3D domain of the given geometry has also been carried out using three mesh sizes: 0.6 million for the coarse grid, 1.5 million for the medium grid, and 3.0 million for the fine grid. For this case also, a constant time step of $1e^{-06}$ s is used. It is observed that the ignition delay for the medium and fine mesh are same but contact time and ignition time for the fine mesh differ from other two meshes. However, the obtained results are in the same order as the published literature, fine mesh (mesh 3) is selected for further analysis (Table 4).

Comparison of the 2D and 3D case

Contour plots for argon mixture fraction and temperature at the contact time for 2D, 3D, and the results obtained by Park et al. [4] are shown in Fig. 6. Contact time obtained for the 2D case is 3.8 ms. High temperature in the impingement zone is noticed that signifies contact time (Fig. 6b). The relative position of the MMH and RFNA species at contact time is also depicted in Fig. 6a. Contact time obtained for the 3D case is 3.5 ms. A high-temperature value in the impingement zone and

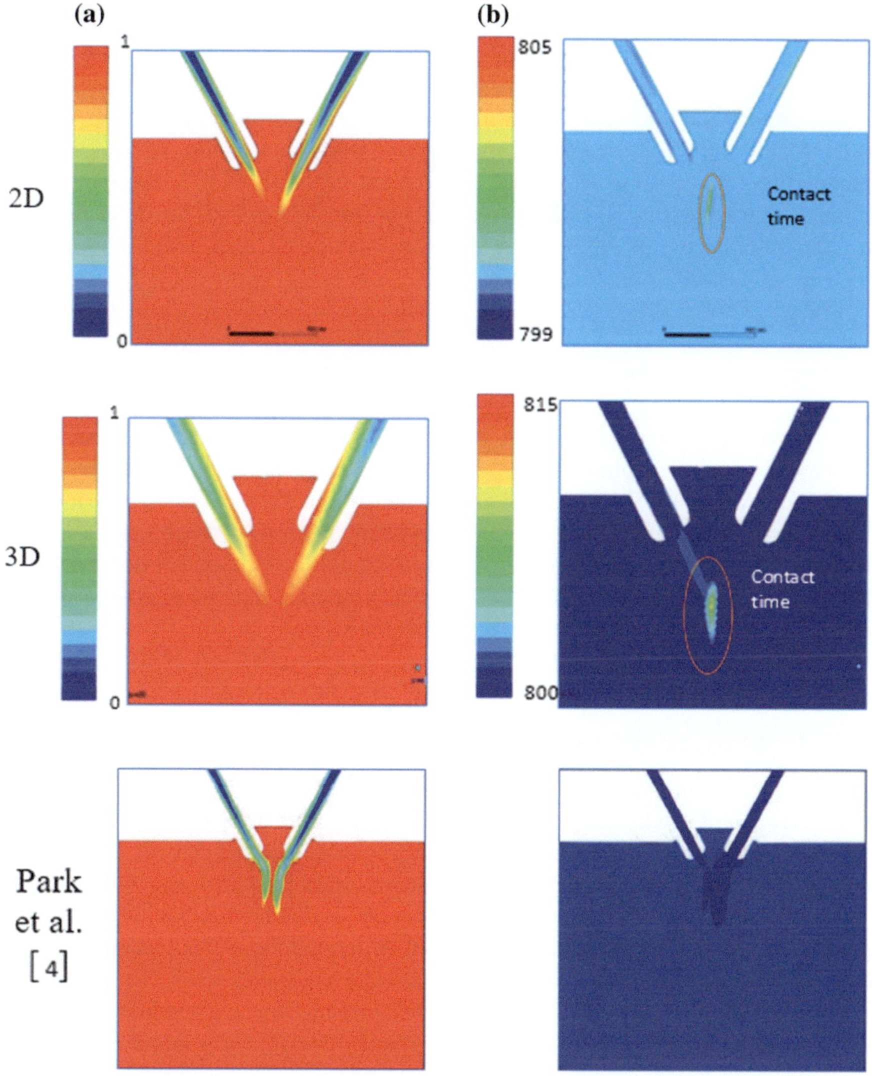

Fig. 6 Comparison of contact time contour plot: **a** Argon mixture fraction **b** Temperature

relative position of the MMH and RFNA species at that time is captured (Fig. 6). The difference of 0.3 ms is noticed in the 3D case with respect to the contact time in the 2D case. It is attributed to the volumetric flow in 3D case whereas, in the case of 2D, flow is considered only in the form of sheets. Contact time contour plots, given in Park et al. [4], are also presented (Fig. 6). No high-temperature zone is noticed since contact time defined there as the time at which 10% mass fraction of MMH and RFNA species comes in contact with each other (Fig. 6b).

Contour plots for argon mixture fraction and temperature at the ignition time for 2D, 3D, and the plots given by Park et al. [4] are presented in Fig. 7. Ignition time obtained for the 2D case is 4.5 ms. A temperature region of 1500 K is reported in the impingement zone that gives a total ignition delay of 0.7 ms. The relative position of the MMH and RFNA species at ignition time is also given (Fig. 7). Ignition time obtained for the 3D case is 4.5 ms. A temperature of 1500 K is obtained in the impingement zone (Fig. 7) that gives a total ignition delay of 1.0 ms. Ignition delay of 1.0 ms in the case of 3D is very close to the value of 1.1 ms given in Park et al. [4] (Fig. 7).

Unsteady Characteristics

A contour plot for time-averaged argon mixture fraction and the temperature is depicted in Fig. 8 for the interval of 4.5 and 5.5 ms for the 2D domain. All newly formed species and temperature profile are developed along the streamwise length that happens because a plane sheet of MMH and RFNA species is coming in contact along the streamwise direction. Different boundary conditions have a considerable impact on the spreading of the data. Ignition delay obtained in the 3D case is higher than that in the 2D case, this along with the volumetric effect of 3D flow results in the higher static temperature of 2250 K for 2D case as compared to 2000 K for 3D case. The time-series plot in Fig. 9 represents the same.

4.3 Liquid-Gas Phase Injection

Hypergolic combustion is characterized by injection and collision of liquid streams of fuel and oxidizer consequently breaking of liquid into small droplet vaporization and finally combustion. Work done so far incorporates only gas–gas combustion to study the combustion characteristic phenomenon. In this section, the discrete phase interaction in the form of liquid fuel droplet with continuous gas phase is studied. Fuel MMH (monomethylhydrazine) is sprayed in the combustion chamber in the form of fine droplets and oxidizer RFNA (red fuming nitric acid) is injected as gas. Computational domain, as shown in Fig. 4, is also used for gas-liquid injection. It has three inlets, MMH inlet which is used as droplet inlet, other is RFNA inlet which is used as an inlet for oxidizer gas, and one is pressure inlet through which argon gas comes inside the domain. Two different types of DPM injection are used (i) uniform droplet diameter and (ii) Rosin-Rammler droplet distribution. Details of injection conditions are depicted in Table 5. The boundary conditions for RFNA inlet and ambient inlet are provided in Table 6.

Setting Mass Flow Rate

RFNA inlet is used as velocity inlet, with the velocity and another boundary parameter is taken from Park et al. [4], but for MMH discrete phase boundary details are not available as only gas–gas injections are used by Park et al. [4]. MMH DPM mass flow rate selected such that time taken by droplet particle to reach the impingement

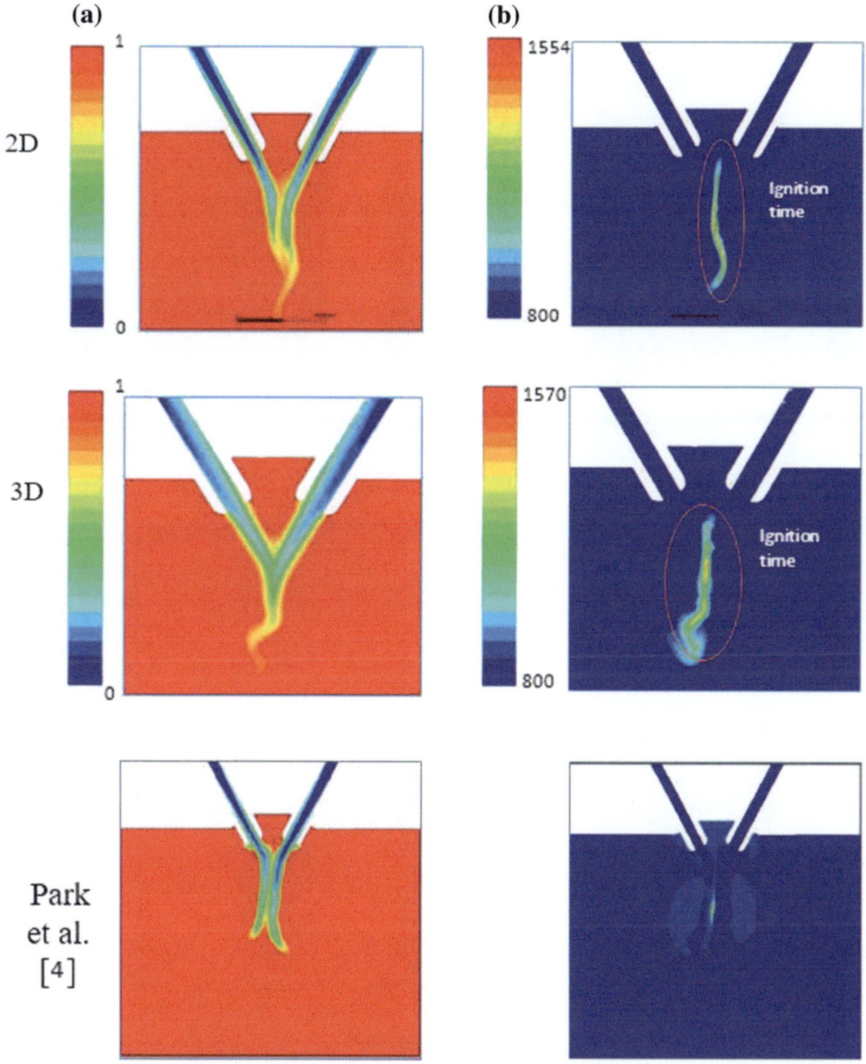

Fig. 7 Comparison of Ignition time contour plots: **a** Argon mixture fraction **b** Temperature

zone should coincide with the time RFNA gas reaches in the impingement zone. To arrive at such a different value, trial has been carried out with a mass flow rate of $2.5e^{-05}$ kg/s, 3.5^{e-05} kg/s, $4.5e^{-05}$ kg/s and finally $4.8e^{-05}$ kg/s is selected for MMH DPM mass flow rate.

MMH droplet properties

Following properties are set for the MMH droplet particle for two-phase combustion modeling (Table 7).

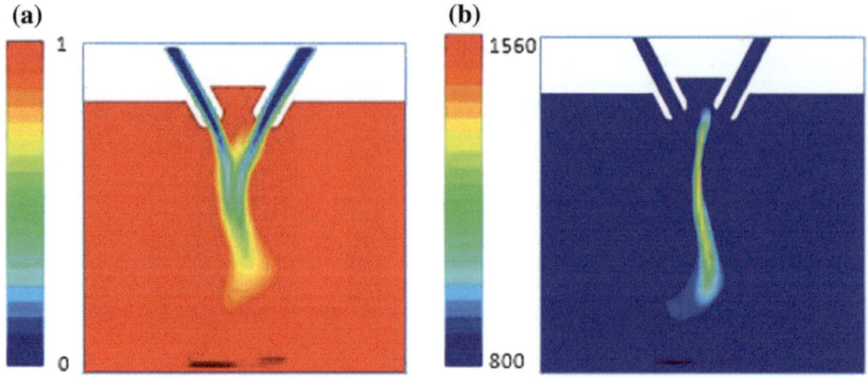

Fig. 8 2D time-averaged contour plot results **a** Argon mixture fraction **b** Temperature

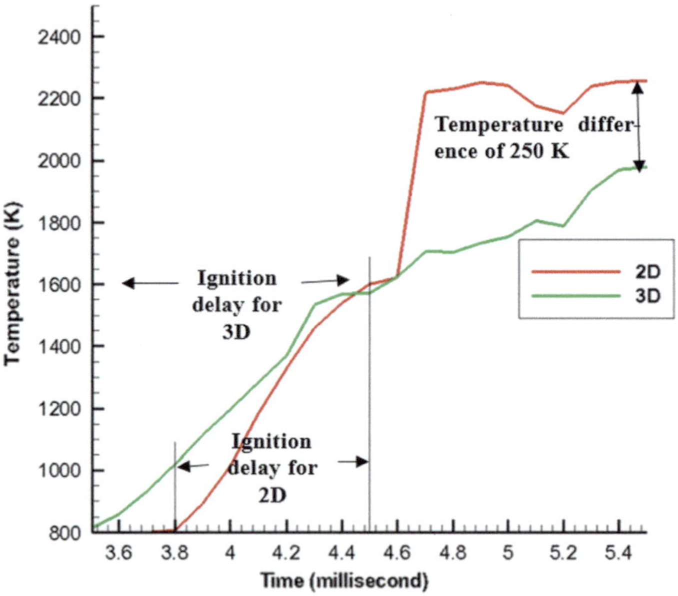

Fig. 9 Transient temperature profiles for 2D and 3D cases

Droplet distribution study

To investigate the effect of different droplet distribution, three cases are studied as given in Table 5 from case 1 to case 3. Contour plot of temperature with DPM injection for case 1 is exhibited in Fig. 10. As given in Table 5, case 1 corresponds to uniform diameter distribution of $1e^{-06}$ m. Figure 10 consists of 4 plots each corresponds to temperature contour at a particular time instant with the MMH droplet particles position. MMH droplets particles are colored as per the diameter. MMH droplets particles come through MMH feeder tube, and small fraction of MMH droplet enters

Table 5 Injection conditions for DPM case

		Case 1	Case 2	Case 3	Case 4	Case 5	Case 6
RFNA gas	Temperature (K)	300			320		
	Velocity (m/s)	15.5					
MMH spray	Temperature (K)	300			400		800
	Velocity (m/s)	18.6					
	Diameter (m)	$1e^{-06}$	0.00132 to $1e^{-06}$	$1e^{-04}$ to $1e^{-06}$	$10e^{-06}$	$20e^{-06}$	$20e^{-06}$

Table 6 Boundary conditions for continuous phase

Inlet gas	Velocity or pressure	Temperature (K)	Turbulent kinetic energy (m^2/s^2)	Turbulent dissipation (1/s)
RFNA	15.5 m/s	300	0.9	18,733
Argon	101,325 Pa	300	0.1	1000

Table 7 MMH droplet properties [13]

Sl No.	Property	Value (unit)
1	Density	880 Kg/m^3
2	C_P	2930 J/kg K
3	Latent heat	87,500 J/kg
4	Vaporisation temperature	265 K
5	Volatile mass fraction	100%

in the impingement zone, where it is getting evaporated into MMH gas phase due to interaction with RFNA hot gas and combustion begins in the impingement zone which can be visualized in Fig. 10a with a maximum temperature of 320 K and presence of MMH droplet near the flame zone.

As the temperature increases due to combustion, it initiates increased rate of evaporation and high rate of generation of MMH gas and consequently flame starts spreading, which is shown in Fig. 10b–d where the maximum temperature in the domain is increased from 300 to 1322 K, and MMH droplets also get distributed near the flame. Small droplet size felicitates quick flame propagation due to quick evaporation of small droplet particles. As soon as droplet comes in contact with RFNA gas flame forms as less heat is required to evaporate small droplet but it fails in quick flame spreading as mass content in these droplets is small. This is the reason it takes around 1.2 ms to reach a high temperature in the impingement zone.

To understand the effect of droplet diameter distribution, case 2 has been studied. Figure 11 gives time wise flame generation with a Rosin-Rammler droplet diam-

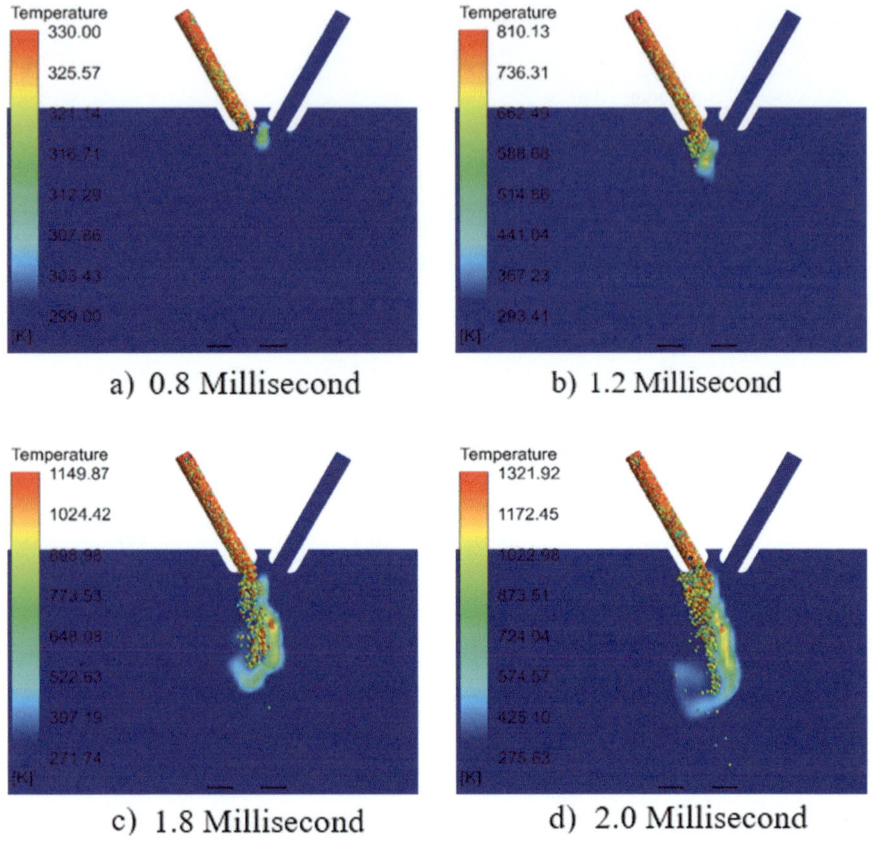

Fig. 10 Temperature contour plot with DPM injection for case 1

eter distribution. In this case, diameter range is selected from a minimum diameter of $1e^{-06}$ m to maximum diameter of MMH feeder tube diameter itself, which is 0.00132 m. Results of selecting Rosin-Rammler diameter distribution in range 0.00132 m to $1e^{-06}$ m are shown in Fig. 11. Figure 11a shows the start of ignition in the impingement zone, as the high temperature is noticed in the zone. With the relative position of MMH droplet particle in the MMH feeder tube and the impingement zone, the first sign of ignition noticed in case 2 is at 0.7 ms, which is close to the case 1 where it happens at 0.8 ms, and that may be due to the same minimum diameter distribution. The large size of maximum diameter causes remaining droplet diameter to shift toward minimum diameter. The presence of diameter distribution near the minimum diameter and availability of slight high mass droplet particle in the vicinity causes rapid combustion, vaporization, and flame spreading as can be seen from the contour plots (Fig. 11). Temperature rise rate is higher compared to the case 1. The presence of very large droplet particle halts the further rise in temperature as seen in

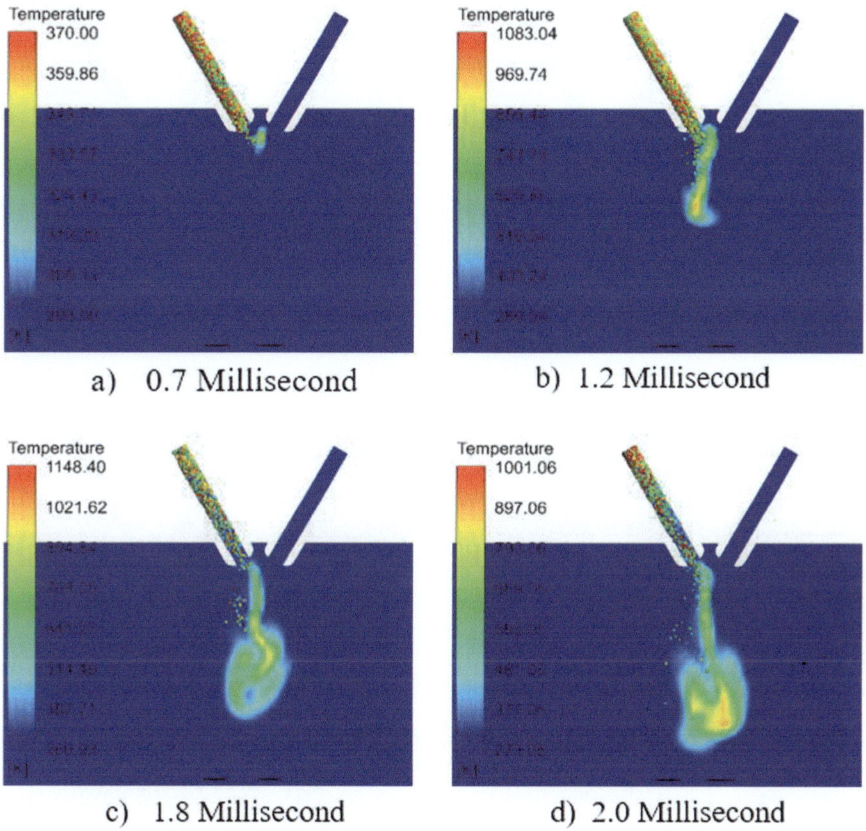

Fig. 11 Temperature contour plot with DPM injection for case 2

Fig. 11d, flame spreading is rapid, but it does not sustain for a more extended period to an extra-large size of maximum diameter particles.

Case 3 is a modified version of case 2 as in this case Rosin-Rammler diameter distribution range is modified to $1e^{-04}$ m to $1e^{-06}$ m to reduce the effect of high maximum diameter (Fig. 12). The first sign of ignition noticed in case 3 is at 1.2 ms which is slightly more compared to the earlier cases where it happens around 0.8 ms for case 1 and 0.7 ms for case 2. Reason for such behavior is attributed to large number of droplet particles are in the higher diameter range, and particle count in the $1e^{-06}$ m diameter range is low as compared to case 1 and case 2. Large diameter particles have large surface area and RFNA gas being the initial heat source for vaporization. It results in a slight delay in the ignition. Slight delay in ignition felicitates more droplet entry into the impingement zone. Here, ignition is delayed due to evaporation delay, but the entry of droplet particles is consistent, the moment combustion begins in the impingement zone due to evaporation and hypergolic effect of RFNA and MMH it causes explosion kind of situation. Since large number of the droplet particle mass

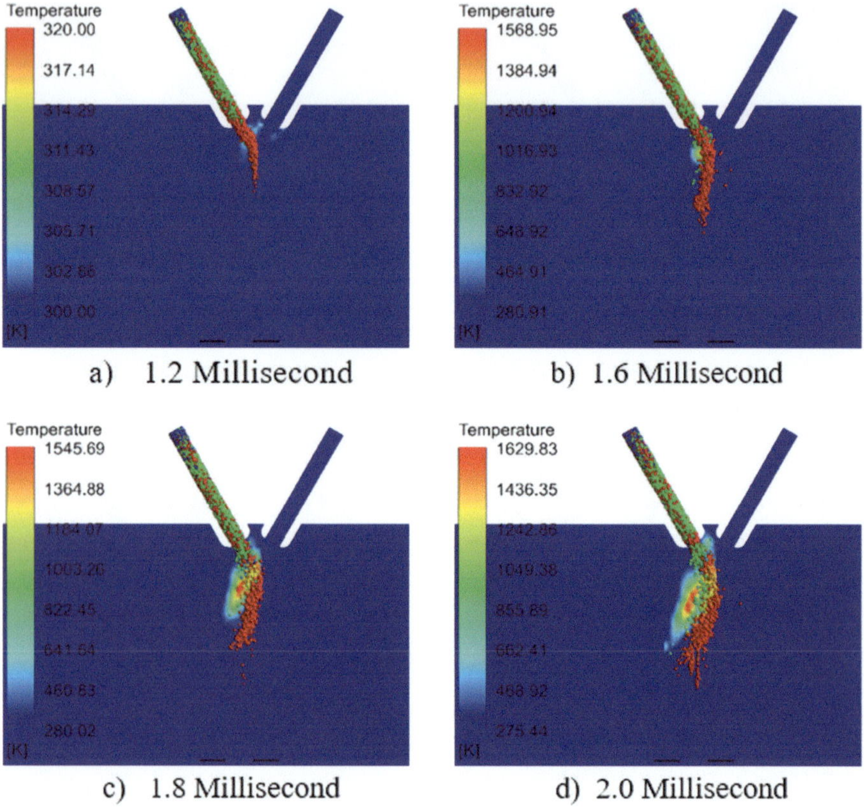

Fig. 12 Temperature contour plot with DPM injection for case 3

source is present in the combustion zone, it results in a very high rate of temperature rise in small time, i.e., rapid flame propagation. It is shown in Fig. 12b where the maximum temperature noticed in the impingement zone is 1569 K within a span of 0.4 ms.

Temperature rise profile for all three cases can be summarized and depicted in Fig. 13. It is a transient temperature rise profile plot. Case 1 which is of uniform diameter shows a steady rise in temperature with minimal temperature rise rate, whereas case 2 which is a Rosin-Rammler-1 distribution has steep initial temperature rise but later the temperature rate is not maintained due to low evaporation, the large surface area of large droplet particle with less volumetric mixing. In case 3, initial lag is more as compared to other cases due to sizeable mean droplet size but the moment ignition starts it propagates with very high rate and sudden rise in the temperature is noticed.

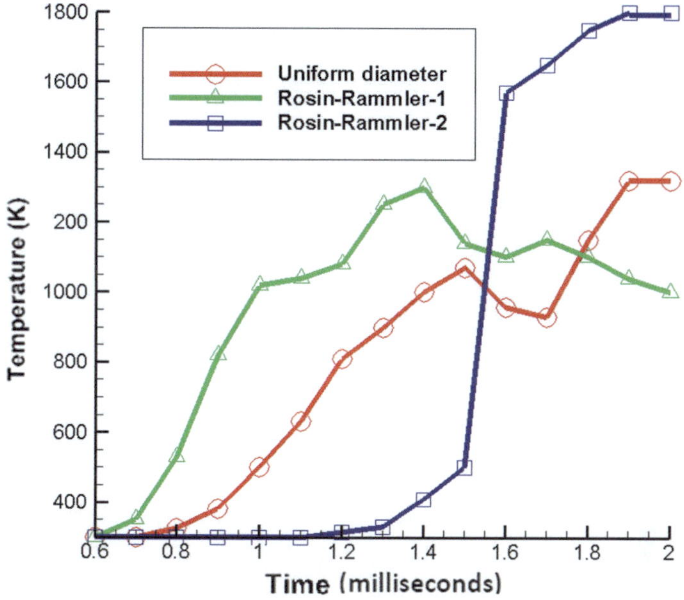

Fig. 13 Transient temperature rise profile for DPM case 1 to case 3

Droplet diameter effect

Studies carried out in the above cases 1, 2, and 3 show even if Rosin-Rammler droplet distribution is close to the droplet distribution which exits in nature. To use this distribution, it requires a better understanding of its distribution parameters which require the presence of experimental results. So for the time being to estimate droplet diameter effects on flame propagation and vortex generation droplet of uniforms diameter 10 and 20 μ is used with the condition as given in Table 5 for case 4 to case 6. Here, temperature and vortices are plotted at a time instant of 2, 3, 4, and 5 ms to understand the effect of droplet diameter on the mixing and combustion mechanism.

Figure 14 depicts the temperature contour plot for the case 4. This case is corresponding to the droplet diameter of 10 μ with a slightly elevated temperature of 320 K for MMH droplet and temperature of RFNA is increased to 400 K to accelerate the vaporization and in turn increase the combustion rate. Flame propagation is seen in the direction of oxidizer flow. Droplet being small in size is pushed through the hot oxidizer stream and in this process combustion as well as evaporation occur. In Fig. 14a, the droplet particles are not in the oxidizer flow path as soon as MMH droplet comes in the hot oxidizer flow path it starts evaporating, and at the same time, gas phase combustion begins that is shown in Fig. 14b, c. Figure 14d shows the burn out gases and fresh burning gases together. Flow field can be better understood by the vortices (Fig. 15). These plots are with velocity in the plane domain to show the geometrical location of vortices with a corresponding velocity component in that zone here vortices are colored with a temperature gradient. The plane domain

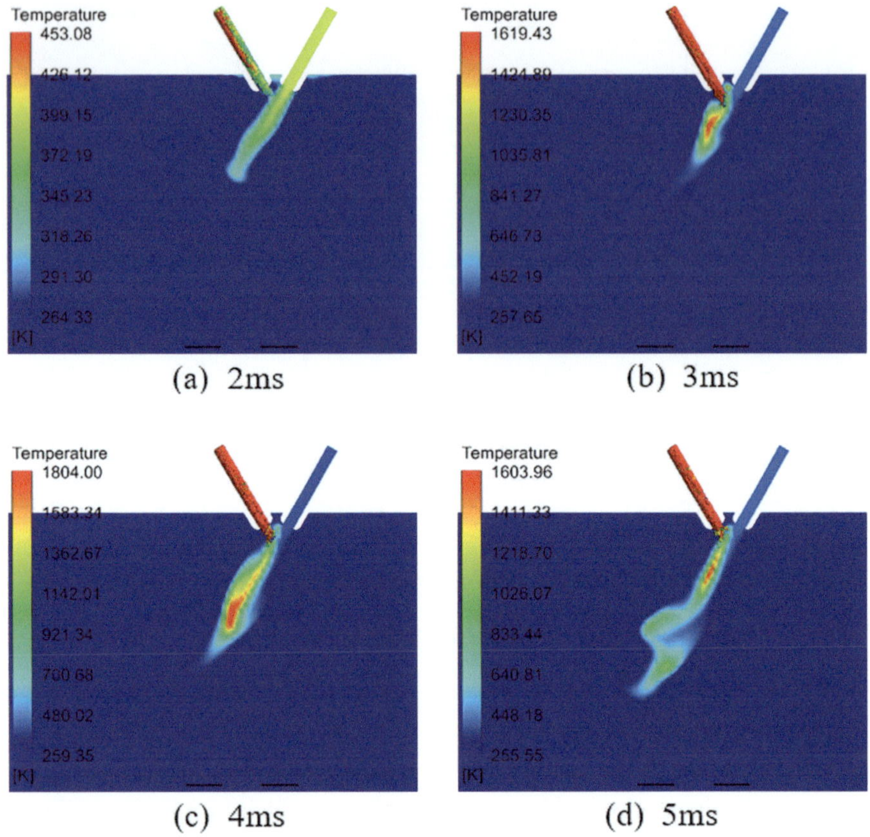

Fig. 14 Temperature contour plot with DPM injection for case 4

is rendered to 0.5 to make it a transparent domain as it helps in clearly visualizing the vortices. It is observed from Fig. 15a–c that a mushroom-shaped vortex forms in the impingement zone. Mushroom-shaped vortex forms due to Rayleigh–Taylor instability which signifies interaction of low-density flow with the high-density flow as happening in this case where MMH discrete particle phase interacts with RFNA continuous gas. Figure 15d shows the formation of complex vortices which is a combination of several vortices. Such complex vortices form due to the interaction of hot combustion product with the fresh RFNA oxidizer gas along with MMH mass source droplet particle. Results in Fig. 15d can be related to the results obtained in Fig. 14d which clearly shows the presence of combustion product and fresh combustion in the zone of interest.

To investigate the effect of larger droplet size, uniform droplets of diameter 20 μ are selected as given in Table 5 as case 5. Temperature contour plot for case 5 with the particle distribution is depicted in Fig. 16. Figure 16a shows the presence of more particle in the impingement zone compared to particle presence for case 4.

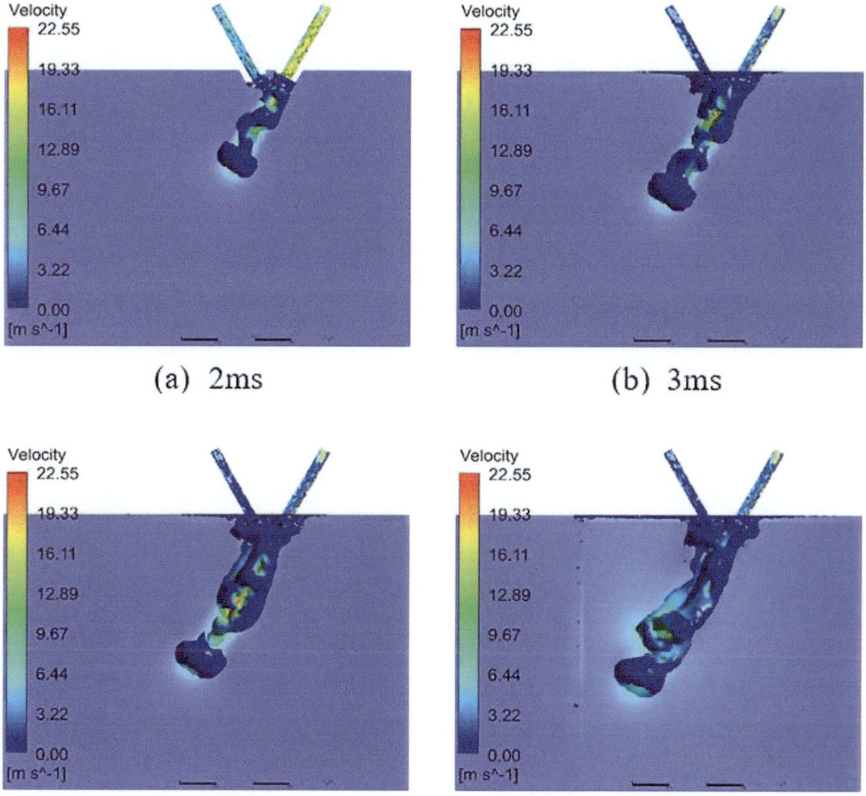

(a) 2ms (b) 3ms

(c) 4ms (d) 5ms

Fig. 15 Vorticity contours for case 4

It may be due to large diameter size that felicitates large combustion area and large combustion masses as well as it could not be pushed away by the oxidizer stream due to considerable momentum. All this effect results in rapid combustion and quick settlement of high temperature in the combustion zone. These large droplet sizes form a barrier to the oxidizer stream and that cause flame deflection from RFNA gas path to the central streamwise perpendicular direction. These shift in flame front direction can be seen in Fig. 16.

Vortex structure corresponding to Fig. 16 is reported in Fig. 17. In Fig. 17a, one small mushroom-shaped vortex is just around the velocity profile ahead of the droplet zone which may be formed due to the high energy level of RFNA gas, and it might have crossed the particle impingement zone before particles reach there. As in the aftermath of particle arrival, the smooth vorticity generation as obtained in case 4 is not possible as it deflects the RFNA flow direction and causes a more vigorous vorticity formation in the impingement zone. This feature is evident from the figures depicted in Fig. 17b, c. These figures also confirm the vorticity characteristics as

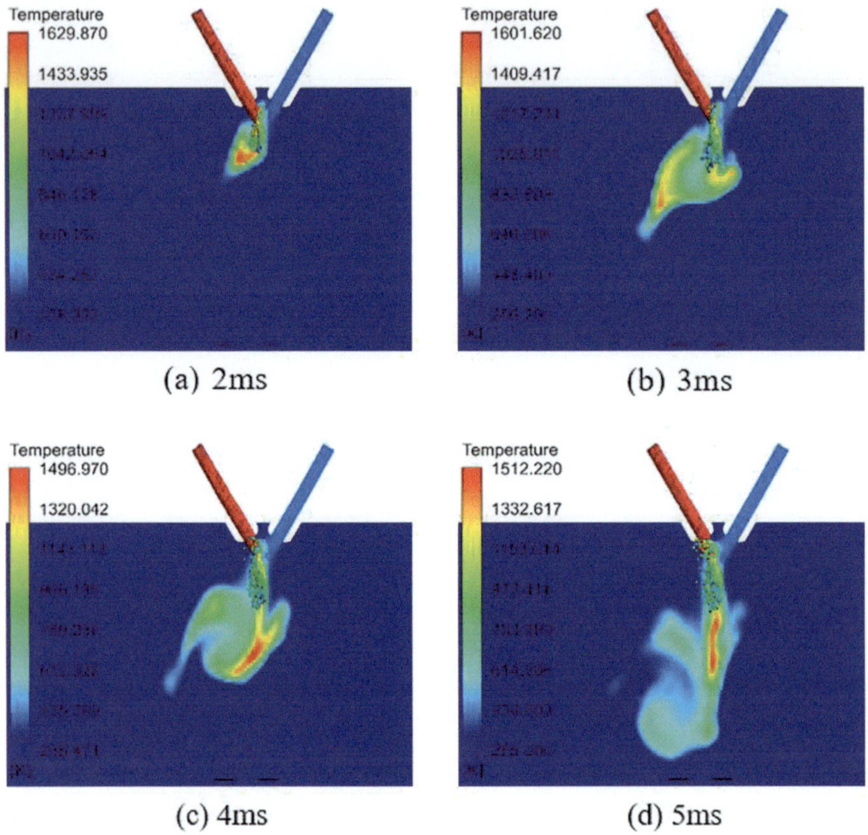

Fig. 16 Temperature contour plot with DPM injection for case 5

well as shifting of flame along the y-axis direction, as opposed to the observation, noticed in case 4 where the flame and vortex form along the RFNA flow path itself. So larger droplet sizes modify the flow field, and it enhances the mixing by creating a complex vortex as seen in the shape of Fig. 17d which is very complicated and close to the perpendicular plane.

Effect of high-temperature oxidizer

As seen in case 5, the large droplet size of diameter 20 μ improves the combustion regime. In case 6, the study is extended to investigate the effect of high-temperature oxidizer stream in the combustion regime. In this case, the oxidizer, i.e., RFNA gas temperature, is increased to 800 K which is 400 K in case 5. Temperature contour with the DPM injection for case 6 is reported in Fig. 18. Compared to the earlier cases, there is a rapid rise in temperature in the impingement zone. As noticed from Fig. 18a, here high-temperature zone of 1781 K is observed. The quick temperature rise may be attributed due to high oxidizer temperature that initiates high vaporization rate of

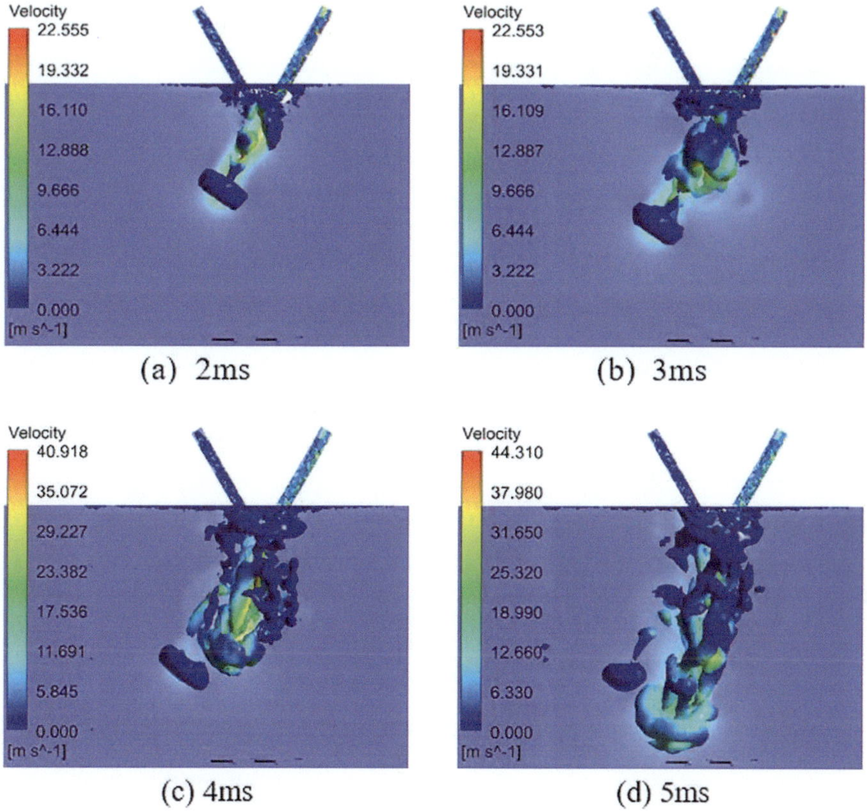

Fig. 17 Vorticity contours for case 5

MMH droplet particle into MMH gas and in turn high combustion rate of MMH gas with the RFNA in the impingement zone. Later the large droplet particles maintain the high mass effect as it is similar to case 5 and starts obstructing the flow of the high-temperature RFNA. Case 6 shows the same effect in deflecting the flame front toward the central streamwise direction as it is shown in case 5. In case 5, due to a comparatively low temperature of the oxidizer stream, the combustion happens while the MMH droplet particles flow past the hot RFNA stream; i.e., it happens downstream of the impingement zone. Whereas in case 6 combustion takes place at the instant, the droplet particles come into contact hence there is no carry forward.

Figure 19 exhibits the 3D vortices for the case 6. As shown in the figure, smooth vortices of RFNA do not form in the initial case due to rapid combustion rate as it is observed for the earlier cases. Since from the beginning of contact itself, combustion process starts taking place in the impingement zone, and that results in the formation of complex vorticity structure (Fig. 19a). Later as the flow and flame get deflected toward the central regime, the vortices appear to shift toward the same region. Here,

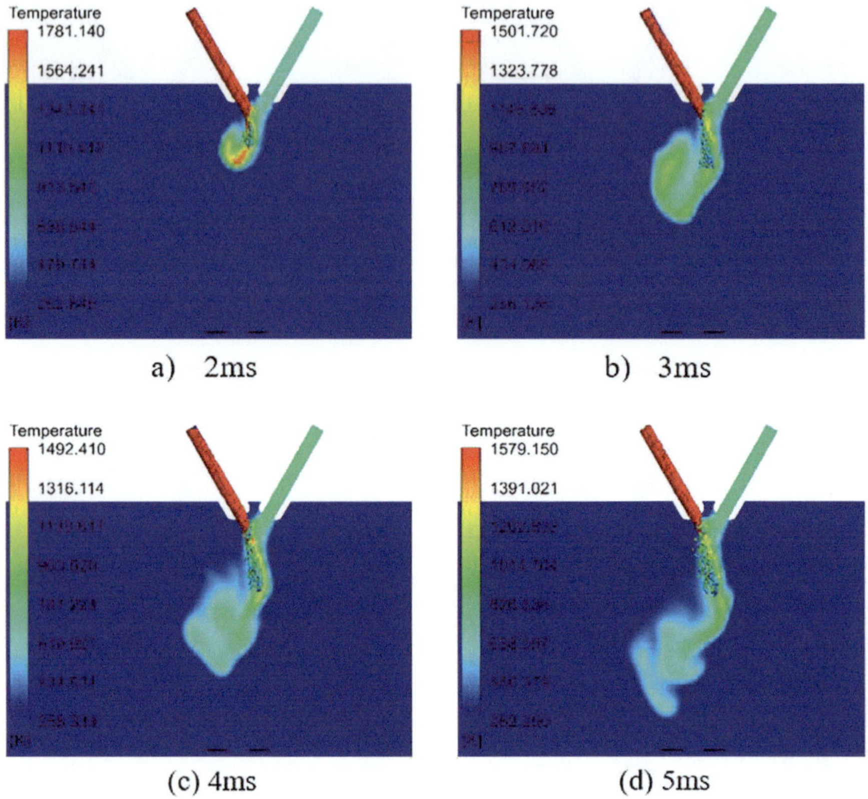

Fig. 18 Temperature contour plot with DPM injection for case 6

vortices formed in Fig. 19 are smoother as compared to that obtained in case 5, and this may be due to smooth combustion and rapid formation of combustion products. The vortices form in this case due to a burnt product which is being pushed by the flame itself.

5 Conclusion

The following conclusions are drawn from the present study:

(a) Equilibrium chemistry modeling with 2D domain shows similar contact time as observed in the literature [4]. However, the contact time obtained for all three cases is different due to a different kinetic energy level at the inlet. Further, different ignition delay observed for all three cases due to different kinetic energy level, which results in different mixing and combustion rate, whereas ignition location obtained for all three cases is again at the same distance of 1.07 cm

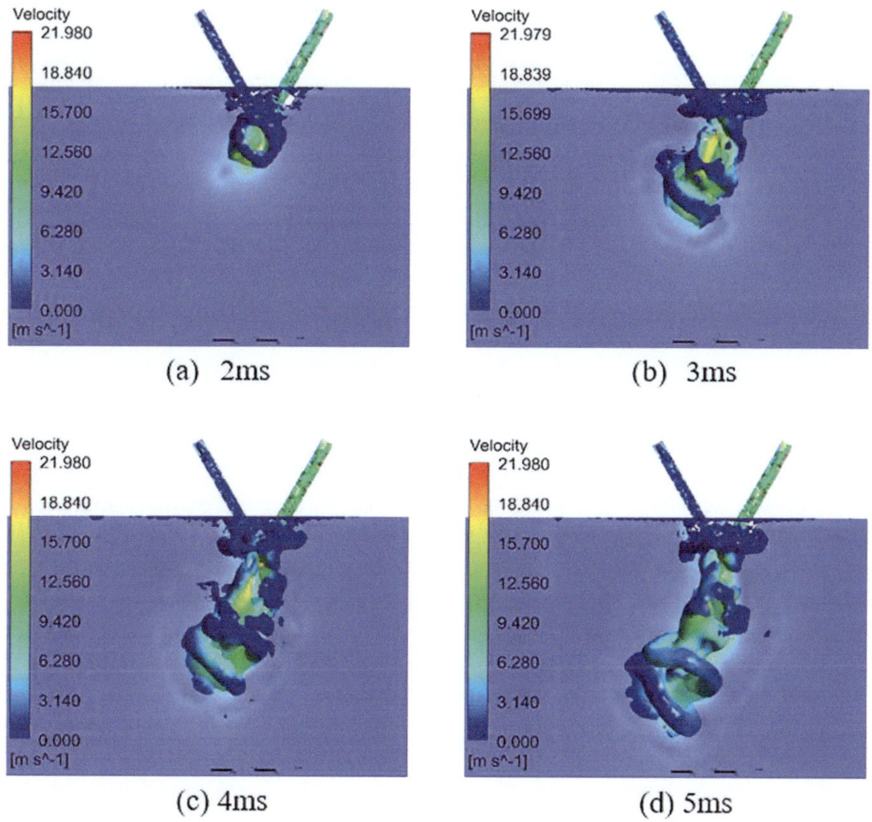

Fig. 19 Vorticity contours for case 6

from the top surface. The discrepancies may be attributed to the equilibrium combustion model.

(b) Comparison of 2D and 3D domain reveals significant differences in predictions of contact time. The difference in contact time is attributed due to volumetric mixing in the 3D domain; this enhances the contact time. Ignition delay obtained for the 2D domain is 0.7 ms whereas for the 3D domain it is 1.0 ms which is very much close to the published results [4]. 2D results also fail in describing the flame spreading behavior whereas in the 3D domain flame spreading characteristics is seen. In addition to that 3D analysis shows different contact time, ignition delay, and less final static temperature as compared to the 2D case which gives different value due to its geometry constraints.

(c) MMH liquid phase injection with gaseous phase RFNA explains hypergolic droplets phase combustion characteristics with gaseous phase. Uniform droplet distribution shows a uniform rise in temperature but flame generation rate is meager due to tiny droplet size as energy content in a droplet of such a small size (1 μ) is very small. Rosin-Rammler droplet distribution with incomplete information about the correct droplet sizes along with low oxidizer temperature does not help in good flame propagation and smooth combustion inside the combustion chamber. Droplet vaporization and combustion happens intermittently as small droplets vaporize early and give high-temperature rise and large droplets have considerable lag time for the vaporization. With much smaller droplet size, the combustion gets delayed, and high temperature observed only after 1.3 ms. Combustion takes place at a delayed time due to the absence of a small droplet and less oxidizer temperature (300 K). Effect of different droplets size is seen to be significant. Uniform temperature rise rate is observed in the case with smaller droplet sizes and flame developed along the oxidizer (RFNA) flow path. This may be due to the high oxidizer stream temperature (400 K) and small droplets size (10 μ). Here, clear vortex of oxidizer stream is seen, and as the time progressed, the vortex moves along the oxidizer stream. At later time instant, multiple vortices are seen due to large combustion products. With 20 μ droplet size, the flame does not propagate along the oxidizer stream, but after initial flame front, it starts shifting toward the center of the chamber. In this case also, at later time instants, very complex vortices are observed due to very high combustion, mixing, and vaporization. Effect of high oxidizer temperature is observed to have some impact on the flame characteristics. Since oxidizer with high temperature (800 K) is used in this case that causes instant ignition and formation of combustion products. The high temperature is seen in the oxidizer flow along the center of the chamber as combustion happens at the moment droplet particle comes in contact with the high-temperature stream. Flame generation is smooth along with the vortices structure due to smooth droplet vaporization and combustion.

Acknowledgements The authors would like to acknowledge the IITK computer center (www.iitk.ac.in/cc) for providing support to perform the computation work, data analysis, and article preparation.

References

1. Anderson WR, McQuaid MJ, Nusca MJ, Kotlar AJ (2010) A detailed, Finite-rate, chemical kinetics mechanism for monomethylhydrazine-red fuming nitric acid systems. Technical report, DTIC Document 2010
2. Labbe N, Kim, YS, Westmoreland P (2010) Computational mechanism development for hypergolic propellant systems: MMH and DMAZ. AIChE Annual Meeting
3. Sardeshmukh S, Heister S, Xia G, Merkle C, Venkateswaran S (2012) Kinetic modeling of hypergolic propellants using impinging element injectors. In: 48th AIAA/ASME/SAE/ASEE joint propulsion conference & exhibit, joint propulsion conferences, Georgia
4. Park KS, Sardeshmukh S, Heister S, Wang H. Numerical simulation of combustion of unlike impinging jets near a wall. In: 49th AIAA/ASME/SAE/ASEE joint propulsion conference, joint propulsion conferences (AIAA 2013-4156)
5. Tani H, Terashima H, Koshi M, Daimon Y (2015) Hypergolic ignition and flame structures of hydrazine/nitrogen tetroxide co-flowing plane jets. Proceed Combust Instit 35:2199–2206

6. Tani H, Terashima H, Kurose R, Kitano T, Koshi M, Daimon Y. Hypergolic ignition and flame structures of hydrazine spray/gaseous nitrogen tetroxide co-flowing jets. In: 53rd AIAA aerospace sciences meeting, AIAA SciTech Forum (AIAA 2015-0422)
7. Daimon Y, Tani H, Terashima H, Koshi M. Three-dimensional structures in hypergolic ignition process and flame holding mechanisms for hydrazine/nitrogen dioxide un-like doublet impinging gas jets. In: 54th AIAA aerospace sciences meeting, AIAA SciTech Forum (AIAA 2016-0691)
8. Wang SQ, Thynell ST (2012) An experimental study on the hypergolic interaction between monomethylhydrazine and nitric acid. Combust Flame 159:438–447
9. Dennis J, Pourpoint T, Son S. Ignition of gelled monomethylhydrazine and red fuming nitric acid in an impinging jet apparatus. In: 47th AIAA/ASME/SAE/ASEE joint propulsion conference & exhibit, joint propulsion conferences
10. Daimon Y, Terashima H, Koshi M (2011) Evaluation of rate constants relevant to the hypergolic reaction of hydrazine with nitrogen dioxide. Chia Laguna, Cagliari, Sardinia, Italy, September 11–15
11. ANSYS Fluent 17.0 User's guide, Canonsburg, PA, USA
12. Saini R, De A (2018) Soot Predictions in higher order hydrocarbon flames: assessment of semi-empirical models and method of moments. Modeling and simulation of turbulent combustion. Springer, Berlin, pp 335–361
13. Toth, Cannon LR, Lewis JC (1997) Monomethlhydrazine propellant/material compatibility investigation and results, California institute of technology jet propulsion laboratory, California, November, AFRPL-TR-77-63

Part II
Renewable Fuels

Green and Clean Upgraded Fuel from Old Landfill Dumpsites for Sustainable Development

Somrat Kerdsuwan and Krongkaew Laohalidanond

Keywords Landfill reclamation · Refuse-derived fuel · Fuel upgrading · Municipal solid waste

1 Introduction

Today's increasing population and economic have led to an increase in municipal solid waste (MSW), which requires proper disposal. Landfill has been considered the most common method of getting rid of this waste due to its lowest operation cost and it is the most used method of waste disposal in many countries. Most developed countries dispose of MSW using sanitary landfill; however, developing countries operate in an opposite manner, using only open dumping or uncontrolled landfill, which has a negative impact on and is harmful to the environment.

Moreover, considering the sustainable waste management hierarchy, it is not sustainable to just bury the waste since valuable materials and energy remain inside the dumpsite. For this reason, the idea of landfill reclamation is taken into consideration since it is a way to recover materials and energy after reclamation. This study focuses on how to recover green and clean energy from dumpsites, the physical and chemical compositions of waste after reclamation, which may vary from one country to another, as well as the life age of dumpsites and the buried depth. After that, technology for upgrading fuel from landfill reclamation in the form of refuse-derived fuel (RDF) will be discussed.

S. Kerdsuwan (✉) · K. Laohalidanond
Department of Mechanical and Aerospace Engineering, King Mongkut's University of Technology, North Bangkok, Bangkok, Thailand
e-mail: somrat_k@yahoo.com

A. K. Gupta et al. (eds.), *Innovations in Sustainable Energy and Cleaner Environment*, Green Energy and Technology,
https://doi.org/10.1007/978-981-13-9012-8_5

2 Energetic Characteristics and Properties of Reclaimed Landfill from Dumpsites

Landfill reclamation or landfill mining is the process to digging up old garbage in a dumpsite in order to obtain empty land that can be reclaimed for reuse. What remains after the reclaimed land consists of soil-like materials, depending on the age of the site, as well as non-degradable materials, which are almost plastics that can be used as fuel after upgrading it to the form of refuse-derived fuel. Landfill reclamation has attracted a lot of interest in the past 50 years in various countries due to its advantages such as [1–3] the following:

- It can increase the potential to reuse the dumpsite to receive new daily garbage.
- RDF can be used as renewable fuel to produce electrical or heat power.
- Some valuable leftover materials after reclamation such as steel and metal can be sold as recycled material.
- Landfill reclamation can reduce the harmful impact on health and the environment.

Since this study focuses on the use of RDF from landfill reclamation to be used for power production, it is essential to study the compositions as well as the characteristics of landfill reclamation.

As mentioned earlier, the most common physical fraction of landfill reclamation is soil-like materials, followed by the non-degradable parts of combustible materials, such as plastic, paper, wood, rags. Table 1 shows the physical compositions of landfill reclamation in various countries.

It has been found that the major fractions of landfill reclamation in various countries are the soil-like materials, which ranges from 40 to 80% by weight. This is due to the biodegradable parts deteriorating, which depend on the life age of the land-

Table 1 Physical composition of landfill reclamation in various countries [4, 5]

	Thailand	Sweden	Belgium	India	USA
Life ages of landfill (years)	3–5	17–25	14–29	NA	NA
Combustible (wt%)					
Plastic	31	4.4	17	11	22
Rubber/leather/foam				14.5	NA
Rags/fabric	7.6	1.8	6.8	2.3	NA
Paper	3.3	7.2	7.5	NA	NA
Wood	8	7.2	6.7	11.6	5
Incombustible (wt%)					
Glass/tile	6.5	0.5	1.3	0.8	8
Metal	6.4	1.6	2.8	0.2	17
Soil-like materials	37.2	77.3	57.9	59.6	48
Total	100	100	100	100	100

fill site. The following fractions are combustible materials, which range from 20 to 50% by weight, and incombustible parts from 10 to 25% by weight. Of course, this depends on the type of waste management, which varies from country to country. For example, Sweden and Belgium employ very good and effective separation of recycled waste at the source, and therefore, there are the fewer amounts of plastic, rubber, and leather than in Thailand and India, where they do not separate waste before discarding material in garbage bins.

The physical composition of landfill reclamation depends not only on the waste management scheme but also on the age of the landfill, which reflects the deterioration of the old garbage inside the dumpsite, as shown in the example of a dumpsite in Thailand in Table 2.

The same trend has been found that most landfill reclamation is still comprised of soil-like materials, where it was observed that the older the landfill is, the greater is the amount of soil-like materials due to the greater portion of deterioration of degradable materials with time. A 5-year landfill reclamation seems to have more soil-like materials than with a 10-year period, which may result from the amount of soil daily cover used where it may depend on the operation scheme. It has also been observed that younger landfill reclamations (2 years) have a higher amount of combustible fractions than older ones, as mentioned earlier.

Over the past decade, The Waste Incineration Research Center (WIRC) has focused its research on landfill reclamation as a fuel for power generation [7–14] and has studied the physical and chemical compositions of reclaimed landfill, as discussed below.

Somboon et al. [13] studied the physical composition of old waste with an age of more than 40 years at the Nongkham solid waste disposal center in Bangkok and compared this with fresh daily waste, as shown in Table 3, while Table 4 shows the study of Somrat et al. [14] on Phuket island.

Table 2 Physical compositions of reclaim landfill from dumpsite in Thailand with various ages [6]

	2 years	5 years	10 years
Combustible (wt%)			
Plastic	36.75	24.64	35.34
Rubber/leather/foam	2.35	1.38	0.60
Rag/fabric	11.51	7.45	1.80
Paper	4.09	0	0
Wood	7.66	3.42	1.20
Incombustible (wt%)			
Glass/tile	2.98	4.86	7.78
Metal	1.79	1.66	4.19
Soil-like materials	32.60	56.59	49.00
Total	100	100	100

Table 3 Comparison of the physical composition of fresh daily garbage with reclaimed landfill from dumpsite: Nongkham solid waste disposal station, Bangkok, Thailand

	Fresh daily waste	Old waste with 40 years old
Combustible (wt%)		
Food waste, vegetables, fruit	44.60	0
Plastic	28.73	38.94
Rubber/leather/foam	1.65	4.65
Rags/fabric	3.70	1.46
Paper	8.99	1.70
Wood	3.99	2.69
Incombustible (wt%)		
Glass/tile	3.31	2.68
Metal	2.10	3.42
Soil-like materials	2.06	0
Total	0.20	0
Glass/tile	0.67	44.46
Metal	100	100

Table 4 Comparison of the physical composition of fresh daily garbage with reclaimed landfill from dumpsite: Phuket island, Thailand

	Fresh daily waste	Old waste
Combustible (wt%)		
Food waste, vegetables, fruit	57.15	0.85
Plastic	18.43	39.98
Rubber/leather/foam	15.79	4.75
Rags/fabric		
Paper	5.26	0.44
Incombustible (wt%)		
Glass/tile	3.36	10.05
Metal	0.01	0.43
Soil-like material	0	43.50
Total	100	100

Additionally, the WIRC has conducted a comparison study of the physical composition of fresh waste with reclaimed landfill from Tha Yang Sub-district Municipality and Phetchaburi Municipality, as shown the detail in Table 5.

The results from Tables 3, 4 and 5 show that the reclaimed landfill contains very little food waste but a higher composition of soil-like material than fresh waste. It can obviously be seen that the fresh waste contains approximately 50% by weight of food waste. When solid waste is disposed of in landfills for a prolonged period, the

Table 5 Comparison of the physical composition of fresh daily garbage with reclaimed landfill from dumpsite: Phetchaburi, Thailand

	Tha Yang Sub-District Municipality		Phetchaburi Municipality	
	Fresh waste	Old waste	Fresh waste	Old waste
Combustible (wt%)				
Food waste, vegetables, fruit	52.04	0	48.73	0
Plastic	20.17	41.73	18.48	37.58
Rubber/leather/foam	2.36	0	1.77	0
Rags/fabric	4.85	3.25	6.62	3.59
Paper	11.92	2.31	17.01	8.01
Wood	5.75	5.41	4	5.72
Incombustible (wt%)				
Glass/tile	2.91	12.06	2.70	5.39
Metal	0	0	0.69	0
Soil-like material	0	35.24	0	39.71
Total	100	100	100	100

fermentation process takes place until the food is degraded into soil-like material. It can be seen that the old waste does not contain food waste. In addition, the older waste contains more plastic than fresh waste, about 30–40% by weight of plastic.

The center has investigated the landfill of the Nonthaburi Provincial Administrative Organization at depths of 3, 7, and 15 m [15] and found that at different depths, there was no significant impact on the physical composition of the solid waste. As shown in Table 6, most of the old solid waste disposed of in landfills is plastic, and soil-like material and paper. This finding is consistent with other studies [4, 5, 9]. The studies are also for the analysis of the chemical composition of reclaimed landfill, as well as other relevant properties, such as bulk density and heating values. Somrat et al. [8] has studied those properties compared with fresh waste derived from Phuket island, as shown in Table 7, and from Tha Yang Sub-District Municipality and Phetchaburi Municipality in Table 8.

It was found that the older waste had lower moisture content than the fresh waste because the old landfill was covered by a final covering of soil, preventing the rain from penetrating inside the dump site. The average moisture content of old waste was about 15% by weight. Other properties were almost the same, and no distinguishing values were observed. However, considering the waste's heating value, the old waste had a higher heating value than the fresh waste, which is about 25 MJ/kg of dry waste.

The effect of the depth of the waste inside the dump site on the chemical and other properties was studied, and Table 9 shows the properties of old the waste from the Nonthaburi Provincial Administrative Organization at depths of 3, 7, and 15 m [15]. It was found that the burial depth had no significant effect on the chemical composition

Table 6 Comparison of the physical composition of reclaimed landfill from dumpsites with various depths of buried material: Nonthaburi, Thailand

	3 m depth	7 m depth	15 m depth
Combustible (wt%)			
Food waste, vegetables, fruit	0	0	0
Plastic	43.18	36.90	39.29
Rubber/leather/foam	6.57	4.19	6.84
Rags/fabric	2.53	6.29	2.65
Paper	4.29	7.55	2.65
Wood	11.11	8.18	12.80
Incombustible (wt%)			
Glass/tile	2.53	4.61	3.31
Metal	4.80	9.22	5.08
Shell/bone	0	0	0
Household hazardous waste	0	0	0
Soil-like material	24.99	23.06	27.38
Total	100	100	100

Table 7 Comparison of the chemical composition of fresh daily garbage with reclaimed landfill from dumpsite: Phuket island, Thailand

	Fresh waste	Old waste
Proximate analysis (wt%, dry and ash-free basis)		
Moisture[a]	46.34	1.23
Volatile matter	77.56	74.82
Fixed carbon	1.53	6.95
Ash	19.92	18.23
Total	100	100
Ultimate analysis (wt%, dry and ash-free basis)		
Carbon	46.10	53.60
Hydrogen	6.38	6.03
Nitrogen	1.10	Not detect
Sulfur	0.12	0.16
Oxygen	46.30	40.21
Total	100	100
Other properties (dry basis)		
Density (kg/m^3)[a]	128.91	90
Heating value (MJ/kg)	8.03[a]	25.82

[a]as received basis

Table 8 Comparison of the chemical compositions of fresh daily garbage with reclaim landfill from dumpsite: Phetchaburi, Thailand

	Tha Yang Sub-District Municipality		Phetchaburi Municipality	
	Fresh waste	Old waste	Fresh waste	Old waste
Proximate analysis (wt%, dry and ash-free basis)				
Moisture[a]	60.87	14.93	58.97	12.12
Volatile matter	84.20	89.51	82.33	74.38
Fixed carbon	14.38	9.84	16.50	25.22
Ash	1.42	0.65	1.17	0.40
Total	100	100	100	100
Ultimate analysis (wt%, dry and ash-free basis)				
Carbon	50.75	66.30	45.40	59.77
Hydrogen	7.78	9.63	6.53	8.75
Nitrogen	0.61	0.08	0.23	0.15
Sulfur	0.16	0.09	0.13	0.16
Chlorine	1.05	2.51	0.58	0.64
Oxygen	39.65	21.39	47.13	30.53
Total	100	100	100	100
Other properties (dry basis)				
Density (kg/m^3)[a]	153.70	124.07	129.59	134.30
Heating value (MJ/kg)	23.31	32.09	18.98	32.63

[a]as received basis

or its heating value since the plastic content in the waste had no variation with the buried depth, as shown earlier.

3 Technology Development of Refuse-Derived Fuel (RDF) Production from Reclaimed Landfill

The disposal of solid waste by using high-temperature technology is often difficult to handle due to the variety in its composition. The changes are based on community and seasonal characteristics. Moreover, this waste has low heating value and high-ash and moisture content. This eliminates many difficulties for technology designers and practitioners, but as a result, environmental control is also difficult. The upgrading of waste as fuel is needed in order to improve the physical and chemical properties of solid waste to make it refuse-derived fuel, (RDF) which can be solved the difficulty as mentioned previously. The RDF can be used as high-class fuel for power

Table 9 Comparison of the chemical compositions of reclaim landfill from dumpsite with various depths of buried: Nonthaburi, Thailand

	3 m depth	7 m depth	15 m depth
Proximate analysis (wt%, dry and ash-free basis)			
Moisture[a]	46.50	44.26	38.26
Volatile matter	76.95	70.12	74.37
Fixed carbon	5.83	6.47	7.07
Ash	17.22	23.42	18.56
Total	100	100	100
Ultimate analysis (wt%, dry and ash-free basis)			
Carbon	59.38	55.33	57.36
Hydrogen	8.19	7.37	7.80
Nitrogen	1.68	0.89	1.09
Sulfur	0.45	0.54	0.70
Chlorine	0.67	0.65	3.22
Oxygen	12.40	11.80	11.27
Ash	17.22	23.42	18.56
Total	100	100	100
Other properties (dry basis)			
Heating value (MJ/kg)	19.17	20.19	20.22

[a]as received basis

generation where it can control the burning process as desired and has less impact on the environment.

RDF refers to waste that has undergone various upgrading processes, such as the separation of non-combustible materials, and tearing or cutting waste into small pieces. The upgraded fuel will have a higher heating value or is better to use as fuel than waste that is collected and used directly. It has both uniformity in its physical and chemical compositions.

The upgrading process of solid waste to high-grade fuel requires a management process that depends on the properties of preferred fuels. The general management process is as follows (Fig. 1):

1. Source separation,
2. Manual or mechanical separation,
3. Size reduction,
4. Size classification,
5. Mixing,
6. Drying and palletization,
7. Packaging and storage.

In the designing fuel, upgrading from waste process shall depend on the preliminary handling process of waste. For example, if there is the preprocess of source

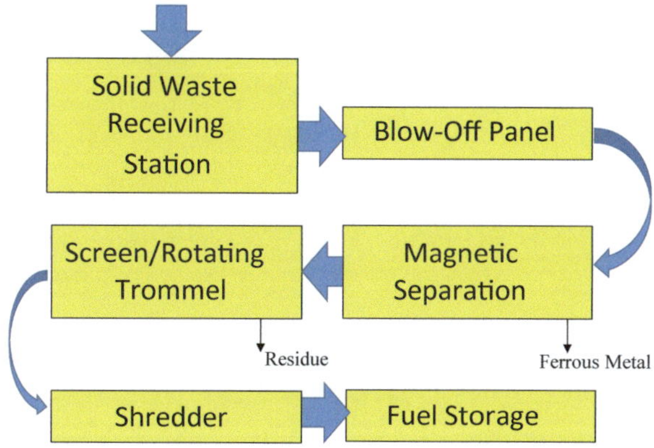

Fig. 1 General process of RDF production

Table 10 Rate of RDF production from solid waste [18]

Country	Process	Rate of RDF production (%)
Austria	MBT	23
Belgium	MBT	40–50
Finland	MT	Indeterminate
The Netherlands	MT	35
The UK	MT	22–50

MBT mechanical biological treatment
MT mechanical treatment

separation in order to separate the recyclable materials, it may not be necessary to have a procedure for separating metal or glass. Generally, the waste is separated into recyclable parts, such as metal and glass, where the organic fractions such as food waste that has high moisture content will be used in the biogas production process or soil conditioners. Other fractions after separation such as paper, wood chips, and plastics will pass through size reduction and be used as direct combustion fuel in the form of coarse RDF (c-RDF) or pass through drying and compacting to form densified RDF (d-RDF) [16, 17]. The selection of which type of fuel to use depends on the technology of the combustion system used, the location between the fuel processing plant and where it will be used, etc. The amount of RDF produced per 1 ton of waste depends on the waste collection, the processes used in waste processing, and the quality of the fuel needed. The amount of RDF produced from solid waste is in the range of 23–50% by weight of processed waste, as shown in Table 10.

The technology for RDF production can be classified as mechanical biological treatment (MBT), biological treatment (BMT), and autoclaving as follows.

Mechanical Biological Treatment (MBT)

This process is a combination of mechanical sorting and biodegradation. Solid waste is treated by mechanical sorting, for example, with the use of conveyor belts, metal sorting machines, air blowers, etc. The separated fractions—waste paper, waste plas-

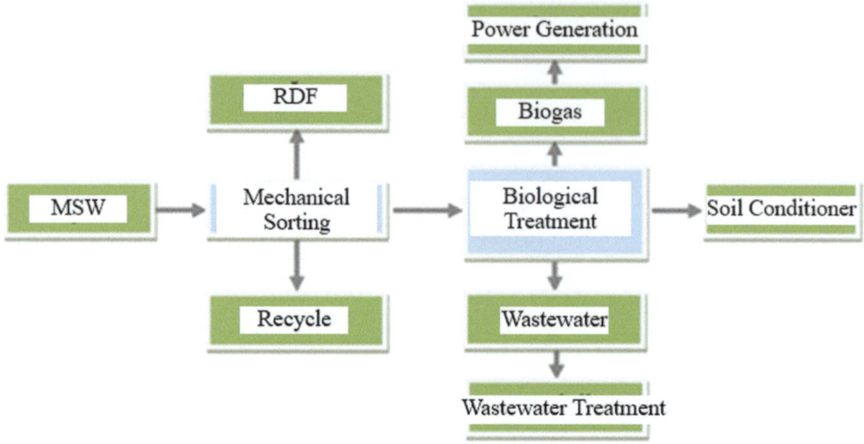

Fig. 2 RDF production by MBT [19]

tic, waste cloth—will be used to produce RDF. The remaining, for example, organic food waste, will be sent to the biodegradation process, which may be either composting or anaerobic digestion, and finally will become soil conditioner. Figure 2 shows the process.

Biological Mechanical Treatment (BMT)

This process is a combination of biodegradation and mechanical separation. Solid waste shall be treated by a composting or anaerobic digestion process before the biodegradable organic waste is biodegradable and become biogas. The remaining waste from biodegradation processes, such as mixed metal, paper, plastic, must be passed through mechanical sorting, such as conveyor belts, metal sorting machine, air blower, and then crushed or pelletized to produce RDF. Figure 3 shows the process of producing RDF by BMT.

Autoclaving RDF

The limitation of the MBT and BMT systems is that the efficiency of the sorting system determines the quality of the RDF, and with the limited availability of various types of waste, low separation efficiency. As a result, the quality of RDF is reduced. Another technology used in RDF production is the replacement of the mechanical waste sorting system by steam. By using steam with high temperature and pressure sprayed into the waste bins. Under appropriate times and conditions, metals, non-metals, plastics, and organic materials are separated. After that, it will be used to make RDF and recycled materials. Because this process is a process that uses steam at high temperatures and pressures, it requires the high energy consumption in the process which resulting in high operating system prices as well. Figure 4 shows the process of autoclaving RDF.

Since there are many types of combustion systems used in power plants and industrial plants, each type of RDF must be selected in accordance with the combustion

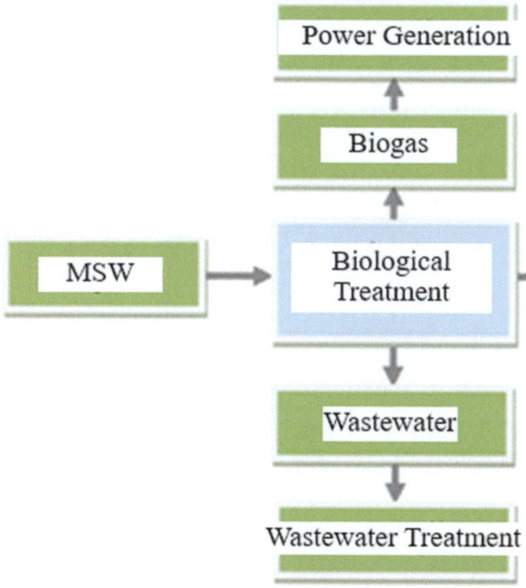

Fig. 3 RDF production by MBT [19]

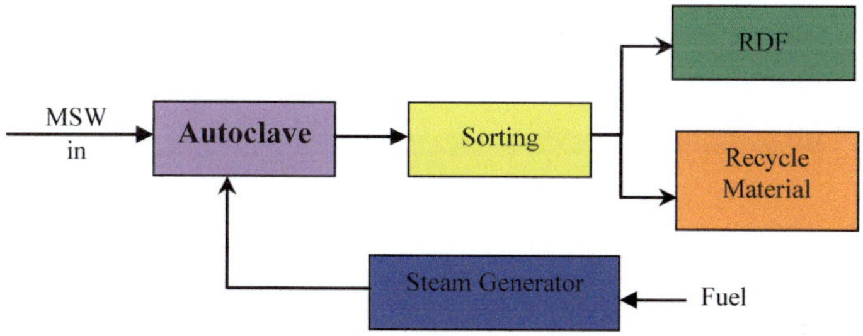

Fig. 4 RDF production by autoclaving [19]

system used. Table 11 shows the characteristics of each type of RDF and the type of combustion system used.

The Waste Incineration Research Center has designed the technology to produce RDF from old dumpsites on Phuket island, as shown in Figs. 5 and 6. The main processes include (1) landfill reclamation, (2) material pre-sorting, and (3) material upgrading.

Landfill Reclamation Process

In the excavation of old waste from a landfill, the dumpsite will be excavated using a hydraulic excavator. The old solid waste is removed from the pit to a height of 0.8–1 m. in order to reduce the moisture content of the old waste to suit the process of sorting out the soil-like materials and other materials. Typically, the time to dehumidify the pit depends on the weather conditions and the rainfall in the project area.

Table 11 Characteristic of each type of RDF and the combustion system used [19]

Type	Manufacturing process	Combustion system used
RDF 1: MSW	Sorting out the incombustible parts as well as large trash	Stoker
RDF2: coarse RDF	Coarsely crushed or cut waste	FBC
RDF 3: fluff RDF	Sorting out the incombustible parts such as metal, glass, crushed or cut until 95% of the sorting waste has a size of less than 2 in.	Stoker
RDF 4: dust RDF	Sorting out of the combustible fractions and grinding into the form of dust	FBC, PF
RDF 5: densified RDF	Sorting out of the combustible fractions and densified until it has a density of more than 600 kg/m^3	FBC
RDF 6: RDF Slurry	Sorting out of the combustible fractions and grinding into the form of slurry	Swirl burner
RDF 7: RDF syngas	Sorting out of the combustible fractions and used to produce syngas through the gasification process	Burner, IGCC

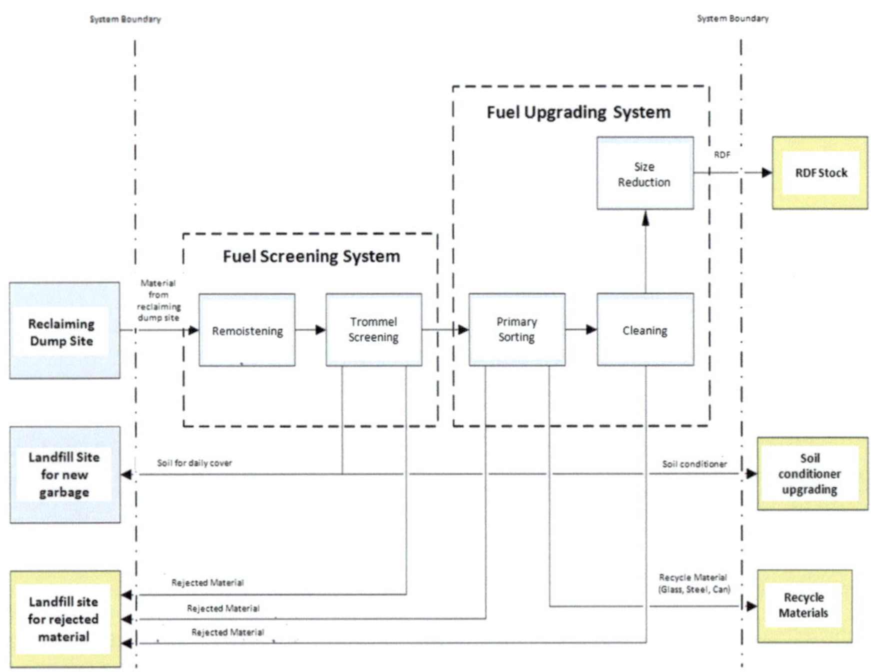

Fig. 5 RDF production from old dumpsite process

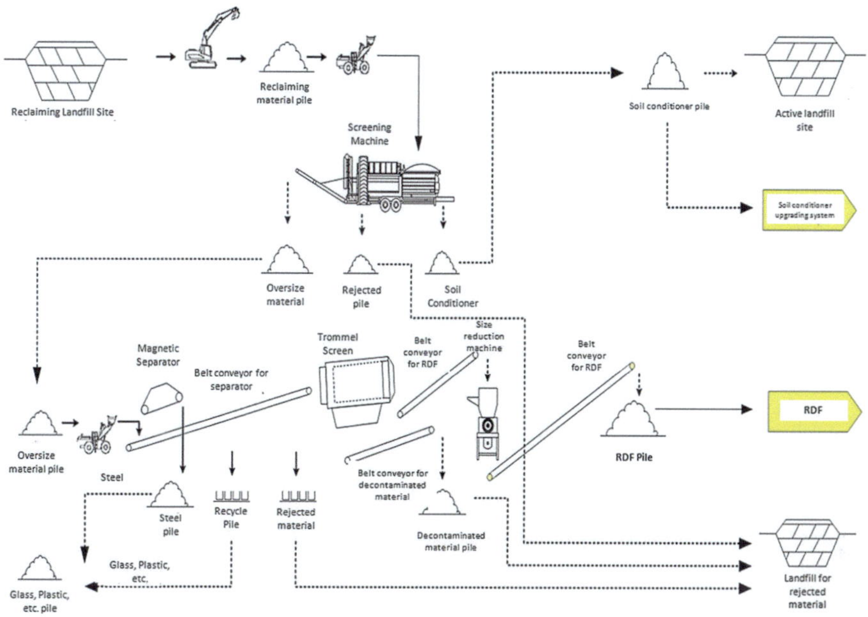

Fig. 6 Machine used for reclaiming process of old dumping sites

Material Pre-sorting Process

The basic process is to use a wheel loader to transport material from the old solid waste pile with proper humidity for entering it into the mobile drum screen. The appropriate amount of the material is fed into the feeder conveyor while larger materials are separated by sieves fitted to the feeder system of the glider before being fed into a separate glider. In order to prevent the splitter from being damaged, the material on the conveyor belt controls the feed rate to suit the performance of the drum screen. The material to be fed into the drum screen is rotated by rotation. At the same time, smaller materials are separated through the drum screen. Small pieces of material that pass through the sieve hole, most of which are soil or soil-like material, fall into the conveyor belt, which is used to remove the small material from the machine to prepare for transportation for further use. While the material which is larger than the grating hole which contains mostly plastic and combustible materials, will move out to the end of the drum screen and is conveyed by the conveyor belt and move to improve the quality of materials which can be used as RDF.

Fuel Upgrading Process

The fuel upgrading process has the purpose of improving the large material obtained from the initial separation process to be suitable for use as fuel. The materials from the initial gliding process are transported by truck to the storage facility. During the material improvement process, the material is fed into the hopper by the wheel loader. The conveyor belt is then conveyed through a metal separator (magnetic separator) to separate the steel from the material. The materials are then sorted by human labor in order to separate the valuable materials and to remove unwanted contaminants, and the valuable materials are collected to prepare them for use. The contaminated goods are collected and prepared to be disposed in the landfill. The material is then transported to the drum screen in order to separate the small particles admixed with the material. The small particles that pass through the grate will fall into the conveyor belt in order to remove the contaminants from the machine, and then, these will be transported and disposed in the landfill. The material larger than the grate is composed of mostly plastic and combustible materials moving toward the end of the drum screen and is conveyed by the conveyor belt to reduce it to a suitable size by the grinder for RDF production. However, if the upgrading process needs the dene fuel, this can be done by adding the compacting/palletizing machines. Materials that are of the correct size and moisture will be transported and prepared for transportation to be used in the form of fuel.

The properties of the RDF received from the green manufacturing of eco-materials by urban mining from landfill dump sites are shown in [20]. It is found that the moisture content of the mined MSW is much lower than that of the received waste, whereas the volatile matter and ash of the mined MSW are higher than those of the received waste since the decomposition of moist kitchen waste into soil-like material and high content of plastic waste in mined MSW, consequently, higher heating value.

4 Conclusion and Recommendation

Since energy recovery from waste is considered green, clean, and renewable energy, energy recovery from mined landfill waste creates benefits in terms of land reclamation and the reduction of the fossil fuel used. Old MSW from landfill reclamation can be considered as an alternative energy source since it contains a high amount of combustible materials, e.g., plastic waste. The landfill process itself behaves as a biological process in terms of decomposing organic waste in MSW, and therefore, only a mechanical process is required for RDF production from old MSW. After the mechanical process, RDF with a high-energy content can be used as an alternative fuel in cement industries or in MSW power plants. Landfill reclamation contributes not only to the production of new energy sources, but also to sustainable waste disposal. Additionally, it provides more landfill space for raw MSW.

Acknowledgements The authors would like to express their gratitude to National Research Council of Thailand (NRCT) for financial support of this project and also the Waste Incineration Research Center (WIRC), Department of Mechanical and Aerospace Engineering, Faculty of Engineering and Science and Technology Research Center (STRI) of King Mongkut's University of Technology, North Bangkok, for the facilities support.

References

1. UA EPA (1997) Landfill reclamation, solid waste and emergency response, USA
2. Centre for Environmental Studies, Dumpsite rehabilitation and landfill mining, Anna University, India
3. Tanha A, Zarate D (2012) Landfill mining: prospecting metal in Gärstad landfill. Linköping University, Sweden
4. Quaghebeur M, Laenen B, Geysen D, Nielsen P, Pontikes Y, Van Gerven T, Spooren J (2013) Characterization of landfilled materials: screening of the enhanced landfill mining potential. J Cleaner Prod 55:72–83
5. Kurian J, Esakku S, Palannivelu K, Selvam A (2003) Study on landfill mining at solid waste dumpsites in India. In: Proceedings Sardinia 2003, Ninth international waste management and landfill symposium, Cagliari, Italy, 6–10 October 2003
6. Chiemchaisri C, Charnnok B, Visvanathan C (2010) Recovery of plastic wastes from dumpsite as refuse-derived fuel and its utilization in small gasification system. Biores Technol 101:1522–1527
7. Laohalidanond K, Cherdphong S, Kreeso P, Kerdsuwan S (2011) Utilization of RDF from dumpsite as alternative fuel for power generation via gasification process. In: Proceedings ISET 2011, international symposium on EcoTopia science, Nagoya, Japan, 9–11 December 2011
8. Kerdsuwan S, Laohalidanond K, Cherdpong S, Uthaikiattikul T (2011) Testing of RDF gasification from old landfill dump site for electricity generation in a downdraft gasifier. In: Proceedings ICAE 2011, third international conference on applied energy, Perugia, Italy, 16–18 May 2011
9. Kreeso P, Kerdsuwan S, Laohalidanond K, Cherdpong S (2012) Testing of 200 hr CONTINUOUS OPERATION of MSW gasification. In: Proceedings GRE 2012, graduate research conference, Khon Kaen University, 2012
10. Laohalidanond K, Kerdsuwan S (2012) Performance evaluation of using producer gas from MSW gasification as supplementary fuel in a modified diesel engine. In: Proceedings ICEAS 2012, international conference on engineering and applied science, Beijing, China, 24–27 July 2012
11. Etutu TG, Cherdphong S, Laohalidanond K, Kerdsuwan S (2012) Air-blown gasification of refuse derived fuel (RDF) in a downdraft gasifier: effect of air flow rate on tar formation. In: Proceedings ICGSI 2012, international conference on green and sustainable innovation, Chiang Mai, Thailand, 24–26 May 2012
12. Etutu TG, Laohalidanond K, Kerdsuwan S (2013) In-depth analysis of tar formation from a downdraft RDF-gasification. In: Proceedings ICAE 2013, international conference on applied energy, Pretoria, South Africa, 1–4 July 2013
13. Chalermcharoenrat S, Sirirermrux N, Laohalidanond K, Kerdsuwan S (2014) Optimization of the production of densified RDF from reclaimed landfill without mixing binding agent using hydraulic hot pressing machine. In: Proceedings on TSME-ICOME 2014, The 5th TSME international conference on mechanical engineering, Chiang Mai, Thailand, 17–19 December 2014
14. The Waste Incineration Research Center (2011) Assessment of waste-to-energy project for Petchaburi, Thailand, Final Report
15. The Waste Incineration Research Center (2015) Study and analysis of old waste from dumpsite of Nontaburi Provincial Organization, Final Report
16. Cheremisinoff NP (2003) Handbook of solid waste management and waste minimization technologies. Elsevier Science, USA
17. White PR, Franke M, Hindle P (2001) Integrated solid waste management: a lifecycle inventory. Blackwell Science, USA

18. European Commission (2003) Refuse derived fuel, current practice and perspectives, Final Report
19. Sommerlad RE, Seeker R, Finkelstein A, Kilgroe JD (1988) Environmental characterization of refuse derived fuel incinerator technology. National Waste Processing Conference
20. The Waste Incineration Research Center, Feasibility Study and Detail Design of the investment and operation of waste disposal by bioreactor landfill for green energy production, Final Report, Phuket Municipality

Nonlinear Synergistic Effects in Thermochemical Co-processing of Wastes for Sustainable Energy

Kiran Raj Goud Burra and Ashwani K. Gupta

Keywords Sustainable energy · Waste and biomass · Thermochemical processing · Pyrolysis · Gasification · Co-processing of biomass and plastics · CO_2 content

1 Introduction

Two hundred years of industrial revolution has provided rapid improvements to the human lifestyle aided by fossil fuel supply from the buried hydrocarbon reservoir formed over a span of million years [1]. The rapid nature of this development has caused rapid depletion of natural hydrocarbon resources as well as the impact of our actions on the use of these resources that are now recognized to be unsustainable, both for energy use and the environment. One such effect is the rapid rise in the atmospheric CO_2 from 280 ppm in the pre-industrial era (~1800 s) to above 400 ppm today due to a rapid increase in fossil fuel combustion for unsustainable energy demands of the industrial society [2, 3]. The consequences were not considered until much later when the impact of this rise was visible in the form of steady increase in global temperature [4]. Before going into the details on the impact of anthropogenic carbon release on the terrestrial climate, we need to understand the working of a stable global carbon cycle and means by which it impacts the global temperature and thus the climate. A brief understanding into the important mechanisms of global carbon cycle is discussed below.

K. R. G. Burra · A. K. Gupta (✉)
The Combustion Laboratory, Department of Mechanical Engineering, University of Maryland, College Park, MD 20742, USA
e-mail: akgupta@umd.edu

A. K. Gupta et al. (eds.), *Innovations in Sustainable Energy and Cleaner Environment*, Green Energy and Technology,
https://doi.org/10.1007/978-981-13-9012-8_6

$$CO_2 + CaSiO_3 \overset{weathering}{\Longleftrightarrow} CaCO_3 + SiO_2 \quad (1) \tag{1}$$

The carbon cycle is stabilized by the weathering–metamorphism equilibrium along with carbonate deposition, given by Eq. (1) which controls the atmospheric CO_2 content [1, 5]. Igneous rocks contain calcium oxide components in mineral phase which form metal carbonates when exposed to atmospheric CO_2 (weathering) [5, 6]. These sediments are then eroded from the continents (land) to the oceans as the carbon sink. Geological carbon source for rise in atmospheric CO_2 is from natural degassing emissions from volcanic gases, and deep sea hot spring fluids. An increase in global temperature increases water vapor pressure causing accelerated hydrological cycles (freshwater runoff from continents into the oceans) [6]. This increased runoff leads to higher erosion rates so that fresh igneous rocks gets constantly exposed to the atmosphere and increased weathering. Thus, the weathering carbon sinks becomes stronger when the temperature increases. An increase in atmospheric CO_2 also leads to increase in weathering rate and to increase the strength of the sink. This negative feedback, also called weathering CO_2 thermostat, is partially responsible for stabilizing the atmospheric CO_2 content and the global temperature [7]. Any variations in the atmospheric CO_2 or global temperature causes the sink to become stronger to cause cooling of the planet and decrease the CO_2 content to a new set point. Over this timescale, the degassing emissions match the weathering to cause stabilized atmospheric CO_2. Note that the response time of this mechanism is in the order of hundred thousand years to reset the carbon balance [7].

An insight into the interaction between faster changes in atmospheric CO_2 on the global temperature can be understood by examining the interglacial cycles on change in the global temperature. The CO_2 levels have been constant over long time periods, averaged of the order of hundred thousand years, suggesting the influence of weathering CO_2 thermostat mechanism, although there have been interglacial cycles in CO_2 concentrations. The polar ice sheets also cycle through formation and melting driven by the change in solar intensity distribution caused by the earth's orbital wobbling, also called Milankovitch cycle [6]. Once ice sheets form, the diffuse reflectance of that surface changes to lower the energy absorption and thus the global temperature, which in turn increases the rate of ice sheet formation. This mechanism has a positive feedback on the temperature change [6]. Decrease in global temperature causes increased ice sheet formation, which increases the reflected solar intensity to further lower the global temperature. The opposite effect occurs during ice melting. Another natural periodic mechanism that controls the atmospheric CO_2 is the atmosphere–ocean gas equilibrium. The temperature and pH dependence on the solubility of CO_2 in seawater are the pathway for control. Increase in ocean temperature leads to lower solubility and thus the dissolved CO_2 leaves the ocean into the atmosphere. This mechanism has the response time of the order of thousand years [6]. Lower temperature increases the solubility of CO_2 so that more CO_2 is captured by the ocean. Proxy global temperature and atmospheric CO_2 data showed that during glacial cycles the atmospheric CO_2 changed in rhythm with the temperature that accounted for a positive feedback mechanism in the carbon cycle causing a direct link between the CO_2

and global temperature [8]. Even though ocean–atmosphere CO_2 equilibrium is a positive feedback mechanism, it has been found that its contribution, based on deep sea data, does not fully account for the interglacial variation in CO_2 when temperature varied [6]. The variation in pH due to the variation in CO_3^{2-} ions also has an effect on ocean's feedback on atmospheric CO_2. CO_2 reacts with water to form carbonate and bicarbonate ions so that increase in CO_2 intake decreases the pH of the ocean, also called ocean acidification. This acidification also increases the solubility of $CaCO_3$, which can then affect the marine life and the sustainability of $CaCO_3$ shells. The increase in acidity can also affect the marine life in other ways due to the changing conditions. All these natural inorganic processes along with significant biological contributions from marine and terrestrial life control the atmospheric concentration of CO_2 and the global temperature. Change in any of these parameters affects the rest of the climate as per the above-described cycle.

2 Carbon Emission

Anthropogenic carbon emissions are mainly contributed from fossil fuel combustion to meet the energy demands. Fossil fuels are carbonaceous deposits in the crust of the earth that have formed over a span of millions of years. While the formation rate of these fuels has been long, their consumption rate, by humans, has been of the order of decades. Such a drastic imbalance in the utilization rate makes anthropogenic carbon emissions so rapid that the only natural carbon sequestration from atmosphere to land, i.e., photosynthesis, has been slow to catch up. Note that even in the case of high wood yield trees such as pine wood, for example, 10 tons/year/acre of wood is harvested which corresponds to net CO_2 absorption of only 3.63 tons of C/year/acre via photosynthesis [9, 10]. This is the main reason for net rise in atmospheric CO_2 content. The carbon emissions of ~9.8 Gt of C year^{-1} (in 2014) have been significant compared to the natural carbon fluxes [11, 12]. The anthropogenic CO_2 flux should be made only with the natural degassing flux since that is the only known natural and significant pathway for carbon movement from the crust into the atmosphere. Table 1 provides carbon pools of the earth. Isotopic studies have shown that the anthropogenic CO_2, which can be counted easily, is the main contributor for increases in the atmospheric CO_2 content. The rate of increase in atmospheric CO_2 was found to be only half that of the anthropogenic carbon emissions rate, which suggests that the rest of the carbon is missing. Carbon flux balancing showed that most possible pathway for missing CO_2 is into the ocean, which the ocean can adjust by increasing its pH [13, 14]. Currently, the global carbon cycle's negative feedback is keeping the atmospheric CO_2 to the present levels. Due to the order of ocean–atmosphere equilibrium timescale, it is expected that around the turn of this century, the ocean carbon pool may saturate causing other positive feedback mechanisms to drive significant climate changes. Predictions into the future are not very clear, but the presence of positive feedback mechanisms and their influence on the sensitive ecosystem makes this rising atmospheric CO_2 an absolutely important and essen-

Table 1 Carbon pools of the earth (from [1])

Carbon pools	Quantity (Gt)
Atmosphere	720
Oceans	38,400
Total inorganic	37,400
Total organic	1000
Surface layer	670
Deep layer	36,730
Lithosphere	
Sedimentary carbonates	$>0.6 \times 10^8$
Kerogens	0.15×10^8
Terrestrial Biosphere (total)	2000
Living biomass	600–1000
Dead biomass	1200
Aquatic biosphere	1–2
Fossil fuels	4130
Coal	3510
Oil	230
Gas	140
Others (peat)	250

tial problem of this century. It also provides a deciding factor for the future on the fate of life on the earth. Significant overhaul of the existing infrastructure and the methodology of problem solving are necessary prior to moving into such uncertain times. Even though the improvement of global lifestyle is necessary, it should not come at the expense of facing the consequences of an unsustainable society, which deteriorated the climate of the planet into an uninhabitable place.

Most of the global population resides in the so-called 'developing' economies. It is very timely for all economies, including industrialized countries, to develop and cultivate sustainable pathways for energy and environment to foster improved lifestyle for all. Energy is the fundamental and most important resource needed for societal development. The developed countries have reached industrialization from utilization of fossil fuel resources that have left large footprint on continuously increasing CO_2 as mentioned above due to lack of understanding on the consequences of carbon cycle disruption along with an underestimation of the population growth. Even in the developed countries such as the USA, some 81% of primary energy consumption was from fossil fuel resources in 2016; see Fig. 1. This represents a significant amount (some 36 million barrels of oil equivalent/day) that needs replacement for sustainable energy and environment in the USA. With the knowledge gained on understanding of the impact of carbon footprint, it is time to foster further investment in sustainable resources for energy production.

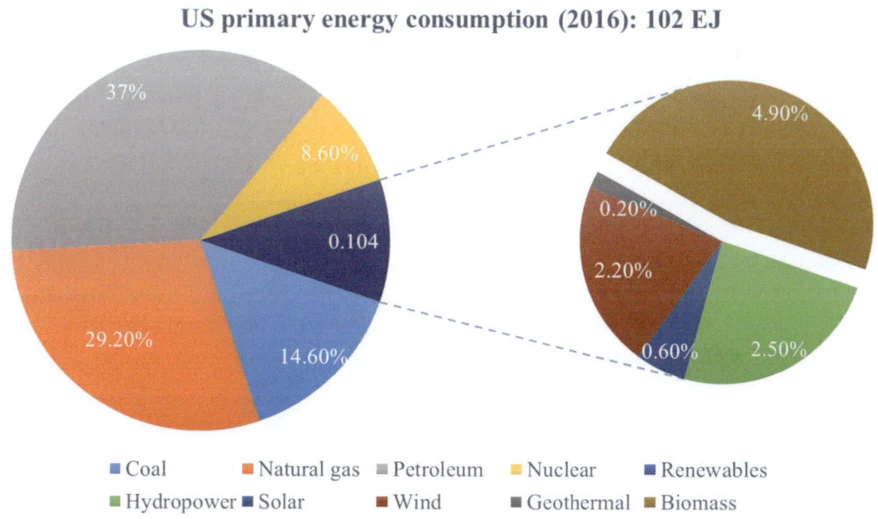

Fig. 1 Primary energy consumption in the USA in 2016 [15]

The fundamental requirement of a sustainable energy resource lies in its ability to recover energy at the same rate as its consumption. Renewable energy resources such as direct solar energy utilization, hydropower and geothermal energy, wind energy are abundant and driven by the solar radiation and earth's energy. Their abundance negates the need for recovery. The authors choose not to include the scalability of nuclear-based energy production, due to our lack of understanding on nuclear waste disposal, or expanding such an energy source globally. France, Slovakia, Ukraine, Belgium, and Hungary use nuclear energy for majority of the country's electricity supply. Other resources, such as biomass energy, need recovery time which is the time taken to grow the same amount of resources as consumed over a given time span. Even though photosynthesis is relatively slow, it extracts solar energy to form well-known form of carbon-based lignocellulosic material at higher efficiency than photovoltaics and the growth of biomass is necessary for the survival of ecosystems. This makes simultaneous plantation and utilization of biomass a sustainable pathway for energy production. The fuels produced from biomass are also carbon based similar to fossil fuels, which means the infrastructure replacement and the understanding needed for their implementation are lower compared to other renewable forms of energy production. In the next section, we will discuss biomass and its potential to meet our goal of greater energy demands globally

3 Biomass as Sustainable Energy Resource

In this chapter, we define biomass to include wood waste (trees, branches, and bark residues), shrubs, agricultural residues (bagasse), and biogenic fraction of commercial and municipal solid wastes (paper, cardboard, food waste, animal manure, and other bio-generated waste components). Currently, world's biomass-based energy, also called bio-energy, contributed to 50.3 EJ/year in 2008 which is ~10% of the global primary energy consumption [16]. Figure 2 shows the contribution of a variety of biomass sources to total primary energy production in 2008 [16]. These bio-energy usage estimates include two generic types of usage—first, bio-energy in the form of direct burning of wood, dung, and straws for cooking, space heating, and lighting in the rural areas of 'developing' countries. This accounts for around 37–43 EJ/year of usage while its low conversion efficiency means that only about 10–20% of this energy reaches the consumer. This shows the availability of resources but lack of available infrastructure for sustainable utilization. Second, bio-energy in the form of electricity and CHP, space heating, and transport fuels (ethanol and biodiesel) from biofuels, such as wood, MSW and biogas, in a developed infrastructure at high efficiency accounts for about 11 EJ/year of primary energy consumption. Note that only up to 60% of this energy reaches the consumer after considering losses from process efficiency in conversion to secondary energy [16]. Table 2 shows the details of these estimates that show the stark differences in the improvements that can be obtained with appropriate infrastructure. Modern bio-energy infrastructure reduced the primary energy requirement by three times for the same amount of secondary energy production. Countries such as India, China, and Brazil were shown to be the major contributors to bio-energy production [17].

The theoretical potential estimates on global bio-energy considering high biomass plantation productivity from global agriculture on land, and considering the need for conservation of a feasible biosphere, one global modeling study showed maximum bio-energy potential of 1548 EJ/year [18]. This is the amount of available biomass for energy considering only biophysical constraints. Assessing the technical potential is difficult to predict that accounts for biomass production with practice limitations, competition with food, fodder, fiber and forest products, area for human infrastructure, along with nature, and biodiversity preservations. This is primarily due to uncertainty in cropping techniques to be implemented, weather conditions, and difficulty in predicting the competition between the demands for forestry and agriculture products (as mentioned above) with the demand for energy. Variations in the assumptions on population, economic and technological growth, variety in social preferences, climate change, and uncertainty of biodiversity cause the broadband in the estimates of technical potential by different studies. Table 3 provides the technical potential estimates by 2050 for each variety of biomass considered for bio-energy. Considering rain-fed lignocellulosic plants on unprotected woodlands and grasslands while accounting for food and fodder requirements (excluding forests for biodiversity purposes), the global technical potential is estimated to be 171 EJ/year [16]. Different plantation and crop management techniques may be considered to

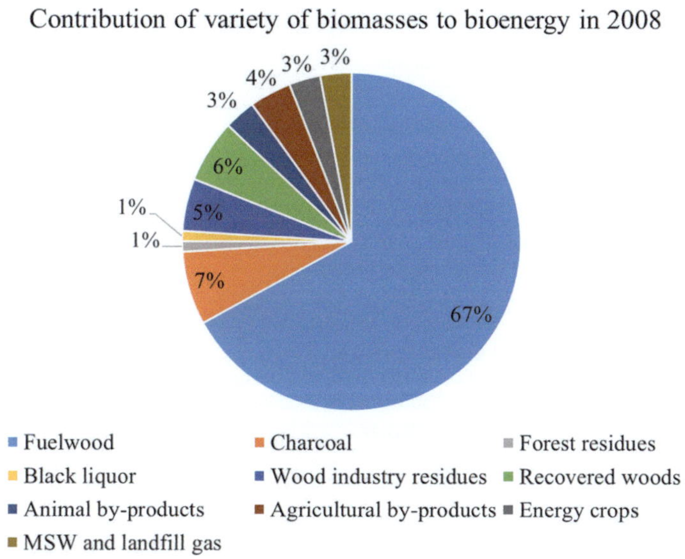

Fig. 2 Contributions of different types of biomass to global bio-energy (from [16])

Table 2 Global bio-energy contributions from different sectors (from [16])

Type	Approx. primary energy (EJ/year)	Approx. average efficiency (%)	Approx. secondary energy (EJ/year)
Traditional biomass			
Accounted for in IEA statistics	30.7	10–20	3.6
Estimated for informal sectors (ex. Charcoal)	6–12		0.6–2.4
Total	37–43		3.6–8.4
Modern bioenergy			
Electricity and CHP from biomass, MSW, and biogas	4	32	1.3
Heat in residential, public/commercial buildings from solid biomass and biogas	4.2	80	3.4
Road transport fuels (ethanol and biodiesel)	3.1	60	1.9
Total	11.3	58	6.6

Table 3 Technical potential estimates of bio-energy from different sources by 2050

Biomass category	Comment	2050 technical potential (EJ/year)
Agricultural residues	By-products of agriculture: primary (cereal straw from harvesting) and secondary (rice husk from mills)	15–70
Dedicated biomass production on surplus agricultural land	Land used: agricultural land not used for food; Conventional agriculture crops and dedicated bioenergy plants Possibility of no surplus land availability from food sector development	0–700
Dedicated biomass production on marginal lands	Usage of deforested, degraded or marginal lands not suited for agriculture to grow biomass by bioenergy schemes, such as reforestation. Lacks appropriate definition of marginal lands and this means addition of surplus agriculture land and marginal lands may be double counting	0–110
Forest biomass	Forest sector by-products-Primary: silvicultural thinning and logging Secondary: Wood processing residues- sawdust, bark Zero potential possible if the demand from other sectors such as paper, wood exceeds the estimates	0–110
Dung	Animal manure	5–50
Organic wastes	Household and restaurant organic wastes; discarded wood including paper, construction and demolition; Waste management critical	5 to >50
Total		<50 to >1000

enhance the potential without disturbing biodiversity. However, this discussion is out of the scope of this chapter and readers are referred to other references [16].

The technical potential estimates provide the importance of different kinds of biomass to meet our energy demands, even though these estimates do not always consider policy and environmental limitations. This means efficient management and utilization of bio-energy are important to negate the continuously rising CO_2 emission. The issue with biomass for energy production lies in its low mass and low energy density as compared to fossil fuels. Their management in terms of harvesting, pretreatment, storage, and transport is estimated to have significant energy costs that can account for some 20–50% of total energy produced from them. The dependence of single type of biomass makes it unreliable due to their limited seasonal availability. This makes the contribution of various secondary bioresidues, such as rice husk, animal manure, and other biogenic wastes from MSW, important to achieve the highest potential to meet the energy demands in a carbon constrained world. Although these secondary residues can potentially provide energy, their conversion efficiency significantly depends on their energy content, which then also depends on their moisture content, chemical composition, and inorganic content. These factors decide the energy consumption for pretreatment of these feedstocks to enhance their compatibility to secondary energy carrier conversion techniques, such as combustion, gasification, pyrolysis, anaerobic digestion, or fermentation.

4 Wastes Supplement to Biomass

Low energy density of secondary biomass along with varying amounts and types of wood production demand causes unreliable availability of biomass feedstocks. Even the supplementary energy feedstocks required for the bio-energy potential mentioned above are not as reliable for biomass feedstocks. It is recognized that there is potential for biomass replacing fossil fuels in the energy sector, but their potential in replacing the petrochemical precursors used in today's world for the production of value-added chemicals and polymers is uncertain and mostly unknown. With reasonable technical assumptions, the bio-energy is only able to supplement coal and natural gas for energy production, but fossil fuels would still be required for transportation and plastics production. This implies that while consideration to only bio-energy consumption along with other conventional renewable energies for carbon emissions, biomass do not account for the potential carbon emissions from fossil fuel usage in materials requirement. Plastic materials are produced from fossil fuels, and their consumption is steadily increasing. Carbon material balance is very important for sustainable utilization of resources. An important solution here is to develop synergy between bio-energy utilization and plastic waste management as a hybrid approach in managing both biomass and plastic wastes.

Material management is necessary for a sustainable society, and this is lacking in the present scenario. Significant portions of energy-intensive materials, such as plastics and food waste, are ending up in landfills. Landfilling is inexpensive option in the short term but unattractive for waste management since significant land wastage along with potential chemical leaching from the landfills makes the soil around these landfills unusable. Furthermore, it often causes hazardous discharges to the water sheds that causes problems to clean water supply. Uncontrolled decomposition of organic wastes in landfills, especially manure waste, causes methane release into the atmosphere leading to greenhouse effect, similar to CO_2 but methane is roughly 30 times more potent as a heat-trapping gas to provide enhanced greenhouse gas emission potential. To mitigate or avoid such influences on climate change, proper disposal practices of these wastes are necessary. Note that, significant portions of these wastes emanate from municipal solid wastes (MSW). As per EPA, the USA produced 262 million tons/annum of MSW in 2015 that consisted of materials shown in Fig. 3 [19]. This is equivalent to producing approximately 6 billion tons of MSW globally, if global lifestyle was equated to that of the USA. While approximately 30% of this waste was recycled or composted (an accepted pathway for waste management), more than half of the total waste still ended into landfills; see Fig. 4 [19]. This represents a significant amount of chemical energy left unused, which could be extracted to meet energy demands. A detailed look into landfilling of several major components in MSW is shown in Fig. 5. Note that significant portions, representing more than 75% of plastics, and more than half of rubber, textiles and leather wastes, end up in landfills. This is hazardous for the environment since organic wastes, such as food wastes, end up three out of four portions in landfills, and their decomposition occurs in an uncontrolled environment. They contribute to significant portions of CH_4 and CO_2 emissions into the atmosphere during their anaerobic digestion, leading to climate risks described above. Waste-to-energy pathway is very critical since significant portions of these wastes cannot be separated out for recycling or composting. Waste components such as plastics, rubber, and textiles are high energy density materials, and their utilization offers necessary supplements to biomass for fossil fuel replacement in both energy and materials production. Note that biomass alone does not show potential for complete fossil fuel replacement when considering both energy and material production. Appropriate conversion techniques chosen or developed to minimize net carbon emissions require sustainable processing of wastes that also can meet demands on energy and material/value-added chemicals.

5 Biomass Conversion Processes

Significant improvements in conversion techniques are necessary for efficient utilization of the available biomass feedstocks. Various conversion techniques summarized in Fig. 6 are based on the feedstocks used, physical pathway, and the products obtained. The conversion processes are classified into three types based on the reactions involved. They include biochemical processing, chemical processing, and thermochemical processing.

Biochemical processing involves utilization of microorganisms under mild conditions for the progress of conversion reactions. Methods of fermentation involve microorganisms such as yeasts, in anoxic conditions to metabolize sugars into ethanol. Since specific feedstock types are required for metabolizable sugars, feedstocks such as sugarcane, sugar beet, sweet sorghum, and other starch crops (such as corn, wheat, or cassava) are the only types used for this type of processing for ethanol production with the goal of substituting/supplementing gasoline as trans-

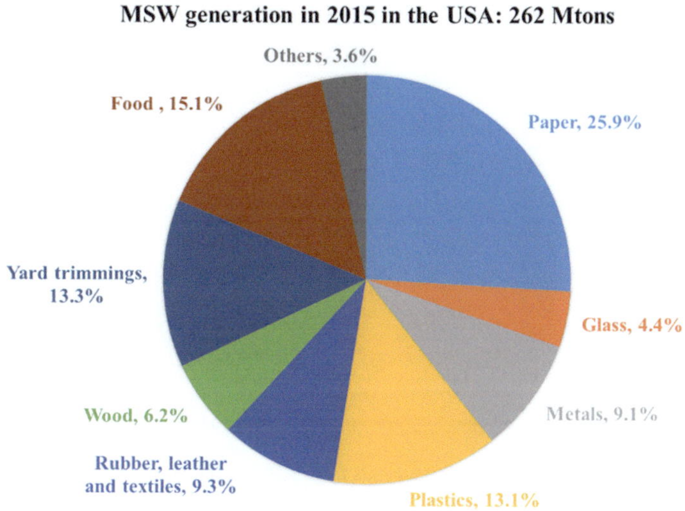

Fig. 3 Contributions in MSW generated in the USA in 2015 (from [19])

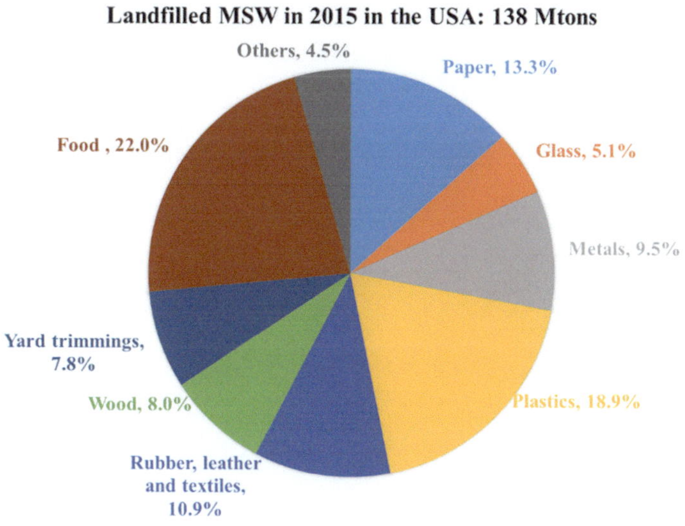

Fig. 4 Materials distribution in landfilled MSW in the USA (from [19])

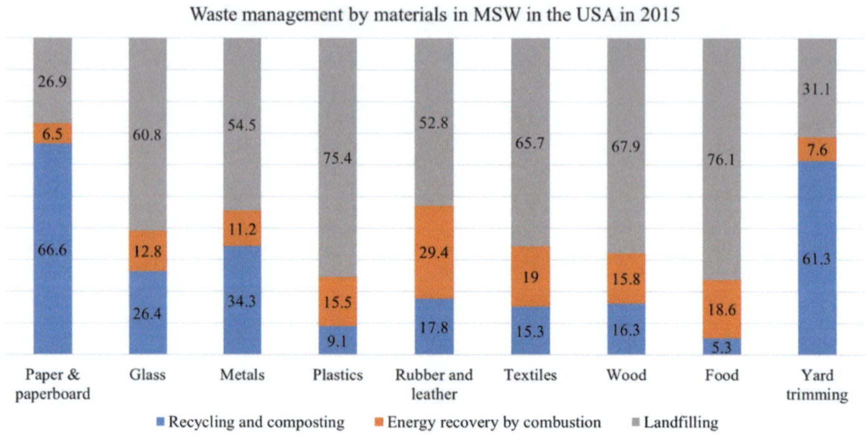

Fig. 5 MSW component pathways by materials from selected process in the USA in 2015 (values in percent) [19]

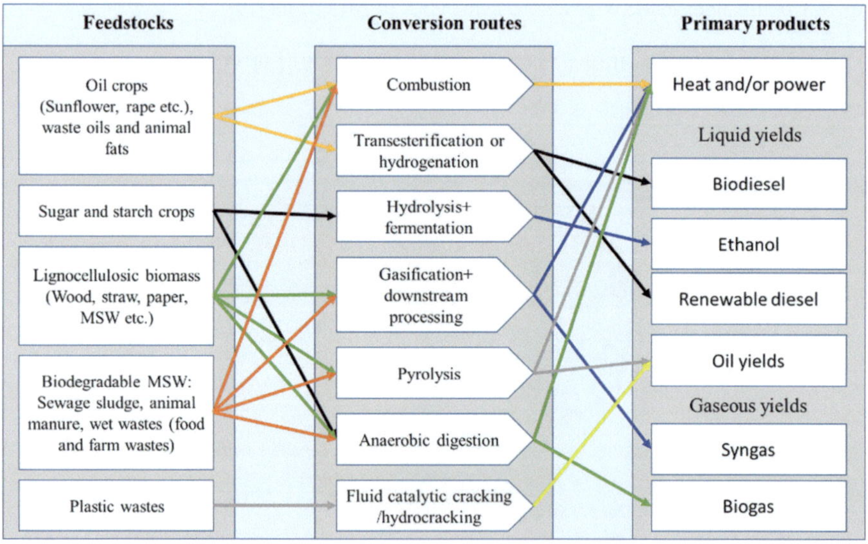

Fig. 6 Conversion routes suitable for each type of feedstock and the desired products (based on [16])

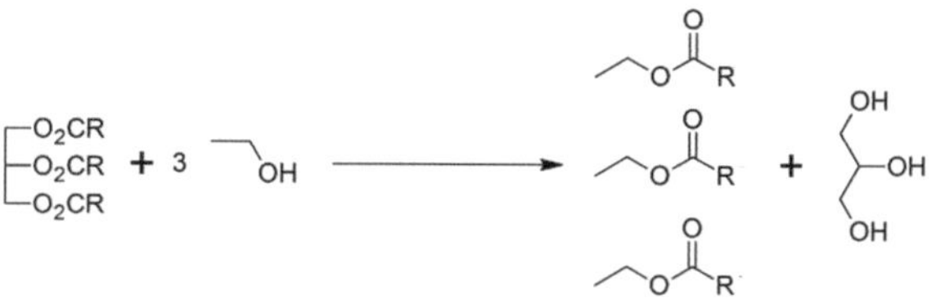

Fig. 7 Generic transesterification reaction

portation fuel [16]. More generic biochemical process is anaerobic digestion, which involves metabolizing a wide variety of feedstocks such as animal manure, leafy feedstocks, and other biogenic solid wastes by methanogenic bacteria to produce methane-rich gaseous mixture containing methane and carbon dioxide, also called biogas. High methane content makes this biogas useful as fuel for energy production by steam cycling, or direct combustion for cooking, heating, and possibly transportation. This gas is suitable when seeking for acceptable natural gas requirements due to high methane content. The residual material also has the potential for usage as soil supplement. While significant contributions can be expected from these processes, their lack of feedstock flexibility and the slow throughput makes these processes unsuitable or inefficient. Most of the biowaste from MSW also needs to be separated out for efficient performance of the biochemical processes, which increases the cost of feedstock management and preparation. Furthermore, lack of feed-flexible pathway for liquid fuel production places limitation on the potential of these biochemical processes.

Chemical processes involve catalytic processing of biomass extracts to produce fuel, energy, or value-added chemicals under mild conditions. Two such common techniques are: transesterification, and hydrogenation. Both these processes use feedstocks that can produce intermediates such as vegetable oil, recycled oils, or animal fat to produce liquid fuel replacements/additives for petroleum-based transportation fuel. Transesterification involves catalytic reaction of methanol with triglycerides obtained from vegetable oil, as shown in Fig. 7, extracted from seeds, to form alkyl esters of fatty acids with glycerol as by-product. These long-chain fatty acid esters, commonly referred to as biodiesel, are useful substitute as transportation fuel. Hydrogenation involves catalytic H_2 saturation/reduction of vegetable oils, recycled oils, or animal fats to form long-chain hydrocarbons, which is a renewable diesel fuel that can be blended with normal diesel. Both these processes involve a very narrow type of feedstock requirement that limits the potential of these processes for scalability and resourceful utilization on available bio-energy.

The thermochemical processes, well known to mankind for centuries, involve conversion of biomass feedstocks into valuable products such as fuel, heat, or electricity at high temperature. Three major thermochemical processes are: direct combustion, pyrolysis, and gasification. Direct combustion is one of the dominantly used processes for biomass utilization in the current scenario, especially in rural areas involving burning of charcoal, wood, or dung cakes for cooking and heat. Pyrolysis involves anaerobic decomposition of biomass at high temperature to produce solid (charcoal), liquid (bio-oil), and gas products whose relative yield depends on the operational conditions such as heating rate, temperature, and vapor residence times. Slow pyrolysis at low heating rates and moderate temperatures predominantly provides charcoal through carbonization, which is dominantly used around the world for char production. Biomass gasification involves partial oxidation of biomass at high temperature to produce gaseous mixture of predominantly H_2 and CO, called producer gas. This gas mixture is a combustible mixture that can be upgraded downstream into syngas of higher heating value, which can then be directly combusted or converted into liquid fuels by Fischer–Tropsch synthesis or into other value-added chemicals such as diethyl ether or methanol [20]. The syngas can be combusted for electricity production in combined heat and power plants without direct biomass combustion. Gasification provides efficient control over the product output that also provides enhanced combustion efficiency over direct combustion. The results have shown higher efficiency in power production due to higher heating value of syngas compared to solid biomass and better mixing in gas-phase combustion. Further details about these thermochemical processes are discussed next along with their choice over biochemical and chemical processes.

6 Thermochemical Processing

Thermochemical conversion is the most versatile pathway for biomass and waste utilization because chemical and biochemical conversion techniques, as mentioned above, are not capable of converting wide range of feedstocks. Since the driving potential in these processes is temperature, unlike difference in chemical potential or microorganism metabolism in chemical and biochemical processes, these reactions provide high throughput along with feed flexibility, which is necessary to utilize wastes for bio-energy production from biomass. Biomass combustion is a well-established technique involving burning of biomass or wastes in the presence of air/oxygen for heat production which can be utilized for combined heat and power. While this technique is well-known, solid combustion, especially biomass and wastes with low heating value, it is inefficient due to lack of mixing and low specific heat output causing low-temperature operation, incomplete combustion, and excessive pollutants' formation and emission. Direct combustion of these heterogeneous oxygenated feedstocks also produces NO_x, SO_x, and other pollutants due to the high reactivity of oxygen, and presence of nitrogen and sulfur in feedstocks [21, 22]. Removal of these pollutants is not only difficult, but also energy intensive that only lowers the net efficiency of direct combustion. Pyrolysis and gasification processes provide better control along with uniformity in the products produced. Lower oxygen content during pyrolysis and gasification results in off-gases produced to contain lower nitrogen oxide and sulfur oxide compounds. The N and S contents leave as NH_3 and H_2S that are easier to separate out using processes such as wet scrubbing [23]. In this chapter, we focus on pyrolysis and gasification as the primary components of thermochemical processing.

Table 4 Thermochemical processes and their product yield by temperature and residence times [24]

Process		Temperature (K)	Vapor residence time	Liquid yield (wt%	Solid yield (wt%)	Gas yield (wt%)
Pyrolysis	Fast pyrolysis	~773	~1 s	75	12	13
	Intermediate pyrolysis	~773	10–30 s	50% in 2 phases	25	25
	Carbonization	~673	~Hours to days	30	35	35
	Torrefaction	~563	Solid residence time ~ 10 to 60 min	5% if condensed	80% solid	20
Gasification		>973	Same as char	5	10	85

Thermochemical processing of carbonaceous materials can be classified based on the operating parameters such as temperature, heating rate, vapor residence time, gasifying agent during pyrolysis, and gasification. Their product yield depends critically on the above parameters. Role of these parameters in pyrolysis and gasification can be seen from Table 4. Solid product produced from these processes is called char, which is a coal-like product with high heating value, high C content, and includes the inorganic residue left after decomposition. Gaseous products formed primarily include CO, CO_2, and H_2 along with low-molecular-weight (C_1, C_2, and C_3) hydrocarbon gases. Their relative yield depends on the above-mentioned operational parameters along with material composition [22]. Liquid yields are called bio-oil/pyrolysis oil in the case of pyrolysis and tar in the case of gasification. They contain wide variety of products including oxygenates, carboxylic acids, aldehydes in the case of bio-oil (from pyrolysis), and PAHs in the case of tar (from gasification) [24].

6.1 Pyrolysis

Pyrolysis is the process of decomposing any material by heating to high temperatures to break its chemical bonds in the absence of oxidizing agents such as air/O_2, steam, or CO_2. This is in contrast to gasification that includes some oxidizing agent. Pyrolyzers can be operated under different conditions for improved yield and increased selectivity of the desired product phase and amounts. Typical kinds of pyrolysis apparatus explored for different amounts of gas/liquid yields are presented in Table 4. Solid biomass upon heating releases vapors from material decomposition. The vapor residence time is the time these vapors evolved from the biomass decomposition stay

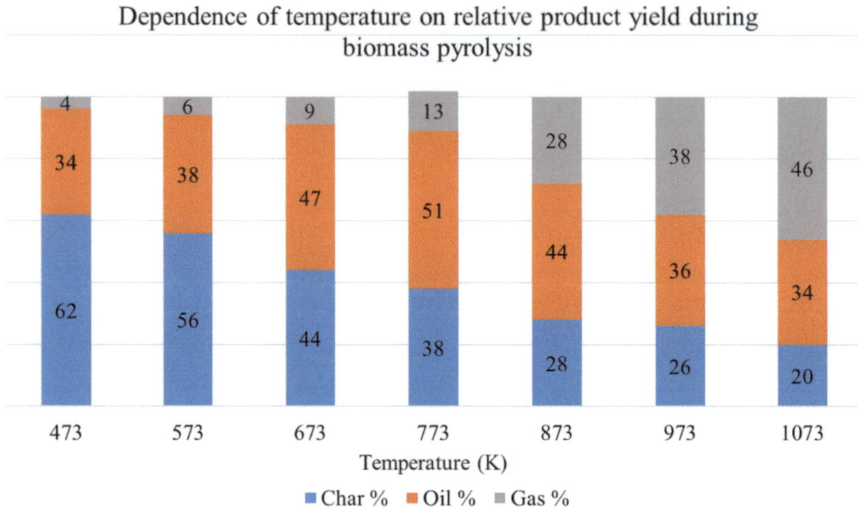

Fig. 8 Effect of temperature on fast pyrolysis and relative product-phase yields [25]

in contact with the solid phase during pyrolysis. This controls the secondary reactions of evolved vapors with the solid components, which further controls the extent of cracking. Temperature controls the thermodynamic extent of reaction. Farther downstream, higher gas yield results from enhanced cracking. Low temperature and long residence times favor solids' formation in the form of char. High temperature and long residence times favor formation of gases as these results in high degree of cracking. However, moderate temperatures and shorter residence time yield liquid formation, usually called bio-oil. The short residence times also consider the contact time on char formed which needs to be reduced to avoid cracking of vapors into low-molecular-weight hydrocarbon gases. The influence of temperature on biomass pyrolysis is shown in Fig. 8 [25]. Optimal temperature for high bio-oil yield is seen to be around 773 K, which balances the repolymerization and the extent of cracking. The low-temperature operation and long residence times lead to enhanced repolymerization to form increased char yields from carbonization. Biomass carbonization or a lower temperature torrefaction is conducted to obtain processed solid fuel with higher heating value and density from a biomass having low density and low heating value. The obtained torrefied biomass can be pelletized into high-density solid fuel for easy transportation and enhanced combustion efficiency when combusted for heat/power. These solid fuels contain high carbon content and can act as substitute for coal in various applications, in particular for power generation.

Fast pyrolysis occurs at vapor residence times as low as 2 s at moderately low temperatures of 773 K with the motive to yield high bio-oils. To avoid secondary vapor-phase reactions and reach thermodynamic non-equilibrium, the collected vapors undergo rapid cooling to obtain high yields of bio-oil. Vapor quenching efficiency can be enhanced by collecting vapors in an immiscible hydrocarbon solvent. The

$$C+0.5O_2 \rightarrow CO;\ \Delta H^o_R=-110.5\ kJ/mol$$
$$CO+O_2 \rightarrow CO_2;\ \Delta H^o_R= -283\ kJ/mol$$
$$H_2+0.5O_2 \rightarrow H_2O;\ \Delta H^o_R=-241.8\ kJ/mol$$
$$CH_4+2O_2 \rightarrow CO_2+2H_2O;\ \Delta H^o_R=-802.23\ kJ/mol$$
Water gas shift reaction: $CO+H_2O \leftrightarrow CO_2+H_2;\ \Delta H^o_R=-41.2\ kJ/mol$
Steam reforming: $CH_4+H_2O \leftrightarrow CO+3H_2;\ \Delta H^o_R= 206.17\ kJ/mol$
Dry reforming: $CH_4+CO_2 \leftrightarrow 2CO+2H_2;\ \Delta H^o_R= 247.37\ kJ/mol$
Boudouard reaction: $CO_2+C \leftrightarrow 2CO;\ \Delta H^o_R= 172.5\ kJ/mol$
Char-steam gasification: $C+H_2O \leftrightarrow CO+H_2;\ \Delta H^o_R= 131.3\ kJ/mol$
Methanation: $C+2H_2 \leftrightarrow CH_4;\ \Delta H^o_R= 74.9\ kJ/mol$
Thermal cracking: $C_nH_m \rightarrow C_{n-a}H_{m-x}+C_aH_b+0.5(x-b)H_2;\ \Delta H^o_R>0$

Fig. 9 Set of equilibrium reactions that follow pyrolytic breakdown of biomass during gasification

high heating rate requirement limits the feedstock properties to moisture content less than 10% and small particle sizes (around 2 mm or less) as heating of this biomass is the rate limiting step in fast pyrolysis. It also helps to reduce aqueous content in the bio-oil. When lignocellulosic biomass is pyrolyzed, the lignin-derived species are inherently more stable due to high aromaticity (lignin being aromatic polymer), and this means low gas yields from lignin. Lignin pyrolysis leads to phenolic derivatives such as guaiacol. Cellulose pyrolysis leads to dehydration and depolymerization leading to levoglucosan if the heating rates are low. At high heating rates, anhydrosugars, such as levoglucosan, can be minimized favoring production of liquid products.

6.2 *Gasification*

Gasification is pyrolysis of solid carbonaceous feedstock in the presence of gasifying/oxidizing agents such as steam, CO_2, or O_2/air at high temperature (>1000 K). This process is developed to obtain high yield of synthesis gas, which is a mixture of H_2, CO, CO_2, H_2O, CH_4, and low-molecular-weight hydrocarbons (C_2, C_3), and minimum yield of tar or char (liquid and solid yields). Typical reaction pathway involves pyrolysis followed by secondary reactions among the vapors formed, reforming of the vapors using gasifying agent, in parallel for the gasification of char formed with the aid of gasifying agent to yield H_2 and CO. Side reactions also occur which involves repolymerization of the vapors to yield polyaromatic hydrocarbons (PAH). Low temperatures favor tar formation while high temperatures (greater than 1300 K) have reported no residual tar yields. The gases yielded at such high temperature adhere to and are controlled by thermodynamic equilibrium. The major secondary reforming reactions, following pyrolytic breakdown, are given in Fig. 9.

The choice of gasifying agent depends upon the desired composition of the syngas yield, which subsequently depends on the downstream applications of the syngas yield. Steam gasification yields syngas with high H_2 content, while (dry) CO_2

gasification can enhance CO content in syngas by utilizing readily available high-temperature CO_2 from flue gases [26]. The endothermicity of these reactions can be maintained by external heat or by use of air/O_2 at very fuel-rich conditions. Operating temperatures vary between 1000 and 1500 K for different types of gasifying reactors used the industrial gasifiers that operate at low to atmospheric pressures to avoid excess methanation. Their operation becomes more favorable at high pressures. However, high pressures pose significant restrictions due to the material handling and high-pressure feeding of the feedstock making industry to avoid such operational conditions.

7 Need for Co-processing

In both fast pyrolysis and gasification, use of biomass as feedstock poses some limitations. Low moisture and high energy contents are required for both these processes. High oxygen content in the feedstock lowers the stability of bio-oil and poses restrictions in its usage due to acidity, high melting point, and high viscosity which leave bio-oil with significant downstream processing requirements before its utilization as fuel [24]. In gasification, high carboxylic acid content in biomass leads to high CO_2 evolution, which dilutes the syngas produced and lowers its energy density. Also, seasonal availability of biomass poses restrictions on the reliability of biomass for gasification [16]. Similarly, conversion of MSW wastes that includes plastic wastes also has some limitations for waste processing by gasification [27]. Significant research on pyrolysis and gasification of plastics has been reported, especially mixed plastic wastes [27–35]. The issues relating to gasification and pyrolysis of plastics alone arise from product composition and the reactor design requirements. Pyrolysis of plastic wastes alone increases the corrosive and toxic properties in bio-oil by yielding products such as HCl, benzoic acid, and significant tar which lower the quality of bio-oil [36]. Plastic feeding also poses technical difficulties in reactor design due to their flowing nature and flammability along with agglomerative behavior on bed materials in fluidized bed reactors [36].

A novel proposed alternative for utilization of biomass and recovery of plastic wastes is to develop synergy between their purposes, and this can be made possible by co-processing blends of biomass and plastic wastes. Significant advantages can be gained from such co-processing in both pyrolysis and gasification [36]. These include:

1. The high heating value of plastic wastes can supplement the secondary biomass (such as rice husk, manure waste) with mediocre heating value, to support gasification without the need for energy-intensive moisture removal from biomass and provide consistent syngas yield.
2. Unreliable biomass availability due to their seasonal production and availability can be minimized as the plastic wastes are often available to blend in and stabilize

with the feedstock supply and thus allow for stable syngas yield throughout the year in all seasons.

3. Addition of biomass to plastic wastes increases the control over feedstock feeding into the reactor by binding it together and allowing for efficient reactor design along with minimizing agglomerative behavior from the sticky and viscous behavior of plastic wastes when heated.
4. Co-processing also decreases the need for waste separation, especially in the case of separating plastic wastes from food, paper, and wood wastes that only decrease the waste management efficiency and increase costs.
5. The low oxygen content in plastic wastes decreases the net oxygen in biomass—plastic blends so that fast pyrolysis of these blends yielded bio-oil with better quality in terms of stability, viscosity, and acidity due to low oxygen content in the oxygenates in the bio-oil. They all help minimize the requirement of downstream processing of bio-oil before their further usage.
6. Coking that occurs during feeding of plastic wastes (polyethylene terephthalate) into air-blown fluidized bed gasifier was found to be reduced by blending the plastic waste with wood and biomass-plastic composite pellets that provided enhanced heat transfer and fluidization.

The above parallel benefits have motivated significant amount of research into co-pyrolysis and co-gasification which will be discussed here with the focus on the effect of feedstock composition [36–44]. While the influence of operational parameters such as temperature, pressure, heating rate, reactor type, catalyst, and gasifier type is crucial for co-pyrolysis and co-gasification, their influence is similar to their effects during individual component pyrolysis and gasification [36]. This motivates us to focus on reviewing the influence of blended feedstock content, whose effect on the product yield is non-trivial, compared to pure feedstock conversion.

8 Co-processing of Biomass with Plastic Wastes

Significant literature is available that provides fundamental studies and reactor scale studies on co-pyrolysis and co-gasification of biomass/biowaste with plastics, and other MSW wastes such as wet sewage sludge [37, 42–47]. Since both pyrolysis and gasification involve pyrolytic decomposition as their starting point, fundamental studies on the reaction mechanism and kinetics involve understanding the influence of feedstock composition on the mechanism involved. Note that plastics are long-chain hydrocarbon polymers as in polyethylene (PE) and polypropylene (PP) that may contain aromatic content as in polyethylene terephthalate (PET) and polystyrene (PS) used in bottling and packaging, halogen content as in polyvinyl chloride (PVC) used for piping. Some characteristic plastics found in the waste and their heating value are given in Table 5. These materials have very high volatile content, and their degradation is preceded, coincided, or succeeded by melting or glass transition.

Table 5 Typical plastics, their structure, and their heating value

Component	HDPE and LDPE	Poly styrene (PS)	Poly carbonate (PC)	Poly propylene (PP)	Polyethylene terephthalate (PET)	Poly hexamethylen eadipamide or Nylon	Poly methyl methacrylate (PMMA)	Polyvinyl chloride (PVC)
Estimated LHV (kJ/g)	40 and 42	40.72	31.3	44.16	24.1	30.9	26.8	16.75
Structure	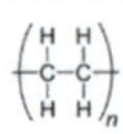	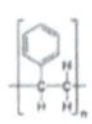	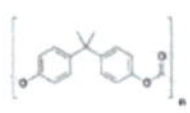	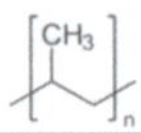	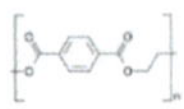	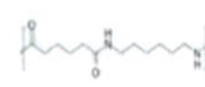	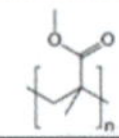	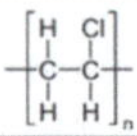

Fundamental studies on these biomass and plastics materials, both separately and as blends, involved micro-scale reactor studies such as thermogravimetric analysis (TGA), micro-pyrolysis which were sometimes equipped with the evolved gas analysis instruments, such as, mass spectroscopy (MS), gas chromatography (GC), and Fourier-transform infrared spectroscopy (FTIR). The individual component studies revealed that polymers such as PE, PP, and PS provide approximately a single sharp TGA peak behavior corresponding to their decomposition [36]. In contrast the plastics follow a free radical mechanism of decomposition involving radical formation from random scission followed by rearrangements and thermal cracking [48]. Low fixed carbon content in plastic wastes mean that majority of plastic wastes produce very low char. Hydrocarbon plastics such as PE and PP decompose via random scission leaving olefins as major products, which in parallel may rearrange into some aromatics as part of tar yield. The olefins during gasification can reform in the presence of H_2O, or CO_2 into H_2, CO, and CO_2 or crack further to yield H_2 and CH_4, depending on the temperature and catalyst present. Lack of carbonization pathway in these plastics mean that no char is produced from these plastics. Tar formation and its dependence on temperature and residence times in plastics is shown in Fig. 10 [27].

Biomass decomposition using TGA peaks shows a complex decomposition behavior that predominantly involves two overlapping large peaks over a wide range. Such behavior corresponds to the complex structure of biomass involving cellulose, hemicellulose, and lignin. Majority of the biomass, that corresponds to energy production, are lignocellulosic and their behavior typically involves with drying around 373 K to remove absorbed moisture. Hemicellulose decomposes in the range of 463–563 K first due to its low thermal stability, followed by a narrow cellulose decomposition in 563–633 K due to its consistent monomeric structure. It is then followed by slow and broadband lignin decomposition over 633–773 K due to its high stability that arises from its aromatic monomers and oligomers [49]. This process is called devolatilization, during which the volatile materials escape the biomass leaving solid residue which carbonizes to form char. Note that the char yield can account for up to 30% by wt. of the product yield [50, 51]. The inorganic content in biomass was also found to have influence on biomass decomposition by acting as catalyst to enhance the tar cracking during biomass gasification [51].

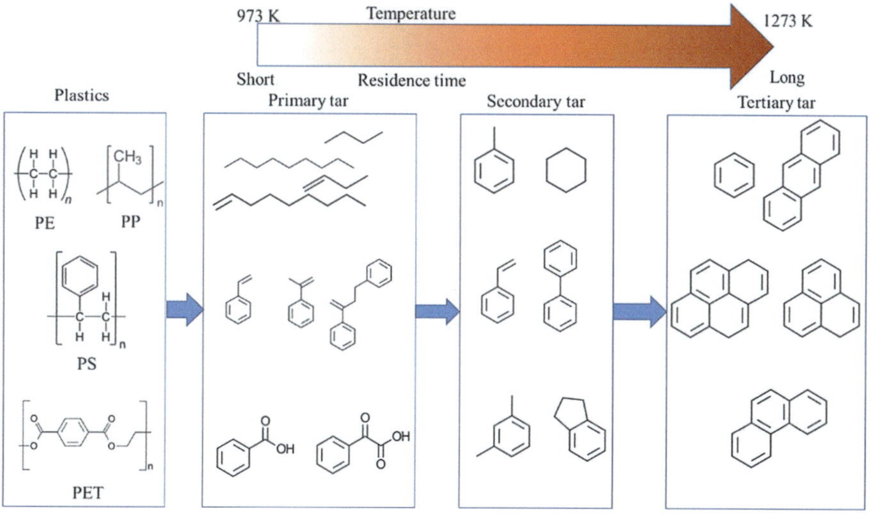

Fig. 10 Tar formation from plastics during gasification (from [27])

Biomass blended with plastics and rubber used for co-pyrolysis and co-gasification showed significant changes to the products yield and their composition. Synergistic and nonlinear effects were also reported when biomass was co-gasified with plastics, which means that the product yields and compositions from co-processing of the blends were not a linear combination of the results from separate processing of the individual components at same respective mass under the same conditions. Next sections discuss the changes in composition and yield (syngas/gaseous yield), bio-oil and tar, and char produced when biomass was co-processed with plastics or rubber.

8.1 *Influence of Feedstock Composition on Gaseous Yield: Synergistic Effects*

Studies on co-gasification and co-pyrolysis carried out by converting the respectively chosen biomass and plastics separately and then comparing those results with the products produced from converting blends of those biomass and plastics for different mass ratios are useful to understand the influence of feedstock composition. Most commonly chosen biomass for these studies was pinewood while plastics was PE.

Co-gasification of pinewood with PE in a steam-rich gasification reactor at 1173 K showed lowering of CO and CO_2 yield with increase in PE content [52]. This was due to lower oxygen content in PE and thus the net feedstock. Figure 11 shows that the presence of PE increased the yield of H_2, C_2H_4, and other hydrocarbon that enhanced the heating value of syngas produced [52]. While this was expected due to the change in C, H, O contents in the feedstock and the type of bonds in the feedstock, another intriguing result was found. The syngas, H_2, and hydrocarbon yield from co-gasification of pinewood and PE was found to be higher than the weighted sum of yields when pinewood and PE were gasified separately. Figure 12 revealed similar results on mole fraction of gaseous species yield (CO, CO_2 and C_1, C_2, C_3, and C_4 hydrocarbons) during co-pyrolysis of beech wood and PP [56]. For similar feedstocks, Dong et al. reported that this non-additive behavior was not apparent in TGA results which showed the decomposition behavior of the blends to be a linear combination of the pure substances [53]. This showed that the interaction between solid particles of pinewood and PE was negligible. The non-additive effects could arise from interaction between the vapor/intermediate species of biomass and plastics among themselves and with the solid species of biomass and plastics during decomposition. Indeed Dong et al. identified the non-additive synergistic behavior from evolved gas emissions which showed non-additive enhancement of hydrocarbon yield and reduction of CO yield [53]. Similar effects were found in gas composition at different conditions and different reactor designs for both pyrolysis and gasification, while relative gas, liquid, and char yield varied with the conditions. This suggests such effects to be intrinsic to the feedstocks involved irrespective of the conditions tested. Berrueco et al. found that gasifying PE with sawdust in a bubbling fluidized bed reactor yielded less tar and more gas yield from co-gasification compared to individual component gasification [54]. Pinto et al. found an optimal blend ratio for pine wood in PE that resulted in maximum H_2 yield, and this was due to favoring of reverse reactions, such as reverse water gas reaction, at high H_2 concentrations [41]. Gasification of high-density polyethylene (HDPE) with pinewood in a conical spouted bed steam gasifier showed synergistic increase in gas yields with simultaneous decrease in tar and char yield [55, 56]. The reason for such synergistic enhancement was proposed to be due to mutual interaction between the H-donor radicals from PE decomposition with the oxygenate volatiles that evolve from biomass to results in synergistic enhancement of both their yields.

Other types of plastic waste additions have also been investigated. A wide variety of other plastics, such as PP, PET (polyethylene terephthalate, also abbreviated as PETE), and PC (polycarbonate), were also blended with pinewood for steam co-gasification and revealed synergistic enhancement of H_2, CO_2 gaseous yields and reduction in CO with the addition of plastic wastes while hydrocarbon yield reduced with increase in these plastics; see Fig. 13 [58]. This was due to the oxygen presence in these plastics that lowered the hydrocarbon yield. The enhancement of CO_2 yield could be due to increased carbonate content from PET and PC along with enhanced water gas reaction. This will lead to higher CO_2 and H_2 with lower CO. Synergistic reduction of char formation can also be found in pinewood blends with PET and PC but absent in blends with PP which suggests that benzyl radicals from these polymers along with their oxygen content enhanced char reduction during co-gasification [58]. Thermogravimetric analysis (TGA) studies on the same pinewood blends with PET, PC, and PP revealed increased synergistic influences showing low-temperature decomposition and decreased high-temperature decomposition in the case of PET and PC but not in the case of PP; see Fig. 14 [59]. Figure 12 showed direct influence

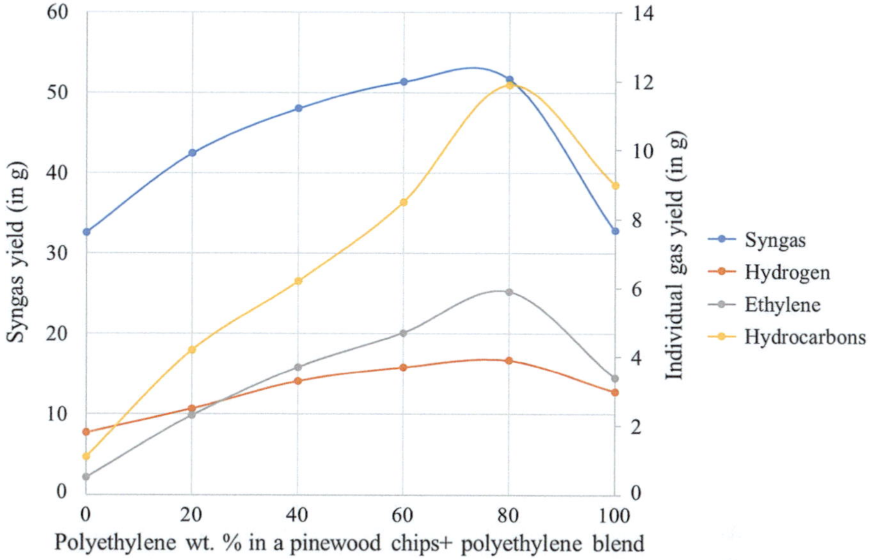

Fig. 11 Synergy in the gaseous yields from steam gasification of pinewood and PE mixtures (mass: 35 g) [52]

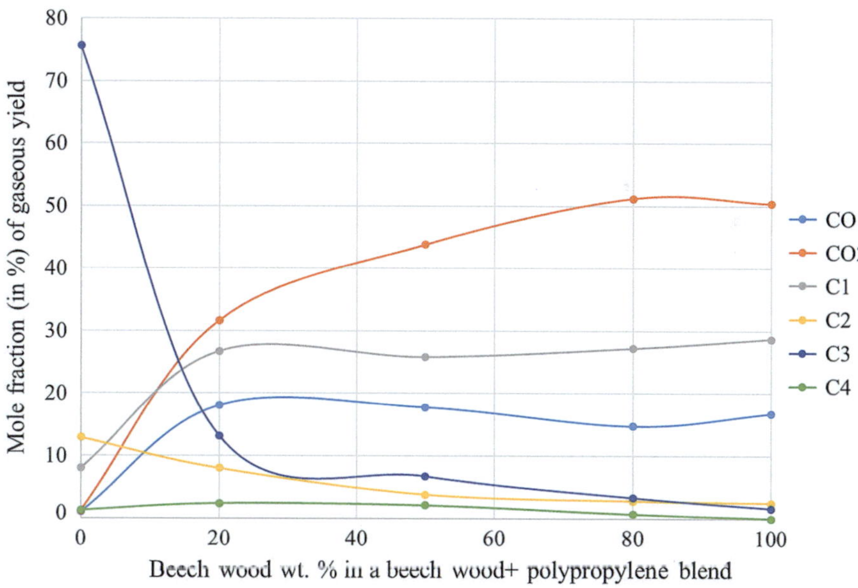

Fig. 12 Effect of composition in beech wood and PP mixture on the gas yield composition [57]

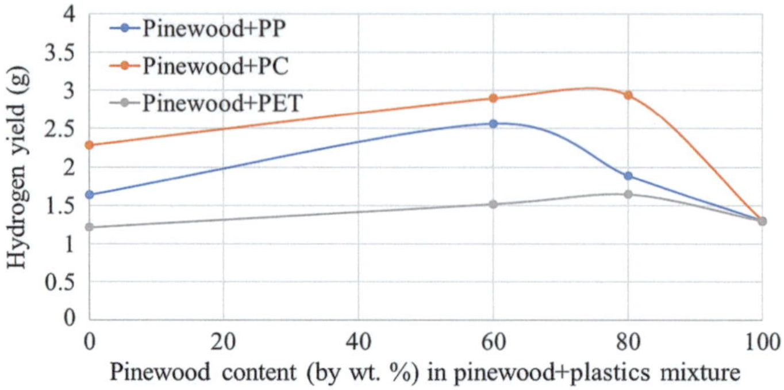

Fig. 13 Synergy in the hydrogen yields from steam gasification of pinewood and different plastic mixtures (mass: 35 g) [58]

of oxygen content in the pyrolysis of beech wood-PP mixtures. Increase in oxygen content increased CO and CO_2 in a thermodynamic equilibrium fashion. From these studies, one can conclude that synergistic influence due to addition of straight-chain hydrocarbon plastics, such as PP and PE, was not apparent from the TGA studies. This suggests that interaction between lignocellulosic content as well as PP and PE melt during pyrolysis to be negligible and that the synergy from these materials is from secondary reactions between their products. Lu et al. investigated co-pyrolysis of pinewood and PVC using a TGA and found that maximum weight loss temperature of pinewood decreased with PVC addition and that it was from HCl evolved during PVC decomposition [60]. TGA results revealed that PS or PE did not affect biomass devolatilization while the char from biomass enhanced their decomposition [40]. Steam gasification of rice straw with PVC in a laboratory-scale reactor also showed H_2 and CH_4 to increase while CO and CO_2 decreased with increase in PVC content [61]. Chen et al. showed that, in pyrolytic conditions at 773 K using waste newspaper blended with HDPE, increase in PE increased the net gas yield but synergistic effect on the gas yield made it lower than the yields expected from individual component pyrolysis [62]. This was compensated by synergistic increase in bio-oil including the tar and aqueous content, while synergistic influence was not observed in the char yield although the net yield lowered with increase in PE. Synergy was also found with coal. Co-gasification of PE with lignite showed non-additive synergistic enhancement of cold gas efficiency. Co-gasification of lignite with wood revealed nonlinear correlations in the product gas yield and composition.

While the synergistic effects were observed, the reason for such contributions is still debated. Some suggest the interaction between the plastic radical intermediates and the oxygenate volatiles of biomass. Others suggest that biomass decomposes earlier leading to biochar along with its alkaline metal content, which can act as catalyst to enhance the plastic component pyrolysis and its hydrocarbon cracking leading to enhanced H_2 yields and lower char yield. The reason for synergy could also be physical and not chemical, meaning that the plastic melt provides a different

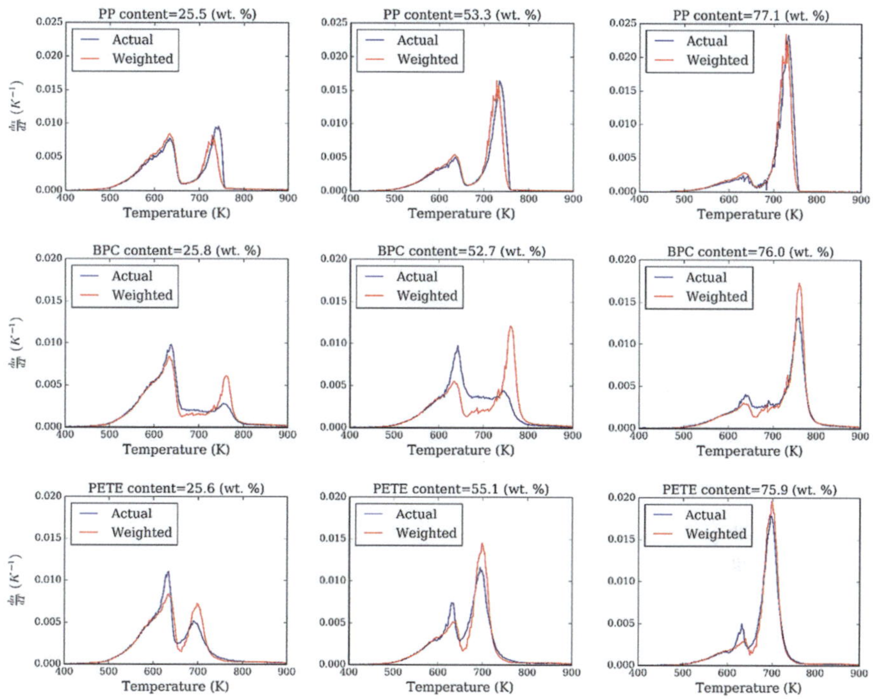

Fig. 14 Mass loss rate behavior of pinewood and plastics mixtures from TGA showing synergy in mixtures of BPC (black PC) and PET with pinewood but not with PP [59]

phase for biomass to decompose and the plastic melt could induce diffusion effects to influence the relative yields [59]. The inorganic content was suggested to be responsible for synergy via catalytic enhancement of thermal cracking which increases the gaseous yield.

8.2 Influence of Feedstock Composition on Bio-oil, Tar, and Char Yield: Synergistic Effects

The idea behind blending lignocellulosic materials with plastic and rubber wastes for bio-oil production lies in the low oxygen/carbon (O/C) of plastic wastes leading to bio-oil with low oxygen content that enhances its heating value along with improving the quality and stability of the oil produced. Sharypov et al. studied different types of wood: pinewood, beech wood, hydrolytic lignin, and cellulose, blended with different types of plastics: medium-density PE (MDPE), isotactic PP, and atactic PP [57]. Similar to other studies, no apparent mutual influence was apparent from TGA results. High-pressure pyrolysis in an autoclave reactor at 0.1 MPa, 633–723 K for ~1 h

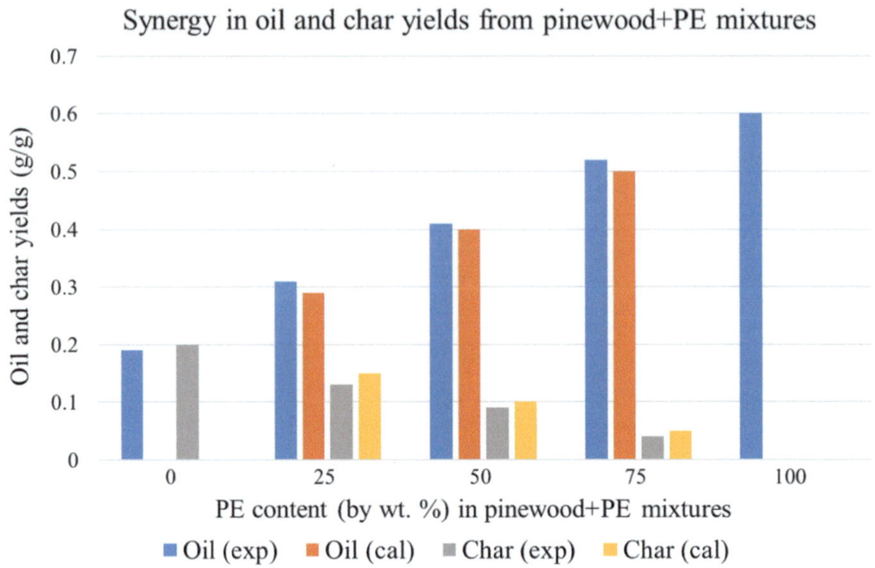

Fig. 15 Synergy in oil and char yields from co-pyrolysis of pinewood with PE (exp: experimental; cal: calculated) [60]

of these blends revealed synergistic evolution of light liquid fractions. Even with similar compositions of beech wood, and pinewood, their influence was noticeable which was interpreted by the differences in chemical properties between softwood and hardwood. Highest light liquid yields were obtained from using PE followed by atactic PP and isotactic PP, which suggested increased stability due to tacticity (that relates to arrangement or steric order of polymer) and high molecular weight. Similar conditions with pinewood and PE, PS, and PP mixture revealed reduction in overall liquid yield and increase in gas and residue yield with increase in pinewood. Reduced tar yields were observed when pinewood was co-gasified with PE [54]. Lu et al. found synergistic increase in oil yields and reduction in char yields from pinewood, PE blends; see Fig. 15. They found the opposite effect using pinewood with PVC during co-pyrolysis in a fixed bed reactor [60]. Waste newspaper co-pyrolysis with HDPE also yielded increased tar, oil, and aqueous yields and thus net liquid yields non-additively to support the synergy [62]. Characterizing this oil yield revealed that the pH was synergistically increased to neutrality with the blends, and the calorific value increased non-additively. The decrease in oxygen content nonlinearly decreased viscosity, density, water content, acid number and increased the calorific value suggesting synergistic improvement in the quality of bio-oil by co-pyrolysis.

Synergistic effects were also reported in bio-oil produced from co-pyrolysis of wood with waste tires, which enhanced the C and H contents along with calorific value [43, 63]. This was found to be due to the reduction in oxygen content via decrease in phenolic products produced from lignin with the addition of tire waste.

Similar effect was found between palm shell and polystyrene but with an increased viscosity of the oil produced [43, 64]. This depends on the plastic/rubber type, since addition of polycarbonate was found to increase phenolic content [43]. The reason for synergy was investigated using micro-pyrolyzer reactors. They showed that enhanced hydrocarbons during co-pyrolysis corresponded to Diels–Alder reaction between oxygenates such as furans from biomass and the olefins obtained from plastics at catalytic site [65–68]. These reactions along with enhanced dehydration led to reduction in oxygen content in the presence of olefins [42, 69].

Char yields and their quality also had synergistic influence from co-processing biomass with plastic wastes. Addition of plastics to biomass decreased char formation non-additively due to enhanced devolatilization and mutual interaction. This was consistent with most of the studies regarding the char yield. Lu et al. also showed that increasing PE or PVC in pinewood blends during co-pyrolysis revealed reduction in H/C in the char produced; see Fig. 16 [60]. Chen et al. studied the co-pyrolysis of PET, PVC, and PP with Paulownia wood and revealed that char derived from co-pyrolysis was found to be morphologically different compared to char from individual components [71]. Char yield from co-pyrolysis showed uniform agglomerative surface as observed using scanning electron microscopy. Figure 17 shows rate of combustion of chars prepared from co-charring/co-pyrolysis of pinewood and PET at 673 K and compared with weighted sum of the combustion rate of chars produced from mono-pyrolysis/mono-charring of the respective individual components. This study on the quality of char yielded from PET, pinewood blends, revealed that the char produced from co-pyrolysis possessed higher heating value and uniform combustion behavior compared to a combination of chars produced separately from the individual components which showed lower heating value and multiple combustion peak; see Fig. 17 [70]. Such uniformity was possibly provided by the plastic melting that yielded in a uniform media for carbonization leading to high-quality char yield with uniform properties.

9 Conclusions and Future Endeavors

Synergistic effects were found to increase the quality of syngas, bio-oil, and char while enhancing their yields according to the operational conditions during co-processing of biomass with plastic and rubber wastes. This builds the potential of developing such blended processing to minimize energy-intensive downstream processing required in pyrolysis and gasification of pure components. While the effects were established for a wide variety of biomass and plastic feedstocks, the reasons for such effects and a generic composite reaction model development are still necessary that will help in the design and implementation of feed-flexible reactors that can operate on multiple composites with minimal changes to the operational conditions. Development of reaction model involves providing improved understanding of the intermediates involved, which could be obtained from in situ vacuum extraction of the intermediates during pyrolysis and analyzing them using GC, FTIR techniques in

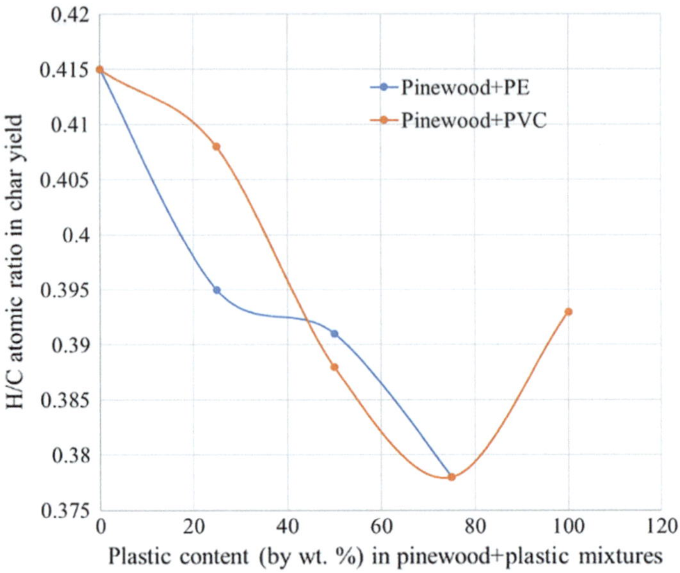

Fig. 16 H/C atomic ratio of char produced from co-pyrolysis of pinewood with PE and PVC [60]

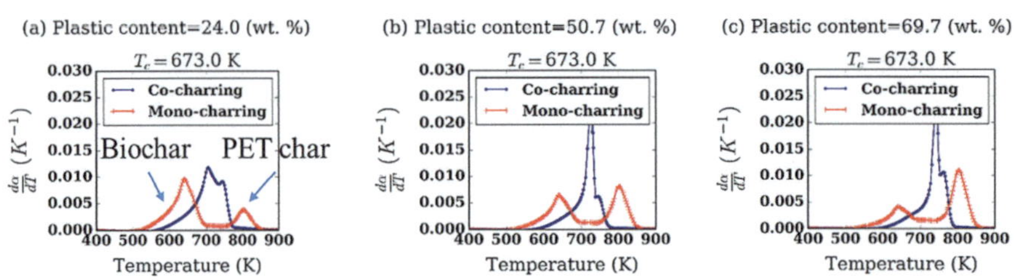

Fig. 17 TGA profiles of oxidation of char produced from co-pyrolysis and mono-pyrolysis at 673 K at different wt% of plastic in the mixture [70]

micro-scale reactors. Upscaling of these composite reactions involves understanding and developing models for thermophysical parameters such as thermal conductivity, absorption coefficient, and density, to build multi-physics CFD model necessary for designing large-scale feed-flexible reactors. This can further be validated using pilot-scale studies before large-scale implementation. Enhanced quality products from co-pyrolysis and co-gasification of biomass with plastic wastes have motivated for necessary research input to utilize the available secondary biomass resources along with the landfilled wastes. This novel approach can synergistically alleviate the problems of energy and material crisis together with waste management. This approach will also help maintain the sensitive balance of carbon cycle necessary to avoid irreversible and uninhabitable climate change.

Acknowledgements This research work was supported by ONR and is gratefully acknowledged.

References

1. Falkowski P (2000) The global carbon cycle: a test of our knowledge of earth as a system. Science 290:291–296. https://doi.org/10.1126/science.290.5490.291
2. Keeling R (2018) Scripps Institute of Oceanography 2018. scrippsco2.ucsd.edu/. Accessed 2 July 2018
3. Tans P. NOAA/ESRL 2018. www.esrl.noaa.gov/gmd/ccgg/trends/. Accessed 2 July 2018
4. Houghton JT, Meira Filho LG, Bruce J, Lee H, Callander BA, Haites E et al (1995) Climate change 1994: radiative forcing of climate change and an evaluation of the IPCC IS92 emission scenarios. Cambridge University Press, Cambridge
5. Kasting JF, Toon OB, Pollack JB (1988) How climate evolved on the terrestrial planets. Sci Am 256:90–97. https://doi.org/10.1038/scientificamerican0288-90
6. Archer D (2010) The global carbon cycle. Princeton University Press, Princeton
7. Archer D (2005) Fate of fossil fuel CO_2 in geologic time. J Geophys Res 110:C09S05. https://doi.org/10.1029/2004jc002625
8. Hansen J, Sato M, Russell G, Kharecha P (2013) Climate sensitivity, sea level and atmospheric carbon dioxide. Philos Trans R Soc A Math Phys Eng Sci 371:20120294. https://doi.org/10.1098/rsta.2012.0294
9. Lee Allen H, Fox TR, Campbell RG (2005) What is ahead for intensive pine plantation silviculture in the south? South J Appl For 29:62–69. https://doi.org/10.1093/sjaf/29.2.62
10. Phyllis2, database for biomass and waste. Energy Res Cent Netherlands 2018. https://www.ecn.nl/phyllis2/Biomass/View/3501. Accessed 28 May 2018
11. Le Quéré C, Raupach MR, Canadell JG, Marland G (2009) Trends in the sources and sinks of carbon dioxide. Nat Geosci 2. https://doi.org/10.1038/ngeo689
12. Boden TA, Marland G, Andres R. Global, Regional, National Fossil-Fuel CO_2 Emissions. Carbon Dioxide Information Analysis Center, Oak Ridge National Laboratory US Department Energy, Oak Ridge, Tenn, USA n.d. https://doi.org/10.3334/cdiac/00001_v2017
13. Sabine CL (2004) The oceanic sink for anthropogenic CO_2. Science 305:367–371. https://doi.org/10.1126/science.1097403
14. Takahashi T (2004) The fate of industrial carbon dioxide. Science. https://doi.org/10.1126/science.1100602
15. Beiter P (2016) Renewable energy data book. https://www.nrel.gov/docs/fy18osti/70231.pdf. Accessed 13 August 2018
16. Berndes G, Dhamija P, Bain R, Paré D, Kahn Ribeiro S, Faaij A et al (2011) Bioenergy. In: IPCC special report on renewable energy sources and climate change mitigation. Cambridge University Press, New York, NY, USA
17. Cannell MGR (2003) Carbon sequestration and biomass energy offset: theoretical, potential and achievable capacities globally, in Europe and the UK. Biomass Bioenergy 24:97–116. https://doi.org/10.1016/S0961-9534(02)00103-4
18. Smeets E, Faaij APC, Lewandowski I, Turkenburg W (2007) A bottom-up assessment and review of global bio-energy potentials to 2050. Prog Energy Combust Sci 33:56–106. https://doi.org/10.1016/j.pecs.2006.08.001
19. EPA. Advancing Sustainable Materials Management: 2015 Fact Sheet Assessing Trends in Material Generation, Recycling, Composting, Combustion with Energy Recovery and Landfilling in the United States. Environ Prot Agency 2015. https://www.epa.gov/sites/production/files/2018-07/documents/2015_smm_msw_factsheet_07242018_fnl_508_002.pdf. Accessed 8 August 2018
20. Aggarwal S (2013) Simulations of combustion and emissions characteristics of biomass-derived fuels. Technol. Convert. Biomass to Useful Energy Combust. Gasification, Pyrolysis, Torrefaction Ferment, CRC Press
21. Saxena RC, Seal D, Kumar S, Goyal HB (2008) Thermo-chemical routes for hydrogen rich gas from biomass: a review. Renew Sustain Energy Rev 12:1909–1927. https://doi.org/10.1016/j.rser.2007.03.005

22. Molino A, Chianese S, Musmarra D (2016) Biomass gasification technology: the state of the art overview. J Energy Chem 25:10–25. https://doi.org/10.1016/j.jechem.2015.11.005
23. Emami-Taba L, Irfan MF, Wan Daud WMA, Chakrabarti MH (2013) Fuel blending effects on the co-gasification of coal and biomass—a review. Biomass Bioenergy 57:249–263. https://doi.org/10.1016/j.biombioe.2013.02.043
24. Bridgwater AV (2012) Review of fast pyrolysis of biomass and product upgrading. Biomass Bioenergy 38:68–94. https://doi.org/10.1016/j.biombioe.2011.01.048
25. Akinola AO (2016) Effect of temperature on product yield of pyrolysis of seven selected wood species in South West Nigeria 9359:176–181
26. Hussein MS, Burra KG, Amano RS, Gupta AK (2017) Temperature and gasifying media effects on chicken manure pyrolysis and gasification. Fuel 202:36–45. https://doi.org/10.1016/j.fuel.2017.04.017
27. Lopez G, Artetxe M, Amutio M, Alvarez J, Bilbao J, Olazar M (2018) Recent advances in the gasification of waste plastics. A critical overview. Renew Sustain Energy Rev 82:576–596. https://doi.org/10.1016/j.rser.2017.09.032
28. The American Chemical Council (2013) Gasification of Non-recycled Plastics From Municipal Solid Waste in the United States The American Chemistry Council
29. Williams PT, Williams EA (1999) Interaction of plastics in mixed-plastics pyrolysis. Energy Fuels 13:188–196. https://doi.org/10.1021/ef980163x
30. Brems A, Baeyens J, Vandecasteele C, Dewil R (2011) Polymeric cracking of waste polyethylene terephthalate to chemicals and energy. J Air Waste Manag Assoc J J Air Waste Manag Assoc 617:1096–2247. https://doi.org/10.3155/1047-3289.61.7.721doi.org/10.3155/1047-3289.61.7.721
31. Apaydin-Varol E, Polat S, Putun A (2014) Pyrolysis kinetics and thermal decomposition behavior of polycarbonate—a TGA-FTIR study. Therm Sci 18:833–842. https://doi.org/10.2298/TSCI1403833A
32. Lopez G, Erkiaga A, Artetxe M, Amutio M, Bilbao J, Olazar M (2015) Hydrogen production by high density polyethylene steam gasification and in-line volatile reforming. Ind Eng Chem Res 54:9536–9544. https://doi.org/10.1021/acs.iecr.5b02413
33. Wong SL, Ngadi N, Abdullah TAT, Inuwa IM (2015) Current state and future prospects of plastic waste as source of fuel: a review. Renew Sustain Energy Rev 50:1167–1180. https://doi.org/10.1016/j.rser.2015.04.063
34. Dayana S, Sharuddin A, Abnisa F, Mohd W, Daud AW, Aroua MK (2016) A review on pyrolysis of plastic wastes. https://doi.org/10.1016/j.enconman.2016.02.037
35. Kunwar B, Cheng HN, Chandrashekaran SR, Sharma BK (2016) Plastics to fuel: a review. Renew Sustain Energy Rev 54:421–428. https://doi.org/10.1016/j.rser.2015.10.015
36. Block C, Ephraim A, Weiss-Hortala E, Pham Minh D, Nzihou A, Vandecasteele C. Co-pyrogasification of plastics and biomass, a review. Waste and Biomass Valorization n.d. 0:3. https://doi.org/10.1007/s12649-018-0219-8
37. Ramos A, Monteiro E, Silva V, Rouboa A (2018) Co-gasification and recent developments on waste-to-energy conversion: a review. Renew Sustain Energy Rev 81:380–398. https://doi.org/10.1016/j.rser.2017.07.025
38. Uzoejinwa BB, He X, Wang S, El-Fatah Abomohra A, Hu Y, Wang Q (2018) Co-pyrolysis of biomass and waste plastics as a thermochemical conversion technology for high-grade biofuel production: Recent progress and future directions elsewhere worldwide. Energy Convers Manag 163:468–492. https://doi.org/10.1016/j.enconman.2018.02.004
39. Kamińska-Pietrzak N, Smoliński A (2013) Selected environmental aspects of gasification and co-gasification of various types of waste. J Sustain Min 12:6–13. https://doi.org/10.7424/jsm130402
40. Wilk V, Hofbauer H (2013) Co-gasification of plastics and biomass in a dual fluidized-bed steam gasifier: possible interactions of fuels. Energy Fuels 27:3261–3273. https://doi.org/10.1021/ef400349k
41. Pinto F, Franco C, André RN, Miranda M, Gulyurtlu I, Cabrita I (2002) Co-gasification study of biomass mixed with plastic wastes. Fuel 81:291–297. https://doi.org/10.1016/S0016-2361(01)00164-8

42. Hassan H, Lim JK, Hameed BH (2016) Recent progress on biomass co-pyrolysis conversion into high-quality bio-oil. Bioresour Technol 221:645–655. https://doi.org/10.1016/j.biortech.2016.09.026
43. Abnisa F, Daud WMAW (2014) A review on co-pyrolysis of biomass: an optional technique to obtain a high-grade pyrolysis oil. Energy Convers Manag 87:71–85. https://doi.org/10.1016/j.enconman.2014.07.007
44. Mallick D, Mahanta P, Suryakant Moholkar V (2017) Co-gasification of coal and biomass blends: chemistry and engineering. Fuel 204:106–128. https://doi.org/10.1016/j.fuel.2017.05.006
45. Peng L, Wang Y, Lei Z, Cheng G (2012) Co-gasification of wet sewage sludge and forestry waste in situ steam agent. https://doi.org/10.1016/j.biortech.2012.03.079
46. Shuang-quan Z, Xiao-ming Y, Zhi-yuan Y, Ting-ting P, Ming-jian D, Tian-yu S (2009) Study of the co-pyrolysis behavior of sewage-sludge/rice-straw and the kinetics. https://doi.org/10.1016/j.proeps.2009.09.104
47. Hu M, Guo D, Ma C, Hu Z, Zhang B, Xiao B et al (2015) Hydrogen-rich gas production by the gasification of wet MSW (municipal solid waste) coupled with carbon dioxide capture. Energy 90:857–863. https://doi.org/10.1016/j.energy.2015.07.122
48. Horton SR, Woeckener J, Mohr R, Zhang Y, Petrocelli F, Klein MT (2016) Molecular-level kinetic modeling of the gasification of common plastics. Energy Fuels 30:1662–1674. https://doi.org/10.1021/acs.energyfuels.5b02047
49. Wang S, Dai G, Yang H, Luo Z (2017) Lignocellulosic biomass pyrolysis mechanism: a state-of-the-art review. Prog Energy Combust Sci 62:33–86. https://doi.org/10.1016/j.pecs.2017.05.004
50. Ahmed I, Gupta AK (2009) Syngas yield during pyrolysis and steam gasification of paper. Appl Energy 86:1813–1821. https://doi.org/10.1016/j.apenergy.2009.01.025
51. Ahmed II, Gupta AK (2010) Pyrolysis and gasification of food waste: syngas characteristics and char gasification kinetics. Appl Energy 87:101–108. https://doi.org/10.1016/j.apenergy.2009.08.032
52. Ahmed II, Nipattummakul N, Gupta AK (2011) Characteristics of syngas from co-gasification of polyethylene and woodchips. Appl Energy 88:165–174. https://doi.org/10.1016/j.apenergy.2010.07.007
53. Dong C, Yang Y, Jin B, Horio M (2007) The pyrolysis of sawdust and polyethylene in TG and U-shape tube reactor. https://doi.org/10.1016/j.wasman.2006.10.021
54. Berrueco C, Ceamanos J, Esperanza E, Mastral F (2004) Experimental study of co-pyrolysis of polyethylene/sawdust mixtures. Therm Sci 8:65–80. https://doi.org/10.2298/TSCI0402065B
55. Lopez G, Erkiaga A, Amutio M, Bilbao J, Olazar M (2015) Effect of polyethylene co-feeding in the steam gasification of biomass in a conical spouted bed reactor. Fuel 153:393–401. https://doi.org/10.1016/j.fuel.2015.03.006
56. Arregi A, Amutio M, Lopez G, Artetxe M, Alvarez J, Bilbao J et al (2017) Hydrogen-rich gas production by continuous pyrolysis and in-line catalytic reforming of pine wood waste and HDPE mixtures. Energy Convers Manag 136:192–201. https://doi.org/10.1016/j.enconman.2017.01.008
57. Sharypov VI, Marin N, Beregovtsova NG, Baryshnikov SV, Kuznetsov BN, Cebolla VL et al (2002) Co-pyrolysis of wood biomass and synthetic polymer mixtures. Part I: influence of experimental conditions on the evolution of solids, liquids and gases. J Anal Appl Pyrolysis 64:15–28. https://doi.org/10.1016/s0165-2370(01)00167-x
58. Burra KG, Gupta AK (2018) Synergistic effects in steam gasification of combined biomass and plastic waste mixtures. Appl Energy 211:230–236. https://doi.org/10.1016/j.apenergy.2017.10.130
59. Burra KG, Gupta AK (2018) Kinetics of synergistic effects in co-pyrolysis of biomass with plastic wastes. Appl Energy 220:408–418. https://doi.org/10.1016/j.apenergy.2018.03.117
60. Lu P, Huang Q, (Thanos) Bourtsalas AC, Chi Y, Yan J (2018) Synergistic effects on char and oil produced by the co-pyrolysis of pine wood, polyethylene and polyvinyl chloride. Fuel 230:359–367. https://doi.org/10.1016/j.fuel.2018.05.072

61. Baloch HA, Yang T, Li R, Nizamuddin S, Kai X, Bhutto AW (2016) Parametric study of co-gasification of ternary blends of rice straw, polyethylene and polyvinylchloride. Clean Technol Environ Policy 18:1031–1042. https://doi.org/10.1007/s10098-016-1092-4
62. Chen W, Shi S, Zhang J, Chen M, Zhou X (2016) Co-pyrolysis of waste newspaper with high-density polyethylene: synergistic effect and oil characterization. Energy Convers Manag 112:41–48. https://doi.org/10.1016/j.enconman.2016.01.005
63. Martínez JD, Veses A, Mastral AM, Murillo R, Navarro MV, Puy N et al (2014) Co-pyrolysis of biomass with waste tyres: Upgrading of liquid bio-fuel. https://doi.org/10.1016/j.fuproc.2013.11.015
64. Abnisa F, Daud WMAW, Sahu JN (2013) Pyrolysis of mixtures of palm shell and polystyrene: an optional method to produce a high-grade of pyrolysis oil. https://doi.org/10.1002/ep.11850
65. Dorado C, Mullen CA, Boateng AA (2015) Origin of carbon in aromatic and olefin products derived from HZSM-5 catalyzed co-pyrolysis of cellulose and plastics via isotopic labeling. Appl Catal B, Environ 162:338–345. https://doi.org/10.1016/j.apcatb.2014.07.006
66. Xue Y, Kelkar A, Bai X (2016) Catalytic co-pyrolysis of biomass and polyethylene in a tandem micropyrolyzer. https://doi.org/10.1016/j.fuel.2015.10.125
67. Mullen CA, Dorado C, Boateng AA (2018) Catalytic co-pyrolysis of switchgrass and polyethylene over HZSM-5: catalyst deactivation and coke formation. J Anal Appl Pyrolysis 129:195–203. https://doi.org/10.1016/j.jaap.2017.11.012
68. Sebestyén Z, Barta-Rajnai E, Bozi J, Blazsó M, Jakab E, Miskolczi N et al (2017) Thermo-catalytic pyrolysis of biomass and plastic mixtures using HZSM-5. Appl Energy 207:114–122. https://doi.org/10.1016/j.apenergy.2017.06.032
69. Zhang H, Cheng Y-T, Vispute TP, Xiao R, Huber GW. Catalytic conversion of biomass-derived feedstocks into olefins and aromatics with ZSM-5: the hydrogen to carbon effective ratio n.d. https://doi.org/10.1039/c1ee01230d
70. Burra KG, Gupta AK (2018) Characteristics of char from co-pyrolysis of biomass and plastic waste. Vol. 1 Fuels, Combust. Mater. Handl. Combust. Turbines Comb. Cycles; Boil. Heat Recover. Steam Gener. Virtual Plant Cyber-Physical Syst. Plant Dev. Constr. Renew. Energy Syst., Lake Buena Vista, Florida, USA: ASME; 2018, p V001T01A008. https://doi.org/10.1115/power2018-7255
71. Chen L, Wang S, Meng H, Wu Z, Zhao J (2017) Synergistic effect on thermal behavior and char morphology analysis during co-pyrolysis of paulownia wood blended with different plastics waste. Appl Therm Eng 111:834–846. https://doi.org/10.1016/j.applthermaleng.2016.09.155

Residual Biomass Resources: An Invaluable Reservoir of Energy and Matter

Biagio Morrone

Keywords Biomass · Energy recovery · Biogas · Matter recovery · Lactic acid

1 Properties of Biomass Feedstock

1.1 Introduction

Biomass represents a complex heterogeneous mix mainly made up of various natural components from growing land- and water-based vegetation such as algae, trees, and crop residues, which are able to use energy from sunlight to convert carbon dioxide and water into carbohydrates via photosynthesis [1, 2]. Animal and municipal residues as well as agricultural and agro-food residues derived by industrial processing can be also included [1, 3]. Therefore, biomass refers to any organic matter, which is available on a renewable or recurring basis, mainly composed of sugar-containing cellulose and hemicellulose and non-sugar lignin, and also of various other organic and inorganic constituents with natural and anthropogenic origin such as lipids, proteins, simple sugars, starches, water, hydrocarbons, ashes and other compounds [2].

The European Union defines '*the biomass as the biodegradable fraction of products, waste and residues from biological origin from agriculture (including vegetal and animal substances), forestry and related industries including fisheries and aquaculture, as well as the biodegradable fraction of industrial and municipal waste*' and '*bio-liquids… liquid fuel for energy purposes other than for transport, including electricity and heating and cooling, produced from biomass*' and '*biofuels … liquid or gaseous fuel for transport produced from biomass*' [4].

B. Morrone (✉)
Department of Engineering, University of Campania "L. Vanvitelli", Aversa, CE, Italy
e-mail: biagio.morrone@unicampania.it

A. K. Gupta et al. (eds.), *Innovations in Sustainable Energy and Cleaner Environment*, Green Energy and Technology,
https://doi.org/10.1007/978-981-13-9012-8_7

Biomass currently constitutes only 3% of primary energy consumption in OECD countries and 22% in non-OECD countries [5]. However, the rural population in developing countries, which represents about 50% of the world's population, is dependent on biomass for fuel, primarily in the form of wood [6].

Biomass of animal origin such as solid and liquid animal manure and livestock slaughter waste like blood and hair can be used to produce biogas by anaerobic digestion without the addiction of any seed bacteria [7]. Manure, in particular, is an optimal substrate to produce biogas with plenty of carbohydrate compounds, an optimal carbon-to-nitrogen ratio and a plethora of microorganisms among which cellulolytic, hydrogen-producing and methanogenic bacteria [8].

Plant biomass such as crops, agricultural crop residues and agro-industrial residues from the food processing industry can be considered a promising renewable source of biofuels and bio-products, too. However, apart from agro-industrial waste, the use of these substrates has relevant drawbacks [2]. The use of crops for producing biogas and bioethanol competes with food production since it subtracts fertile lands and water from growing food crops. While the removal of residual agricultural biomass from soil can deplete the Soil Organic Carbon (SOC) stock [9] reducing soil quality and agronomic productivity [10].

However, in general, the controlled use/re-use of residual biomasses has the double benefit to reduce the amount of waste allocated for landfills and decrease the request for fossil fuels, thus reducing the greenhouse gas emissions, especially CH_4, CO_2, NO_X, SO_X, and trace elements. They can therefore provide a renewable energy source sustainably exploitable in the future, able to improve the economy and energy security with an environment-friendly approach.

1.2 Biomass Properties

According to the intrinsic properties of the different biomass categories, as wood residues, agricultural wastes, dedicated energy crops or municipal solid waste, the choice of the conversion processes and subsequent difficulties are evaluated [11].

The properties can be classified by Proximate analysis and Elemental (or Ultimate) analysis. Samples are analysed for their elemental and/or isotopic composition using the Elemental analysis which gives quantitative analysis of various elements, such as carbon, hydrogen, sulphur, oxygen, and nitrogen. Elemental analysis can be qualitative or quantitative. Instead, Proximate analysis gives more comprehensive and global parameters including moisture, total solids, volatile matter, ash and fixed carbon.

A large number of biomass properties need to be defined considering which kind of matter should be referred to. The properties can be referred to the so-called fresh matter, which is the biomass '*As received: including humidity and ashes*' for brevity indicated with **ar**. Once the humidity is removed, the sample is called '*Dry: without humidity*' **dry**. After removing moisture and ashes, which are the inorganic part of the biomass, the substrate is '*Dry no ashes: without humidity and ashes*' reported as **daf**.

The concentration and behaviour of elements such as Ca, Cl, K, Na, P, S, Si and heavy metals are able to cause many technological and environmental problems during biomass-to-energy conversion process [12]. Moreover, it is essential to consider other properties of biomass such as moisture content (dependent or not on weather effects), calorific value, proportions of fixed carbon and volatiles, ash/residue content, and cellulose/lignin ratio [13].

Moisture/water content can range from approximately 5 to 60% ar. Higher moisture content can negatively impact on energy conversion processes, in particular, gasification process due to the energy request during the vapourization process and the subsequent quality of the extracted syngas. Even though, biomass drying is too energy-consuming making the process not economically worth doing [14]. The greater the moisture content, the lesser the heating value.

The heating value is the energy content of matter released when fully oxidized and represents its chemical energy potential content. It is usually expressed in MJ/kg for solids and liquids, or MJ/Nm^3 for gases. The Lower (LHV) and Higher Heating Values (HHV) are calculated when the combustion process occurs completely, i.e. all carbon evolves in carbon dioxide (CO_2), hydrogen in water (H_2O) and sulphur in sulphuric acid (SO_2), at stoichiometric conditions[1] and with the reactants entering and products released at the Standard Temperature (298 K). The two values differ because the water released with the products can be either as vapour phase or liquid. The LHV is calculated considering a complete combustion reaction with stoichiometric air conditions and the H_2O released as vapour. The HHV is calculated considering the water condensed after the combustion process, thus releasing its latent heat of condensation. The two quantities are strictly related and from the knowledge of one of the two, the second can be easily calculated. In fact, the relationship between the two HVs is:

$$\mathrm{HHV} = \mathrm{LHV} + m_{\mathrm{w}} h_{\mathrm{fg}}(T_{\mathrm{fg}}) \quad (1)$$

In this equation, m_{w} is the mass of water present in the products after the combustion and h_{fg} represents the latent heat of vapourization at specific temperature, T_{fg}. It has been shown that calorific values can vary between 6 and 8 MJ/kg for fresh biomass with a 50–60% moisture content to 15–17 MJ/kg for air-dried biomass with a moisture content of 10–20%, up to 19 MJ/Kg for dried biomass with a moisture lower than 8% [15] (Fig. 1).

The solid combustible residue remaining after biomass is heated and the volatile matter expelled is called fixed carbon. It is calculated by subtracting the percentage of moisture/water, volatile matter and ashes from a biomass sample. It is clear that a biomass that has a lower fixed carbon residual amount is convenient for being converted into syngas, while higher fixed carbon residues are advantageous to maximize

[1]Stoichiometric condition means that the oxygen content is exactly that required by the chemistry.

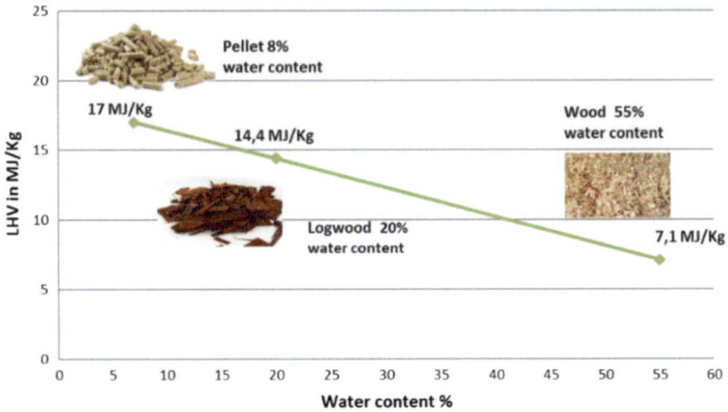

Fig. 1 Lower heating value as a function of the water content for some biomass feedstock

biochar yield [16]. However, a very high fixed carbon content is an indication of recalcitrance of the biomass to both aerobic and anaerobic degradations [17].

The biomass volatile matter is the condensable and non-condensable vapour released when the biomass is heated, and its content depends on the rate and temperature of heating. A typical biomass material has a volatile matter content varying from 60 to 90%, while a usual coal has a volatile matter ranging from 20 to 50% [18].

The ashes, composed mainly of silica, aluminium, iron and calcium, are the inorganic solid residue left after the complete combustion of biomass. Its content is generally very small (<1%) but may play an important role in the biomass conversion to energy, especially if it contains alkali metals that reduce the biomass usability. In fact, these latter react with silica producing a liquid and sticky phase, that can block airways in the furnace and boiler plants. Ashes therefore constitute a standard and important measurement parameter for solid and liquid fuels, influencing both the handling and processing costs [19–21] (Table 1).

The cellulose/lignin ratio is an important issue for biomass energy use. Simple sugars originated from the degradation of cellulose and hemicellulose in plant cell wall is easily convertible to biofuel. However, a main fraction of biomass can be lignin, which is constituted by a diverse group of polymers formed by the polymerization of hydroxycinnamoyl alcohol (monolignol) derivatives which are easily polymerized but impossible to deal with after polymerization. The strong recalcitrance of lignin to degradation, which implies very long retention times or high-cost processes, hampers lignocellulosic biomass use in energy production [22, 23].

The biomasses, as fuel resources, may be divided into different groups and subgroups based on their biological diversity. The interplay between the biomass biological diversity and the form in which the energy is required to determine the choice of the conversion process and any subsequent processing difficulties that may arise [13].

Table 1 Some proximate and ultimate analysis for different types of biomass (https://www.ecn.nl/phyllis2/)

Biomass	Proximate analysis				Ultimate analysis					HHV (MJ/kg)
	Moist. % (ar)	Ash % (ar)	Volatile matter % (ar)	Fixed carbon % (ar)	C	H	N	S	O	
Cattle manure fresh #1882	13.88	13.67	60.52	11.94	39.09	4.61	0.83	0.25	26.68	15.96
Cattle manure part. composted #1883	18.00	34.11	38.79	9.10	27.02	4.02	0.57	0.64	14.33	12.40
Peat #1407	29.80	4.42	47.81	17.97	37.42	3.93	1.00	0.11	23.27	14.99
Wood #908	50.30	10.01	30.91	8.78	19.98	2.04	0.34	0.06	17.19	8.45
Sunflower husk #1911	9.10	1.73	62.81	26.36	45.83	4.46	0.54	–	38.34	16.83

2 Biomass Sustainability

One main reason for pushing forward biomass resources use is their claimed **carbon neutrality**, which sometimes is difficult to prove. It means that in a reasonable lapse of time (usually 100 years), the CO_2 emitted in combustion products is then reabsorbed by plants and animals showing a net CO_2 balance equal to zero. In fact, photosynthesis converts water and CO_2 into sugars and oxygen, therefore transforming solar energy into chemical energy, then stored in plant tissue biomass.

Dealing with biomass is a matter of sustainability. The concept of sustainability is quite controversial, anyway, a wide definition has been given by Bruntlandt Commission of the World Commission on Environment and Development:

> Development that meets the needs of the present without compromising the ability of future generations to meet their own needs. [24]

The European Union identified in 2010, four main key issues in biomass resources exploitation about their 'sustainability', i.e. how natural systems function, remain diverse and produce everything it needs for the ecology to remain in balance [25]:

1. Sustainability in production refers to the conservation of biodiverse ecosystems and carbon stocks (land management, cultivation and harvesting)
2. Land use, land use change and forestry (LULUCF) accounting deforestation, forest degradation and a number of other practices can result in a significant loss of biogenic carbon and/or significant changes in productivity
3. Life cycle greenhouse gas (GHG) performance The potential environmental benefits, including in terms of GHG savings that can be obtained from replacing fossil fuels with biomass sources
4. Energy conversion efficiency The energy conversion efficiencies of household.

The evaluation of sustainability is obtained, then, by accounting for the carbon dioxide equivalent emissions for kilogram of biomass harvested, pre-processed, processed, transported and ready for use.

$$E = e_{\mathrm{ec}} + e_{\mathrm{l}} + e_{\mathrm{p}} + e_{\mathrm{td}} + e_{\mathrm{u}} - e_{\mathrm{sca}} - e_{\mathrm{ccs}} - e_{\mathrm{ccr}} \tag{2}$$

The terms that account for positive emission values are related to:

1. Extraction or cultivation of raw materials, e_{ec}
2. Carbon stock exchange caused by land use change (these two are zero for residual Biomass), e_{l}
3. Processing of biomass, e_{p}
4. Transport and distribution, e_{td}
5. Use of fuel derived from biomass, e_{u}.

Three terms can account for negative values of emissions (saving emissions):

1. Due to soil carbon accumulation because of improved agricultural management, e_{sca}

Table 2 Examples of typical and default values of greenhouse gases equivalent emissions per unit of energy available, for different biomass resources and pathways [25]

Primary solid and gaseous biomass pathway	Typical greenhouse gas emissions (gCO_2 eq/MJ)	Default greenhouse gas emissions (gCO_2 eq/MJ)
Bagasse briquettes-wood as process fuel	14	17
Bagasse briquettes—natural gas as process fuel	29	35
Biogas from wet manure	7	8
Biogas from dry manure	6	7
Biogas from wheat and straw (wheat whole plant)	18	21

2. Carbon capture and storage, e_{ccs}
3. Carbon capture and replacement, e_{ccr}.

As an example, Table 2 shows some values of emission for different types of biomass resources.

These values do not take into account the end use as fuel since it can be very different and the equipment employed can vary largely.

3 Biomass Resources Processes

Biomass can be converted into different energy forms by several processing steps using three main different conversion technologies: (i) the biochemical conversion includes anaerobic digestion (to biogas) and fermentation (to lactic acid, ethanol and so on); (ii) the thermochemical conversion routes include pyrolysis, gasification, combustion and hydrothermal processing; (iii) the physical–chemical conversion process pertains to the extraction of oil by esterification from oilseeds [26–28]. The different processes which a biomass feedstock can undergo mainly depend on their physical and chemical composition. The three main classes of processes are listed below:

- Thermochemical
- Biochemical
- Physical–chemical

The pathways from biomass to energy and/or matter recovery are strongly dependent on the chemical and physical composition of the biomass, as detailed in the previous paragraph. Figure 2 sketches the main pathways from biomass feedstock to energy/matter recovery, highlighting each class with coloured frames: Green is for biochemical, Red for thermochemical and Blue for physical–chemical. Very often,

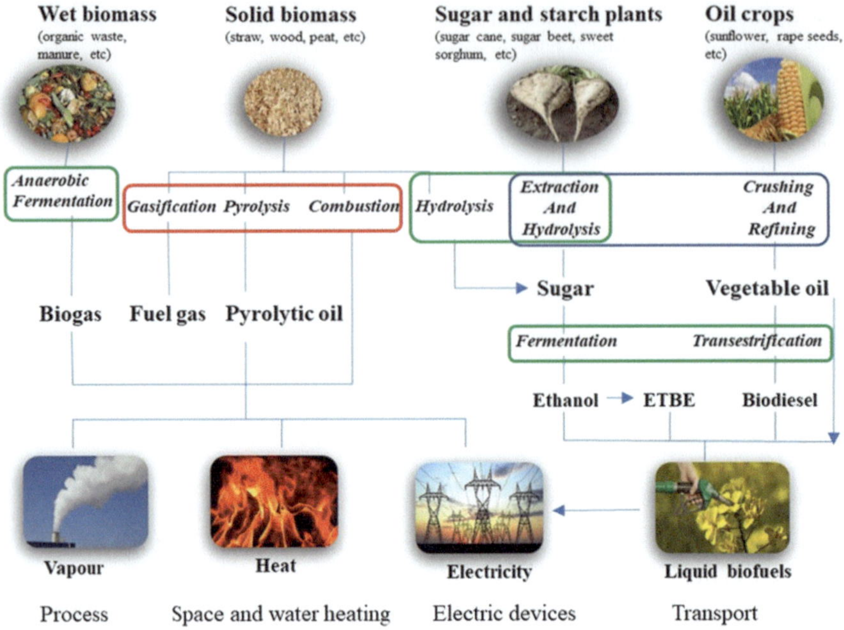

Fig. 2 Schematic diagram of typical pathways biomass-to-energy as a function of the biomass source

the mechanical extraction (physical process) is followed by a biochemical or thermochemical conversion process.

Thermal conversion processes include combustion, pyrolysis, torrefaction and gasification processes. Combustion accounts for the decomposition of the feedstock using sufficient oxygen at high temperature. Pyrolysis is the decomposition of biomass in the absence of oxygen. Torrefaction is a pyrolysis attained at low temperature, and gasification is the conversion of solid biomass into various gases using heat and varying amounts of oxygen. Biochemical conversion involves the use of enzymes of bacteria or other microorganisms to breakdown biomass through the processes of anaerobic digestion, fermentation or composting [29].

In the following, only the main thermochemical and biochemical processes will be described in detail.

3.1 *Thermochemical Processes*

Thermochemical processes involve the degradation and transformation of complex molecules into simpler ones by means of high temperature conversions that can be usefully converted either directly to energy or to some solid, liquid or gaseous fuels.

Combustion

The simplest way of managing solid biomass is combustion, which is a typical process with sufficient oxygen to oxidize the hydrocarbon molecules. In this process, the oxidation of the biomass feedstock occurs to generate directly heat without producing high-value gaseous, liquid or solid fuels. In the case of combustion of municipal solid waste (MSW), the process is called **Incineration**.

The combustion process is a fast oxidation exothermic process, which is initiated once the three main elements are present: the fuel, the oxidizer and high temperature. The fuel is the biomass, mainly composed of hydrocarbon molecules, the oxidizer is the oxygen present in the air and the temperature is provided by any ignition mean that allows the combustion process to start up, increasing the temperature so that the activation energy barrier is overcome [30]. The products of the combustion process are: CO_2, water, sulphur oxide and heat. Almost ever, other undesired products are produced, such as the carbon monoxide, nitric oxides and unburned hydrocarbons. All these products pollute the environment.

The biomass Heating Value (HV) is a very important datum about its use in the combustion process. As already detailed in a previous paragraph, it is only a potential measure of how much heat can be delivered when a fuel is burnt, but this value allows a comparison between different feedstocks.

The efficiency of the combustion process is strongly dependent on the air/fuel ratio, maximum temperature attained in the process, combustion chamber shape and many other parameters, and it is usually calculated by using the so-called 'smokestack heat losses', which account for how much energy the exhaust gases bring with them flowing out of the combustion chamber. The relationship is:

$$\eta = \frac{Q_{\text{avail}}}{m_{\text{c}}\,\text{LHV}} = 1 - \frac{Q_{\text{loss}}}{m_{\text{c}}\,\text{LHV}} \tag{3}$$

where Q_{avail} is the available heat deployed to users, m_{c} the fuel mass burnt, LHV the Lower Heating Value of the fuel and Q_{loss} the heat loss by exhaust gases.

As far as the biomass is concerned, the wood feedstock can be usually dealt with combustion in order to decompose the long chains of lignin and/or cellulose and hemicellulose, which are recalcitrant to be decomposed by many other treatments.

Pyrolysis

Pyrolysis is the main reaction precursor of gasification process which decomposes the organic matter without (or using a very low content of) oxygen. It represents the basic and most important step to convert the biomass feedstock to a useful fuel, breaking it down into charcoal (biochar), tar gases and liquids. The thermal decomposition starts when its temperature rises above around 500 K and usually takes place at temperatures around 750 and 900 K.

The gases and liquids produced during lower temperature pyrolysis are partially decomposed fragments of the original biomass that break off with heat. These fragments contain mainly H, C and O molecules collectively referred to as volatiles, which are reactive. The solids that remain are usually called *charcoal*, or sometimes

Table 3 Classification of pyrolysis/gasification processes and typical yields for the different products

Type of process	Main characteristics	Liquid bio-oil (%)	BioChar (%)	SynGas (%)
Fast/Flash	Short residence times, intermediate temperatures	50	25	15
Slow	Very long residence times, low temperatures	25	35	25
Gasification	Long residence times, very high temperatures	5	10	80

called *biochar* when the solid is biomass, whereas the gases and liquids released are called *tars* [31].

Different heating rates, temperatures attained in the process and residence time give rise to the diverse composition of the products. Schematic indications of different yields are given in Table 3, highlighting that the target products determine the choice of the process. The sum of the products does not account for 100% since there are always some losses [32].

Two different processes can be identified: Fast pyrolysis is currently the most widely used method yielding around 60% of bio-oil, the rest biochar and syngas plus losses and takes seconds for completing the pyrolysis. Fast pyrolysis processes imply that biomass feedstock has very short residence time attaining temperatures around 850 K producing mainly liquid products, with a very high heating rate (K/s) of the feedstock, very short residence time of vapours and a subsequent rapid cooling of the same vapours. On the other hand, slow pyrolysis takes several hours to complete and charcoal is the main product [33].

Feedstock moisture content, required to be around 10%, strongly influences the pyrolysis process. In fact, when the moisture content is higher, much water is produced, whereas for smaller levels, the process could produce only dust instead of oil. Thus, waste streams with large moisture figures, such as sludge and meat processing wastes require drying process before subjecting to pyrolysis [34].

The pyrolysis process is also strongly affected by the particle size of feedstock. In fact, the greatest majority of the pyrolysis technologies can process only small particles to a maximum of 2 mm because of the requirements for very fast heating. To obtain small particle size, the feedstock needs to be reduced by grinding, milling or pulverizing before being used for pyrolysis [35].

Gasification

Gasification is an incomplete combustion of carbonaceous biomass feedstock, through different intermediate steps, which produces a gaseous fuel product together with other solid and liquid by-products. The process requires feedstock with high

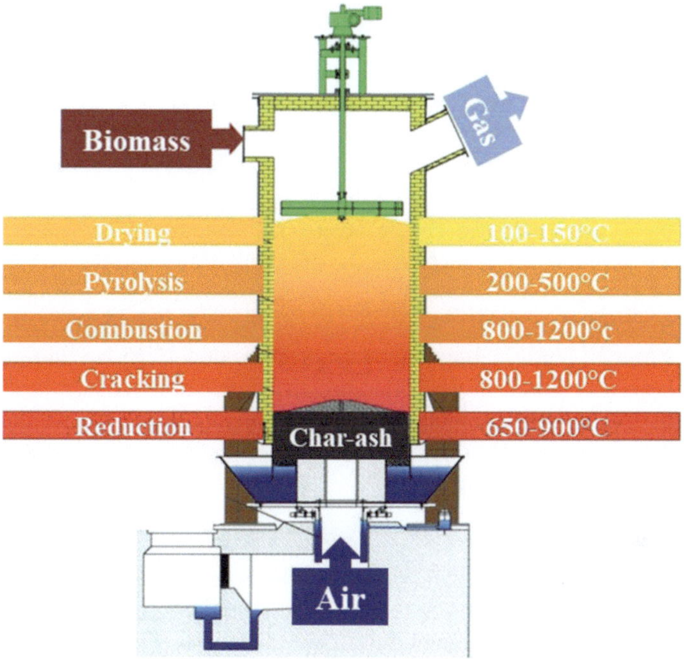

Fig. 3 Main gasification processes and illustrative temperatures occurring at each stage

levels of carbon, such as wood biomass, coal and municipal solid waste. The chemical reactions and processes occur at elevated temperatures, generally larger than 700 °C. The main stages of the process can be summarized as (i) evaporation or drying of moisture, (ii) release of volatile higher organics by pyrolysis, (iii) combustion, (iv) cracking of hydrocarbons and (v) reduction (or steam reforming) as schematically reported in Fig. 3. There are different types of gasifier in which gasification processes occur:

- Fixed bed (updraft or downdraft)
- Fluidized bed
- Different others with moving grate beds

Drying

Drying is the phase in which the moisture in the biomass is removed before entering the reactor because it needs to be around 10% of the total fresh weight. During the increasing temperatures, the water in the biomass is vapourized out of the fuel. This is an important step that should be correctly accomplished to produce clean gas.

Pyrolysis

This step has been already described in the previous subparagraph.

Cracking

During this crucial phase, large complex molecules depolymerize into gases using heat. Cracking is required to guarantee proper combustion because it occurs only with combustible gases thoroughly mixed with oxygen. During the combustion, the high temperatures attained at this stage decompose the large tar molecules that pass through the combustion zone.

Combustion

Combustion is the only exothermic process of the processes of gasification; the heat arises directly from combustion or recovered from combustion by heat exchange in a gasifier. Combustion can be fuelled by either the tar gases or char from Pyrolysis. Different reactor types can be used for each type of process.

Reduction

The stripping oxygen atoms of the combustion products of hydrocarbon (HC) molecules takes place during the reduction phase. Combustion releases heat and as products water vapour and carbon dioxide. During the reduction stage, the removal of oxygen from combustion products at high temperature to produce again fuel gases occurs. Combustion and reduction in most burning environments operate simultaneously.

Reduction is accomplished by passing water vapour or carbon dioxide, the so-called gasifier agents, across hot charcoal. The carbon is highly reactive with oxygen, stripping the oxygen off water vapour and carbon dioxide. No free oxygen is left in its diatomic form. The reduction process leads to CO and H_2, which are both combustible fuel gases.

A bit of chemistry of gasification process

The main chemistry of gasification process can be summarized as follows:

$$\text{Biomass} + O_2 \rightarrow CO, CO_2, H_2O, H_2, CH_4 + (CH) + \text{tar} + \text{char} + \text{ashes} \tag{4}$$

The amounts of carbon monoxide, carbon dioxide, other gases, char and tar produced depend on the process parameters. As far as the gases are concerned, their relative amount is dictated by the used gasifying agent and the air/fuel ratio, when air is used as agent.

The most important thermochemical reactions are shown here:
Methane Reforming (Steam and Methane)

$$CH_4 + H_2O \rightleftharpoons CO + 3H_2 \tag{5}$$

Water gas shift

$$CO + H_2O \rightleftharpoons CO_2 + H_2 \tag{6}$$

Carbon char to methane

$$C + 2H_2 \rightleftharpoons CH_4 \tag{7}$$

Carbon char oxides

$$C + CO_2 \rightleftharpoons 2CO \tag{8}$$

The process can be very efficient with a conversion fraction over 95% of biomass carbon to syngas [28].

3.2 Biochemical Process

Anaerobic Digestion Process

Anaerobic digestion process is of paramount importance when dealing with a large number of biomass resources, either dry or wet, but always with a large content of fermentescible matter. In addition, indigenous bacteria population is usually required to start the process without external inoculum.

Anaerobic digestion is a complex biochemical process carried out in a number of steps by several types of microorganisms, which produces biogas, a mixture of carbon dioxide, methane and small amounts of other gases, as H_2S and hydrogen being present. Inside a heated, sealed airless container the process takes place in order to attain ideal conditions for the microflora to digest the organic material in anoxic conditions. Anaerobic digesters produce conditions that favour the natural breakdown of organic matter [27, 29].

Three types of anaerobic digestion processes can be industrially set up as a function of the process temperature:

- *psicrophilic digestion* the digester is heated to 10–25 °C, and the feedstock remains in the digester typically for long time, usually more than 30 days. The main characteristics are:
  - No heating required
  - Low reaction rates

  - Usually used in wastewater treatments
  - High tolerance to toxicity
  - Very long hydraulic retention time (HRT) >30 days

- *mesophilic digestion*, in which the digester is heated to 30–37 °C, and the feedstock resides in the digester usually for 20–35 days.
  - Most commonly used
  - Municipal and industrial wastewater and co-digestion
  - Medium tolerance to toxicity
  - HRT around 20–35 days

- *thermophilic digestion,* in which the digester is heated to around 55 °C, and the residence time is typically around 15 days.
  - More compact digesters in terms of volume
  - Higher degradation rates
  - High temperature kills most pathogens and bacteria
  - Slow and difficult start-up
  - Can miss some bacteria fit to these temperatures
  - Low tolerance to toxicity
  - HRT around 10–15 days

Whereas in terms of the biomass total solids (TS) property, the process can be identified as:

- **Wet**
  - When TS < 10%
  - Mainly for animal manure
  - Low organic loads 2–4 kg VS/m^3 day
  - Biogas yield 100–150 m^3/t waste
  - Specific biogas yield 0.4–0.5 m^3/kg vs
  - Methane 50–70%

- **Semi-Dry**
  - When TS around 10% (conveniently diluted)
  - Mainly for organic fraction of municipal solid waste (MSW)
  - CSTR reactor
  - Low organic loads 8–12 kg VS/m^3 day
  - Biogas yield 100–150 m^3/t waste
  - Specific biogas yield 0.3–0.5 m^3/kg vs
  - Methane % 55–60%

- **Dry**
  - when TS > 20% (conveniently diluted)
  - Mainly for organic fraction of MSW

- Plug flow reactor (or batch)
- Low organic loads 8–10 kg VS/m^3 day
- Biogas yield 90–150 m^3/t waste
- Specific biogas yield 0.2–0.3 m^3/kg vs

In the anaerobic fermentation process, 30–60% of the digestible solids are usually converted into biogas [36], which typically consists of 50–70% CH_4, 30–50% CO_2 [37] and smaller amounts of other gases.

The full process can be summarized in four stages; each of them has a bacteria population fitted for the specific environmental conditions [38, 39]:

1. *hydrolysis*, during this stage, polymers like carbohydrates and lipids are converted into glucose, glycerol and other substances. Hydrolytic microorganisms eliminate hydrolytic enzymes, converting biopolymers into simpler compounds. The hydrolytic bacteria, responsible for this step of anaerobic digestion, belong to facultative anaerobes like *Cellulomonas* and *Lactobacillus* [40, 41].
2. *acidosis,* at this stage, simple compounds are converted to volatile fatty acids (VFA). The products of hydrolysis, simple sugars, amino acids and fatty acids are converted by acidogenic bacteria into VFA, such as propionate, butyrate and lactate or neutral compounds (ethanol, methanol) [39]. The acidogenesis dominant community members are *Bacillus*, *Clostridium*, *Micrococcus*, *Pseudomonas* and *Ruminococcus* [42].
3. *acetogenesis*, VFA is converted by hydrogen-producing acetogenic bacteria to acetic acid, hydrogen and carbon dioxide. *Clostridium* is mainly responsible for the production of hydrogen [8], while *Acetobacterium* is responsible for acetate formation [43].
4. *methanogenesis*: methane and carbon dioxide are produced from acetic acid and hydrogen. Produced methane originates for about 70% from acetate, while the remaining 30% is obtained from the conversion of hydrogen and carbon dioxide. The bacteria responsible for this last conversion are acetotrophic or acetoclastic methanogens, which are obligate anaerobes [44].

The digestion is not complete until the substrate has undergone all the four stages. The consortium of the first types of bacteria enables the production of intermediate products, such as hydrogen, acetic acid and carbon dioxide, which are then used by methanogenic bacteria to produce methane. When the symbiotic relationship is disturbed, there is an accumulation of volatile fatty acids (VFA) which may cause the reduction of pH value with a decrease or arrest of methane production [45].

Temperature, pH and nutrients are the most important process parameters that severely affect the stability of anaerobic processes, mainly due to the fragile nature of microorganisms, especially methanogens [46]. In fact, these latter are sensitive to changes in temperature and pH and are inhibited by a high level of volatile fatty acids and other process intermediates, such as hydrogen, ammonia and hydrogen sulphide [47]. Moreover, pH and temperature are the discriminating factors that can cause a metabolic shift, which can affect the type of reaction that bacteria accomplish or the bacterial population involved in it [48].

A bit of chemistry of anaerobic fermentation

Three typical idealized reactions can be listed to represent the fermentation process from the substrate to the biogas. Each reaction considers only one type of substrate and the products obtained in the whole process. When starch is considered as substrate, the products methane and carbon dioxide are the main components of biogas with a 50:50 (1:1) composition. Instead, dealing with fat/proteins as substrate, the ratio is a little more favourable to methane 44:39 (1.13:1). When triglycerides represent the substrate, the methane/carbon dioxide ratio is around 70:30 (2.33:1). Of course, when the actual biochemical reactions are taken into account, some other products such as hydrogen sulphide in ppm are present and should be stripped off to protect metallic parts in contact with biogas.

From glucose (starch)

$$C_6H_{12}O_6 \rightarrow 3CH_4 + 3CO_2 \tag{9}$$

From fats or proteins

$$C_{10}H_{10}O_6N_2 + 3.5\ H_2O \rightarrow 5.25\ CH_4 + 4.75\ CO_2 + 2\ NH_3 \tag{10}$$

From triglycerides

$$C_{54}H_{106}O_6 + 24.5\ H_2O \rightarrow 38.75\ CH_4 + 15.25\ CO_2 \tag{11}$$

Recovery and use of Biogas

Biogas composition can vary because of the biomass substrate organic material, process parameters, retention time and climatic conditions. As previously shown, gas composition can be around 50–70% CH_4 and 30–50% CO_2 with very small amounts of NH_3 (80–100 ppm), H_2S (400–3000 ppm) and hydrocarbons (<100 ppm). Biogas may also contain dust particles and siloxanes, usually below to 100 mg/m^3 [49]. The most important post-treatments can be: desulphurization, dehydration, carbon dioxide and nitrogen/oxygen stripping. With post-treatments, biogas shows greater methane content, thus a larger calorific value. By considering the pure methane LHV, on a mass basis, equal to 13.9 kWh/kg when the methane content is about 75% of the gaseous mixture, the LHV decreases to about 11 kWh/kg, with a 20% reduction.

Some results for biogas production

It has been shown that a starting pH value of about 7.0–7.5 with a fermentation temperature of 37 °C seems to be the optimal conditions for producing methane from fermenting buffalo manure, as it is shown in Fig. 4, where methane production is reported for three biological replicates, with an initial pH = 7.0 and a process temperature $T = 37$ °C. It is to observe that three different replicates show the same typical behaviour of such processes. The first phase is called lag phase, in which the bacteria start to acclimate before the process starts. Once the fermentation is started, the production growth is very intense, followed by a plateau phase in which the cumulated

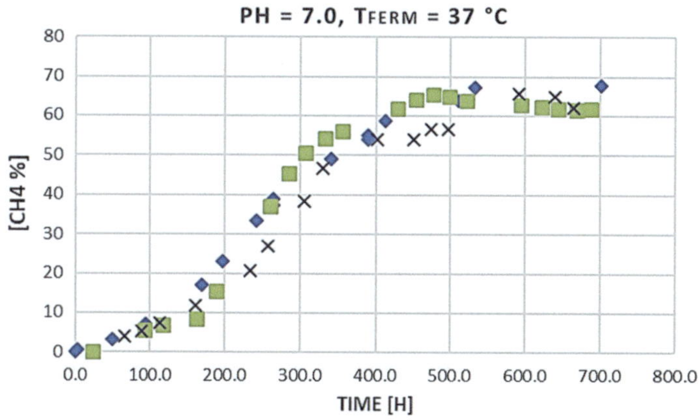

Fig. 4 Methane production from buffalo manure with initial pH = 7.0 and processing temperature $T = 37$ °C, for three replicates

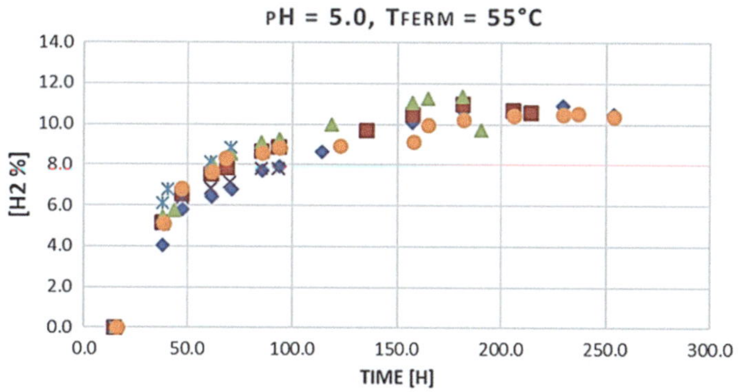

Fig. 5 Hydrogen production from buffalo manure with initial pH = 5.0 and processing temperature $T = 55$ °C, for three replicates

methane production reaches its maximum value. The Gompertz law describes very well these three phases, indicating a sigmoid growth of the biogas production. In the case reported in Fig. 4, the methane yield is around 65%, indicating that the substrate digested by bacterial community is mainly composed of triglycerides.

In some cases, from anaerobic fermentation bio-hydrogen can be produced. Peculiar conditions, anyway, are required because hydrogen is an intermediate product of the process. In fact, thermal pre-treatments at about 90 °C for about three hours, with an initial pH of 5.0 and a process temperature of 55 °C have shown that hydrogen production can be maximized from the same substrate, Fig. 5 [50, 51].

Some results for lactic acid recovery

Lactic acid (LA) represents one of the many possible chemical building blocks that can be recovered from organic waste. It is a valuable matter since its use is largely

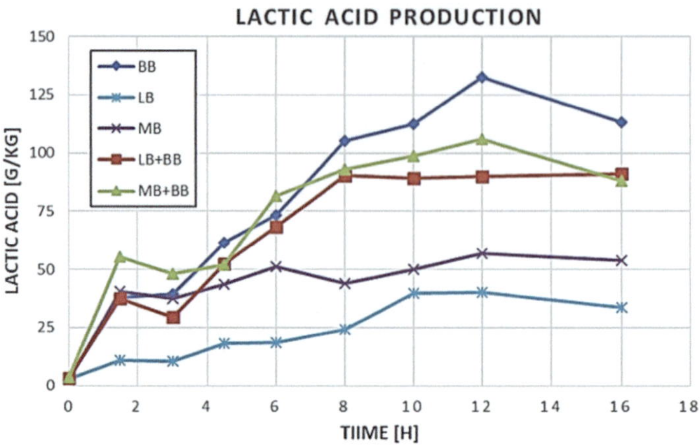

Fig. 6 Lactic acid production by anaerobic fermentation for different bacteria population process temperature $T_{\text{ferm}} = 55$ °C and initial pH = 5.0

widespread in chemical, pharmaceutical, nutraceutical and bio-plastic industries. The volume of LA market in 2008 was estimated around 260 Mt per year [52] with a forecast demand for 2020 of about 830 Mt per year [53].

Organic residues from agro-food sector, such as tomato processing waste, are the optimal substrate to produce LA avoiding problems of their disposal in landfills, which contribute to cause water pollution, rotten odours, as well as release into the atmosphere of greenhouse gases [54].

Organic acids, as LA, are formed during the breakdown of the biomass substrate by acidogens microbial populations. They represent the intermediate products of the anaerobic fermentation.

Some illustrative results about the production of lactic acid (LA) from fermentable sugars using as substrate tomato pomace are presented, Fig. 6.

LA was produced from tomato pomace fermentable sugars using bacteria consortia obtained from water buffalo manure in comparison with the most efficient commercial LA producing bacteria, as a complementary process to biogas production [55].

Manure of Mediterranean water buffalo was collected in a commercial buffalo farm [8]. Bacteria from buffalo dung (BB) were grown and selected on a specific clostridial nutrient medium. LA producing bacteria (LB) Lactobacillus rhamnosus GG (ATCC 53103) and bacteria from mozzarella cheese wastewater (MB).were cultivated and enriched. Then, the three types of bacteria consortia were used as inoculum alone or in combination (BB, LB, MB, LB + BB, MB + BB) for the anaerobic fermentation of tomato pomace as a substrate for the production of fermentable sugars and their conversion to LA. The initial value of the pH of the substrate was set to 5.0.

As displayed in Fig. 6, LA production increases in the first hours of the anaerobic fermentation. The largest LA production recorded after 16 h digestion is obtained by tomato pomace inoculated with BB, while the two mixtures containing BB bacteria, i.e. LB + BB and MB + BB show a slightly lower maximum production. The pure BB population shows a maximum after 12 h, with a production of 125 g/kg, whereas LB + BB and MB + BB had a maximum production around 100 g/kg at the same time (12 h). The other two single populations display a quite lower production, less than half the maximum obtained using the BB population.

4 Conclusions

Biomass is a complex mix mainly made up of various natural constituents originated algae, plant, animal and municipal residues as well as agricultural and agro-food residues derived by industrial processing are also included.

Their different chemical and physical composition requires distinct processes to obtain either directly energy or biofuels and biomaterials. The thermochemical, biochemical or physical-chemical conversion pathways are the possible process to produce energy. Biochemical methods include the anaerobic digestion, which is an efficient way of producing biogas, a mixture of methane and carbon dioxide. The main process parameters, which affect the biogas production, are the pH of the substrate, temperature, partial pressure of biogas. It has been displayed herein that buffalo manure fermentation can produce biogas with a methane concentration of around 65%. Pre-treating the substrate and changing temperature and pH of the process allow to extract hydrogen up to a 15% of concentration from the same substrate.

Not only energy but also matter can be recovered from organic waste. In fact, tomato waste residues have been demonstrated to be the optimal substrate for the lactic acid extraction using anaerobic fermentation. Using different bacterial species or mixtures of them, diverse results are obtained. The best lactic acid production is obtained using buffalo manure bacteria with a maximum production of about 150 g/kg.

These results highlight the dramatic potential of residual biomasses, which represent a giant step to the future both in terms of energy and matter recovery.

References

1. Vassilev SV, Baxter D, Andersen LK, Vassileva CG, Morgan TJ (2012) An overview of the organic and inorganic phase composition of biomass. Fuel 94:1–33. https://doi.org/10.1016/j.fuel.2011.09.030
2. Carillo P, Morrone B (2017) Converting residual biomasses into energy through anaerobic digestion. In: Tzortzakis N (ed) Municipal solid waste: management strategies, challenges and future directions. Nova Science Publisher, New York, p 21
3. Anwar Z, Gulfraz M, Irshad M (2014) Agro-industrial lignocellulosic biomass a key to unlock the future bio-energy: a brief review. J Radiat Res Appl Sci 7(2):163–173. https://doi.org/10.1016/j.jrras.2014.02.003
4. Parliament TE (2009) Directive 2009/28/EC of the European Parliament on the promotion of the use of energy from renewable sources and amending and subsequently repealing Direc-

tives 2001/77/EC and 2003/30/EC. The European Parliament and the Council, Brussels, BE: Bruxelles
5. Macqueen D, Korhaliller S (2011) Bundles of energy: the case for renewable biomass energy 24:1–86
6. Balat M, Ayar G (2005) Biomass energy in the world, use of biomass and potential trends 27:931–940
7. Yokoyama H, Waki M, Ogino A, Ohmori H, Tanaka Y (2007) Hydrogen fermentation properties of undiluted cow dung. J Biosci Bioeng 104(1):82–85. https://doi.org/10.1263/jbb.104.82
8. Carillo P, Carotenuto C, Di Cristofaro F, Kafantaris I, Lubritto C, Minale M, Morrone B, Papa S, Woodrow P (2012) DGGE analysis of buffalo manure eubacteria for hydrogen production: effect of pH, temperature and pretreatments. Mol Biol Rep 39(12):10193–10200. https://doi.org/10.1007/s11033-012-1894-3
9. Blanco-Canqui H, Lal R (2009) Crop residue removal impacts on soil productivity and environmental quality. Crit Rev Plant Sci 28(3):139–163. https://doi.org/10.1080/07352680902776507
10. Muth DJ Jr, Bryden KM, Nelson RG (2013) Sustainable agricultural residue removal for bioenergy: a spatially comprehensive US national assessment. Appl Energy 102:403–417. http://dx.doi.org/10.1016/j.apenergy.2012.07.028
11. Popa VI (2018) Biomass for fuels and biomaterials. In: Popa V, Volf I (eds) Biomass as renewable raw material to obtain bioproducts of high-tech value. Elsevier, Amsterdam, pp 1–37
12. Ali Abdoli M, Golzary A, Hosseini A, Sadeghi P (2018) Biomass as a source of energy, pp 1–31
13. McKendry P, McKendry P (2002) Energy production from biomass (part 1): overview of biomass. Bioresour Technol 83:37–46
14. Korus A, Szlęk A (2015) The effect of biomass moisture content on the IGCC efficiency. Biomass Bioenerg 80:222–228. https://doi.org/10.1016/j.biombioe.2015.05.014
15. Kamińska A (2011) Method for increasing the calorific value of fragmented wood biomass. Czasopismo Techniczne. Chemia. Available at http://yadda.icm.edu.pl/baztech/element/bwmeta1.element.baztech-article-BGPK-3545-3403. Accessed 2 August 2018
16. Demirbas A (2004) Effects of temperature and particle size on bio-char yield from pyrolysis of agricultural residues. J Anal Appl Pyrol 72(2):243–248. https://doi.org/10.1016/j.jaap.2004.07.003
17. Oginni O, Singh K (2016) Assessment of soil amendment and carbon sequestration potential of Arundo, Miscanthus and Switchgrass Biochars
18. Gaqa S, Mamphweli S, Katwire D, Meyer E (2014) The properties and suitability of various biomass/coal blends for co-gasification purposes 04:175–182
19. Mascia PN, Scheffran J, Widholm JM (2010) Plant biotechnology for sustainable production of energy and co-products 66
20. James A, Thring R, Helle S, Ghuman H (2012) Ash management review—applications of biomass bottom ash. Energies 5(10):3856
21. Basu P (2013) Biomass characteristics (Chap. 3). In: Basu P (ed) Biomass gasification, pyrolysis and torrefaction, 2nd edn. Academic Press, Boston, pp 47–86
22. Himmel ME, Ding S-Y, Johnson DK, Adney WS, Nimlos MR, Brady JW, Foust TD (2007) Biomass recalcitrance: engineering plants and enzymes for biofuels production. Science 315(5813):804–807. https://doi.org/10.1126/science.1137016
23. Lockhart J (2015) Altering lignin composition to improve biofuel production. Plant Cell 27(8):2082. https://doi.org/10.1105/tpc.15.00668
24. United Nations (1987) Our common future. World Commission on Environment and Development, p 300
25. Report Commission to the Council and the European (2010) Parliament on sustainability requirements for the use of solid and gaseous bio-mass sources in electricity, heating and cooling. Bruxelles
26. Adams P, Bridgwater T, Lea-Langton A, Ross A, Watson I (2018) Biomass conversion technologies. In: Thornley P, Adams P (eds) Greenhouse gas balances of bioenergy systems. Academic Press, New York, pp 107–139

27. Ramachandra TV, Joshi NV, Subramanian DK (2000) Present and prospective role of bioenergy in regional energy system. Renew Sustain Energy Rev 4(4):375–430. https://doi.org/10.1016/S1364-0321(00)00002-2
28. Demirbas A (2009) Biofuels. Securing the planet's future energy needs. Green energy and technology. Springer, London, p 331
29. Dahiya A (2015) Bioenergy. Biomass to biofuels. Academic Press, London
30. Atkins P, de Paula J (2009) Physical chemistry, 9th ed. W. H. Freeman, San Francisco
31. Ronsse F, van Hecke S, Dickinson D, Prins W (2012) Production and characterization of slow pyrolysis biochar: influence of feedstock type and pyrolysis conditions. GCB Bioenergy 5(2):104–115. https://doi.org/10.1111/gcbb.12018
32. Cantrell KB, Hunt PG, Uchimiya M, Novak JM, Ro KS (2012) Impact of pyrolysis temperature and manure source on physicochemical characteristics of biochar. Biores Technol 107:419–428. https://doi.org/10.1016/j.biortech.2011.11.084
33. Al Arni S (2018) Comparison of slow and fast pyrolysis for converting biomass into fuel. Renewable Energy 124:197–201. https://doi.org/10.1016/j.renene.2017.04.060
34. Tag AT, Duman G, Ucar S, Yanik J (2016) Effects of feedstock type and pyrolysis temperature on potential applications of biochar. J Anal Appl Pyrol 120:200–206. https://doi.org/10.1016/j.jaap.2016.05.006
35. Suliman W, Harsh JB, Abu-Lail NI, Fortuna A-M, Dallmeyer I, Garcia-Perez M (2016) Influence of feedstock source and pyrolysis temperature on biochar bulk and surface properties. Biomass Bioenerg 84:37–48. https://doi.org/10.1016/j.biombioe.2015.11.010
36. Warburton B (1997) Anaerobic digestion of farm and food processing residues: the development of a sustainable Industry. Environment Council. Environmental Resolve Editor
37. Nijaguna BT (2002) Biogas technology. New age international (P) Ltd., publishers, Daryayanj, New Delhi, India
38. Demirel B, Neumann L, Scherer P (2008) Microbial community dynamics of a continuous mesophilic anaerobic biogas digester fed with sugar beet silage. Eng Life Sci 8(4):390–398. https://doi.org/10.1002/elsc.200800010
39. Zieminski K, Frac M (2012) Methane fermentation process as anaerobic digestion of biomass: transformat ions, stages and microorganisms. Afr J Biotechnol 11:4127–4139
40. Bagnara C, Toci R, Gaudin C, Belaich JP (1985) Isolation and characterization of a cellulolytic microorganism, *Cellulomonas fermentans* sp. nov. Int J Syst Bacteriol 35(4):502–507
41. Heilig HG, Zoetendal EG, Vaughan EE, Marteau P, Akkermans AD, de Vos WM (2002) Molecular diversity of lactobacillus spp. and other lactic acid bacteria in the human intestine as determined by specific amplification of 16S ribosomal DNA. Appl Environ Microbiol 68(1):114–123. https://doi.org/10.1128/aem.68.1.114-123.2002
42. Zhao Y, Ren N, Wang A (2008) Contributions of fermentative acidogenic bacteria and sulfate-reducing bacteria to lactate degradation and sulfate reduction. Chemosphere 72(2):233–242. https://doi.org/10.1016/j.chemosphere.2008.01.046
43. Ljungdhal LG (1986) The autotrophic pathway of acetate synthesis in acetogenic bacteria. Annu Rev Microbiol 40(1):415–450. https://doi.org/10.1146/annurev.mi.40.100186.002215
44. Al Seadi T, Rutz D, Prassl H, Köttner M, Finsterwalder T, Volk S, Janssen R (2008) Biogas handbook. Published by University of Southern Denmark Esbjerg, Niels Bohrs Vej 9-10, DK-6700 Esbjerg, Denmark
45. Vavilin VA, Lokshina LY, Rytov SV, Kotsyurbenko OR, Nozhevnikova AN, Parshina SN (1997) Modelling methanogenesis during anaerobic conversion of complex organic matter at low temperatures. Water Sci Technol 36(6):531–538. https://doi.org/10.1016/S0273-1223(97)00565-9
46. Khanal SK (2008) Anaerobic biotechnology for bioenergy production. Wiley-Blackwell, Oxford
47. Zeikus JG (1977) The biology of methanogenic bacteria. Bacteriol Rev 41(2):514–541
48. Ali Shah F, Mahmood Q, Maroof Shah M, Pervez A, Ahmad Asad S (2014) Microbial ecology of anaerobic digesters: the key players of anaerobiosis. Sci World J 2014: 183752. https://doi.org/10.1155/2014/183752

49. Leonzio G (2016) Upgrading of biogas to bio-methane with chemical absorption process: simulation and environmental impact. J Clean Prod 131:364–375. https://doi.org/10.1016/j.jclepro.2016.05.020
50. Di Cristofaro F, Carotenuto C, Carillo P, Woodrow P, Morrone B, Minale M (2014) Evaluation of CH_4 and H_2 yield with different mixtures of digested and fresh buffalo manure. Chem Eng Trans 37:283–288
51. Carotenuto C, Guarino G, Morrone B, Minale M (2016) Temperature and pH effect on methane production from buffalo manure anaerobic digestion. Int J Heat Technol 34, Special(2):425–429
52. Dusselier M, Van Wouwe P, Dewaele A, Makshina E, Sels BF (2013) Lactic acid as a platform chemical in the biobased economy: the role of chemocatalysis. Energy Environ Sci 6(5):1415–1442. https://doi.org/10.1039/C3EE00069A
53. PlasticsToday. Bioplastics demand spurs global growth in lactic acid production. PlasticToday 2011 30/07/2018]; Available from: http://www.plasticstoday.com/
54. Carillo P, D'Amelia L, Dell'Aversana E, Faiella D, Cacace D, Giuliano B, Morrone B (2018) Eco-friendly use of tomato processing residues for lactic acid production in campania. Chem Eng Trans 64:223–228. https://doi.org/10.3303/CET1864038
55. Tang J, Wang X, Hu Y, Zhang Y, Li Y (2016) Lactic acid fermentation from food waste with indigenous microbiota: effects of pH, temperature and high OLR. Waste Manag 52:278–285. https://doi.org/10.1016/j.wasman.2016.03.034

Next-Generation Biofuels—Opportunities and Challenges

Naveen Kumar, Ankit Sonthalia, Harveer S. Pali and Sidharth

Keywords Non-edible crops · Second generation · Algae · Biofuel policy · Biofuel cost · Hydroprocessing

1 Introduction

The human population is projected to exceed 9 billion by 2050 which will increase the calorie intake and push the production of food crops on scarce arable land to its limit. The demand of the energy will also increase. Figure 1 shows a comparison of the world total primary energy supply (TPES) in 1973 and 2016. The energy consumption of the world increased from 6101 Million tons of oil equivalent (Mtoe) in 1973 to 13,699 Mtoe in 2016 [1], an increase of 124%.

The fossil fuels provide a major chunk of the energy requirement, however, depleting reserves along with price volatility coupled with environmental issues preclude

N. Kumar (✉) · A. Sonthalia · H. S. Pali · Sidharth
Centre for Advanced Studies and Research in Automotive Engineering, Delhi Technological University, Bawana Road, Delhi 110042, India
e-mail: naveenkumardce@gmail.com

A. Sonthalia
Department of Automobile Engineering, SRM Institute of Science and Technology, NCR Campus, Ghaziabad 201204, Uttar Pradesh, India

H. S. Pali
Department of Mechanical Engineering, NIT Srinagar, Srinagar 190006, Jammu & Kashmir, India

Sidharth
Department of Mechanical and Automation Engineering, Maharaja Agrasen Institute of Technology, Rohini, Delhi 110086, India

A. K. Gupta et al. (eds.), *Innovations in Sustainable Energy and Cleaner Environment*, Green Energy and Technology,
https://doi.org/10.1007/978-981-13-9012-8_8

their use as a primary energy source [2]. Most of the countries are moving toward renewable and cost-effective options which are biodegradable and abundantly available. Renewable sources of energy, as shown in Fig. 2, namely wind, rain, solar, hydroelectric, and wave energy and bioenergy such as biomass, and biofuels have the potential to provide a holistic solution to growing energy needs [3].

The most effective and efficient form of renewable energy is biofuels. They can be easily produced from the biomass, they are biodegradable and are environment-friendly [4]. Their combustion is almost similar to fossil fuels [5] and they produce less toxic compounds and emissions [6, 7]. Also, their net carbon emissions are significantly low [8]. With no or little modifications to the engine, they can easily replace fossil fuels. Some of the available biofuels are bioethanol [9, 10], renewable methanol, biogas, biobutanol, biodiesel [11], and biohydrogen [12].

The biofuels can be classified as first generation [13], second generation [14], third generation [15], and fourth generation. Corn, wheat, barley, sugarcane, palm, and

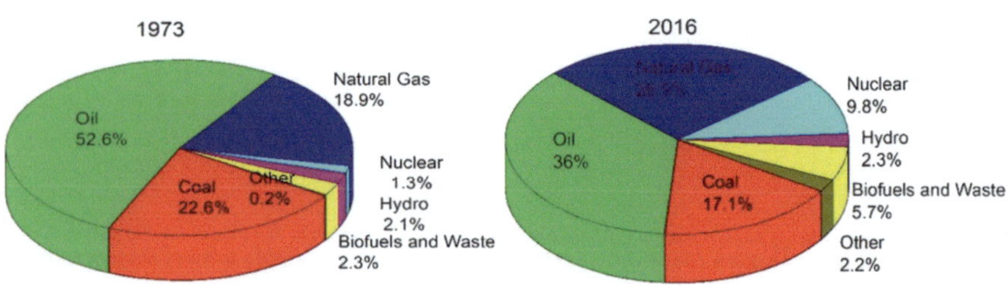

Fig. 1 Comparison of total primary energy supply in 1973 and 2016 [1]

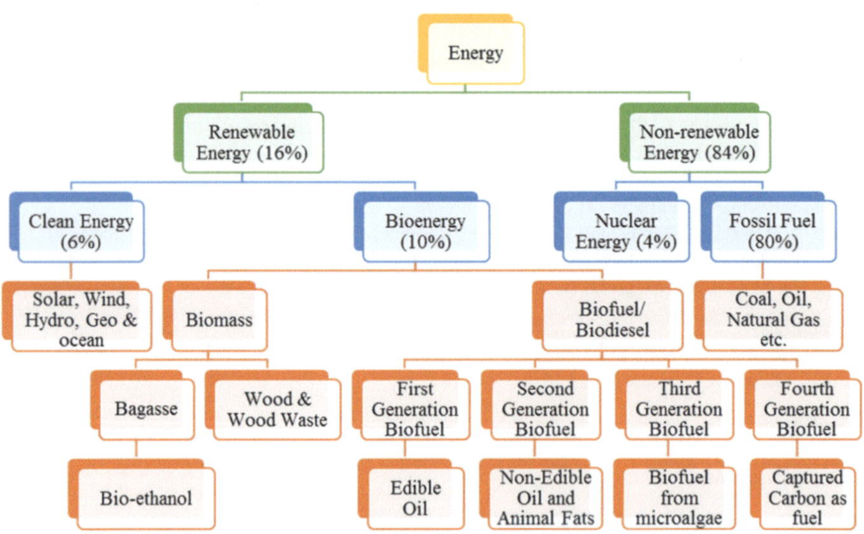

Fig. 2 Classification of energy by source type [3]

rapeseed are the major feedstock for the first-generation biofuels. Since they are also a source of food, it results in competition between fuel and food and may also increase the commodity prices of food [16]. Use of non-edible crops, agricultural and forest wastes can resolve this issue. Second-generation biofuels comprise of by-products, wastes, and dedicated feedstock. However, they also need land causing competition with food crops [17]. Therefore, a search was initiated to explore alternatives so that there is less competition between the food crop and the non-food crop. The quest directed the search toward the microbial world where a number of microbes had the potential to produce biofuels and also have certain advantageous traits such as faster growth rate, simple nutrient uptake and less complex bio-machinery associated with them. Among the wide pool of microbes, algae are of main interest due to its richness in biofuels, its inept ability to grow in various types of aquatic environment and ability to be grown any time of the year [18]. Biofuel produced from algae is known as third-generation biofuel. The abovementioned fuels cannot completely replace fossil fuels since they have limitations such as economic performance, geographical location, and dependence on sunlight [19]. Genetic engineering of algae for providing clean fuel is the solution to this problem and they are considered as the fourth generation of biofuels [20].

This chapter presents the brief overview of production mechanism, policies of Government of India and other countries pertaining to biofuels, use of these fuels in internal combustion engines. The availability of biomass in India along with the cost of the biomass for biofuel production is also presented.

2 Worldwide Biofuel Scenario

Both developed and developing countries are emphasizing the use of biofuels to meet their energy demands. These countries have framed their own policies to regulate and improve the use of biofuels. Brazil, the USA, and the European Union were the first one to acknowledge the importance of biofuels [21]. In the year 2007, the European Union set a target of meeting 20% of its energy demand by using renewables by 2020. Out of this 20%, 10% should be fulfilled by biofuels [22]. European Union has also laid a road map for Competitive Low Carbon Economy wherein the greenhouse gases need to be reduced to 40, 60, and 80% by 2030, 2040, and 2050, respectively, using technologies which emit less carbon [23, 24]. To achieve this target, several initiatives have been taken by the member countries such as improvement in biofuels production, reducing the taxes, and allowing the blending of higher percentages of biofuels with conventional fuels [25]. Germany, for instance, also notified a National Renewable Energy Action Plan in the year 2010 which aimed at increasing the renewable energy share to 18% by 2020 [23].

The USA endorsed a sequence of renewable energy legislation for biofuel plants construction, it comprises policies for tax reduction, financial aids, and loan assurance [23]. Energy policy was approved in 2005 with the aim of contributing 4 billion gallons of biofuels to the transport sector by 2006 [26]. They also enacted the Energy

Independence and Security Act in 2007 that targets 18% renewable to be used in the transport sector by 2022 [27]. In 2008, a biomass program was started to increase the corn-based ethanol production such that the gasoline consumption could be reduced to 30% by 2030. Canada sanctioned the Environmental Protection Act, 1999 (BillC-33) which sets a target of blending 2% biofuel in conventional diesel. Argentina in 2006 legislated regulation and promotion of the production and sustainable use of biofuels with the aim of blending 5% biofuels in conventional diesel [26]. Thailand has also set a target of 5% blending of biodiesel (from palm oil) in diesel and 10% blending of ethanol (from cassava) in petrol [28].

3 Energy Scenario in India

The Indian economy is one of the fastest growing, and energy is building block in its development. The electricity generating capacity of India is 5th largest in the world, and India is the 6th largest energy consumer which is nearly 3.4% of the world energy consumption [29]. However, the per capita energy consumption is only one-third of the world average because a large part of the Indian population does not have access to reliable power, despite the fact that the power systems are rapidly expanding in recent years. The residential per capita electricity consumption shows a broad range between urban and rural population. Delhi has per capita consumption higher than Organisation for Economic Co-operation and Development (OECD) average whereas Bihar has per capita consumption ten times lower than OECD levels [30].

Power generation in India comprises of diversified resources such as coal, natural gas, hydro, oil, nuclear, and renewable sources such as solar, wind, and domestic and agricultural waste. The total installed electricity generation capacity in 1947 was 1362 MW and it increased to 344,002 MW in March 2018. The installed capacity, as on March 2018, for coal-based is 197,171 MW (57.3%), for gas-based is 24,897 MW (7.23%), for diesel-based is 837 MW (0.24%), for nuclear-based is 6780 MW (1.97%), for hydro-based is 45,293 MW (13.17%), and for renewable-based is 69,022 MW (20%). The generation of power from renewables can be subdivided into small hydropower (4485 MW), wind power (34,046 MW), biomass power (8700 MW), waste to energy (138 MW), and solar power (21,651 MW) [31]. In 2011, the renewable sector amounted to only 11% of the total installed capacity which has increased to 20% of the total installed capacity in 2018. This shows that India is on track to fulfill its commitment to CO_2 reduction.

The international energy agency (IEA) in its report on world energy outlook predicted that oil demand in India will increase four times by 2040. It also predicted that the growth in oil demand of developed countries will be offset by a growth in demand of developing countries, half of that will come from India. The gas demands in India will be tripled as more gas-based power plants will mushroom and the increasing use of CNG vehicles. By 2040, power generation from coal will drop to half from current figures and power generation from solar will increase 100-folds and wind will increase 9-folds. This prediction is based on 6.5% gross domestic product (GDP) growth rate for the entire period, i.e. 2016–2040 [32].

The crude oil production in India in the year 2016–17 is 36.01 million metric tonnes (MMT), and 213.93 MMT of crude oil was imported in the same year. The oil imports increased by 5.46% in comparison with the previous year. It is estimated that by 2021 the demand for diesel and petrol will be 110 and 31.1 MMT, respectively [33].

3.1 Biofuel Policy in India

As discussed earlier, the use of biofuels as a supplement to the conventional fuels can provide the much-needed energy security and reduce the dependence of imported fossil fuels. In 2002, a five percent ethanol blending program in petrol was started and in 2003, a National Biodiesel Mission (NBM) with an aim of blending twenty percent biodiesel in diesel by 2011–12 was formulated. The ethanol blending program mandated 5% blending of ethanol with gasoline from January 2003 in the nine major sugarcane producing states and Union Territories, [34]. In November 2006, all the states were included in the mandate except for some northeastern states and Jammu and Kashmir. However, the biofuel program was unsuccessful due to the non-availability of fuel, pricing formula, delayed procurement, and other reasons [35]. The government also started the ethanol blending program under which the oil marketing companies (OMCs) have to sell 10% ethanol blended petrol. The companies could procure 66.51 crore liter of petrol in the year 2016–17 as compared to 111 crore liter in the previous year due to a shortage of supply of ethanol [33].

With the aim of promoting the biofuels industry, a National Policy on Biofuels was adopted by India in December 2009 [36]. The main aim of the policy was to encourage the use of renewables by blending 20% of biofuels (biodiesel and ethanol) by the end of 2017. This policy was expected to stimulate rural development and generate rural employment. The policy also had the provision of minimum support price for the growers which will be periodically revised. However, the biofuel manufacturers were not permitted to directly sell the fuel in open market; instead, the fuel was to be procured by oil marketing companies. At oil marketing company's collection centers, the quality standards will be verified based on the Bureau of Indian Standards (BIS) specification for biofuels. National policy also laid down the price and minimum quality requirements of the biofuels.

The National Biodiesel Mission (NBM) report focused especially on Jatropha which can be cultivated in wastelands. The report reckoned that Jatropha can be cultivated on 13.4 million hectares of land. The goals of the mission were divided into two phases. In phase 1, the experimentation and demonstration had to be implemented by 2006–07 [37]. In phase 2, nursery development, seeds procurement, setup of plants for oil extraction, transesterification plants installation, and marketing of biodiesel had to be carried out [37, 38]. To achieve the targets for the second phase of the mission, the government also encouraged the state agencies, the private sector, and foreign institutions. The government is also offering subsidy and tax concessions to boost the feedstock production for biofuels. As a result, during April–November 2017, 43,551 kl of biodiesel was procured compared to 34,910 kl procured in the same period in the previous year, i.e. an increase of 24% [33].

The Government of India took a decision on biofuels in June 2015 which allowed private biodiesel manufacturers to sell pure biodiesel in open market, provided the fuel met the standards laid down by BIS. The price of the fuel was also determined by the policy. The government is also exploring new avenues for use of biodiesel such as railways and defense establishments [39].

The Indian Government approved the National Policy on Biofuels in 2018. The policy allowed production of ethanol from sugarcane juice, sugar-containing materials like sugar beet, sweet sorghum, and starch-containing materials like corn, cassava. The policy also included damaged food grains like wheat, broken rice, and rotten potatoes which are unfit for human consumption. During surplus production years, the price of the product may not be appropriate and to safeguard the farmer's interests, the policy allows the use of surplus for ethanol production. Setting up of a supply chain management system for collection of non-edible oilseeds, used cooking oil and short gestation crops are also encouraged by the policy. The expected benefits from this policy are savings in foreign currency reserves (especially US$), reduction in CO_2 emission, health benefits as cooking oil will not be used multiple times, employment generation, and extra income to farmers [40].

Ministry of New and Renewable Energy (MNRE) started a "Biomass power and Bagasse Co-generation" program whose aim is to recover energy from the biomass which includes bagasse, de-oiled cakes, agricultural residues, and wood from dedicated plantations for power generation. Till December 2017, 500 such projects with a capacity of 8414 MW have been installed. A biogas-based power generation (off-grid) program was also started by the ministry. Under this program, 409 biogas projects with power generation capacity of 7.04 MW have been set up till December 2017 [41].

4 Biofuel in India

Biomass energy is derived from two sources, one is the primary bioenergy which uses fertile lands and forests, and the other is biomass residue which is generated as a by-product. Agricultural residues, by-products, and organic wastes are considered as second-generation biofuels. They have characteristics similar to today's fuels, and they can be produced using the infrastructure of petroleum and sugar industries. The second-generation fuels can be integrated into first-generation biofuel markets and after grueling technological investigations, they will become the predominant fuel [42].

India has started the search for alternative feedstocks for ethanol production due to feedstock unavailability and economic unavailability of molasses-based ethanol [43, 44]. Sweet sorghum could be a potential source due to advantages such as its adaptability to saline and alkaline soils, drought, and water logging conditions. Also, it has four months growing period and it has low water requirements [45]. The second-generation biofuels are advantageously based on the total residues annually available, their ease of use and the surplus available for energy use. The following sections evaluate the potential availability of biomass residues for their cost, the production methods used for biofuel production, and the cost of production of biofuels.

4.1 Biomass Availability in India

Understanding the area under crops, their pattern and residues utilization are the important factors to assess the crop residues potential for biofuel production. 43% of India's geographic area is cropped, and the cropped area has been stable at 140 Mha since 1970. The area under which multiple crops are cultivated per year has increased from 132 Mha in 1950–51 to nearly 200.9 million for the year 2013–14. Two cropping seasons are possible in India, first when the southwest monsoon enters India (Kharif season) and second when the northeast monsoon enters India (Rabi season). The net irrigated area has increased from 21 Mha in 1950–51 to 68.2 Mha in 2013–14. Rice and wheat cultivation account for 41% of the cropped area, whereas pulses account for 13.8%, oilseeds account for 15.9%, and other commercial crops account for 10.2% of the cropped area.

Different crops produce a different quantity of residues. Straw, a dominant residue for cereals, is low in density. The ratio of straw to grain varies from 2.5 for maize to 1.6 for wheat. The by-product from milling of rice grains is called rice husk and it accounts for 20% of the paddy. Woody (ligneous) residues are produced from crops such as rapeseed and cotton. As shown in Table 1, the total crop residue produced during the year 2015–16 is estimated to be 816.4 Mt. Rice, wheat, sugarcane, and cotton cultivation produced the maximum quantity of residues which accounted for 75.8% of the total residue [42]. The specific ratios of residue to crop production shown in Table 1 are taken from [42, 46, 47].

Different regions have different use of crop residues and their use mainly depends on the lignin and energy content, density and nutritive value for livestock. Pulses and cereals residues have fodder value. Whereas residues of some food crops are woody in nature, hence, they are suitable for use as a fuel. The predominant use of crop residues is cattle fodder, fuel for cooking, and thatching material for houses. The non-woody leguminous residue is placed in composting pit to produce bio-fertilizers. Also, the residue leaves from the sugarcane are burnt in the field. Other uses are consumption of rice husk by the rice mills, cement plants, paper, and brick kilns [50].

The cattle population in India is very big and the grazing land available is also huge. However, the produce of grass on grazing land is low because of climate and degradation of land resulting in total dependence of fodder on crop residues. Table 2 shows that nearly 301 Mt of crop residue was utilized as fodder in 1996–97 and it is expected to increase by 20% every year. As seen in Table 2, cereal residues are

Table 1 Area under different crops and their respective residue production [48, 49]

Crop	Economic produce	Type of residue	Residue to crop ratio	Area (Mha) 2015–16	Total economic production (Mt) 2015–16	Total residue production (Mt) 2015–16
Rice	Foodgrains	Straw + husk	1.8	43.39	104.32	187.8
Wheat	Foodgrains	Straw	1.6	30.23	93.5	149.6
Jowar (sweet sorghum)	Foodgrains	Stalk	2.0	5.65	4.41	8.82
Bajra	Foodgrains	Straw	2.0	6.98	8.06	16.12
Maize	Foodgrains	Stalk + cobs	2.5	8.69	21.81	54.5
Other cereals	Foodgrains	Stalk	2.0	2.46	3.66	7.3
Gram	Foodgrains	Waste	1.6	8.35	7.17	11.5
Tur (ahar)	Foodgrains	Shell + waste	2.9	3.75	2.46	7.1
Lentil (masur)	Foodgrains	Shell + waste	2.9	1.47	1.04	3
Other Pulses	Foodgrains	Shell + waste	2.9	11.69	5.8	16.8
Groundnut	Oilseeds	Waste	2.3	4.56	6.77	15.57
Rapeseed and mustard	Oilseeds	Waste	2.0	5.76	6.82	13.6
Other oilseeds	Oilseeds	Waste	2.0	15.81	11.71	23.4
Cotton	Fiber	Seeds + waste	3.5	11.87	30.15	105.5
		Cotton gin trash	0.1			3
Jute and mesta	Fiber	Waste	1.6	0.78	10.47	16.75
Sugarcane	Sugar	Bagasse + leaves	0.5	4.95	352.16	176
Total				166.39	670.31	816.4

Table 2 Biofuel potential from the net availability of agricultural residues [51]

Crop residue	Agricultural residue used for fodder, fuel, and other purposes			Net agri-residue availability for biofuels (Mt)	Net ethanol availability (billion liters)
	Fodder	Fuel	Other		
Rice straw and husk	80.8	11.1	8.0	13.8	3.0
Wheat straw	86.4	0.0	13.6	0.0	0.0
Jowar stalk	100.0	0.0	0.0	0.0	0.0
Bajra straw	89.8	0.0	10.2	0.0	0.0
Maize stalk and cobs	81.0	19.0	0.0	7.4	1.6
Other cereals stalk	100.0	0.0	0.0	0.0	0.0
Gram waste	0.0	100.0	0.0	9.5	2.0
Tur shell and waste	3.5	48.5	48.0	2.9	0.5
Lentil shell and waste	3.5	48.5	48.0	1.0	0.2
Other pulses shell and waste	3.5	48.5	48.0	6.4	1.3
Groundnut waste	0.0	13.2	86.8	1.8	0.4
Rapeseed and mustard waste	0.0	100.0	0.0	11.8	2.5
Other oilseeds waste	0.0	100.0	0.0	23.1	4.9
Cotton seeds and waste	0.0	100.0	0.0	14.1	3.0
Cotton gin trash	0.0	100.0	0.0	0.3	0.1
Jute and mesta waste	0.0	100.0	0.0	2.2	0.5
Sugarcane bagasse and waste	11.8	41.0	47.2	35.9	7.7

mostly used as fodder and only ligneous residues can be used as an energy source. It is worth noting that if the entire cereals residues used as fodder is converted to ethanol then nearly 120 billion liters of bioethanol can be produced [51].

The residues left after utilization as fodder or fertilizer is used in biomass power and bagasse cogeneration. MNRE implemented the program to optimally utilize the biomass resources for grid power generation. The states and center have made policies which provide an encouraging environment for the program. Rajasthan announced an exclusive policy in 2010 for the program's promotion and it continues till date [52]. The policy provides a package of fiscal concessions, like accelerated depreciation, concessional customs duty, income tax exemption, and excise duty exemption on power generation plants for 10 years. As of December 2017, 8414 MW of biomass power and bagasse cogeneration power plants has been installed. Therefore, the net biofuel potential of agricultural residues can only be estimated after factoring in the power generation from biomass. Table 2 shows that nearly 20.7 billion liters of bioethanol can be produced which is less than the actual requirement of ethanol for 5% blending with gasoline.

4.2 Cost of Available Biomass

Use of agricultural residues as an energy source depends upon its availability, the characteristics of the produced fuel, the financial availability as compared to other options and lastly the accessibility of the bio-refinery. It is assumed that the agricultural residues are available free of cost to the user which is an invalid assumption. Since the producers may themselves use the agricultural residues so there is no need to determine its cost. However, when a user needs to use these residues, then the cost may be determined on the basis of: crop from which it is produced, its farm-handling technique and its transportation and handling at its place of utilization. Also, it is difficult to ascertain its market costs due to the absence of established markets.

The price of rice husk in Chhattisgarh is $74 per truck which makes it viable to transport it to a distance of 100 km, and the rice mill owners can still earn $9 per ton [53]. In Uttar Pradesh, the rice husk cost varies from $37 to $46 per ton whereas coal price varies from $83 to $92 per ton [54]. Sharma et al. [55] observed the price (pan India) of rice husk varied from US$18 to US$74 per ton whereas rice straw price varied from US$11 to US$13 per ton. This is due to the various use of straw ranging from fodder for cattle to strawboard, construction material, and packaging material for glass wares, paper, and hardboard units [53].

4.3 Methods of Biofuel Production

Biomass can be converted into liquid biofuels either by thermochemical processing or by biochemical processing. In thermochemical processing, the biomass is converted into a range of products by thermal decay and chemical reformation. The main advantage of the thermochemical process over the biochemical process is the complete conversion of the biomass's organic components whereas the latter focuses only on polysaccharides [56]. In this section, different types of thermochemical conversion processes are discussed.

Agricultural and forest residues and waste biomass from industries cannot be easily converted to biofuels due to their uneven bulky and troublesome characteristics. Therefore, they are densified into compact regular shapes called briquettes. Briquetting is performed by compressing and squeezing out the moisture and breaking the elasticity of the residues. If the elasticity is not removed, then the residues may regain their pre-compression volume. The densification is performed in a compression chamber at high temperature and pressure. The natural lignin present can easily bind to form a solid piece [57].

Different processes for converting biomass into energy are shown in Fig. 3. Some of them are direct combustion, gasification, liquefaction, and pyrolysis.

In indirect combustion, the fuel is combusted in the presence of oxygen (present in air) to form carbon dioxide, water vapor along with heat release. The heat can be used in domestic stoves or in industrial boilers. They are a good substitute for fossil fuels as they are typically free of sulphur and particulate formation.

Biomass gasification to produce fuel has been under investigation for the past thirty years. The biomass is gasified using oxygen, steam, or air to produce carbon monoxide, carbon dioxide, hydrogen, methane, and nitrogen mixture which is either called synthesis gas (syngas) or producer gas, based on the relative proportions of the gases produced [58]. Stationary power generation can be carried using producer gas whereas a range of fuels and chemical intermediates can be made using syngas. For producing fuels used in transportation, syngas using the water gas shift reaction can be converted to hydrogen [59], by using Fischer–Tropsch synthesis or methanol synthesis to hydrocarbons [60]. Syngas can be produced from biomass either in the presence of a catalyst or without a catalyst. Non-catalytic gasification requires operation temperature as high as 1300 °C, whereas a lower temperature is required for catalytic gasification [61].

Liquefaction of the biomass to biofuel is carried out in water at a moderate temperature (280–370 °C) and high pressure (10–25 MPa). The process produces liquid bio-crude along with other by-products [62]. The process produces high-energy-content product along with enhanced heat recovery. The process can be carried out for both lignocellulose (dry biomass) and algal (wet biomass) biomass. The liquefaction is done by decomposing and depolymerizing the biomass [63]. The compounds formed are highly reactive, and they can be easily polymerized into bio-crude, gas, and solid components under suitable conditions such as temperature and residence time [64].

Biomass can also be thermally degraded in the absence of oxygen resulting in the formation of char, oil, and gas. The process is called pyrolysis. The process, based on operating conditions, can be of three types namely conventional, fast, and flash pyrolysis [65]. Conventional pyrolysis is done for massive wood pieces with

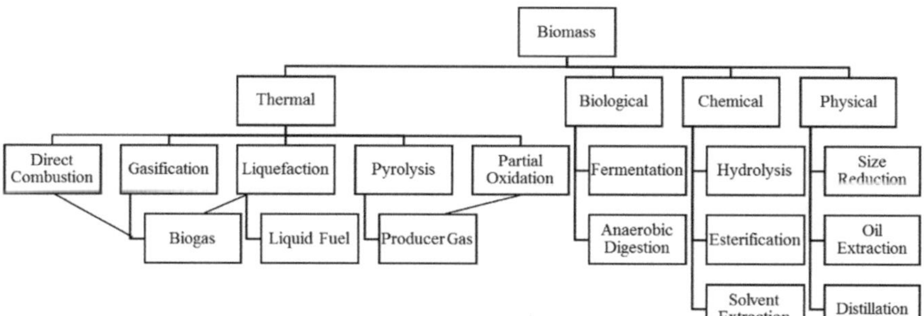

Fig. 3 Conversion processes [14]

a heating rate of 0.1–1 K/s and 45–550 s of residence time. In the first stage of decomposition of biomass, internal rearrangement takes place. In the second stage, the decomposition occurs at a high rate and pyrolysis products are formed. In the final stage, the slow decomposition of charge takes place and residue rich in carbon is formed. Fast pyrolysis is used for fine particles at high temperatures (850–1250 K), with a heating rate of 10–200 K/s and residence time of 0.5–10 s. The biomass decomposes to form vapor, aerosols, and char. The condensation of vapors and aerosol results in the formation of a dark brown liquid which has a heating value half that of conventional fuel [66]. For very fine particles, flash pyrolysis is used. The process can be done in the temperature range of 1050–1300 K, with heating rates greater than 1000 K/s and residence time less than 0.5 s. The process can achieve up to 70% efficiency for crude oil production from biomass. The crude oil produced can be directly used in turbines and engines, and they can also be used as a feedstock for refineries [67].

Hydrotreating the vegetable oil in the presence of a catalyst is a relatively new technology. The petroleum industry uses this technology to remove unwanted atoms as well as for cracking and saturation of olefins and aromatics. The process can also be used to produce diesel-like fuel from vegetable oil with better properties in comparison with biodiesel. The process consists of catalytic saturation, deoxygenation, decarboxylation, and isomerization. By changing the process parameters, any kind of feedstock can be used to produce a sulphur-free fuel with high cetane index. The fuel has good storage stability and it can easily be blended with conventional fuel. The fuel can be easily transported using the existing infrastructure [68].

Supercritical fluid processing of biomass is an alternate method for converting biomass. It can easily and quickly convert biomass into a mixture of oils, organic acids, alcohol, and methane. In a supercritical state (300–644 K, 200–250 bar) and near critical state (523–573 K), water is converted into acid (H^+) and basic components (OH^-) which dissolve in the biomass. These dissolved components break the biomass rapidly into small sugar molecules, glucose, xylose, and oligosaccharide [69]. Thus, the supercritical water processing is a potential method for converting biomass into value-added products without using any catalyst.

5 Algal Biofuel

If the entire diesel used in transport is to be replaced with biodiesel, then the annual requirement of biodiesel is nearly 200 billion gallons. If the entire requirement of biodiesel is to be fulfilled by jatropha oil, then jatropha needs to be grown in an area of 384 Mha, which is more than the geographical area of India [70]. Thus, it is obvious that neither edible nor non-edible crops can entirely replace fossil fuels in the near future. Microalgae, however, requires much smaller area to produce the same amount of diesel. For example, in a well-designed production system, annual microalgal biomass production can be around 1.5 kg/m^3/day [71]. If the oil content in the dry algae biomass is considered to be 30%, then the yield of oil can be around

123 m^3 per hectare of the total land area for most of the year, resulting in 98.4 m^3/ha of biodiesel yield. At this rate, to produce 200 billion gallons of fuel, the microalgal needs to be grown in only 5.4 Mha land which is nearly 2% of the Indian geographical area. The scenario is feasible even if 15% of oil is considered in the dry weight of the biomass.

Algae both in micro- or macro-form can produce large quantities of biomass and oil which can then be converted to biodiesel and other fuels. Algae can easily grow in extreme conditions. They can grow at high cell densities up to 10^7 cells/ml [72] in nutrient-rich waters. They can strip the common pollutants of the polluted water by absorbing a large amount of polluting plant nutrients. Some algae variants grow well in a variety of salt concentrations [73, 74]. They can grow well even in wastewaters. In India, it is well known that microalgae clean the domestic wastewaters in oxidation ponds and ditches; however, the function and growth of algae are poorly controlled and the biomass is not even economically utilized. Since India has a long sea line, macro-algae (seaweed) grown in coastal areas can also be used for biodiesel production.

Photoautotrophic algae can be cultivated either in raceway pond systems or in photobioreactors. Raceway ponds are oval in shape, 0.25–0.4 m deep, open to air and fitted with a paddle wheel to avoid sedimentation and proper circulation of water. As the penetration of light is low due to optical adsorption and self-shading of the algae, the pond's depth is kept low. In photobioreactors, the culture is placed in a transparent array of plates or tubes, and they are circulated from a central reservoir. Thus, allowing better control of the environment but they are expensive and their demand for auxiliary power is also high [75].

In developing countries like India, the wastewater generation is very high and the cost to treat them is very expensive [76]. The municipal wastewater of the large Indian metros has a concentration of nitrogen and phosphorus ranging from 30–100 ppm to 10–45 ppm, respectively, [77–79], and agricultural wastewaters from farms have nearly 1000 ppm [80–82]. Today only small parts of agricultural wastewater are recycled for farm use and rest of it is discharged, example coffee pulping wastewater [83]. Thus, a large amount of wastewater is available with the required nutrients for the algae to grow. This solves two problems simultaneously, first low-cost wastewater treatment [84–86] and second biofuel generation [87]. This way the nutrients are available at low cost for the algae and quite often the wastewater generator will pay for wastewater treatment. The nutrients, in this case, will be available at a negative price, at least for the present [88].

The feedstock available from algae can be used directly or processed further to form a gas or liquid biofuels by various biochemical and thermochemical processes [89, 90]. The algae can be dried and directly burnt for recovery of energy [91]. Various thermochemical methods previously described can also be used to convert algal biomass to oil or gaseous biofuels [92–94]. In the biochemical conversion process, the algal biomass is fermented and then anaerobic digestion is done to produce bioethanol/biobutanol, methane, and biohydrogen [92, 95]. Biodiesel can be produced from lipids extracted from the algae by transesterification process [71,

96]. Biodiesel produced from algae has higher oil production potential in comparison with oilseeds [71, 97].

6 Issues with Biofuels

In the early 2000s, it looked as if biofuels were the best solution to meet the energy demands. However, later it was realized that due to their own problems they cannot completely replace the conventional fuels. This section briefly describes the issues that are being faced with biofuels.

6.1 Cost of Biofuel Production

Bioethanol is used by many countries as an alternative fuel by blending it with gasoline as it is compatible with gasoline and has high octane value. The blends of gasoline and ethanol help in reduction of greenhouse gases as the number of carbon atoms is less; however, the energy content of ethanol is low which results in higher consumption of fuel to produce the same power [98]. Moreover, ethanol is still produced by a fermentation process which includes pretreatment costs, high-charge enzymes to get better yields which make the production costs of ethanol higher than gasoline [99]. In India, the ethanol production depends upon the availability of molasses whose prices fluctuate every year. Raju et al. [100] reported the price of molasses ranging from $18 to $37 per ton. This can have serious implications on the cost of conversion of molasses to ethanol. The Indian Government raised the prices of ethanol procurement by oil marketing companies from $0.47 (ex-mill) to $0.61 (ex-mill). This resulted in the procurement of 83 crore liters of ethanol per annum as compared to 32 crore liters per annum in the previous year [101]. In Brazil, the cost of ethanol has dropped to $0.18 per liter from new plants of ethanol production [102]. In the USA and EU, the grain-based ethanol costs 50% and 100% more as compared to sugarcane-based ethanol in Brazil. Transportation, blending, and distribution costs add nearly $0.20 per liter in the retail price [5].

The production cost of biodiesel from algae is higher than conventional fuels and other biomass-derived fuels. All the stages of production require high-energy inputs which increase the production costs. A comparison between microalgae and other biomass-derived fuels is shown in Table 3.

Table 3 Comparison of microalgae with other biomass-derived fuels for biodiesel production [103]

Raw material	Oil content (% in dry weight biomass)	Output (L oil/ha year)	Land used (m^3 year/kg biodiesel)	Water footprint (m^3/ton)	Production cost (US$/L)	Acid value of oil	Biodiesel yield (%)
Soybean	18	636	18	4200	0.40–0.66	0.2	90
Rapeseed	41	974	12	4300	0.99	2.0	87
Sunflower	40	1070	11	6800	0.62	0.1	90
Palm	36	5366	2	5000	0.68	6.1	95
Castor	48	1307	9	24,700	0.92–1.56	4.6	89
Microalgae	50	97,800	0.1	591–3276	3.96–10.56	8.9	60

6.2 *Limitations to the Use of Biofuels in Internal Combustion Engines*

Today, bioethanol and biodiesel are the major biofuels used in internal combustion (IC) engines. Ethanol is blended with gasoline for use in gasoline-fueled engines. 20% blending of ethanol in gasoline gives higher engine brake power, torque, volumetric and brake thermal efficiencies and reduction in carbon monoxide and unburned hydrocarbon emissions whereas the fuel consumption increases [104]. However, the engine exhaust emissions contain toxic aldehydes namely formaldehyde, acetaldehyde, etc. [105]. 100% of ethanol cannot be used in unmodified gasoline engines because of the corrosion of the components used in the fuel injection system. Also, the valve seat should be made of resistant material due to the low lubricating effect of ethanol [106].

To run the compression ignition engine on biodiesel, its properties must be comparable to diesel. Some of the important properties are density, viscosity, calorific value, flash point, moisture content, acid value, oxidation stability, cetane number, and cold flow properties [107, 108]. Since biodiesel produced from different feedstock have different properties, it is difficult to maintain a standard such that the engine in which the fuel is used sustains its efficiency and emissions as per the regulations. Therefore, countries have set up their own standards which the biodiesel manufacturer has to uphold.

One of the major issues with biodiesel is its oxidation stability due to the presence of a mixture of unsaturated and saturated fatty acid esters resulting in degradation and formation of undesired by-products [109]. The important properties of biodiesel produced by using non-edible feedstock are shown in Table 4. The other problems are higher oxides of nitrogen emission [110], some aldehydes and hazardous aromatic compounds which are unregulated [105]. The increase in oxides of nitrogen emission is due to higher combustion temperature which also helps in a decrease of carbon monoxide and unburned hydrocarbon emissions [110].

Table 4 Properties of biodiesel as per ASTM standards [111]

Properties	Jatropha curcas	Pongamia pinnata	Madhucaindica	Neem	Rice barn	Rubber seed	Polanga seed
Density @ 15 °C (g/cm^3)	0.8642	0.8690	0.875	0.8850	–	–	0.878
Viscosity @ 40 °C (mm^2/s)	4.3	4.65	5.10	5.21	3.5	3.89	5.5
Oxidation stability (h, @ 110 °C)	2.8	2.5	–	–	1.7	–	–
CFPP (°C)	–	−7	6	11	0	0	11
Pour point (°C)	3	−1	4	14.4	−10	3.2	12
Cloud point (°C)	3	−6	–	2	−11	−2	13
Flash point (°C)	169	180	159	–	169	152	163
Calorific value (kJ/kg)	39,698	38,960	36,914	37,810	38,853	39,700	39,513

7 Conclusions

India is now the sixth largest economy of the world and the industrial along with the transportation sector has seen a phenomenal growth in the last two decades. The energy consumption has increased multi-fold and Government of India is pushing renewable options for a sustainable energy mix with minimal carbon footprints. Biofuels are an emerging option for India, and major initiatives are taken by government for their commercialization and in this context, National Biofuel Policy is adapted to harness environmental, societal, and economic benefits. The program can be successful only when the stakeholders engage with the government to tackle the various issues arising during the program implementation.

The bioethanol program cannot be sustained only by using sugarcane molasses since the costs involved are very high and a sufficient amount of ethanol cannot be produced. Also, the biodiesel program is faced with obstacles like slow progress in the planting of trees and problems associated with processing and marketing. A thrust in research is required for the development of the next feedstock which can address the existing and future energy requirements of the country.

The land is intensively used in India for agriculture, and the exploitation of water is beyond the sustainable limits. Rising population and increase in spending power will raise the food demand further. Thus, the biofuel production will depend upon feedstock derived from agricultural and forest residues or rain-fed biomass produced on the non-food land.

Presently, technology for large-scale conversion of crop residue to fuel is in various stages of development in India. Also, the crop residue is available in abundance at affordable prices; however, its conversion cost is high. The infrastructure required for collecting, transporting, and handling of biomass is in under developing stage. The ramping up of the infrastructure will provide an impetus to the use of biofuels in India, and the country will enter into next generation of biofuels production after reducing or eliminating the technical and costs barriers.

References

1. IEA (2017) Key energy world statistics. https://doi.org/10.1017/cbo9781107415324.004
2. Monari C, Righi S, Olsen S (2016) Greenhouse gas emissions and energy balance of biodiesel production from microalgae cultivated in photobioreactors in Denmark: a life-cycle modeling. J Clean Prod 112:4084–4092
3. Azad AK, Rasul MG, Khan MMK, Sharma SC, Hazrat MA (2015) Prospect of biofuels as an alternative transport fuel in Australia. Renew Sustain Energy Rev 43:331–351. https://doi.org/10.1016/j.rser.2014.11.047
4. Gaurav N, Sivasankari S, Kiran GS, Ninawe A, Selvin J (2017) Utilization of bioresources for sustainable biofuels: a Review. Renew Sustain Energy Rev 73:205–214. https://doi.org/10.1016/j.rser.2017.01.070
5. Oumer AN, Hasan MM, Baheta AT, Mamat R, Abdullah AA (2018) Bio-based liquid fuels as a source of renewable energy: a review. Renew Sustain Energy Rev 88:82–98. https://doi.org/10.1016/j.rser.2018.02.022
6. Surriya O, Saleem SS, Waqar K, Kazi AG, Öztürk M (2015) Bio-fuels: a blessing in disguise. In: Phytoremediat Green Energy. Springer, Berlin, pp 11–54
7. Nigam PS, Singh A (2011) Production of liquid biofuels from renewable resources. Prog Energy Combust Sci 37:52–68
8. Razzak SA, Hossain MM, Lucky RA, Bassi AS, de Lasa H (2013) Integrated CO_2 capture, wastewater treatment and biofuel production by microalgae culturing—a review. Renew Sustain Energy Rev 27:622–653
9. Rosegrant MW, Zhu T, Msangi S, Sulser TG (2008) Global scenarios for biofuels: impacts and implications. Appl Econ Perspect Policy 30:495–505
10. Msangi S, Sulser T, Rosegrant M, Valmonte-Santos R, Ringler CG (2006) Global scenarios for biofuels: impacts and implications. Int Food Policy Research Inst, 1–20
11. Antizar-Ladislao BTJ, Turrion-Gomez T (2008) Second-generation biofuels and local bioenergy systems. Biofuels Bioprod Biorefining 2:455–469
12. Cherubini F, Bird ND, Cowie A, Jungmeier G, Schlamadinger BW-GS (2009) Energy- and greenhouse gas-based LCA of biofuel and bioenergy systems: key issues, ranges and recommendations. Resour Conserv Recycl 53:434–437
13. Demirbas M (2009) Biorefineries for biofuel upgrading: a critical review. Appl Energy 86:S151–S161
14. Naik SN, Goud VV, Rout PK, Dalai AK (2010) Production of first and second generation biofuels: a comprehensive review. Renew Sustain Energy Rev 14:578–597. https://doi.org/10.1016/j.rser.2009.10.003
15. Lee RALJ-M, Lavoie J-M (2013) From first-to third-generation biofuels: challenges of producing a commodity from a biomass of increasing complexity. Anim Front 3:6–11

16. OECD/IEA (2008) From 1st to 2nd generations of biofuel technologies, overview of current industry and RD&D activities
17. Sims REH, Mabee W, Saddler JN, Taylor M (2010) An overview of second generation biofuel technologies. Bioresour Technol 101:1570–1580. https://doi.org/10.1016/j.biortech.2009.11.046
18. Ziolkowska JR, Simon L (2014) Recent developments and prospects for algae-based fuels in the US. Renew Sustain Energy Rev 29:847–853
19. Dutta K, Daverey A, Lin JG (2014) Evolution retrospective for alternative fuels: first to fourth generation. Renew Energy 69:114–122. https://doi.org/10.1016/j.renene.2014.02.044
20. Lu J, Sheahan C, Fu P (2011) Metabolic engineering of algae for fourth generation biofuels production. Energy Environ Sci 4:2451–2466
21. Mohan MR, Phillippe GT, Shiju M (2006) Biofuel laws in Asia: instruments for energy access, security, environmental protection and rural empowerment. Asian Biotechnol Dev Rev 8:51–75
22. Hat D (2009) Directive 2009/139/EC of the European Parliament and of the Council
23. Su Y, Zhang P, Su Y (2015) An overview of biofuels policies and industrialization in the major biofuel producing countries. Renew Sustain Energy Rev 50:991–1003
24. Bastos Lima MG, Gupta J (2014) The extraterritorial dimensions of biofuel policies and the politics of scale: live and let die? Third World Quaterly 35:392–410
25. Pelkmans L, Portouli E, Papageorgiou A Georgopoulos P (2006) Impact assessment of measures towards the introduction of biofuels in the European Union
26. Sorda G, Banse M, Kemfert C (2010) An overview of biofuel policies across the world. Energy Policy 38:6977–6988
27. Yacobucci BD, Bracmort K (2010) Calculation of lifecycle greenhouse gas emissions for the renewable fuel standard (RFS). R40460
28. Kumar S, Shrestha P, Salam PA (2013) A review of biofuel policies in the major biofuel producing countries of ASEAN: production, targets, policy drivers and impacts. Renew Sustain Energy Rev 26:822–836
29. Kumar S, Bhattacharyya B, Gupta VK (2014) Present and future energy scenario in India. J Inst Eng Ser B 95:247–254. https://doi.org/10.1007/s40031-014-0099-7
30. IEA (2015) India energy outlook
31. Central Electricity Authority (2018) Government of India ministry of power central electricity authority New Delhi executive summary power sector
32. Simi Thambi (2017) Highlights for India from the latest World Energy Outlook. http://indiaenergy.gov.in/highlights-for-india-from-the-latest-world-energy-outlook/. Accessed 30 May 2018
33. Petroleum and Natural Gas Ministry (2018) Annual report 2017–18
34. Aradhey A (2010) India-biofuels annual
35. Ray S, Miglani S, Goldar A (2011) Ethanol blending policy in India: demand and supply issues
36. MNRE (2009) National policy on biofuels. New Delhi
37. Biswas PK, Pohit S, Kumar R (2010) Biodiesel from jatropha: can India meet the 20% blending target? Energy Policy 38:1477–1484
38. Bandyopadhyay KR, Das K (2014) Biofuels in South Asia: connecting the dots. Renew Energy Law Policy Rev 5:145–161
39. GoI (2015) India's intended nationally determined contribution: working towards climate justice
40. GoI P (2018) Cabinet approves national policy on biofuels - 2018. http://pib.nic.in/newsite/PrintRelease.aspx?relid=179313. Accessed 10 Jun 2018
41. Ministry of New and Renewable Energy (2018) Annual report 2017–18
42. Ravindranath NH, Somashekar HI, Nagaraja MS, Sudha P, Sangeetha G, Bhattacharya SC, Abdul Salam P (2005) Assessment of sustainable non-plantation biomass resources potential for energy in India. Biomass Bioenerg 29:178–190. https://doi.org/10.1016/j.biombioe.2005.03.005

43. Findlater KM, Kandlikar M (2011) Land use and second-generation biofuel feedstocks: the unconsidered impacts of Jatropha biodiesel in Rajasthan, India. Energy Policy 39:3404–3413
44. Sasmal S, Goud VV, Mohanty K (2012) Characterization of biomasses available in the region of NorthEast India for production of biofuels. Biomass Bioenerg 45:212–220
45. Basavaraj G, Parthasarathy Rao P, Basu K, Reddy CR, Kumar AA, Srinivasa Rao P, Reddy BVS (2013) Assessing viability of bio-ethanol production from sweet sorghum in India. Energy Policy 56:501–508. https://doi.org/10.1016/j.enpol.2013.01.012
46. Kumar A, Purohit P, Rana S, Kandpal TC (2002) An approach to the estimation of value of agricultural residues used as biofuels. Biomass Bioenerg 22:195–203
47. Purohit P, Michaelowa A (2007) CDM potential of bagasse cogeneration in India. Energy Policy 35:4779–4798
48. Ministry of Agriculture and Farmers Welfare (2017) Agricultural Statistics at a Glance 2016
49. Purohit P, Fischer GDI (2014) Second-generation biofuel potential in India: sustainability and cost considerations
50. Ghosh SK (2016) Potential of economic utilization of biomass waste in India—implications towards SDGs
51. Purohit P, Dhar S (2015) Biofuel roadmap for India
52. GoR (2010) Policy for promoting generation of electricity from biomass. Rajasthan, India
53. Pandey R, Sar SK, Bhui AK (2012) Feasibility of installing rice husk power plant in Chhattisgarh to meet sustainable energy demands. Int J Adv Eng Res Stud 1:57–60
54. Yadav JP, Singh BR (2011) Study on future prospects of power generation by bagasse, rice husk and municipal waste in Uttar Pradesh. SAMRIDDHI-A J Phys Sci 2:63–75
55. Sharma SP (2010) Mini-thermal power projects turning unviable in Malwa Raw material, agriculture waste, prohibitively expensive. Trib News Serv
56. Gomez LD, Clare GS, McQueen-Mason J (2008) Sustainable liquid biofuels from biomass: the writing's on the walls. New Phytol 178:473–485
57. Khan BH (2017) Non-conventional energy resource, 3rd edn. Tata McGraw-Hill Education
58. Rowlands WN, Masters A, Maschmeyer T (2008) The biorefinery-challenges, opportunities, and an Australian perspective. Bull Sci Technol Soc 28:149–158
59. Balat M (2006) Sustainable transportation fuels from biomass materials. Energy, Educ Sci Technol 17:83–103
60. Steen EVCM, Claeys M (2008) Fischer-Tropsch catalysts for the biomass-to-liquid (BTL) process. Chem Eng Technol 31:650–660
61. Lee S, Speight JG, Loyalka S (2007) Hand book of alternative fuel technologies. CRC Taylor and Francis Group, USA
62. Toor S (2010) Modeling and optimization of CatLiq biofuel process. Ph.D. thesis, Aalborg University
63. Sohail S, Rosendahl L, Rudolf A (2011) Hydrothermal liquefaction of biomass: a review of subcritical water technologies. Energy 36:2328–2342
64. Gollakota ARK, Kishore N, Gu S (2018) A review on hydrothermal liquefaction of biomass. Renew Sustain Energy Rev 81:1378–1392. https://doi.org/10.1016/j.rser.2017.05.178
65. Demirbas A (2004) Current technologies for thermo-conversion of biomass into fuels and chemicals. Energy Sources 26:715–730
66. Shafizadeh F (1982) Introduction to pyrolysis of biomass. J Anal Appl Pyrolysis 3:283–305
67. Mohan D, Pitman CU, Steele P (2006) Pyrolysis of wood/biomass for bio-oil: a critical review. Energy Fuels 20:848–889
68. Sonthalia A, Kumar N (2017) Hydroprocessed vegetable oil as a fuel for transportation sector: a review. J Energy Inst. https://doi.org/10.1016/j.joei.2017.10.008
69. Sasaki M, Kabyemela B, Malaluan R, Hirose S, Takeda N, Adschiri T, Arai K (1998) Cellulose hydrolysis in subcritical and supercritical water. J Supercrit Fluid 13:261–268
70. Khan SAR (2008) Algae: a novel source of renewable energy and carbon sequestration. Renew Energy (Akshay Urja) 2:14
71. Chisti Y (2007) Biodiesel from microalgae. Biotechnol Adv 25:294–306

72. Chow CWK, Drikas M, House J, Burch MD, Velzeboer R (1999) The impact of conventional water treatment processes on cells of the cyanobacterium Microcystis aeruginosa. Water Res 33:3253–3262
73. Rodolfi L, Zittelli GC, Bassi N, Padovani G, Biondi N, Bonini G, Tredici M (2009) Microalgae for oil: strain selection, induction of lipid synthesis and outdoor mass cultivation in a low-cost photobioreactor. Biotechnol Bioeng 102:100–112
74. Takagi M, Karseno YT (2006) Effect of salt concentration on intracellular accumulation of lipids and triacylglyceride in Marine Microalgae Dunaliella cells. J Biosci Bioeng 101:223–226
75. Slade R, Bauen A (2013) Micro-algae cultivation for biofuels: cost, energy balance, environmental impacts and future prospects. Biomass Bioenerg 53:29–38. https://doi.org/10.1016/j.biombioe.2012.12.019
76. Gasperi J, Garnaud S, Rocher V, Moilleron R (2008) Priority pollutants in wastewater and combined sewer overflow. Sci Total Environ 407:263–272
77. Chanakya HN, Karthick B, Ramachandra T (2006) Nitrogen and carbon flows through Bellandur Lake—role of Bellandur lake as a natural wetland treating Bangalore wastewater. In: Lake 2006-National Seminar on Wetlands and Solid Waste management. IISC Bangalore
78. Chanakya HN, Sharatchandra H (2008) Nitrogen pool, flows, impact and sustainability issues of human waste management in the city of Bangalore. Curr Sci 94:1447–1454
79. Mahapatra DM, Chanakya HN, Ramachandra T (2011) Assessment of treatment capabilities of Varthur Lake, Bangalore. Int J Environ Technol Manag 14:84–102
80. de la Noue J, Laliberte GPD (1999) Algae and waste water. J Appl Phycol 4:247–254
81. Gonzalez LE, Canizares RO, Baena S (1997) Efficiency of ammonia and phosphorus removal from a Colombian agroindustrial wastewater by the microalgae Chlorella vulgaris and Scenedesmus dimorphus. Bioresour Technol 60:259–262
82. Wilkie AC, Mulbry W (2002) Recovery of dairy manure nutrients by benthic freshwater algae. Bioresour Technol 84:81–91
83. Chanakya HN, de Alwin A (2004) Environmental issues and management in primary coffee processing. Process Saf Environ Prot 82:1–10
84. Hoffmann JP (1998) Wastewater treatment with suspended and nonsuspended algae. J Phycol 34:757–763
85. Khan SA, Rashmi Hussain MZ, Prasad S, Banerjee UC (2009) Prospects of biodiesel production from microalgae in India. Renew Sustain Energy Rev 13:2361–2372. https://doi.org/10.1016/j.rser.2009.04.005
86. Chinnasamy S, Bhatnagar A, Hunt RW, Das K (2010) Microalgae cultivation in a wastewater dominated by carpet mill effluents for biofuel applications. Bioresour Technol 101:3097–3105
87. Benemann JR, Weissman JC, Koopman BL, Oswald W (1977) Energy production by microbial photosynthesis. Nature 268:19–23
88. Chanakya HN, Mahapatra DM, Sarada R, Abitha R (2013) Algal biofuel production and mitigation potential in India. Mitig Adapt Strateg Glob Chang 18:113–136. https://doi.org/10.1007/s11027-012-9389-z
89. Rittmann B (2008) Opportunities for renewable bioenergy using microorganisms. Biotechnol Bioeng 100:203–212
90. Amin S (2009) Review on biofuel oil and gas production processes from microalgae. Energy Convers Manag 50:1834–1840
91. Kadam K (2002) Environmental implications of power generation via coal-microalgae cofiring. Energy 27:905–922
92. McKendry P (2002) Energy production from biomass (part 1): overview of biomass. Bioresour Technol 83:37–46
93. McKendry P (2002) Energy production from biomass (part 2): conversion technologies. Bioresour Technol 83:47–54
94. Miao X, Wu Q (2004) High yield bio-oil production from fast pyrolysis by metabolic controlling of Chlorella protothecoides. J Biotechnol 110:85–93

95. Melis A (2002) Green alga hydrogen production: progress, challenges and prospects. Int J Hydrogen Energy 27:1217–1228
96. Miao X, Wu Q (2006) Biodiesel production from heterotrophic microalgal oil. Bioresour Technol 97:841–846
97. Sheehan J, Dunahay T, Benemann J, Roessler P (1998) A look back at the U.S. department of energy's aquatic species program: biodiesel from Algae
98. Masum BM, Masjuki HH, Kalam MA, Palash SM, Habibullah M (2015) Effect of alcohol—gasoline blends optimization on fuel properties, performance and emissions of a SI engine. Renew Sustain Energy Rev 86:230–237
99. Moka S, Pande M, Rani M, Gakhar R, Sharma M, Rani JBA (2014) Alternative fuels: an overview of current trends and scope for future. Renew Sustain Energy Rev 32:697–712
100. Raju SS, Parappurathu S, Chand R, Joshi PK, Kumar P, Msangi S (2012) Biofuels in India: potential, policy and emerging paradigms. In: National Centre for Agricultural Economics and Policy Research (NCAP), New Delhi, India
101. Sharma VP, Pal S, Sharma S (2016) Price policy for sugarcane
102. Institute W (2007) Biofuels for transport: global potential and implications for sustainable energy and agriculture. Earhscan, London
103. González-Delgado Á-D, Kafarov V (2011) Microalgae based biorefineries: issues to consider. Ciencia Technol y Futur 4:5–22
104. Al-Hasan M (2003) Effect of ethanol-unleaded gasoline blends on engine performance and exhaust emission. Energy Convers Manag 44:1547–1561. https://doi.org/10.1016/S0196-8904(02)00166-8
105. Dhillon R, von Wuehlisch G (2013) Mitigation of global warming through renewable biomass. Biomass and Bioenergy 48:75–89
106. De Simio L, Gambino M, Iannaccone S (2012) Effect of ethanol content on thermal efficiency of a spark-ignition light-duty engine. ISRN Renew Energy 2012:1–8. https://doi.org/10.5402/2012/219703
107. Serrano M, Martinez M, Aracil J (2013) Long term storage stability of biodiesel: influence of feedstock, commercial additives and purification step. Fuel Process Technol 116:135–141
108. Yakoob Z, Narayanan BN, Padikkaparambil S, Unii SK, Akbar M (2014) A review on the oxidation stability of biodiesel. Renew Sustain Energy Rev 35:136–153
109. Joshi G, Lamba BY, Rawat DS, Mallick S, Murthy K (2013) Evaluation of additive effects on oxidation stability of Jatropha curcas biodiesel blends with conventional diesel sold at retail outlets. Ind Eng Chem Res 52:7586–9592
110. Varuvel EG, Govindan N, Nagalingam B (2011) Experimental study on the performance, emission and combustion characteristics of rubber seed oil-diesel blends in a DI diesel engine. Int J Renew energy Technol 2:306–323
111. Joshi G, Pandey JK, Rana S, Rawat DS (2017) Challenges and opportunities for the application of biofuel. Renew Sustain Energy Rev 79:850–866. https://doi.org/10.1016/j.rser.2017.05.185

Part III
Flames and Reacting Systems

Laminar Burning Speed Study of Alternative Fuel Air Diluent Mixtures at High Pressures and Temperatures

Ziyu Wang, Guangying Yu and Hameed Metghalchi

Keywords Alternative fuel · Diluent · Laminar burning speed · High temperature and pressure

Nomenclature

A_b	Burned gas area
a	Fitted constant
b	Fitted constant
C	Hydrogen percentage
D	Diluent concentration
E_b	Burned gas energy
E_i	Initial energy
E_u	Unburned gas energy
E_{eb}	Electrodes boundary energy defect
E_{ph}	Preheat zone energy defect
E_{wb}	Wall boundary energy defect
e_b	Burned gas specific energy
e_u	Unburned gas specific energy
e_{bs}	Isentropically compressed burned gas specific energy
e_{us}	Isentropically compressed unburned gas specific energy
I	Radiation intensity
j	Number of narrow bands

Z. Wang (✉) · G. Yu · H. Metghalchi
Mechanical and Industrial Engineering Department, Northeastern University, Boston, MA 02115, USA
e-mail: wang.ziyu2@huksy.neu.edu

A. K. Gupta et al. (eds.), *Innovations in Sustainable Energy and Cleaner Environment*, Green Energy and Technology,
https://doi.org/10.1007/978-981-13-9012-8_9

k_n	K-distribution absorption coefficient
M	Number of directions
m	Total mass
m_b	Burned gas mass
m_u	Unburned gas mass
$\dot{m}_b$	Mass burning rate
N	Number of Gaussian quadrature points
P	Mixture pressure
P_0	Reference pressure
p	Pressure
p_i	Initial pressure
Q_e	Energy loss to electrodes
Q_r	Radiation energy loss
Q_w	Energy loss to wall
R	Specific gas constant
r	Flame radius
r_e	Electrode radius
S_u	Laminar burning speed
S_{uo}	Reference laminar burning speed
T	Temperature
T_b	Burned gas temperature
T_i	Initial temperature
T_u	Unburned gas temperature
T_{u0}	Reference temperature
t	Time
V	Energy source volume
V_b	Burned gas volume
V_c	Chamber volume
V_e	Electrodes volume
V_i	Initial volume
V_u	Unburned gas volume
V_{eb}	Electrodes boundary displacement volume
V_{ph}	Preheat zone displacement volume
V_{wb}	Wall boundary displacement volume
v_b	Burned gas specific volume
v_u	Unburned gas specific volume
v_{bs}	Isentropically compressed burned gas specific volume
v_{us}	Isentropically compressed unburned gas specific volume
x_b	Burned gas mass fraction
$\dot{x}_b$	Rate of burned gas mass fraction
α	Fitted constant
α_0	Fitted constant
α_1	Fitted constant
α_2	Fitted constant
α_3	Fitted constant

β Fitted constant
β_0 Fitted constant
β_1 Fitted constant
β_2 Fitted constant
β_3 Fitted constant
γ Fitted constant
γ_1 Fitted constant
γ_2 Fitted constant
θ Fitted constant
θ_0 Fitted constant
θ_1 Fitted constant
θ_2 Fitted constant
θ_3 Fitted constant
ν Wave number
ρ_u Unburned gas density
ϕ Equivalence ratio
ω_n Weight function

1 Introduction

Syngas mainly consists of carbon monoxide and hydrogen [1]. Syngas can be derived through various methods such as steam reforming of coke, gasification of biomass or coal, and recycling of organic waste emissions. It has gained considerable attention recently due to its potential use as an alternative fuel for natural gas in the gas turbine industry [2–5]. Syngas is also considered as a reformer gas that can be used to reduce the amount of hydrocarbon emissions during engine cold start [6]. Furthermore, syngas and its mixtures can reduce pollutions such as SO_x, NO_x, and CO_2 in different combustion systems [7].

GTL fuel includes iso-alkanes, normal-alkanes, and cyclic-alkanes produced from natural gas through Fisher–Tropsch process. It attracts big focus as an alternative fuel because of supply security as well as environmental issues [8, 9]. Recently, many groups have paid attention to the benefits of using GTL fuel in different fields [10–12]. The combustion properties of GTL fuel for various types of engines have been compared with conventional fuels [13–29].

The laminar burning speed represents the fundamental characteristic of a fuel which is widely used to validate chemical kinetic mechanisms and analyze turbulent flame propagation. It is very useful to measure the laminar burning speed in gas turbines, power plants, and internal combustion engines relevant conditions (high temperatures and pressures) with various diluents. One typical diluent with the same specific heat as the burned gases in internal combustion engines includes 14% carbon dioxide and 86% nitrogen. Additionally, replacing nitrogen with helium in the air can strongly increase flame stability and extend temperature and pressure range due to its higher thermal diffusivity and smaller specific heat than nitrogen.

There are many studies for the laminar burning speed measurement of syngas and GTL without and with diluent. McLean et al. [30] collected the burning speed data of syngas air mixture at atmospheric condition using constant pressure spherical flame method. Hassan et al. [31] determined the laminar burning speed for different H_2 percentages (3–50%) and equivalence ratios varied from 0.6 to 5.0 in a spherical vessel at various pressures (0.5–4 atm.) and temperature of 298 K. Sun et al. [3] measured the laminar burning speed of different H_2 percentages (1–50%) at pressures up to 4 MPa and temperature of 298 K for different equivalence ratios (0.5–5.0) and diluted with helium to extend pressure range of the smooth flame using a dual-cylindrical chamber. Laminar burning speeds of syngas/air/CO_2 were measured by Natarajan et al. [1, 5] for different hydrogen concentrations at high temperatures (up to 700 K) and pressures of 0.1 and 0.5 MPa. The laminar burning speed of syngas is studied by Prathap et al. [2, 32] with CO_2 and N_2 dilution in a constant pressure vessel. Laminar burning speeds of syngas were measured by Dong et al. [33] for various fuel compositions (0–100% hydrogen concentrations) in a Bunsen burner. The effects of He, CO_2, and N_2 diluents on burning speed of syngas were studied by Vu et al. [34] at elevated pressures and temperature of 298 K. Bouvet et al. [35, 36] investigated the laminar burning speed for different hydrogen percentages in syngas mixtures (5–50%) and equivalence ratios (0.4–5.0) at pressure of 0.1 MPa and temperature of 298 K using the Bunsen burner setup and constant pressure method. Burbano et al. [37, 38] studied the effects of N_2 and CO_2 diluents on the laminar burning speed of syngas for different equivalence ratios (0.6–4.3) at the temperature of 298 K and pressure of 0.1 MPa. Laminar burning speeds of syngas diluted with methane and carbon dioxide were measured by Lapalme and Seers [39] at temperature up to 450 K. Singh et al. [40] determined the laminar burning speed of different H_2 concentrations (5–75%) in syngas mixture with H_2O addition (up to 40%) at high temperatures (up to 500 K) and atmospheric pressure in a constant pressure setup. Fu et al. [41] focused on studying the laminar burning speed of syngas with H_2/CO ratios (25–80%) for different equivalence ratios (0.5–1.2). Ai et al. [42] collected the laminar burning speed data of H_2/CO/air mixtures at room and high pressures and temperatures by constant pressure method. Li et al. [43] investigated the stoichiometric and lean burning speed data of syngas for different H_2/CO ratios at normal temperature. The laminar burning speeds were determined by Han et al. [44] at different temperatures, pressures, and CO_2 diluent percentages using a dual-cylindrical chamber setup. Wang et al. [45] studied the laminar burning speed of syngas for various hydrogen percentages (5–75%) in the fuel mixture and equivalence ratios (0.6–5.6) diluted with N_2 and CO_2 (0–60%) using an OH-PLIF method. Li et al. [46] collected laminar burning speed data of syngas with N_2 and CO_2 dilution for lean and stoichiometric mixtures. The laminar flame speed of syngas/air mixtures (H_2 fractions from 20 to 80%) was measured by Xie et al. [47] for lean and stoichiometric mixtures at elevated pressures from 0.4 to 0.6 MPa. Askari et al. [48, 49] studied the laminar burning speed of syngas air mixture and the effects of diluent on laminar burning speed for various hydrogen concentrations (5–25%), equivalence ratios (0.6–5.0), pressures (0.5–5.5 atm.), and temperatures (298–617 K) using a new differential based multi-shell model. Askari et al. [50] also investigated the laminar burning speed of syngas/O_2/He flame for various hydrogen concentrations (5–25%), equivalence ratios (0.6–3.0), pressures (0.5–7.3 atm.), and temperatures (298–650 K) using the same method.

Laminar burning speeds of S-8/air mixtures were measured by Kumar and Sung [24, 25] at high temperatures and pressure of 0.1 MPa using the counterflow method. Singh et al. [51] collected the burning speed data of S-8 for equivalence ratios varied from 0.7 to 1.4 at a temperature of 400 K and pressure of 0.1 MPa. Kick et al. [52] studied the burning speed of GTL experimentally and computationally for various equivalence ratios (0.9–1.4) at pressure of 0.1 MPa and temperature of 473 K. Laminar burning speeds of traditional and alternative jet fuels were determined by Ji et al. [53] using the counterflow method at pressure of 1 atm and temperature of 403 K. Laminar burning speeds of alternative and traditional jet fuels were also compared by Hui et al. [54, 55] at temperatures of 400 and 470 K and atmospheric pressure. Laminar burning speeds of GTL and its blend were determined by Vukadinovic et al. [56] in a cubic vessel at pressures of 0.1, 0.2, and 0.4 MPa and temperatures of 373, 423, and 473 K using the optical laser method. Kinetic simulations and experimental measurements of the burning speed for GTL/air have been done by Dagaut et al. [57, 58] at pressure of 1 atm and temperature of 473 K. Laminar burning speeds of GTL/air mixtures (equivalence ratios from 0.7 to 1.4) were computationally calculated by Yu et al. [59] at different pressures (1.0–25 atm.) and temperatures (400–800 K). Wang et al. [60, 61] studied the laminar burning speed of GTL/air and the effects of diluent on the laminar burning speed of GTL/air for different equivalence ratios (0.7–1.2) at pressures (0.5–4.3 atm.) and temperatures (490–620 K).

In the present work, laminar burning speeds of syngas/air and GTL/air mixtures have been reviewed. Laminar burning speed correlations for syngas/air mixtures have been developed for pressures (0.5–5.5 atm.), temperatures (298–617 K), equivalence ratios (0.6–5.0), and hydrogen percentages (5–25%) in the fuel mixture. Laminar burning speed correlations for GTL/air mixtures have been developed for pressures (0.5–4.3 atm.), temperatures (490–620 K), and equivalence ratios (0.7–1.2). A blend of 43% n-dodecane, 25% n-decane, and 32% iso-octane is considered as the surrogates of GTL fuel [62].

In the present work, the effects of diluent addition to syngas/air and GTL/air mixtures on laminar burning speed have been reviewed. The diluent with the same specific heat as the burned gases includes 14% carbon dioxide and 86% nitrogen. Laminar burning speed correlations for syngas/air/diluent mixtures have been developed for pressures (0.5–5.5 atm.), temperatures (298–450 K), equivalence ratios (0.6–3.0), diluent percentages (0–10%), and various hydrogen concentrations of the fuel (5–25%). Laminar burning speed correlations for GTL/air/diluent mixtures have been developed for pressures (0.5–3.2 atm.), temperatures (490–610 K), equivalence ratios (0.7–1.2), and diluent concentrations (0–10%).

The effects of replacing nitrogen with helium in the air on the laminar burning speeds of syngas/O_2/He flames have been reviewed in this study. Laminar burning speed correlations for syngas/O_2/He mixtures have been developed for various hydrogen fractions (5–25%), pressures (0.5–7.3 atm.), temperatures (298–650 K), and equivalence ratios (0.6–3).

2 Experimental Facilities

The experimental system consists of a spherical chamber and a cylindrical chamber. Laminar burning speed results were determined using a spherical chamber. The stainless-steel spherical chamber withstands pressure up to 400 atm. The whole chamber is placed into an oven which can elevate the vessel temperature up to 500 K. Three ionization probes were placed into the chamber wall. They were used to find when the flame hit the wall. Flame structures were observed using a cylindrical chamber. Two fused quartz windows are sealed to both ends of the cylindrical vessel with high-temperature O-rings. Two band heaters are used to elevate the initial temperature of the cylindrical chamber up to 500 K. A high-speed CMOS camera with a Z-type Schlieren system was used for taking flame images up to 40,000 frames per second [63–66].

Two spark plugs with a very small ignition gap are used in both chambers for central point ignition. The spark energy is tuned to be sufficiently low to reduce the effect of spark discharge on flame propagation [67, 68]. A manifold supply system has been used to fill both chambers by the partial pressures method (Dalton's law). A heated liquid line was used to evaporate liquid fuels. The premixed composition in both chambers has been verified by a gas chromatograph (GC). The pressure rise data were collected by a Kistler's high-sensitivity pressure transducer. The pressure rise as a function of time and flame propagation images was recorded by a data acquisition system.

Figure 1 shows the experimental facilities in the laboratory. The burning speed experiments have been operated at each condition using both vessels. The flame seems to be affected by the geometry of the cylindrical vessel since the flame is spherical. Due to the gas motion in the corner of the cylindrical vessel, laminar burning speed measurement may have some unwanted errors [69]. Therefore, the cylindrical vessel is only used for taking flame pictures. The spherical vessel collects the pressure signal and checks its reproducibility for the same operational condition. Only totally spherical and smooth flames were used for laminar burning speed calculation. Each condition was operated at least three times to guarantee the accurate level is higher than 95% [70]. More researches can be found in previous papers regarding the experimental facilities [48–50, 60, 61].

3 Burning Speed Model

The integral-based model with multi-shell assumption described in the present study was previously created by Metghalchi and Keck and their groups [71–79]. The pressure as a function of time inside the combustion vessel is the input of this model. Figure 2 shows that the unburned and burned gases in the combustion vessel are separated by a reaction layer with negligible thickness. The unburned gases are surrounded by a preheat zone and a wall boundary layer having a uniform temperature.

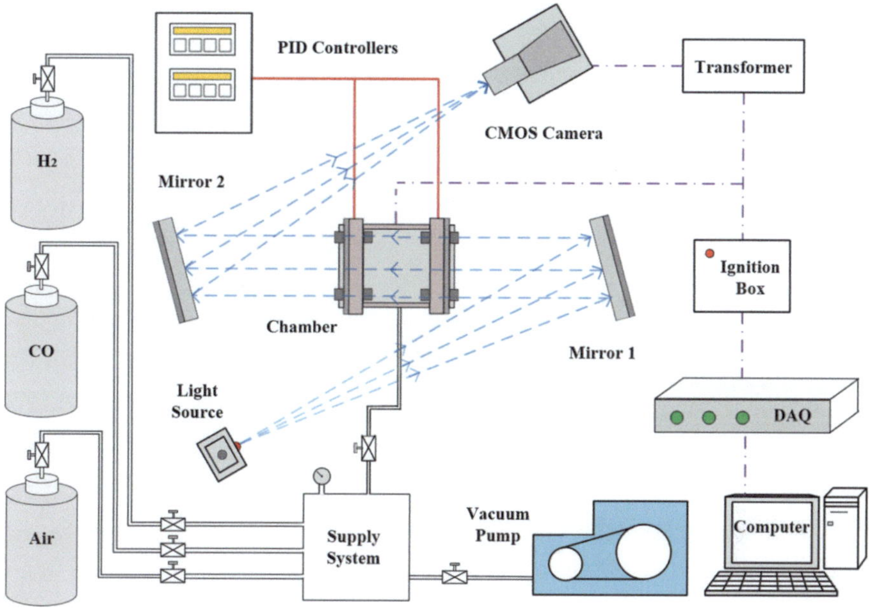

Fig. 1 Experimental facilities in the laboratory

The burned gases are separated into a lot of shells having different temperatures surrounded by a preheat zone. The spark electrodes boundary layers and the wall boundary layer due to the energy transfer from the unburned and burned gases are considered as the thermal boundary layer. The other assumptions include that the pressures in both vessels are uniform, both unburned and burned gases behave as ideal gases and are compressed isentropically.

The mass fraction of burned gases and the temperature distribution of unburned and burned gases are calculated from the measured pressure rise using the mass and energy balances.

The ideal gas law is

$$pv = RT \tag{1}$$

where p and T are the pressure and temperature, and v and R are the specific volume and specific gas constant.

The mass balance equation is

$$m = m_{\mathrm{u}} + m_{\mathrm{b}} = p_{\mathrm{i}}(V_{\mathrm{c}} - V_{\mathrm{e}})/RT_i \tag{2}$$

where m is the mass of the gases in the chamber, m_{u} and m_{b} are the mass of unburned and burned gases, and V_{c} and V_{e} are the volume of the vessel and spark electrodes. Subscripts i indicate the initial conditions.

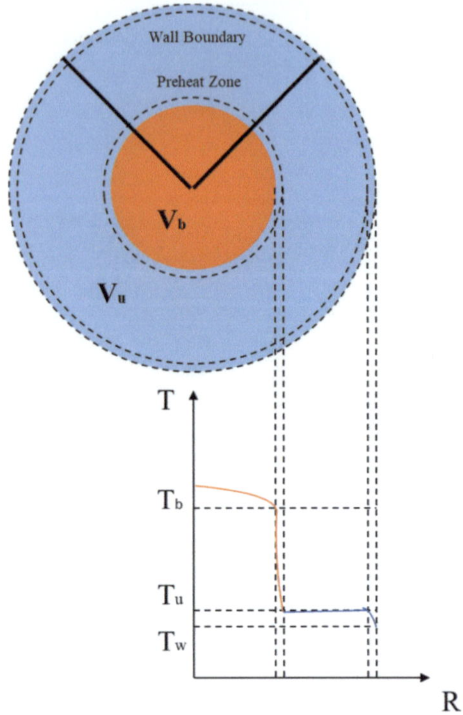

Fig. 2 Schematic of burning speed model [60]

The volume balance equation is

$$V_{\mathrm{i}} = V_{\mathrm{u}} + V_{\mathrm{b}} = V_{\mathrm{c}} - V_{\mathrm{e}} \tag{3}$$

where

$$V_{\mathrm{b}} = \int_{0}^{m_{\mathrm{b}}} v_{\mathrm{b}} \mathrm{d}m = \int_{0}^{m_{\mathrm{b}}} v_{\mathrm{bs}} \mathrm{d}m - V_{\mathrm{eb}} \tag{4}$$

is the volume of the burned gases, v_{bs} is the specific volume of burned gases, and V_{eb} is the displacement volume for the electrode boundary layer.

$$V_{\mathrm{u}} = \int_{m_{\mathrm{b}}}^{m} v_{\mathrm{us}} \mathrm{d}m = m(1 - x_{\mathrm{b}}) v_{\mathrm{us}} - V_{\mathrm{ph}} - V_{\mathrm{wb}} \tag{5}$$

is the volume of the unburned gases, x_{b} is the burned gases mass fraction, v_{us} is the specific volume of unburned gases, V_{ph} is the displacement volume for the preheat zone, and V_{wb} is the displacement volume for the wall boundary layer.

The energy balance equation is

$$E_\mathrm{i} - Q_\mathrm{e} - Q_\mathrm{w} - Q_\mathrm{r} = E_\mathrm{u} + E_\mathrm{b} \tag{6}$$

where E_i is the energy of the gas mixture, Q_e and Q_w are the energy losses to the electrodes surface and vessel wall by conduction, and Q_r is the energy loss by radiation,

$$E_\mathrm{b} = \int_0^{m_\mathrm{b}} e_\mathrm{b}\mathrm{d}m = \int_0^{m_\mathrm{b}} e_\mathrm{bs}\mathrm{d}m - E_\mathrm{eb} \tag{7}$$

is the burned gas energy, e_bs is the specific energy of the burned gas, and E_eb is the defective energy due to the electrode boundary layer,

$$E_\mathrm{u} = \int_{m_\mathrm{b}}^{m} e_\mathrm{u}\mathrm{d}m = m(1 - x_\mathrm{b})e_\mathrm{us} - E_\mathrm{wb} - E_\mathrm{ph} \tag{8}$$

is the unburned gas energy, e_us is the specific energy of the unburned gas, E_wb is the defective energy due to the wall boundary, and E_ph is the defective energy due to the preheat zone.

The energy loss by radiation is determined by a FSNB-CK model [80, 81]. The effects of spectral radiation absorption on energy loss were calculated in this model.

$$Q_\mathrm{r} = V \sum_{j=1}^{N\text{ - band}} \Delta\nu \sum_{n=1}^{N} \omega_n k_n \left(\sum_{m=1}^{M} I_{j,n,m} - 4\pi I_{b,j} \right) \tag{9}$$

where M and N are the number of directions and Gaussian quadrature points, k_n is the k-distribution of absorption coefficient, I is the radiation intensity, ω_n is the weight function, ν is the wave number, N - band is the number of narrow bands, and V is the total volume of the energy source.

Combining Eqs. (3–5) gives the volume equation

$$\begin{aligned} &\int_0^{x_\mathrm{b}} v_\mathrm{bs}(T, p)\mathrm{d}x + (1 - x_\mathrm{b})v_\mathrm{us} \\ &\quad = (V_\mathrm{o} - V_\mathrm{e})/m + \left(V_\mathrm{eb} + V_\mathrm{wb} + V_\mathrm{ph}\right)/m \end{aligned} \tag{10}$$

and combining Eqs. (6–8) gives the energy equation

$$\int_0^{x_\mathrm{b}} e_\mathrm{bs}(T, p)\mathrm{d}x + (1 - x_\mathrm{b})e_\mathrm{us}$$

$$= \frac{E}{m} + \left(E_{\text{eb}} + E_{\text{wb}} + E_{\text{ph}} - Q_{\text{e}} - Q_{\text{w}} - Q_{\text{r}}\right)/m \quad (11)$$

Two unknowns in Eqs. (10) and (11) are the temperature of the last shell in burned gas, $T_{\text{b}}(r, t)$ and the mass fraction of the burned gas, $x_{\text{b}}(t)$. Those two equations can be solved numerically when the pressure is given $p(t)$.

Finally, the laminar burning speed is

$$S_{\text{u}} = \frac{\dot{m}_{\text{b}}}{\rho_{\text{u}} A_{\text{b}}} = \frac{m\dot{x}_{\text{b}}}{\rho_{\text{u}} A_{\text{b}}} \quad (12)$$

where

$$A_{\text{b}} = 4\pi r^2 - 2\pi r_{\text{e}}^2 \quad (13)$$

is the area of the flame, $\dot{x}_{\text{b}}$ is the rate of burned gases mass fraction, r_{e} is the radius of the electrode, and r is the radius of the burned gas.

4 Results and Discussion

4.1 *Laminar Burning Speed of Syngas*

All laminar burning speeds were determined in regions where the stretch rate is low (less than 100 s^{-1}) and flame radii greater than 4 cm [82]. The influence of spark energy discharge and stretch on laminar burning speed can be neglected in these regions. The experimental burning speed results of syngas/air/diluent were fitted to a power-law correlation

$$S_{\text{u}} = S_{\text{u0}}\left(1 + a(\phi - 1) + b(\phi - 1)^2\right)\left(\frac{T_{\text{u}}}{T_{\text{u0}}}\right)^{\alpha}\left(\frac{P}{P_0}\right)^{\beta}(1 - D)^{\gamma} \quad (14)$$

where S_{u0} is the laminar burning speed for $\phi = 1.0$ in reference state ($T_{\text{u0}} = 298$ K, $P_0 = 1$ atm.), T_{u} is the unburned temperature, P is the mixture pressure, ϕ is the equivalence ratio, D is the diluent concentration, and a, b, α, β, and γ are fitted constants.

$$\alpha = \alpha_1 + \alpha_2(\phi - 1) \quad (15)$$

$$\beta = \beta_1 + \beta_2(\phi - 1) \quad (16)$$

$$\gamma = \gamma_1 + \gamma_2(\phi - 1) \quad (17)$$

Table 1 Power-law fitting coefficients for syngas/air/diluent

C (%)	S_{u0}	a	b	α_1	α_2	β_1	β_2	γ_1	γ_2
5	33.351	1.313	−0.410	1.884	−0.008	−0.136	−0.003	3.130	0.818
10	47.433	1.337	−0.444	1.843	0.004	−0.173	0.019	3.579	0.429
25	76.209	1.351	−0.510	1.842	0.008	−0.175	0.035	3.631	0.347

The least square fitted coefficients $S_{u0}, a, b, \alpha_1, \alpha_2, \beta_1, \beta_2, \gamma_1$, and γ_2 for various hydrogen percentages (C) are shown in Table 1. This power-law correlation is valid for smooth flames only in the temperature range of 298–450 K, pressure range of 0.5–5.5 atm., equivalence ratio range of 0.6–3.0, and diluent concentration range of 0–10%.

The calculated laminar burning speeds of syngas/O_2/He were used to develop a power-law correlation

$$S_u = \alpha \left(\frac{T_u}{T_{u0}}\right)^{\beta} \left(\frac{P}{P_0}\right)^{\theta} \tag{18}$$

$$\alpha = \alpha_3\phi^3 + \alpha_2\phi^2 + \alpha_1\phi + \alpha_0 \tag{19}$$

$$\beta = \beta_3\phi^3 + \beta_2\phi^2 + \beta_1\phi + \beta_0 \tag{20}$$

$$\theta = \theta_3\phi^3 + \theta_2\phi^2 + \theta_1\phi + \theta_0 \tag{21}$$

The least square fitted coefficients $\alpha_0, \alpha_1, \alpha_2, \alpha_3, \beta_0, \beta_1, \beta_2, \beta_3, \theta_0, \theta_1, \theta_2$, and θ_3 for various hydrogen percentages (C) are reported in Table 2. This power-law correlation is valid for smooth flames only in the temperature range of 298–650 K, pressure range of 0.5–7.3 atm., and equivalence ratio range of 0.6–3.0.

The laminar burning speed correlations were compared with the computational calculation by different detailed kinetic mechanisms and available experimental results for syngas [83–85]. The laminar burning speeds were calculated by the 1-D steady-state flame code from CANTERA [86]. Figure 3 shows the comparison over a wide range of equivalence ratios for three different hydrogen concentrations at a pressure of 1 atm and temperature of 298 K. The results calculated by Davis kinetic mechanism [83] is very close to the selected experimental data.

Figure 4 shows a comparison of experimental results versus different detailed kinetic mechanisms (Davis [83] and Li [84] mechanisms) for syngas/air/diluent mixture ($C = 25\%$) for various diluent concentrations (0, 5, and 10%) at atmospheric pressure and temperature. The results calculated by Davis kinetic mechanism is very close to selected experimental data.

Figure 5 shows a comparison of experimental results versus different detailed kinetic mechanisms (Davis [83] and Li [84] mechanisms) for syngas/O_2/He mixture ($C = 25\%$) at different temperatures and pressures. In this case, the Li mechanism

Table 2 Power-law fitting coefficients for syngas/O_2/He

C (%)	α_3	α_2	α_1	α_0	β_3	β_2	β_1	β_0	θ_3	θ_2	θ_1	θ_0
5	11.27	−78.18	187.08	−47.32	−0.245	1.551	−3.048	3.229	0.0265	−0.202	0.492	−0.335
10	15.54	−102.5	236.49	−51.92	−0.167	0.994	−1.831	2.378	0.0326	−0.226	0.516	−0.289
25	−0.413	−50.52	249.65	−54.18	−0.0116	0.267	−0.961	2.173	0.0492	−0.253	0.422	−0.183

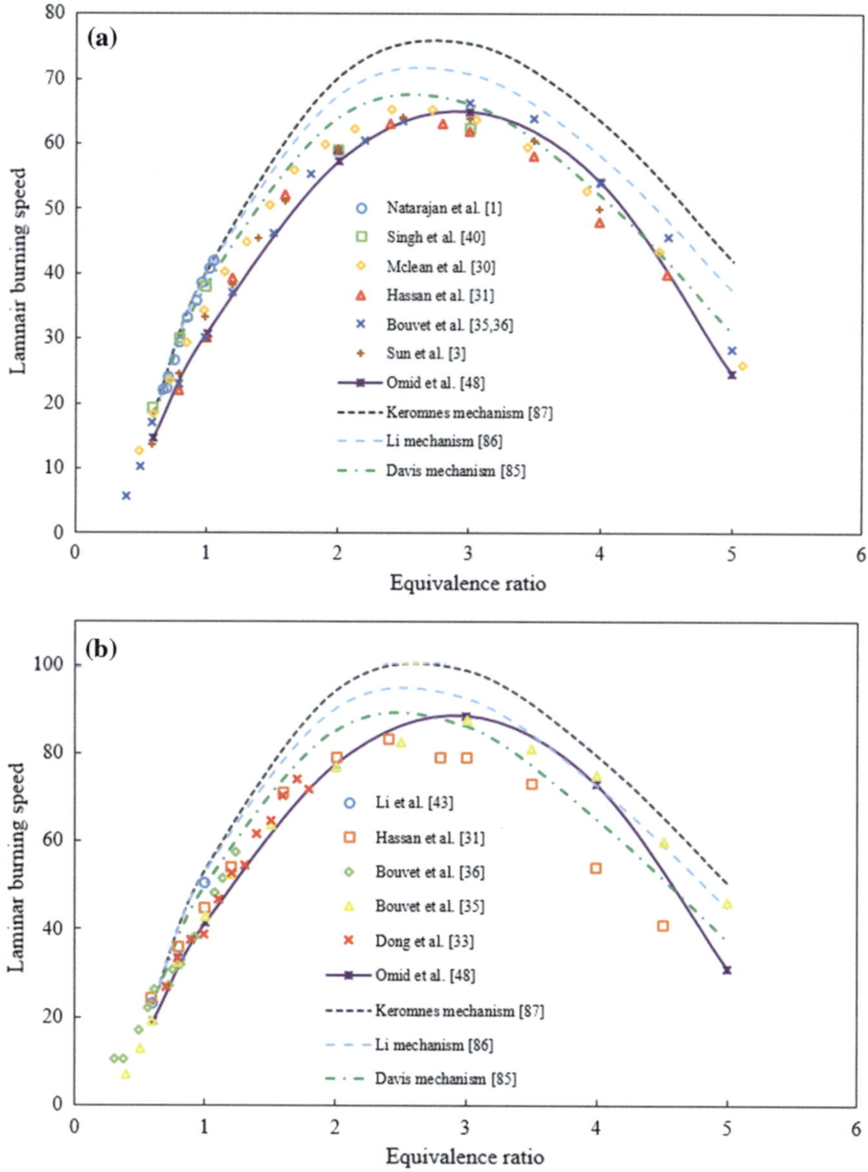

Fig. 3 Comparison of experimental data and kinetic simulations for syngas/air mixture at normal conditions **a** $C = 5\%$, **b** $C = 10\%$, **c** $C = 25\%$, and **d** $C = 50\%$

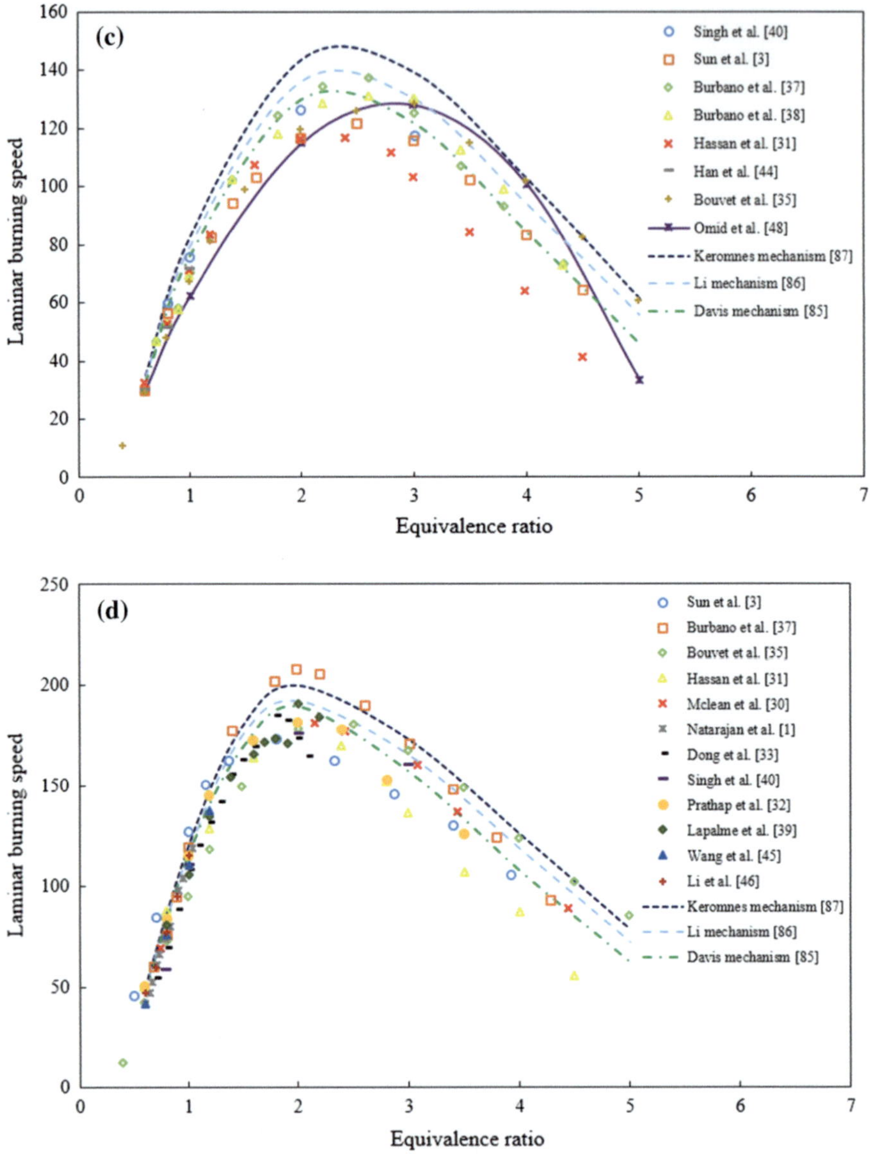

Fig. 3 (continued)

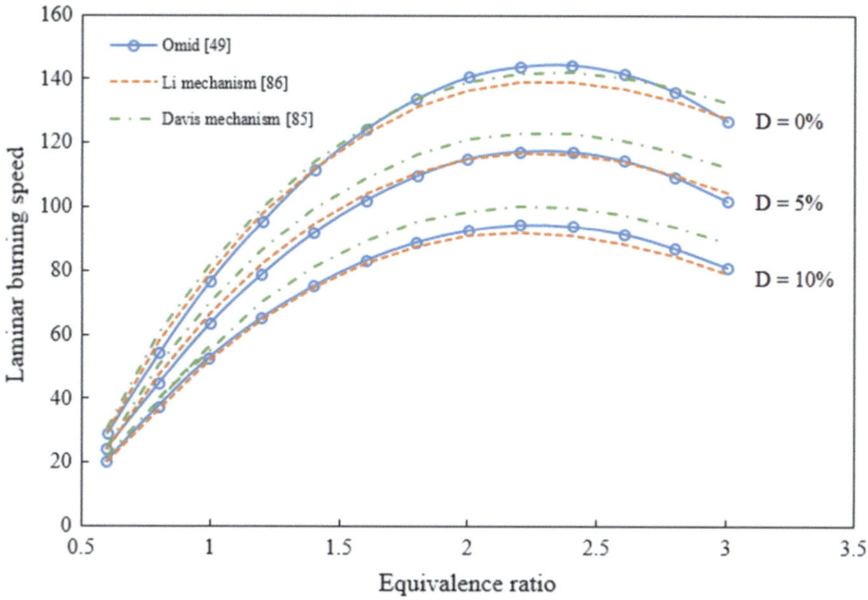

Fig. 4 Comparison of syngas/air/diluent ($C = 25\%$) experimental data versus kinetic simulations for various diluent concentrations at normal temperature and pressure

has a better agreement than the Davis mechanism with the selected experimental results.

4.2 *Laminar Burning Speed of GTL*

The experimental burning speeds of GTL/air/diluent were fitted to a power-law correlation

$$S_u = S_{u0}\left(1 + a(\phi - 1) + b(\phi - 1)^2\right)\left(\frac{T_u}{T_{u0}}\right)^{\alpha_0+\alpha_1\phi}\left(\frac{P}{P_0}\right)^{\beta_0+\beta_1\phi}(1 - \gamma D) \quad (22)$$

where S_{u0} is laminar burning speed ($\phi = 1.0$) in the reference state, $T_{u0} = 500$ K and $P_0 = 1$ atm are the temperature and pressure at the reference point, T_u is the unburned temperature, P is the mixture pressure, ϕ is the equivalence ratio, D is the diluent concentration, and a, b, α_0, α_1, β_0, β_1, and γ are the least square fitted coefficients shown in Table 3. These coefficients were fitted over pressure range of 0.5 atm. $\leq P \leq$ 3.2 atm., temperature range of 490 K $\leq T \leq$ 610 K, diluent concentration range of 0% $\leq D \leq$ 10%, and equivalence ratio range of 0.7–1.2.

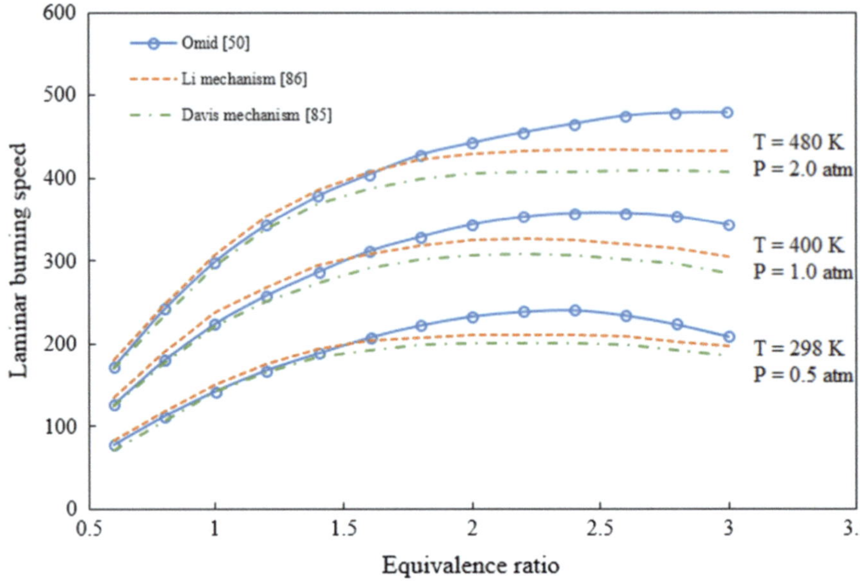

Fig. 5 Comparison of syngas/O_2/He ($C = 25\%$) experimental data versus different kinetic simulations at different temperatures and pressures for various equivalence ratios

The laminar burning speed power-law correlations of GTL were compared with the available computational calculation by different detailed kinetic mechanisms and experimental data. Figures 6 and 7 present the comparison of measured burning speed versus three kinetic simulations [62, 87, 88] at normal pressure and temperatures of 400 and 473 K for various equivalence ratios.

Figure 8 presents the comparison of measured burning speed for various equivalence ratios of GTL/air mixtures and diluent concentrations at temperatures of 550 K and atmospheric pressure. As it can be seen, the Ranzi mechanism [88] including 484 species and 19,341 chemical reactions is very close to the experimental data.

5 Summary

Laminar burning speeds of syngas/air/diluent, syngas/O_2/He, and GTL/air/diluent were determined by the pressure rise of the combustion process. Laminar burning speeds were measured for smooth flames and with radii larger than 4 cm having min-

Table 3 Power-law fit coefficients for GTL/air/diluent

S_{u0}	a	b	α_0	α_1	β_0	β_1	γ
89.257	0.4727	−2.2685	1.5740	0.4312	−0.3112	0.1658	2.156

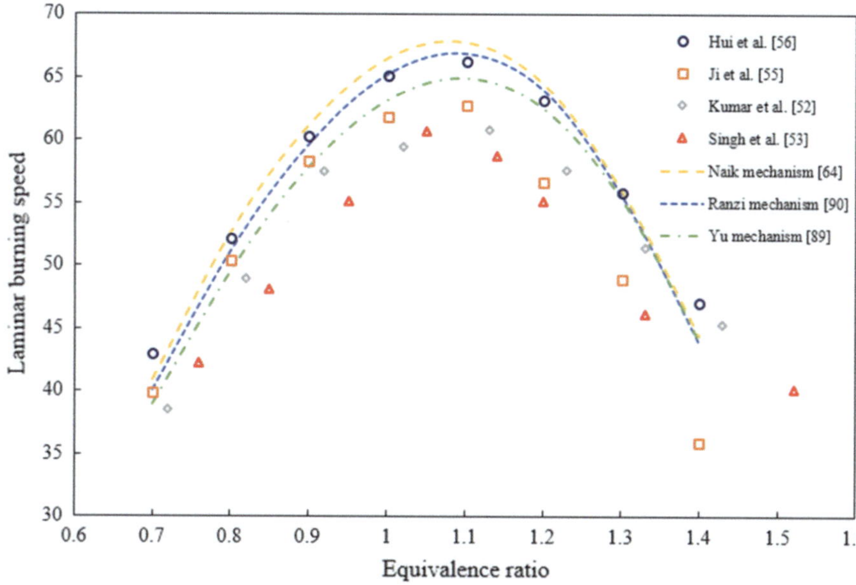

Fig. 6 Comparison of experimental burning speed versus kinetic simulations for GTL/air mixtures at a temperature of 400 K and normal pressure

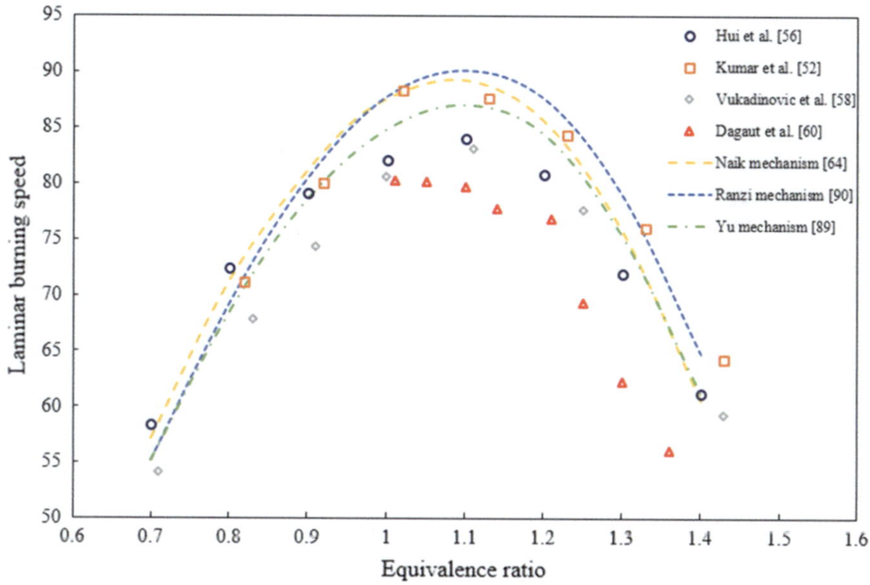

Fig. 7 Comparison of measured burning speed versus kinetic simulations for GTL/air mixtures at a temperature of 473 K and atmospheric pressure

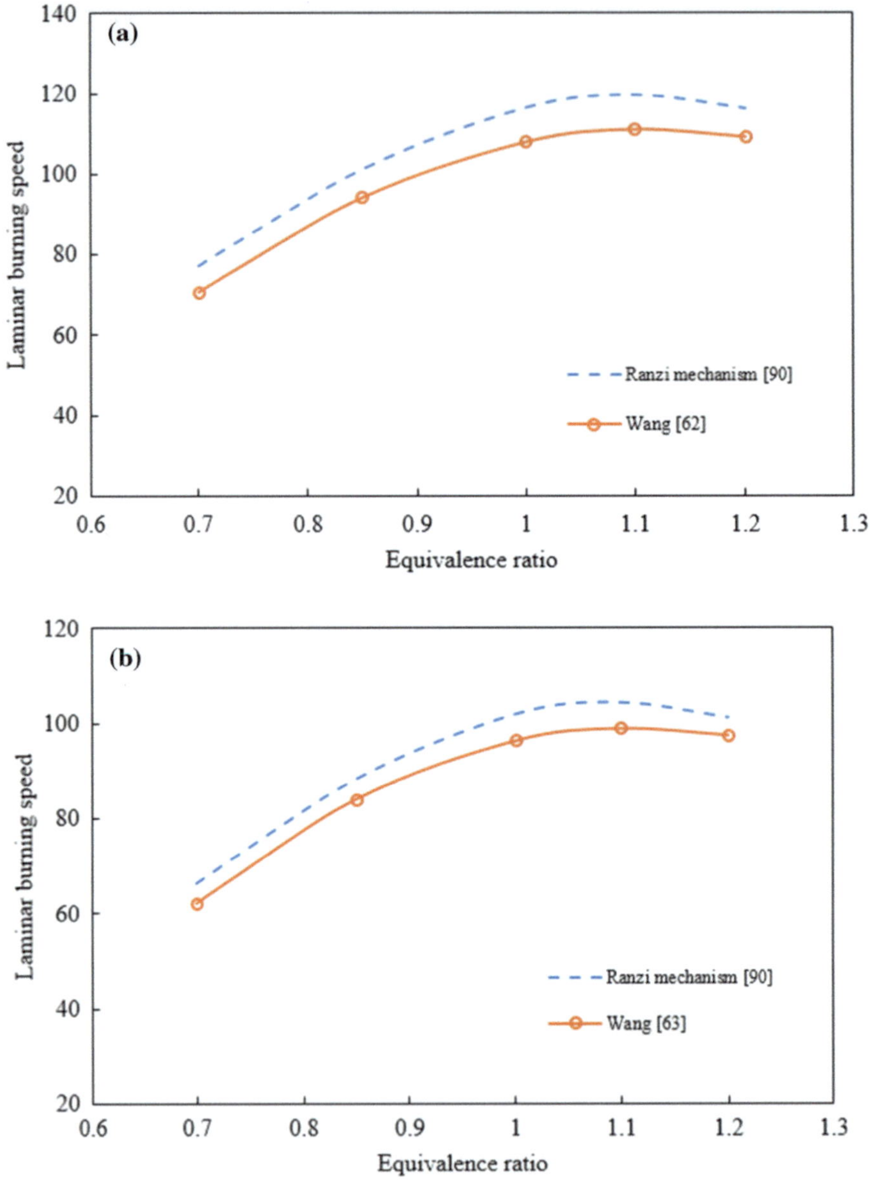

Fig. 8 Comparison of measured burning speed versus kinetic simulations for GTL/air/diluent mixtures at pressure of 0.1 MPa and temperature of 550 K **a** $D = 0\%$, **b** $D = 5\%$, and **c** $D = 10\%$

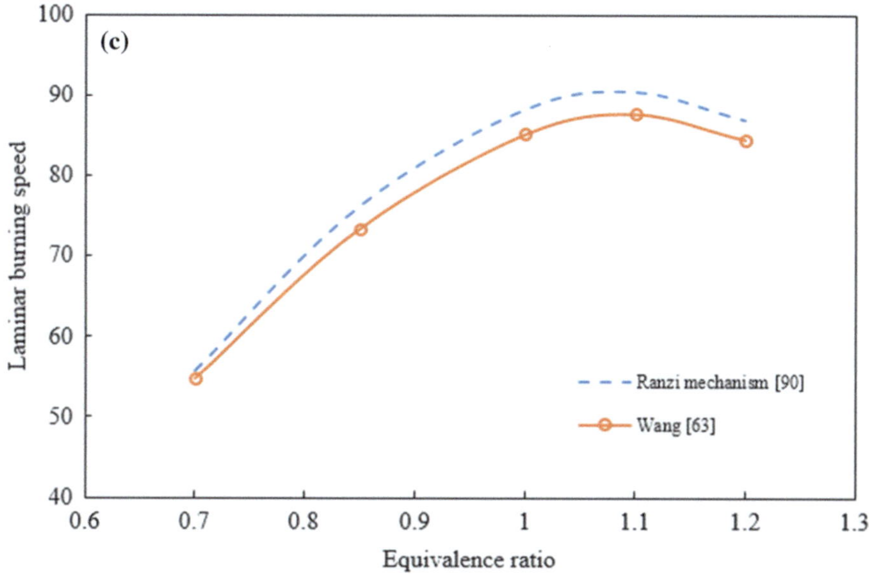

Fig. 8 (continued)

imum stretch effects. Power-law correlations are reported for experimental burning speed whose fitting constants are presented in Tables 1, 2 and 3. As diluent percentage increases, the laminar burning speed decreases. The laminar burning speed decreases as mixture pressure increases. As unburned gases temperature increases, the laminar burning speed increases. The maximum burning speed of syngas happens at equivalence ratio range of 2.5–3.0. The maximum burning speed of GTL happens at an equivalence ratio of 1.1. Simulation results have been compared with available experimental data for various temperatures and pressures. Measured laminar burning speed results of syngas are very close to the kinetic simulations by using Davis mechanism. Measured laminar burning speed results of GTL have very good agreement with the kinetic simulations as well especially by using the Ranzi mechanism.

References

1. Natarajan J, Lieuwen T, Seitzman J (2007) Laminar flame speeds of H_2/CO mixtures: effect of CO_2 dilution, preheat temperature, and pressure. Combust Flame 151:104–119
2. Prathap C, Ray A, Ravi MR (2008) Investigation of nitrogen dilution effects on the laminar burning velocity and flame stability of syngas fuel at atmospheric condition. Combust Flame 155(1–2):145–160
3. Sun H, Yang SI, Jomaas G, Law CK (2007) High-pressure laminar flame speeds and kinetic modeling of carbon monoxide/hydrogen combustion. Proc Combust Inst 31:439–446
4. Som S, Ramirez AI, Hagerdorn J, Saveliev A, Aggarwal SK (2008) A numerical and experimental study of counterflow syngas flames at different pressures. Fuel 87:319–334
5. Natarajan J, Kochar Y, Lieuwen T, Seitzman J (2009) Pressure and preheat dependence of laminar flame speeds of H_2/CO/CO_2/O_2/He mixtures. Proc Combust Inst 32(1):1261–1268
6. Huang Y, Sung CJ, Eng JA (2004) Laminar flame speeds of primary reference fuels and reformer gas mixtures. Combust Flame 139:239–251
7. Park J, Lee DH, Yoon SH, Vu TM, Yun JH, Keel SI (2009) Effects of Lewis number and preferential diffusion on flame characteristics in 80%H_2/20%CO syngas counterflow diffusion flames diluted with He and Ar. Int J Hydrogen Energy 34(3):1578–1584
8. Bao B, El-Halwagi MM, Elbashir NO (2010) Simulation, integration, and economic analysis of gas-to-liquid processes. Fuel Process Technol 91(7):703–713
9. Abdel-Kreem M, Bassyouni M, Abdel-Hamid SMS, Abdel-Aal H (2009) The role of GTL technology as an option to exploit natural gas resources. Open Fuel Cells J 2(1):5–10
10. Udaeta MEM, Burani GF, Arzabe Maure JO, Oliva CR (2007) Economics of secondary energy from GTL regarding natural gas reserves of bolivia. Energy Policy 35(8):4095–4106
11. Economides M (2005) The economics of gas to liquids compared to liquefied natural gas. World Energy 8(1):136–140
12. Hobbs HOJ, Adair LS (2012) Gas-to-liquids plants offer great ROI. The American Oil & Gas Reporter, Haysville
13. Moon G, Lee Y, Choi K, Jeong D (2010) Emission characteristics of diesel, gas to liquid, and biodiesel-blended fuels in a diesel engine for passenger cars. Fuel 89(12):3840–3846
14. Hassaneen A, Munack A, Ruschel Y, Schroeder O, Krahl J (2012) Fuel economy and emission characteristics of gas-to-liquid (GTL) and rapeseed methyl ester (RME) as alternative fuels for diesel engines. Fuel 97:125–130
15. Abu-Jrai A, Rodriguez-Fernandez J, Tsolakis A, Megaritis A, Theinnoi K, Cracknell RF, Clark RH (2009) Performance, combustion and emissions of a diesel engine operated with reformed EGR. Comparison of diesel and GTL fueling. Fuel 88(6):1031–1041
16. Li X, Huang Z, Wang J, Zhang W (2007) Particle size distribution from a GTL engine. Sci Total Environ 382(2–3):295–303
17. Alleman TL, Eudy L, Miyasato M, Oshinuga A, Allison S, Corcoran T, Chatterjee S, Jacobs T, Matthey J, Cherrillo RA, Clark R, Virrels I, Nine R, Wayne S, Lansing R (2004) Fuel property, emission test, and operability results from a fleet of class 6 vehicles operating on gas-to-liquid fuel and catalyzed diesel particle filters south coast air quality management district. SAE technical paper no. 2004-01-2959
18. Alleman TL, Tennant CJ, Hayes RR, Barton G, Rumminger M, Duggal V, Nelson C, May M, Cherrillo RA (2005) Achievement of low emissions by engine modification to utilize gas-to-liquid fuel and advanced emission controls on a class 8 truck. SAE technical paper no. 2005-01-3766
19. Alleman TL, Barnitt R, Eudy L, Corcoran T, Cherrillo RA, Clark N, Wayne WS (2005) Final operability and chassis emissions results from a fleet of class 6 trucks operating on gas-to-liquid fuel and catalyzed diesel particle filters. SAE technical paper no. 2005-01-3769
20. Mancaruso E, Vaglieco BM (2012) Premixed combustion of GTL and RME fuels in a single cylinder research engine. Appl Energy 91(1):385–394

21. Hajialimohammadi A, Ahmadisoleymani S, Abdullah A, Askari O, Rezai F (2012) Design and manufacturing of a constant volume test combustion chamber for jet and flame visualization of CNG direct injection. Appl Mech Mater 217–219:2539–2545
22. Kahandawala MSP, DeWitt MJ, Corporan E, Sidhu SS (2008) Ignition and emission characteristics of surrogate and practical jet fuels. Energy Fuels 22(6):3673–3679
23. Balagurunathan J, Flora G, Saxena S, Kahandawala M, DeWitt M, Sidhu S, Corporan E (2011) Ignition delay times of a range of alternate jet fuels and surrogate fuel candidate hydrocarbons under fuel-lean conditions: a shock tube study. AIAA paper no. 2011-696
24. Kumar K, Sung CJ (2010) A comparative experimental study of the autoignition characteristics of alternative and conventional jet fuel/oxidizer mixtures. Fuel 89(10):2853–2863
25. Kumar K, Sung CJ, Hui X (2011) Laminar flame speeds and extinction limits of conventional and alternative jet fuels. Fuel 90(3):1004–1011
26. Wang H, Oehlschlaeger MA (2012) Autoignition studies of conventional and Fischer-Tropsch jet fuels. Fuel 98:249–258
27. Kannaiyan K, Sadr R (2014) Experimental investigation of spray characteristics of alternative aviation fuels. Energy Convers Manage 88:1060–1069
28. Fyffe D, Moran J, Kannaiyan K, Sadr R, Al-Sharshani A (2011) Effect of GTL-like jet fuel composition on GT engine altitude ignition performance: part I combustor operability. ASME paper no. GT2011-45487
29. Mosbach T, Gebel GC, Le Clercq P, Sadr R, Kannaiyan K, Al-Sharshani A (2011) Investigation of GTL-like jet fuel composition on GT engine altitude ignition and combustion performance: part II detailed diagnostics. ASME paper no. GT2011-45510
30. Mclean IC, Smith DB, Taylor SC (1994) The use of carbon monoxide/hydrogen burning velocities to examine the rate of the CO + OH reaction. In: 25th symposium on combustion, pp 749–757
31. Hassan MI, Aung KT, Faeth GM (1997) Properties of laminar premixed CO/H_2/air flames at various pressures. J Propuls Power 13(2):239–245
32. Prathap C, Ray A, Ravi MR (2012) Effects of dilution with carbon dioxide on the laminar burning velocity and flame stability of H_2/CO mixtures at atmospheric condition. Combust Flame 159(2):482–492
33. Dong C, Zhou Q, Zhao Q, Zhang Y, Xu T, Hui S (2009) Experimental study on the laminar flame speed of hydrogen/carbon monoxide/air mixtures. Fuel 88(10):1858–1863
34. Vu TM, Park J, Kwon OB, Bae DS, Yun JH, Keel SI (2010) Effects of diluents on cellular instabilities in outwardly propagating spherical syngas-air premixed flames. Int J Hydrogen Energy 35(8):3868–3880
35. Bouvet N, Chauveau C, Gökalp I, Halter F (2011) Experimental studies of the fundamental flame speeds of syngas (H_2/CO)/air mixtures. Proc Combust Inst 33:913–920
36. Bouvet N, Chauveau C, Gökalp I, Lee SY, Santoro RJ (2011) Characterization of syngas laminar flames using the Bunsen burner configuration. Int J Hydrog Energy 36(1):992–1005
37. Burbano HJ, Pareja J, Amell AA (2011) Laminar burning velocities and flame stability analysis of H_2/CO/air mixtures with dilution of N_2 and CO_2. Int J Hydrog Energy 36(4):3232–3242
38. Burbano HJ, Pareja J, Amell AA (2011) Laminar burning velocities and flame stability analysis of syngas mixtures at sub-atmospheric pressures. Int J Hydrog Energy 36(4):3243–3252
39. Lapalme D, Seers P (2014) Influence of CO_2, CH_4, and initial temperature on H_2/CO laminar flame speed. Int J Hydrogen Energy 39(7):3477–3486
40. Singh D, Nishiie T, Tanvir S, Qiao L (2012) An experimental and kinetic study of syngas/air combustion at elevated temperatures and the effect of water addition. Fuel 94:448–456
41. Fu J, Tang C, Jin W, Thi LD, Huang Z, Zhang Y (2013) Study on laminar flame speed and flame structure of syngas with varied compositions using OH-PLIF and spectrograph. Int J Hydrogen Energy 38:1636–1643
42. Ai Y, Zhou Z, Chen Z, Kong W (2014) Laminar flame speed and Markstein length of syngas at normal and elevated pressures and temperatures. Fuel 137:339–345
43. Li H, Li G, Sun Z, Zhai Y, Zhou Z (2014) Measurement of the laminar burning velocities and Markstein lengths of lean and stoichiometric syngas premixed flames under various hydrogen fractions. Int J Hydrog Energy 39(30):17371–17380

44. Han M, Ai Y, Chen Z, Kong W (2015) Laminar flame speeds of H_2/CO with CO_2 dilution at normal and elevated pressures and temperatures. Fuel 148:32–38
45. Wang ZH, Weng WB, He Y, Li ZS, Cen KF (2015) Effect of H_2/CO ratio and N_2/CO_2 dilution rate on laminar burning velocity of syngas investigated by direct measurement and simulation. Fuel 141(55):285–292
46. Li H, Li G, Sun Z, Zhou Z, Li Y, Yuan Y (2016) Investigation on dilution effect on laminar burning velocity of syngas premixed flames. Energy 112:146–152
47. Xie Y, Wang J, Cai X, Huang Z (2016) Self-acceleration of cellular flames and laminar flame speed of syngas/air mixtures at elevated pressures. Int J Hydrogen Energy 41:18250–18258
48. Askari O, Moghaddas A, Alholm A, Vein K, Alhazmi B, Metghalchi H (2106) Laminar burning speed measurement and flame instability study of H_2/CO/air mixtures at high temperatures and pressures using a novel multi-shell model. Combust Flames 168:20–31
49. Askari O, Vien K, Wang Z, Sirio M, Metghalchi H (2016) Exhaust gas recirculation effects on flame structure and laminar burning speeds of H_2/CO/air flames at high pressures and temperatures. Appl Energy 179:451–462
50. Askari O, Wang Z, Vien K, Sirio M, Metghalchi H (2017) On the flame stability and laminar burning speeds of syngas/O_2/He premixed flame. Fuel 190:90–103
51. Singh D, Nishiie T, Qiao L (2010) Laminar burning speeds and Markstein lengths of n-Decane/Air, n-Decane/O_2/He, Jet-A/Air and S-8/Air Flames. AIAA paper no. 2010-951
52. Kick T, Herbst J, Kathrotia T, Marquetand J, Braun-Unkhoff M, Naumann C, Riedel U (2012) An experimental and modeling study of burning velocities of possible future synthetic jet fuels. Energy 43(1):111–123
53. Ji C, Wang YL, Egolfopoulos FN (2011) Flame studies of conventional and alternative jet fuels. J Propul Power 27(4):856–863
54. Hui X, Kumar K, Sung CJ, Edwards T, Gardner D (2012) Experimental studies on the combustion characteristics of alternative jet fuels. Fuel 98:176–182
55. Hui X, Sung CJ (2013) Laminar flame speeds of transportation-relevant hydrocarbons and jet fuels at elevated temperatures and pressures. Fuel 109:191–200
56. Vukadinovic V, Habisreuther P, Zarzalis N (2012) Experimental study on combustion characteristics of conventional and alternative liquid fuels. ASME J Eng Gas Turbines Power 134(12):121504
57. Dagaut P, Karsenty F, Dayma G, Dievart P, Hadj-Ali K, Mze-Ahmed A, Braun-Unkhoff M, Herzler J, Kathrotia T, Kick T, Naumann C, Riedel U, Thomas L (2014) Experimental and detailed kinetic model for the oxidation of a gas to liquid (GtL) jet fuel. Combust Flame 161(3):835–847
58. Dagaut P, Dayma G, Dievart P, Hadj-Ali K, Mze-Ahmed A (2014) Combustion of a gas-to-liquid based alternative jet fuel: experimental and detailed kinetic modeling. Combust Sci Technol 186(10–11):1275–1283
59. Yu G, Askari O, Hadi F, Wang Z, Metghalchi H, Kannaiyan K, Sadr R (2017) Theoretical prediction of laminar burning speed and ignition delay time of gas-to-liquid fuel. ASME J Energy Resour Technol 139(2):022202–022206
60. Wang Z, Alswat M, Yu G, Allehaibi MO, Metghalchi H (2017) Flame structure and laminar burning speed of gas to liquid fuel air mixtures at moderate pressures and high temperatures. Fuel 209:529–537
61. Wang Z, Bai Z, Yelishala SC, Yu G, Metghalchi H (2018) Effects of diluent on laminar burning speed and flame structure of gas to liquid fuel air mixtures at high temperatures and moderate pressures. Fuel 231:204–214
62. Naik CV, Puduppakkam KV, Modak A, Meeks E, Wang YL, Feng Q (2011) Detailed chemical kinetic mechanism for surrogates of alternative jet fuels. Combust Flame 158(3):434–445
63. Parsinejad F, Matlo M, Metghalchi H (2004) A mathematical model for Schlieren and Shadowgraph images of transient expanding spherical thin flames. J Eng Gas Turbines Power 126(2):241–247
64. Parsinejad F, Keck JC, Metghalchi H (2007) On the location of flame edge in Shadowgraph pictures of spherical flames: a theoretic and experimental study. Exp Fluids 43(6):887–894

65. Askari O, Metghalchi H, Hannani SK, Moghaddas A, Ebrahimi R, Hemmati H (2012) Fundamental study of spray and partially premixed combustion of methane/air mixture. J Energy Res Technol 135(2):021001
66. Askari O, Metghalchi H, Hannani SK, Hemmati H, Ebrahimi R (2014) Lean partially premixed combustion investigation of methane direct-injection under different characteristic parameters. J Energy Resour Technol Trans ASME 136(2):1–7
67. Bradley D, Gaskell PH, Gu XJ (1996) Burning velocities, Markstein lengths, and flame quenching for spherical methane-air flames: a computational study. Combust Flame 198:176–198
68. Eisazadeh-Far K, Parsinejad F, Metghalchi H (2010) Flame structure and laminar burning speeds of JP-8/air premixed mixtures at high temperatures and pressures. Fuel 89(5):1041–1049
69. Burke MP, Chen Z, Ju Y, Dryer FL (2009) Effect of cylindrical confinement on the determination of laminar flame speeds using outwardly propagating flames. Combust Flame 156(4):771–779
70. Trochim WM, Donnelly JP (2006) The research methods knowledge base, 3rd edn. Atomic Dog, Cincinnati
71. Metghalchi M, Keck JC (1980) Laminar burning velocity of propane-air mixtures at high temperature and pressure. Combust Flame 38:143–154
72. Metghalchi M, Keck JC (1982) Burning velocities of mixtures of air with methanol, isooctane, and indolene at high pressure and temperature. Combust Flame 48:191–210
73. Elia M, Ulinski M, Metghalchi M (2001) Lamianr burning velocity of methane-air-diluent mixtures. J Eng Gas Turbines Power 123(1):190–196
74. Parsinejad F, Arcari C, Metghalchi H (2006) Flame structure and burning speed of JP-10 air mixtures. Combust Sci Technol 178:975–1000
75. Rahim F, Eisazadeh-Far K, Parsinejad F, Andrews RJ, Metghalchi H (2008) A thermodynamic model to calculate burning speed of methane-air-diluent mixtures. Int J Thermodyn 11:151–161
76. Eisazadeh-Far K, Moghaddas A, Rahim F, Metghalchi H (2010) Burning speed and entropy production calculation of a transient expanding spherical laminar flame using a thermodynamic model. Entropy 12(12):2485–2496
77. Eisazadeh-Far K, Parsinejad F, Metghalchi H, Keck JC (2010) On flame kernel formation and propagation in premixed gases. Combust Flame 157(12):2211–2221
78. Eisazadeh-Far K, Moghaddas A, Metghalchi H, Keck JC (2011) The effect of diluent on flame structure and laminar burning speeds of JP-8/oxidizer/diluent premixed flames. Fuel 90(4):1476–1486
79. Rokni E, Moghaddas A, Askari O, Metghalchi H (2015) Measurement of laminar burning speeds and investigation of flame stability of acetylene (C_2H_2)/air mixtures. ASME J Energy Resour Technol 137(1):012204
80. Chen Z, Qin X, Xu B, Ju Y, Liu F (2007) Studies of radiation absorption on flame speed and flammability limit of CO_2 diluted methane flames at elevated pressures. Proc Combust Inst 31:2693–2700
81. Riviere P, Soufiani A (2012) Updated band model parameters for H_2O, CO_2, CH_4 and CO radiation at high temperature. Int J Heat Mass Transf 55:3349–3358
82. Chen Z, Burke MP, Ju Y (2009) Effects of compression and stretch on the determination of laminar flame speeds using propagating spherical flames. Combust Theor Model 13(2):343–364
83. Davis SG, Joshi AV, Wang H, Egolfopoulos F, Troe J, Li J, Kumaran SS, Smith D, Dryer FL (2005) An optimized kinetic model of H_2/CO combustion. Proc Combust Inst 30(1):1283–1292
84. Li J, Zhao Z, Kazakov A, Chaos M, Dryer FL, James J, Scire J (2007) A comprehensive kinetic mechanism for CO, CH_2O, and CH_3OH combustion. Int J Chem Kinet 39:109–136
85. Keromnes A, Metcalfe WK, Heufer KA, Donohoe N, Das AK, Sung CJ, Herzler J, Naumann C, Griebel P, Mathieu O, Krejci MC, Petersen EI, Pitz WJ, Curran HJ (2013) An experimental and detailed chemical kinetic modeling study of hydrogen and syngas mixture oxidation at elevated pressures. Combust Flame 160(6):995–1011

86. Goodwin D, Moffat H, Speth R (2015) Cantera: an object-oriented software toolkit for chemical kinetics, thermodynamics, and transport processes. http://www.cantera.org
87. Yu J, Wang Z, Wang W, Gou X (2016) Surrogate definition and chemical kinetic modeling for Fischer-Tropsch fuels. Energy Fuels 30(2):1375–1382
88. Ranzi E, Frassoldati A, Grana R, Cuoci A, Faravelli T, Kelley AP (2012) Hierarchical and comparative kinetic modeling of laminar flame speeds of hydrocarbon and oxygenated fuels. Prog Energy Combust Sci 38:468–501

Hydrogen and Hydrogen-Rich Fuels: Production and Conversion to Electricity

Jens Klingmann and Martin Andersson

Keywords Hydrogen · Fuel cells · Gas turbines · Combined and hybrid cycles

Abbreviations

CC	Combined cycle
CGO	Gadolinia doped ceria
GT	Gas turbine
NG	Natural gas
HTE	High-temperature electrolysis
SOEC	Solid oxide electrolysis cell
SOFC	Solid oxide fuel cell
PtG	Power-to-Gas
PtL	Power-to-Liquids

1 Introduction

The use of fossil fuels generates pollutions, for example, carbon dioxide, nitrogen oxide and particles. Alternate energy sources would be required for the survival of the present industrial and fast-paced world. Several alternatives have been suggested, but only few can replace conventional combustion engines or gas-fired power plants. Hydrogen may also be used as a "clean" fuel in "conventional" gas-fired power

J. Klingmann (✉) · M. Andersson
Department of Energy Sciences, Lund University, Lund, Sweden
e-mail: jens.klingmann@energy.lth.se

A. K. Gupta et al. (eds.), *Innovations in Sustainable Energy and Cleaner Environment*, Green Energy and Technology,
https://doi.org/10.1007/978-981-13-9012-8_10

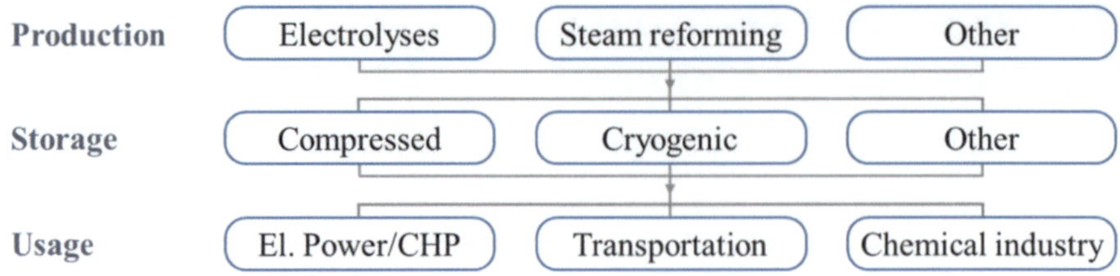

Fig. 1 Overview of production, storage and usage of hydrogen

plants. The opportunity for renewable energy sources is dependent on research breakthroughs improving efficiency, decreases cost and increases life length. Among many, fuel cells and electrolyzes have emerged as solutions with advantages such as compactness, fuel flexibility and environment-friendly operation [1]. Roughly 80% of the energy resources traded today is of fossil origin (natural gas, coal and oil). The fossil energy reserve will sooner or later be depleted, starting with oil [2]. Hydrogen has advantages as an energy carrier since it can be created from every primary energy resource, i.e., renewable energy sources, fossil fuels as well as nuclear power, and also from synthesized hydrogen carriers like ammonia, methanol as well as synthetic fuels. Presently, hydrogen production is dominated by steam reforming or partial oxidation of fossil fuels.

Figure 1 shows an overview of production methods, storage possibilities and usage of hydrogen. Here production—other includes thermolysis, photosynthetic hydrogen production and storage other includes materials capable of binding and releasing hydrogen, usually adding heat. There are different methods of hydrogen storage, e.g., cryogenic storage or chemical reactions with other materials that can be reversed, typically by heat addition. Applications of the former are essentially limited to space industry due to the high (energy) costs and the latter is mainly applicable for the transport sector. For larger scales, stationary application and simple pressurized storage have been most used. The disadvantage of this is that hydrogen has low energy density (J/m^3) in comparison with hydrocarbons, meaning that larger volumes or pressures have to be used to store the energy. Examples of storages range from steel or carbon bottles to underground caverns. A specific pressurized storage is the injection of hydrogen into a natural gas grid leading to a mixture of hydrogen and natural gas.

Power-to-Liquid (PtL) and Power-to-Gas (PtG) are expected to become important parts of a forthcoming renewable energy system. PtG processes links between the power grid and the gas grid by converting surplus power-to-gas via a two-step process: (1) water electrolysis and (2) conversion to H_2 with an external CO_2 or CO source to CH_4 via methanation (see Fig. 2). The integration of fluctuating renewables like solar and wind power in the energy system is supported, including power balancing. Various operating strategies, for example, the absorption of surplus energy at limited annual operating periods, arbitrage trading (exploitation of the highly fluctuating electricity prices) or providing grid services are possible, which also could

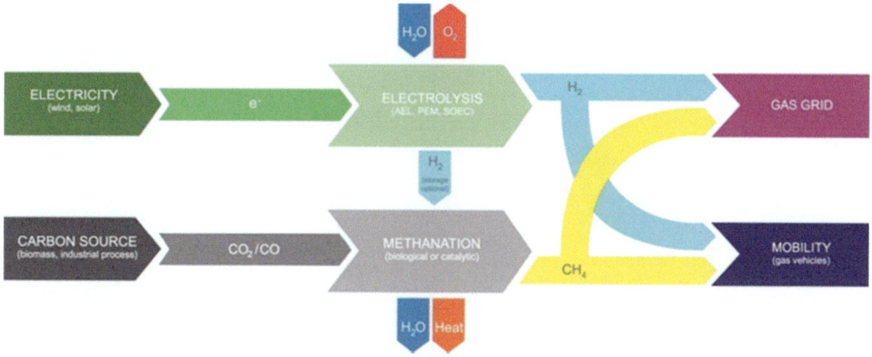

Fig. 2 Examples of power-to-gas process streams. With publisher permission from [4]

be combined. This postures original requirements regarding flexibility, efficiency, part-load as well as stand-by performance [3, 4].

Besides hydrocarbon reforming, hydrogen can also be produced from "green energy" approaches, such as water splitting (thermochemical water splitting, electrolysis or photolysis) or from biomass. Currently at a large scale, water electrolysis is a feasible route for hydrogen production and contains the splitting of water into oxygen and hydrogen by applying electrical energy. An increased temperature increases the amount of energy required for electrolysis slightly, while the amount of required electrical energy decreases (see Fig. 3), i.e., high-temperature electrolysis may be an interesting route to consider when there is an availability of waste heat from other processes. As seen in Fig. 3, the overall energy demand (ΔH) changes only marginally with temperature, i.e., varies just between 283.5 and 291.6 kJ/mole H_2, in a temperature range between 0 and 1000 °C. However, the fraction of useable heat integration (ΔQ) increases with increasing temperature, decreasing the electrical demand (ΔG). Beside enhanced kinetics, the possible high heat utilization of internal losses is a key motivation of high-temperature electrolysis, i.e., operated at 700–900 °C. Furthermore, a fraction of the heat demand related to the latent heat of vaporization is supplied by feeding water vapor, instead of liquid water, as it is the situation for low-temperature electrolysis [3]. Essentially, solid oxide electrolysis cells (SOECs) may use the industrial waste heat to reduce the endothermic electrochemical reaction heat demand. Furthermore, there is also a growing interest in combining electrolysis cells with nuclear energy to improve the hydrogen production efficiency [5].

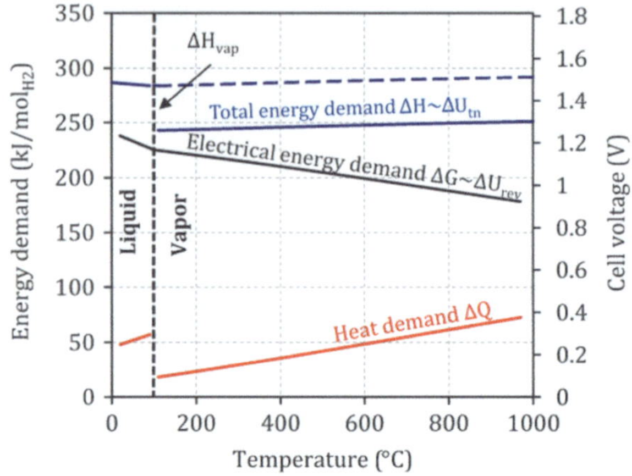

Fig. 3 Total, electrical and heat energy demands as a function of temperature. With publisher permission from [3]

2 Methods for Hydrogen Production

2.1 *High-Temperature Electrolysis*

High-temperature electrolysis (HTE) of water occurs at 700–1000 °C, and the technology may be based on SOECs. SOECs can be understood as solid oxide fuel cells (SOFCs) operating in reversed mode due to its similarity (see Figs. 4 and 5). An SOEC consists of the identical components as an SOFC, i.e., air and fuel channels, an anode, a cathode and an electrolyte. The material of the components is normally the same as for an SOFC, i.e., YSZ for the electrolyte, Ni-YSZ cermet for the hydrogen/steam electrode as well as LSM for the oxygen-side electrode. The transport behavior in SOECs differs notably from the ones in SOFCs, especially those connected to kinetic models, gradients involved and potential ranges, affecting long-term stability and performance in various ways and thus, SOFCs cannot normally be directly used to SOECs. This accentuates the limited amount of SOEC research and the requirement for further development of the technology. Typically, SOC electrodes contain three-phase composites (an electron conducting, an ion conducting as well as a gas phase) forming a porous microstructure allowing the transport of electrons, ions and gases simultaneously within the electrodes. The electrochemical reactions occur at the triple-phase boundaries (TPBs), where the electronic and ionic pathways convene with the gas phase. The performance of the electrode is determined by multiple factors, the primary factors are the transport properties and the abundance of TPBs. Additionally, the electrode microstructure is fundamental for improving the performance of the electrochemistry. The electrode TPB length has been increased by porous composite structure development. In order to advance cell performance, a substantial understanding of the microstructure of the electrode is necessary in quantitative terms. Steam (H_2O) participates in water splitting inside the electrolysis cell, giving oxygen and hydrogen as products. While hydrogen is

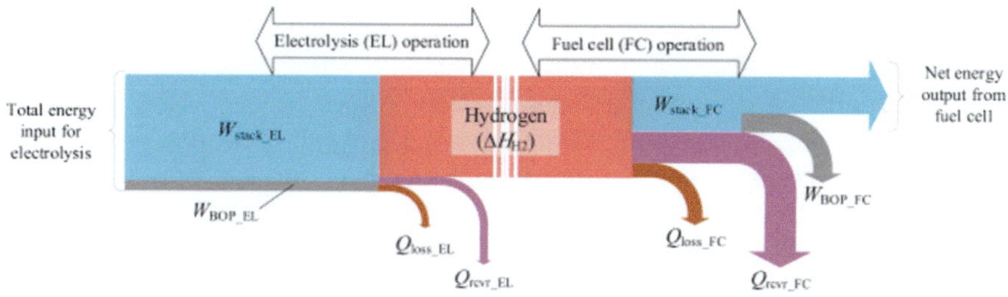

Fig. 4 Schematic of energy flow, including both electrolyzer and fuel cell. With publisher permission from [6]

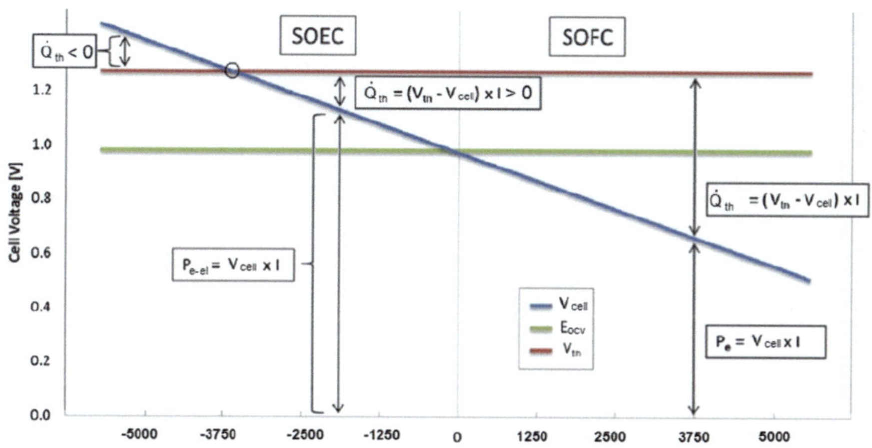

Fig. 5 Polarization curve for fuel cell as well as electrolyzer mode. From [7] (CC-BY)

the main product obtained, oxygen could be counted as a co-product since it is a commodity with a value and can be used for various applications [5].

2.2 *Reforming of Hydrocarbons*

The hydrocarbon conversion, i.e., from methane, ethanol and methanol into carbon monoxide and hydrogen, is enabled by reforming reactions. SOFC system reforming reaction may be conducted either internally or externally. External reforming (ER), for example a SOFC system pre-reforming, means that the reformer is located outside the cell, which is feasible for all types of FCs. Internal reforming (IR) means that the reforming reactions occur inside the cell. IR can be indirect or direct and in the direct internal reforming (DIR) approach, the reforming reactions occur together with electrochemical reactions inside the anode. On the other hand, in the indirect internal reforming (IIR) approach the reforming reactions occur in a reformer within the cell and in close contact with the anode where the exothermic electrochemical reactions take place. IR reactions reduce the need for cell cooling, i.e., less surplus of air is required. A decreased amount of steam is then needed and it also gives benefits concerning the capital cost of the system. Up to half of the heat produced by the oxidation reaction can be "consumed" by the methane steam reforming reaction, which improves the system electrical efficiency [6].

3 Methods of Converting Hydrogen into Electricity

3.1 Fuel Cells

With the concern about global warming, energy security and air pollution, the possibility of using fuel cells (FCs) in future sustainable and renewable energy conversion systems has achieved noteworthy momentum [7]. FCs convert directly the chemical reactant free energy to electrical energy and heat, which differ from a conventional thermal power plant, where fuel is oxidized in a combustion process combined with a conversion process (thermal-mechanical-electrical energy), which takes place after the combustion. A FC consists of two electrodes: one cathode for oxidant and one anode for fuel. An electrolyte separates the electrodes and is connected into an electrically conducting circuit. A gas or liquid flow, as the oxidant or fuel, is transported to a specific electrode, which is permeable via a porous structure. Multiple cells are structured to stacks. The FC is not a new innovation, with the electrochemical process revealed already in 1838–39 [2, 6]. In the 1960s, the use of FCs was limited to niche applications, such as space missions, because of their high costs. But with the evolution in technology and science in the past decades, it is expected that FCs will become a significant source of energy in future. FCs have shown a dramatic performance in terms of electrical efficiency of 40–60% and when they are used with CHP they may reach efficiencies of 75–80%, after future significant upscaling. Additionally, having no moving parts makes them suitable for longer operating periods. The FC emissions, while working, are low because the output of an FC is pure water if hydrogen is used as fuel. Furthermore, due to the absence of moving parts and conversion of energy via electrochemical reactions, as opposed to burning, they have a silent operation. All these promising features make FCs a choice for future power generation [1].

FC systems on specific niche markets, i.e., where FCs bring an supplementary worth, are now competitive compared to traditional (fossil-based) energy systems. A high per weight energy content was the vital advantage in the American space program and a low noise level is the vital advantage for FC development aiming for military use. A rising technology, i.e., FCs, is normally introduced into the market in specific niches, where non-traditional and new characteristic of the technology provides significant added value to reward for a higher (capital) cost. Recently, demonstration project used FCs as APUs (auxiliary power units) in onboard vehicle applications. In the short term, hydrogen production onboard makes the current fuel refilling system possible to use. Current fuels, such as diesel, gasoline or kerosene may still be supplied as fuel and only a single fuel storage system is then required onboard the vehicle. The vehicle industry is recognized to be conservative concerning fuels and the tradition of gasoline or diesel as fuel may support FC commercialization. The hydrocarbon fuel may then be reformed onboard to hydrogen in a pre-reformer. FC APUs may be seen as a worthy transition state aiming of a vehicle applications hydrogen economy [6].

Current drivers for FC development can be summarized as [8]:

- Political "push" for significantly reduced global warming emissions is expected to increase after the Paris agreement.
- "Low" carbon today and possibility for zero carbon in the future.
- Demand for an increased energy security (i.e., local production).
- Requirement of increased (electrical) energy efficiency.
- Development of local and regional energy markets, i.e., there will be multiple solutions and with flexible production pathways.
- Energy can be stored as hydrogen, enabling more renewable energy such as wind power.
- Air quality (significantly lower emissions compared to a combustion engine).
- Japanese government is pushing the development to make the 2020 Tokyo Olympics a showpiece for its hydrogen energy strategy, which can be compared to the Shinkansen (high-speed train) development for the previous Tokyo Olympics. It should be noted that also the Chinese society used the Summer Olympics 2008 and is planning to use the Winter Olympics in 2022 as showcase for their hydrogen and fuel cell technologies.
- Reduced water consumption (compared to, e.g., coal-based power plants).
- Electrification (for vehicles).

Some examples of the current status of FC development [8]:

- The number of companies developing fuel cell systems based on multiple technologies is increasing. Also in public funded programs (e.g., Callux in Germany and the ENE-FARM in Japan) multiple fuel cell technologies are being installed.
- Honda, Toyota and Hyundai launched FC cars.
- Daimler, GM and BMW are expected to launch FC cars around 2020.
- FC as a range extender, for example, to La Poste (France)
- First prototypes of FC trains are developed by Alstom.
- FCs for trams (in China) are being developed.
- Over 10,000 FC powered material handling devices are operating (mostly in USA).
- Significant potential for unmanned aerial vehicles (UAVs), due to high system energy density, low heat signature and less vibration.
- Hydrogen infrastructure roll-out is coming, starting with Japan, California, Europe (starting with Germany) and South Korea.

High-temperature fuel cells. SOFC has a higher electrical efficiency compared to other types of fuel cells. It is possible to achieve an electrical efficiency of 60%

(LHV) for systems as small as 1.5 kW. Among different types of FCs, the SOFC has attained noteworthy attention due to its high electrical efficiency. A high-temperature operation gives many advantages, such as flexibility of using various fuels, a high electrochemical reaction rate and toleration of impurities [6, 9].

Some examples of the current status of SOFC development [8]:

- Lifetimes over 78,000 h reached for stack in laboratory environment (FZ Jülich). It should be noted that this test is with technology from nine years ago and even longer lifetime with technology of today is expected. However, no new similar long-time tests have been started.
- The electrical efficiency is over 60% for a 1.5 kW SOFC CHP system (fueled with natural gas) that is available for purchase (Solid Power).
- Large-scale Bloom Energy SOFC installations in the USA have achieved significant deployment by large companies. Unfortunately, the heat is not used. The motivation for most companies to install their own electricity generation is to ensure reliable power supply.

Low-temperature fuel cells. The first PEFC was developed by General Electric, for the two persons Gemini Space Vehicle in the early 1960s. However, instead of demonstrating the breakthrough in history of PEFC, this excursion to space limited further PEFC development. One of the major reasons was the water management inside the cell and for later space missions alkaline FCs were developed. In the mid-1960s, Dupont developed the Nafion™ membrane, which showed an enhanced lifetime and performance and, PEFCs were taken to space again, this time in a biosatellite mission in 1968. However, the water management problem was still too difficult to deal with. Again, the space program used Alkaline FCs for several missions. In the 1970s and early 1980s, further development in PEFCs was on hold. However, in the late 1980s and early 1990s, the research of Los Alamos National Laboratory and Ballard Power Systems resulted in the advance in new catalyst loading techniques and membrane properties. As of today, the PEFC technology is commercializing slowly. Significant activities in PEFC research and development have taken place in the last decade to increase the electrical efficiency [1, 10]. The PEFC has developed as a suitable FC type for automotive and some portable applications and has also potential for backup power unit applications due to its high power density, low operating temperature and comparative simplicity of construction [7].

The PEFC has potential for producing high power densities, making the technology appealing for mobile and portable applications. Especially many automobile manufacturers have seen the PEFC as a choice replacing the gasoline or diesel engine. In PEFCs, a solid polymer membrane works a separator which distinguishes it from other types of the fuel cells. Due to the polymer membrane, the PEFC typically operates at low temperatures, 60–80 °C, i.e., having faster start-up characteristics compared to high-temperature FCs. One of the restraints of the PEFC is that it operates on pure hydrogen only, but (on-board) fuel processors have been developed, which allows conventional fuels, such as gasoline or natural gas to be used in the cell. Additionally, extended development starting from the PEFCs has been developed, which uses methanol without the fuel processor. This is the direct methanol fuel cells (DMFC), which is prime candidate for use in notebook computers, cameras and other portable electronics. Still extensive and in-depth research is required to extend the PEFC commercialization [1, 10].

Despite engineering progress and scientific advances over the last few decades, the PEFC commercialization remains unrealized, due to: (1) technical problems relating to water management; (2) high prices of materials and components; (3) membrane fragility; and (4) membrane hydration issues. The challenge in understanding the water management processes lies mainly in the two-phase multi-component flow, which involves phase change within the porous media, coupled mass and heat transfer, interactions between the gas channel and the porous layers as well as the complex relationship between cell performance and water content. In PEFCs, the electrochemical reactions are comprehensively coupled to the transport of species (in liquid as well as gas phase), charges, momentum and heat transport. All transport and reaction processes within a PEFC occur simultaneously at different time and length scales. Because of the low operating temperature, the water from the electrochemical reactions may condense, i.e., possibly flooding the microporous layer, catalyst layer, gas diffusion layer as well as gas channel. Understanding of the vital processes of liquid water evolution and transport is still absent, which prevents further enhanced FC development [7, 10].

3.2 Thermal Methods

Combustion of hydrogen and natural gas in gas turbines. Most modern commercial combustion systems are premixed and developed for NG or liquid fuel combustion. The premixing at lean conditions prevents the very high combustion temperatures responsible for most of the NOx formation, but also leads to a number of instability issues. The very high reaction rates of hydrogen combustion lead to high flame speed. Hydrogen's high diffusivity also contributes, but to a lesser degree [11]. At standard conditions, methane/air laminar flame speed is around 0.4 m/s close to stoichiometric conditions whereas peak hydrogen/air laminar flame speed is around 3 m/s at an equivalence ratio of approximately 1.7. Also autoignition delay times are significantly lower [12] reducing the available time for premixing.

Increasing the hydrogen content of the fuel significantly increases the risk for the flame to propagate into (or auto-ignite in) the premixing section upstream of the combustor, an event usually referred to as flashback. Most larger machines have flashback detectors that will immediately trip the machine if this event is detected.

In the development of new combustors for combustion of hydrogen or hydrogen-rich mixtures, a few questions need answers:

- Fuel flexibility: Is the combustor aimed at being used for a hydrogen or should it be capable of combusting a large range of mixtures between hydrogen and other species?
- If the combustor should be capable of handling both hydrogen and natural gas, is the fuel supplied in separate streams or as a mixture with varying composition in time?

Combustor modifications to reduce the flashback propensity of hydrogen-containing fuels can be done, e.g., by increasing the combustor throat velocity. However, most manufacturers are trying to develop combustors that are capable of handling a variety of fuel mixtures. These fuel-flexible combustors pose a significantly larger design challenge than optimizing for a specific fuel. A possible strategy pursued by some manufacturers is to have separate stages for the case where hydrogen and natural gas are supplied separately. In this case, the fuel/air mixing and combustor inlet may be optimized separately.

Gas turbines and combined cycles. Of the present thermal methods to convert fuel into electricity, combined GT/Rankin cycles show the highest efficiencies. Up to 63% of gas has been reported for very large power plants [13]. Combined cycles represent heavy investments and for smaller plants, often only a simple cycle gas turbine is used. Efficiencies for these are typically below 40%.

Size is an important factor. The efficiency of turbomachinery is size dependent so that the compressor and turbine efficiencies of a gas turbine limit the cycle efficiency. Going down to very small engines (for instance the Ansaldo T100 100 kW engine) will provide efficiencies of around 30%. These are typically low-pressure ratio, recuperated machines and at this size, piston engines provide severe competition.

Stoichiometric hydrogen/oxygen combustion in steam cycles. Assuming both hydrogen and oxygen can be stored opens new possibilities. Since the products of hydrogen combustion with pure oxygen are water (without nitrogen that is present in combustion with air) the applicable cycles can be more similar to steam cycles than gas turbine cycles. In fact, there is the possibility to use this combustion in existing steam power plants, see below.

Since the temperatures are moderated by steam, the combustion product, net reaction rates are slightly reduced, but from a very high level. Diffusion flames have been used in space applications and premixed combustion has been demonstrated successfully [14].

Use of hydrogen/oxygen combustion in existing steam cycles. Generally, the maximum temperature of a steam cycle is set by the outer temperature of the tubes in the superheaters that has to be low enough to avoid high-temperature corrosion.[1] Adding a combustion chamber for hydrogen/oxygen in front of the turbines in a steam cycle

---

[1]Gas turbines may have considerably higher temperatures because they are internally fired engines and thus, the heat does not have to be transferred into the cycle through a heat exchange process. This makes it possible to cool the most endangered parts.

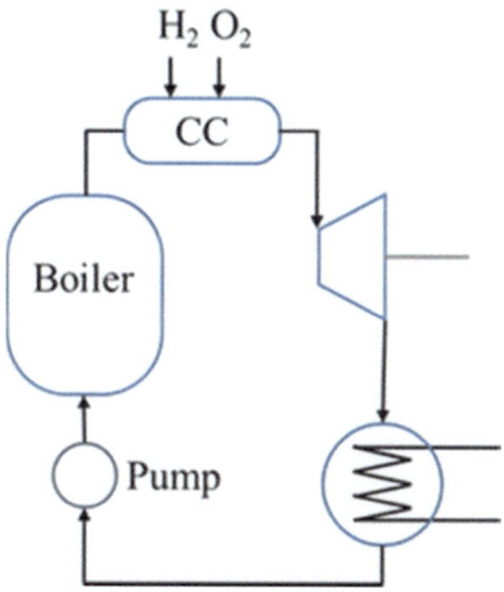

Fig. 6 Simple schematic of supplementary H_2/O_2 firing in a steam cycle

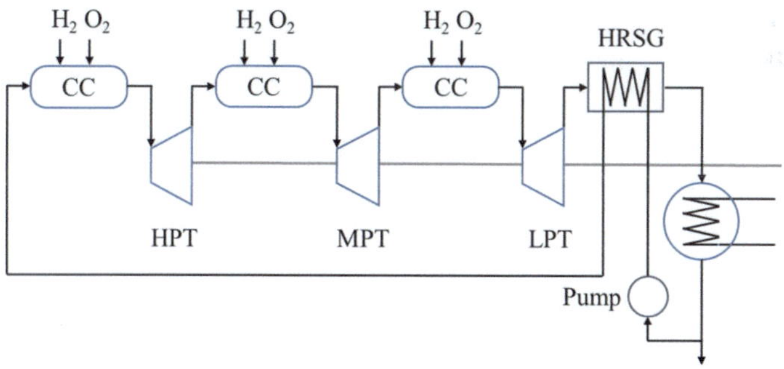

Fig. 7 Schematics of the MNRC cycle

has the potential of increasing the steam temperatures to almost any value. Obviously existing steam turbines are designed for a maximum turbine inlet temperature, meaning that the potential for retrofit is limited. However, there is a clear potential for development in terms of increasing this temperature (Fig. 6).

The efficiency of this process can be discussed in terms of a marginal efficiency defined as the ratio of added power to the cycle to the addition heat provided by the hydrogen/oxygen combustion. Values of this efficiency are highly dependent on the possible temperature increase. Possible candidates for this technology include boiling water nuclear reactors, which generally have fairly low turbine inlet temperatures.

New proposals. There are a few proposals for hydrogen/oxygen fired steam cycles with potentially very high efficiencies. An example of these is the Modified New Rankine Cycle (MNRC) [15], schematically shown in Fig. 7. For turbine inlet temperatures at 1700 °C, the theoretical efficiency is 79%, but inlet temperatures for steam turbines at this level most likely require significant development.

Hybrid concepts. Academic studies developing hybrid cycles, i.e., combining SOFCs and gas turbines may be found. Here, one recent study is chosen. Four representative configurations of hybrid SOFC systems were selected in a recent study at Delft University, which was compared to a stand-alone SOFC system. The flow sheets of the investigated hybrid cycle configurations are shown in Fig. 8. Note that in this chapter, FCs are not included in "combined cycle," as denoted in Fig. 9; instead, "hybrid system" is used to indicate a system that includes both a FC and a gas turbine. As a reference system was a stand-alone SOFC system defined, i.e., to compare with the different proposed SOFC hybrid cycles. A combine cycle, including a SOFC as well as a steam turbine, is designed that stipulates a quite simple method to recover the "waste heat" that is available in the exhaust from the SOFC, into supplementary electricity using power plant technology that is established. Additionally, a combined cycle with a SOFC-reciprocating engine is introduced, where a lean burn spark ignited engine is used to combust anode off-gas, i.e., using chemical instead of thermo-mechanical energy from the SOFC. Two combined cycle SOFC-gas turbine layouts are also investigated, which allow a synergetic coupling using cathodic air, i.e., the cooling medium and oxygen source in the SOFC, as the gas turbine working fluid. It should be noted that the SOFC stack is either operated at ambient or at elevated pressures in the two different systems.

To our knowledge, the only commercially available SOFC-gas turbine hybrid is the Mitsubishi Hitachi Power Systems at 250 kW rated power. The pressurized-type SOFC-MGT hybrid system brought about by combining the 200-kW-class SOFC with a micro gas turbine [17].

4 Results and Discussion

Fuel cells are still fairly small, and upscaling is hindered mainly by costs and partly by life length. Electrical efficiencies as high as 60% for high-temperature FC systems as small as 1.5 kW (with natural gas as fuel) is impressive, comparing to competing technologies. Large-scale conventional power production, such as gas turbines and combined cycles, on the other hand suffers from reduced efficiencies at lower powers. Development work is done to handle hydrocarbons in fuel cells as opposed to GT-based methods where the development is toward including more hydrogen in systems optimized for NG or liquid hydrocarbons. Figure 9 show some sample efficiencies of commercially available power production plants.

Gas turbine based refers to recuperated in the sub-MW scale, simple cycle and combined cycles for the larger plants. FCs are (high temperature) SOFCs and hybrid refer to the abovementioned SOFC/gas turbine from Mitsubishi Hitachi Power Systems. Data is taken from manufacturers homepages and technical notes. Included for reference are two internally fired reciprocating engines in the 1–2 MW class (Siemens SGE-EM and GE Jenbacher J316). The gas turbine-based methods gen-

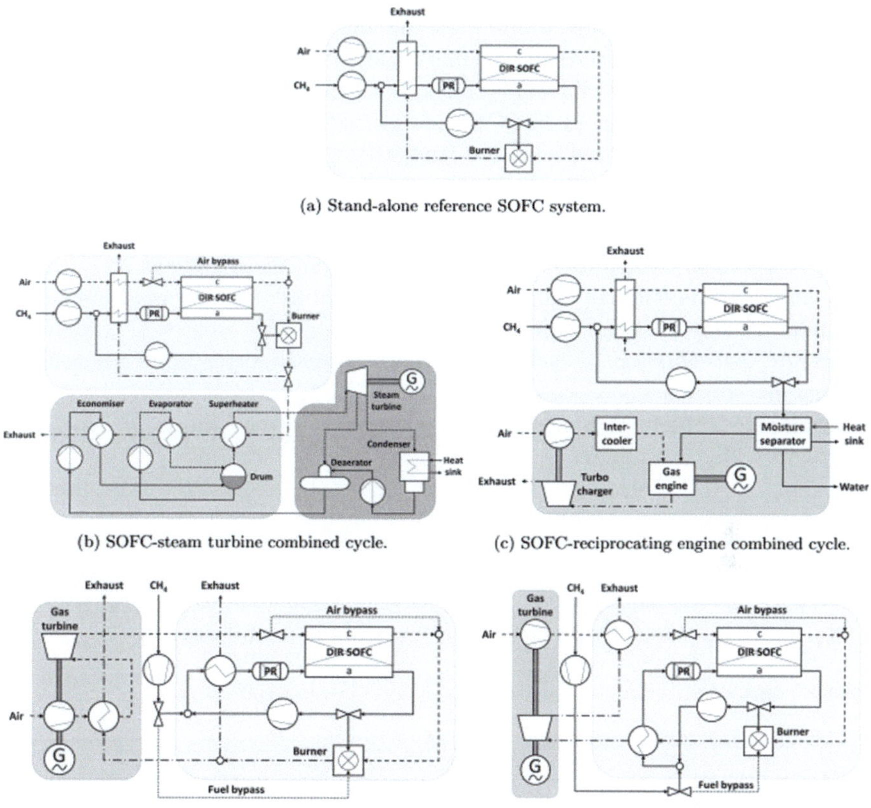

(a) Stand-alone reference SOFC system.

(b) SOFC-steam turbine combined cycle.

(c) SOFC-reciprocating engine combined cycle.

(d) Ambient SOFC-gas turbine combined cycle.

(e) Pressurised SOFC-gas turbine combined cycle.

Fig. 8 Flow sheets of a stand-alone and various combined cycle configurations. From [16] (CC-BY)

Fig. 9 Examples of currently available technologies

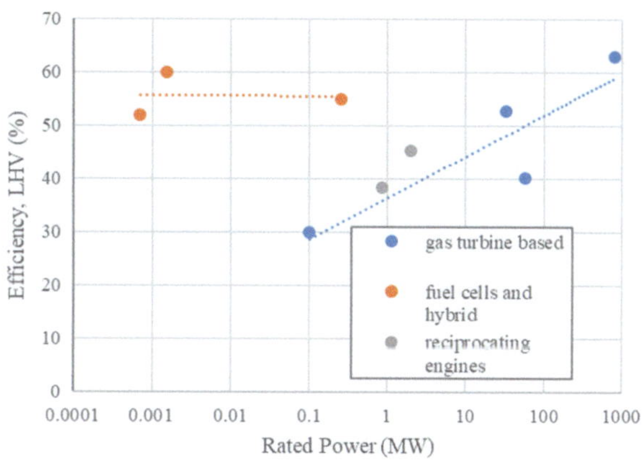

erally have fuel specifications limiting the amount of hydrogen whereas fuel cells operate on natural gas, which can be either internally or externally reformed.

The world largest fuel cell plant is currently in Hwaseong City, South Korea, The 59 MW facility consists of 21 FuelCell Energy DFC3000 power plants, each rated at 2.8 MW, and occupies totally 2 ha of land [18].

A number of developments are ongoing, most mentionable

- Upscaling of fuel cells: It is expected that electrical efficiencies (LHV) of 75–80% can be reached for hybrid systems of 100 MW, but these systems are far from commercialization and upscaling from today's 250 kW systems need to be done in multiple steps, solving various challenge.
- Hybrids, also with steam cycles and internally fired reciprocating engines, are being investigated.
- Developing fuel cells able to work with higher pressures, enabling improved hybrid systems with gas turbines.
- Developing electrolyzer systems that can (easily) be integrated, for example, to industries where large amounts of waste heat are available.

Finally, the reader may note that the speed of development/implementation of future power producing hybrid system with a significant improved electrical efficiency would increase if the price of energy would rise significantly.

Acknowledgements The Åforsk foundation (project 17-331) is acknowledged.

References

1. Khan MA (2011) Multiphysics modelling of PEFCs-with reacting transport phenomena at micro and macroscales. Doctoral dissertation, Lund University
2. Andersson M, Nakajima H, Kitahara T, Shimizu A, Koshiyama T, Paradis H, Yuan J, Bengt S (2014) Comparison of humidified hydrogen and partly pre-reformed natural gas as fuel for solid oxide fuel cells applying computational fluid dynamics. Int J Heat Mass Transf 77:1008–1022
3. Butler A, Spliethoff H (2018) Current status of water electrolysis for energy storage, grid balancing and sector coupling via power-to-gas and power-to-liquids: a review. Renew Sustain Energy Rev 82(3):2440–2454
4. Götz M, Lefebvre J, Mörs F, Koch AM, Graf F, Bajohr S, Reimert R, Kolb T (2016) Renewable power-to-gas: a technological and economic review. Renew Energy 85:1371–1390
5. Navasa M (2016) Three dimensional multiphysics modeling of reversible solid oxide electrochemical cells for degradation studies. Doctoral dissertation, Lund University
6. Andersson M (2011) Solid oxide fuel cell modeling at the cell scale—focusing on species heat, charge and momentum transport as well as the reaction kinetics and effects. Doctoral dissertation, Lund University
7. Andersson M, Beale S, Espinoza M, Wu Z, Lehnert W (2016) A review of cell-scale multiphase flow modeling, including water management, in polymer electrolyte fuel cells. Appl Energy 180:757–778
8. Andersson M, Sundén B (2017) Technology review—solid oxide fuel cell. Energiforsk AB
9. Andersson M, Yuan J, Sundén B (2010) Review on modeling development for multiscale chemical reactions coupled transport phenomena in SOFCs. J Appl Energy 87:1461–1476

10. Andersson M, Beale S, Reimer U, Lehnert W, Stolten D (2018) Interface resolving two-phase flow simulations in gas channels relevant for polymer electrolyte fuel cells using the volume of fluid approach. J Hydrog Energy 43:1961–2976
11. Tang CL, Huang ZH, Law CK (2011) Determination, correlation, and mechanistic interpretation of effects of hydrogen addition on laminar flame speeds of hydro-carbon-air mixtures. Proc Combust Inst 33:921–928
12. Schönborn A, Sayad P, Konnov A, Klingmann J (2014) OH*-chemiluminescence during autoignition of hydrogen with air in a pressurised turbulent flow reactor. Int J Hydrog Energy 39(23):12166–12181
13. Siemens, [Online]. Available: https://www.turbomachinerymag.com/siemens-will-deliver-one-of-the-worlds-most-efficient-combined-cycle-power-plants-to-uk/. Accessed 10 Sept 2018
14. Sebastian S, Stathopoulos S, Tom T, Oliver PC (2015) Blue combustion: stoichiometric hydrogen-oxygen combustion under humidified conditions, pp GT2015-43149
15. Milewski J (2015) Hydrogen utilization by steam turbine cycles. J Power Technol 95:258–264
16. Ito H, Miyazaki N, Ishida M, Nakano A (2016) Efficiency of unitized reversible fuel cell systems. Int J Hydrog Energy 41:5803–5815
17. Kobayashi Y et al (2015) Mitsubishi heavy industries technical review, vol 52, no. 2
18. FCE wraps up world's largest fuel cell park in Korea, new Seoul one. Fuel Cell Bull, pp 6 (2014)
19. van Biert L, Woudstra T, Godjevac M, Visser K, Aravind P (2018) A thermodynamic comparison of solid oxide fuel cell-combined cycles. J Power Sour 397, 382–396

On Soot Reduction Using Oxygenated Combustion in Counterflow Diffusion Flames

Krishna C. Kalvakala and Suresh K. Aggarwal

1 Introduction

Major advances have been reported in fundamental and applied research pertaining to combustion and pollutants formation. However, fundamental understanding and modeling of soot formation continue to pose challenges due to the complex chemical and physical processes involved [1]. Soot also presents significant health and environmental hazards. Fine particles with diameters on the order of micrometers (PM10, PM2.5, and smaller) can penetrate human lung tissue and subsequently enter into bloodstream, resulting in serious health issues, such as cardiovascular and lung diseases [2–4]. Soot from the combustion of fossil fuels accounts for a significant portion of ambient particulate matter [5], with a significant negative impact on life expectancy. Also, soot (aerosol) particles ubiquitous in earth's atmosphere contribute to the global warming, and as reported in the literature [6], the efficacy of soot climate forcing is twice that of CO_2. Major processes associated with soot production include the formation of soot precursors, particularly PAHs (polycyclic aromatic hydrocarbons) species, particle formation through nucleation and polymerization, particle surface growth through surface reactions and coagulation, particle agglomeration and aggregation, and the oxidation of PAHs and soot. Traditional modeling approaches have been mostly empirical [7]. However, significant progress has been made in the modeling of soot formation processes, and reasonably reliable predictive capabilities have been reported in recent years, particularly for laminar flames [8, 9]. Simultaneously, several innovative strategies have been developed for reducing both NO_x and soot in combustion systems. Such strategies include:

1. Mild combustion (subset of HiTAC);

K. C. Kalvakala · S. K. Aggarwal (✉)
Department of Mechanical & Industrial Engineering,
University of Illinois at Chicago, Chicago, IL 60607, USA
e-mail: ska@uic.edu

A. K. Gupta et al. (eds.), *Innovations in Sustainable Energy and Cleaner Environment*, Green Energy and Technology,
https://doi.org/10.1007/978-981-13-9012-8_11

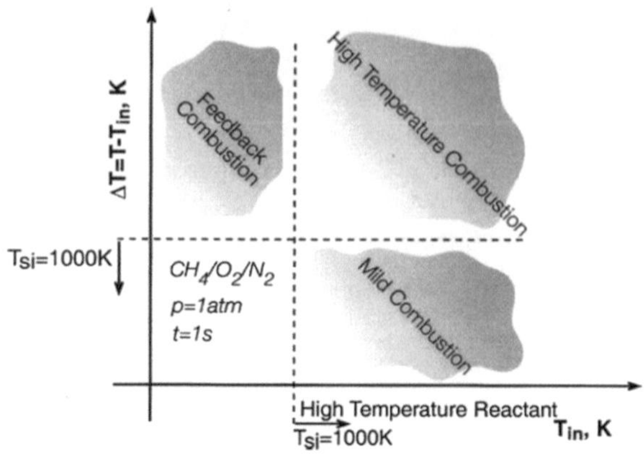

Fig. 1 T_{si}—Self-Ignition temperature; T_{in}—inlet temperature; ΔT—difference between the peak reactor temperature and the temperature of reactants. From Ref. [13]

2. **Low-t**emperature **c**ombustion (LTC);
3. Oxy-fuel/oxygen-enhanced combustion.

HiTAC (**hi**gh-**t**emperature **air c**ombustion) strategy has been utilized in industrial furnaces [10–12]. The basic idea is to use the waste heat from hot products to preheat the inlet air. This results in higher thermal efficiency and fuel saving due to heat recovered from hot products and higher furnace temperatures. This approach can also be used for the combustion of low calorific fuels or to operate the furnace under lean conditions. However, NO_x and soot emissions are generally increased due to higher temperatures, and there are issues due to nonuniform heat flux and temperature distribution in the furnace. Many of these issues can be addressed using the mild combustion strategy [13], which is an extension of HiTAC approach. As indicated conceptually in Fig. 1, the maximum temperature is controlled by diluting the air stream with certain amount of exhaust gases, which lowers the amount of oxygen in the air stream while increasing its temperature. In general, the reactant stream is preheated to a temperature higher than the mixture self-ignition temperature (T_{si}), and the maximum allowable temperature increase during combustion with respect to inlet temperature is lower than T_{si} (Fig. 1). For instance, as shown in Fig. 1, for propane combustion, with air preheated to temperatures of 1000 °C or higher along with O_2 concentrations between 5 and 15%, this approach produces flames that are predominantly blue in color (nearly soot-free) with significantly lower NO_x [11, 13]. As discussed in Ref. [13], these conditions can also produce stable flames at very low equivalence ratios that are not possible with normal combustion air. Thus, mild combustion, which is characterized by both an elevated temperature of reactants and low temperature increase during combustion, can significantly reduce both NO_x and soot emissions.

Low-temperature combustion (LTC) is a very promising technology that is currently being widely researched to achieve ultra-low NO_x and soot emissions in inter-

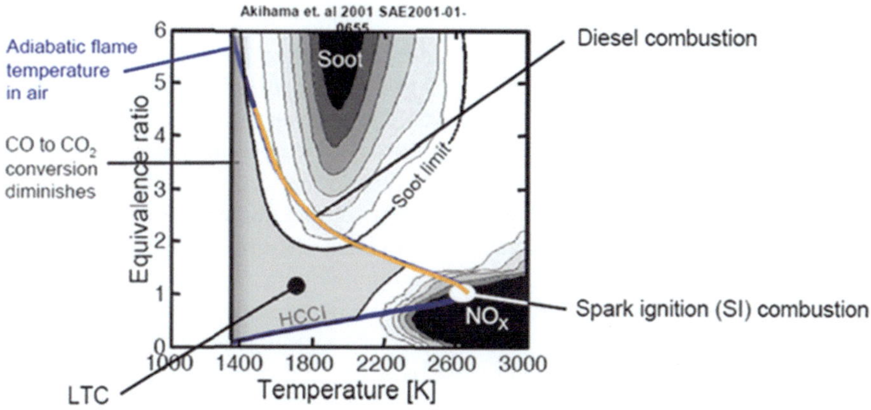

Fig. 2 Low-temperature combustion (LTC) regime shown in $T - \phi$ coordinates. Adopted from [17]

nal combustion engines and gas turbine combustors. The basic idea is to deploy highly diluted, nearly homogeneous fuel–air mixtures by using combustion products and other means [14]. Figure 2 shows qualitatively the soot and NO_x formation regimes in temperature–equivalence ratio ($T - \phi$) space relevant to IC engines. As indicated, LTC approach aims to operate engine in the low temperature, low equivalence regime, thus achieving significantly low soot and NO_x emissions. Homogenous charge compression ignition (HCCI), premixed charged compression ignition (PCCI), and reactivity controlled compression ignition (RCCI) are some of the known variants of this strategy [15, 16]. The challenges faced by this approach include precise control of ignition event in compression ignition engines and flame stability under fuel lean conditions in gas turbines [17].

Oxygen-enhanced combustion (OEC) or oxy-fuel combustion is another approach that has been extensively investigated for reducing soot in combustion systems. Traditionally, the approach has employed increased amount of oxygen (>21% by volume) in the air stream along with a flue gas to regulate the temperature [18]. Another approach to reduce soot emissions is to use oxy-fuels such as alcohols and biodiesels [19, 20]. However, both of these approaches lead to higher flame temperatures and NO_x emissions. An alternate strategy for oxygenated combustion is to use oxygen-enriched air stream along with nitrogen-diluted fuel stream. This modifies the stoichiometric mixture fraction (ζ_{st}) while maintaining flame temperature nearly constant [21–25]. Chen and Axelbaum [23] showed experimentally that increasing ζ_{st} makes a counterflow flame more resistant to extinction, with the extinction strain rate increasing with ζ_{st}. Du and Axelbaum [21] examined the effect of this approach on soot inception in ethylene flames, while Skeen et al. [24] demonstrated its effectiveness in reducing PAH species in ethylene flames. It was shown that by increasing ζ_{st}, one can significantly reduce PAH emissions, and the effect was attributed to the changes in flame structure, resulting in higher oxygen concentration for PAH oxidation in high-temperature regions. However, Sugiyama [26] and Lin and Faeth [27] attributed the PAH inhibiting effect of this approach to hydrodynamic effects. Thus, the previous studies have examined some important aspects of this oxy-combustion strategy, especially its effect on PAH emissions. However, the effect of this strategy on soot emission has not been reported. Moreover, there is limited understanding on the extent to which flame structure and hydrodynamics effects play roles in reducing PAHs and soot formation. In addition, the effects of fuel properties and pressure on PAHs and soot emissions in oxygenated flames have not investigated.

The major objective of this chapter is to examine in detail the oxygen-enhanced strategy of increasing ζ_{st} in reducing PAH and soot formation in counterflow laminar diffusion flames. ζ_{st} is increased continuously by simultaneously using nitrogen-diluted fuel stream and oxygen-enriched oxidizer stream, and its effect on PAH and soot formation and oxidation processes is characterized. Results also focus on identifying the effects of hydrodynamics and changes in flame structure in inhibiting soot emission in these flames. In addition, results from the reaction path analysis are presented in order to identify the dominant reactions associated with the formation of PAH precursors at different oxygenation levels. Finally, the effects of fuel molecular structure (i.e., the presence of C=C bond) and pressure on PAHs and soot formation are analyzed. Note that, most of the results presented in this chapter are taken from our two recent studies concerning the aforementioned aspects [22, 28].

2 Numerical Model

Counterflow flames all fuels and different pressures are simulated using the CHEMKIN-PRO software [29]. A counterflow configuration involves two opposing jets issuing from two identical coaxial nozzles. The distance between the nozzles is 20 mm and the inlet temperatures of fuel and oxidizer jets are maintained at 300 K. The velocities at the two nozzle exits are calculated by using the given strain rate and equating the momenta of the two jets. The global strain rate (a_G), as defined in Eq. (1), is kept constant at 100 s^{-1} [30]. Other boundary conditions, such as axial velocities and species mole fractions at inlet boundaries for ethylene flames are listed in Table 1. As the global strain rate is fixed, the inlet axial velocities can be specified independent of pressure. The grid independence was established by using successively finer grids and varying the GRAD and CURV parameters such that the solution is nearly independent of the grid system.

$$a_G = \frac{2v_o}{L}\left(1 + \frac{v_f}{v_o}\frac{\sqrt{\rho_o}}{\sqrt{\rho_f}}\right) \tag{1}$$

As shown in Table 1, five cases are considered for each fuel at atmospheric pressure with varying stoichiometric mixture fraction (ζ_{st}). The effect of pressure is examined by considering ethylene flames at different pressures. The mixture fraction as defined in Ref. [22] is written as:

Table 1 Species mole fractions and axial velocities for ethylene diffusion flames at a strain rate of 100 s^{-1}

Cases—percentage of N_2 removed from air (%)	Fuel stream			Air stream			Stoichiometric mixture fraction
	C_2H_4 mole fraction	N_2 mole fraction	Velocity (cm/s)	Velocity (cm/s)	O_2 mole fraction	N_2 mole fraction	
0	1	0.0	50.745	50	0.21	0.79	0.064
25	0.262	0.738	50.926	50	0.262	0.738	0.243
50	0.151	0.849	51.225	50	0.347	0.653	0.423
75	0.106	0.894	51.808	50	0.515	0.485	0.602
100	0.081	0.919	53.452	50	1.0	0.0	0.782

From Ref. [22]

$$Z(F) = \sum\left[N_C \times Y_i \times \left\{\frac{M_C}{M_i}\right\}\right] + \sum\left[N_h \times Y_i \times \left\{\frac{M_h}{M_i}\right\}\right] \tag{2}$$

where N_C and N_h are the number of carbon and hydrogen atoms in species i, Y_i is the mass fraction of species i, M is the molecular weight, and the sum is over all the species. The elemental mass fraction here is thus based on C and H atomic species. In order to accommodate a broader set of conditions, Z is normalized as:

$$\zeta = \frac{[Z(F) - Z(F,l)]}{[Z(F,r) - Z(F,l)]} \tag{3}$$

where subscripts r and l correspond to fuel and oxidizer streams, respectively. Since the stoichiometric mixture fraction (ζ_{st}) requires only the fuel and oxidizer mass fractions in stoichiometric proportions, ζ_{st} can also be calculated using [15]:

$$\zeta_{st} = \left(1 + \left(\frac{Y_f M_o \vartheta_o}{Y_o M_f \vartheta_f}\right)\right)^{-1} \tag{4}$$

where Y_f and Y_o are the mass fractions of fuel and oxygen at the respective boundaries, and ν_f and ν_o are the stoichiometric coefficients for fuel and oxygen.

In order to compute the flame structure and PAH formation accurately, the detailed Ranzi mechanism is used in the present study. The mechanism consists of 197 species and 4909 reactions with coronene ($C_{24}H_{12}$) being the largest species. It has been extensively validated in our previous studies [16, 27, 31]. As discussed in Ref. [22], the soot model considers one soot nucleation reaction based on pyrene ($C_{16}H_{10}$) recombination and HACA surface growth reactions, while the soot oxidation reactions are based on hydroxyl radical and oxygen.

Nucleation

$$2C_{16}H_{10} \leftrightarrow 32C + 20C_{soot}{-}H + 28.75C_{soot}^{\cdot} \tag{R1}$$

HACA mechanism

$$H + C_{soot}{-}H \leftrightarrow C_{soot}^{\cdot} + H_2 \tag{R2}$$

$$OH + C_{soot}{-}H \leftrightarrow C_{soot}^{\cdot} + H_2O \tag{R3}$$

$$C_{soot}^{\cdot} + H \leftrightarrow C_{soot}{-}H \tag{R4}$$

$$C_{soot}^{\cdot} + C_2H_2 \leftrightarrow C_{soot}{-}H + 2C + H \tag{R5}$$

Oxidation

$$C_{soot}^{\cdot} + C + OH \leftrightarrow C_{soot}\text{–}H + CO \tag{R6}$$

$$C_{soot}^{\cdot} + 2C + O_2 \leftrightarrow C_{soot}^{\cdot} + 2CO \tag{R7}$$

During nucleation process, the pyrene recombination reaction forms one soot nucleus containing 32C atoms. Further, the soot surface growth is modeled by the HACA mechanism [32]. The soot model has also been validated in previous studies [16, 27]. The dynamics of soot formation process is modeled by solving the particle size distribution functions (PDSFs) using a detailed lumping technique—sectional approach [33].

3 Results and Discussions

Results are provided in two subsections. The first one discusses the effects of oxygenation and fuel unsaturation, while the second discusses the effect of oxygenation at higher pressures on flame structure, PAHs, and soot formation.

3.1 Effect of Oxygenation and Fuel Unsaturation on PAH and Soot Formation

3.1.1 Flame Structure

Figure 3a and b depict the structures for two propene flames with $\zeta_{st} = 0.064$ and 0.423, respectively. Profiles of temperature and important species including soot precursors (acetylene, benzene, and pyrene) are shown as a representation of the flame structure. As indicated, an increase in ζ_{st} shifts the flame toward the fuel side. For instance, at $\zeta_{st} = 0.064$, the flame is located on the oxidizer side and the peaks in soot precursor profiles occur in fuel pyrolysis zone, while for $\zeta_{st} = 0.423$, the flame shifts toward the fuel side with significantly reduced amounts of soot precursors formed. Note that, this behavior is observed for all fuels under consideration.

Figure 4 plots the peak flame temperatures and peak acetylene mole fractions for the five flames for different stoichiometric mixture fractions (ζ_{st}) and the three fuels—ethylene, propane, and propene. As discussed above, for a given fuel, the adiabatic flame temperature is held constant across the five cases. However, Fig. 4 shows small variation in the peak flame temperature. As discussed in Ref. [21], this variation in peak temperature with ζ_{st} can be attributed to the amount of N_2 available at the flame location. It may also be attributed to the particular configuration used. In short, adiabatic flame temperatures are calculated at equilibrium conditions,

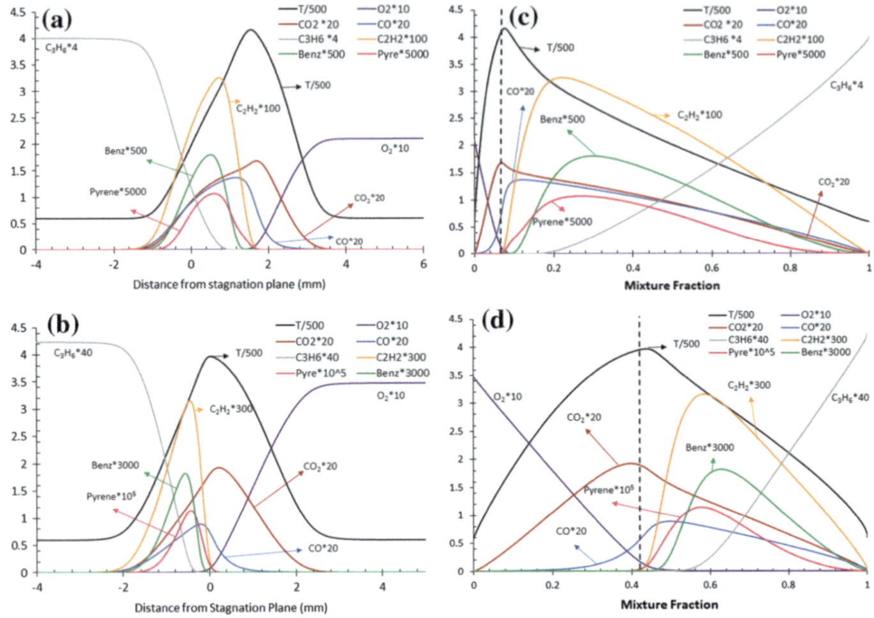

Fig. 3 Structures of two propene flames for $\zeta_{st} = 0.064$ (**a**, **c**) and 0.423 (**b**, **d**). Profiles of temperature and mole fractions of important species are shown in both spatial and mixture fraction coordinates. From Ref. [22]

while counterflow flames have transport processes involved in establishing the flame. Further, as shown in Fig. 4, for a given fuel, the differences in peak temperatures are relatively small (<60 K), and thus, the concept of maintaining a constant flame temperature can be considered acceptable. Also, at a given ζ_{st}, the differences in peak flame temperatures for the three fuels are related to the difference in enthalpies of formation of these fuels.

3.1.2 Effect on Soot Precursors—Acetylene, Benzene, and Pyrene Formation

The effect of oxygenation on PAH and soot formation can be attributed to two factors—residence time that occurs due to change in hydrodynamics and the flame structure effect. However, these two effects are coupled and together represent the effect of oxygenation in reducing PAH and soot.

In order to understand the effect of oxygenation and fuel unsaturation (C=C bond) on PAH formation, a detailed reaction path analysis has been carried out for all three fuels. Figure 4 shows that the highest amount of acetylene (C_2H_2) is formed in ethylene flames, followed by propene and propane flames. Reaction path analysis indicates that for a given fuel, the reduction of ethylene into vinyl, which is fol-

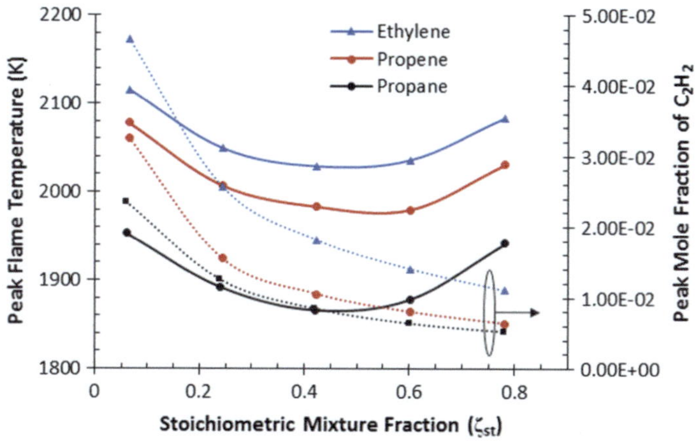

Fig. 4 Peak flame temperature (solid lines) and peak C_2H_2 mole fraction (dotted) versus ζ_{st} for three fuels. From Ref. [22]

lowed by the H abstraction reaction of vinyl, is the major reaction for the formation of acetylene. Consequently, ethylene flames have the highest amount of acetylene formed. Comparing propane and propene flames, we identified additional routes in propene flames (absent in propane flames), which lead to higher amounts of acetylene in propene flames. These reactions (R8, R9, and R10) involve allyl, propyne, and C_3H_5 (structure CH_2CCH_3).

$$\text{Allyl} + \text{H} \leftrightarrow \text{Vinyl} + \text{CH}_3 \tag{R8}$$

$$\text{Propyne} + \text{H} \leftrightarrow \text{C}_2\text{H}_2 + \text{CH}_3 \tag{R9}$$

$$\text{CH}_2\text{CCH}_3 \leftrightarrow \text{C}_2\text{H}_2 + \text{CH}_3 \tag{R10}$$

Figure 7 presents the emission indices (calculated using the formula below) of benzene and pyrene for all fuels at various ζ_{st} conditions.

Emission Index of species P:

$$\text{EIP} = \frac{\int_0^L M_\text{P}\dot{\omega}_\text{P}\text{d}x}{-\int_0^L M_\text{fuel}\dot{\omega}_\text{fuel}\text{d}x}$$

where M_P is the molecular weight of species P, L is the distance between the two nozzles, and $\dot{\omega}$ is the net species production/consumption rate. We notice that the amounts of benzene and pyrene formed are the highest in propene flames, followed by ethylene and propane flames.

Figure 5 presents the important paths for benzene formation in ethylene flames ($\zeta_{st} = 0.064$). Benzene formation in ethylene flames involves two paths, one through

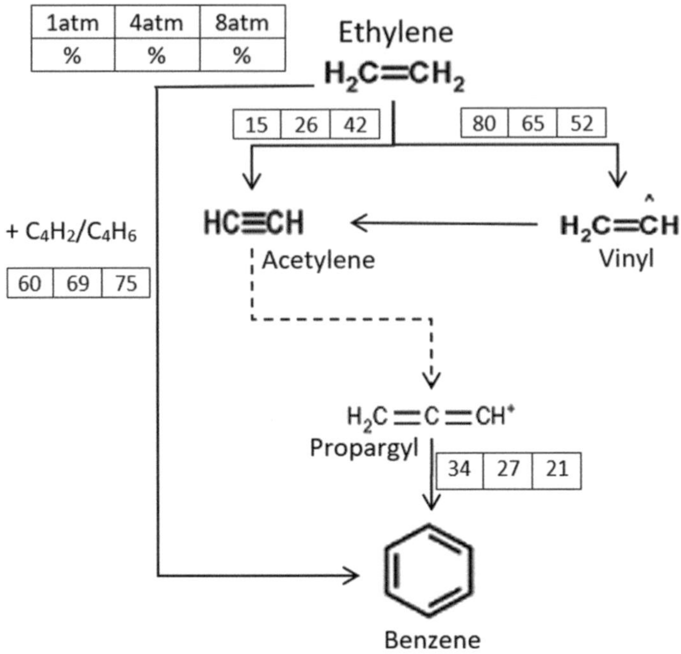

Fig. 5 Key benzene formation routes in ethylene flames ($\zeta_{st} = 0.064$) at 1, 2, 4, and 8 atm. From Ref. [22]

recombination of propargyl and other through pyrolysis reaction path (C2/C4 path). The C2/C4 path involves reactions with cyclobutadiene (C_4H_4), diacetylene (C_4H_2), and butadiene (C_4H_6) species, where these species are mainly formed from acetylene and ethylene. Further, high amounts of acetylene make the C2/C4 path dominant for benzene formation in ethylene flames. The effect of pressure on path (shown in Fig. 5) is discussed in a later section.

Figure 6 presents the major intermediates and benzene formation path in propane and propene flames. It is noticed that propargyl recombination is the dominant path for benzene formation in both propane and propene flames. Propargyl is mainly formed through H abstraction reactions involving allene and propyne isomers. Further, reaction between acetylene and methylene to form propargyl is a minor route in propane and propene flames (but is important in ethylene flames). Propargyl thus formed undergoes recombination reaction to form benzene.

While benzene formation routes are similar in propane and propene flames, the amount of benzene formed is much higher in propene flames. This is due to the fact that propene is an intermediate species in propane flames, but is the fuel itself in propene flames. In propane flames, propene is formed through two paths. One involves propane pyrolysis, while the other involves the reduction of isopropyl. Also note that about 30% of propene produced is converted back into n-propyl, which leads to the production of ethylene but reduces the amount of propene available

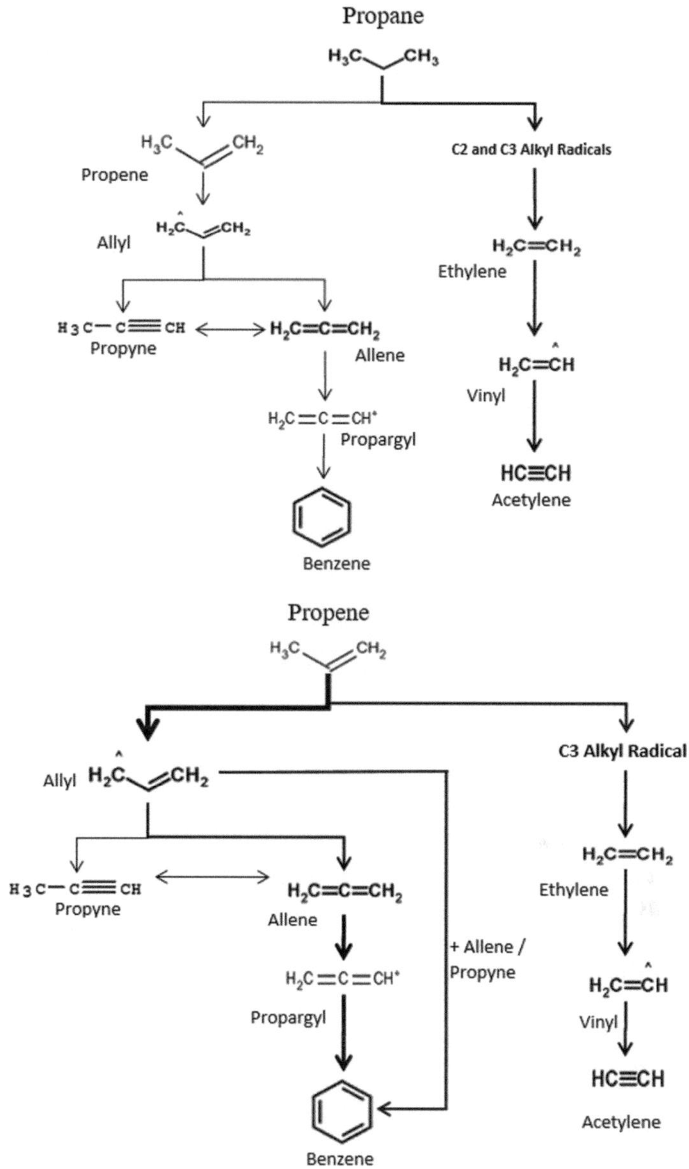

Fig. 6 Dominant acetylene and benzene formation routes in propane and propene diffusion flames at a strain rate of 100 s^{-1}. The air stream contains 21%O_2 and 79%N_2. Line thickness indicates the relative contribution of each path. From Ref. [22]

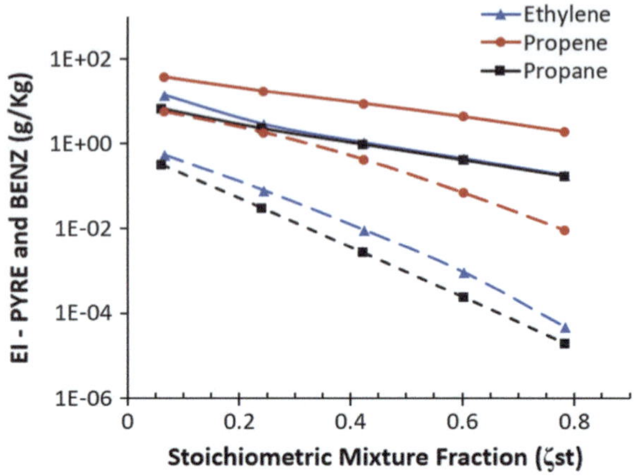

Fig. 7 Emission indices of benzene and pyrene versus ζ_{st}. Solid and dashed lines indicate EIBENZ and EIPYRE, respectively. From Ref. [22]

for conversion into PAHs. Another reason for the increased benzene formation in propene flames is due to higher amounts of allene, propyne, and allyl, which undergo reactions R11 and R12 to form benzene directly (without forming propargyl). This path contributes about 20% of the total benzene produced. Hence, we notice significantly higher amounts of benzene in propene flames compared to propane and ethylene flames. Further, as shown in Fig. 7, ethylene flames produce higher amounts of benzene than propane flames, but smaller amounts than propene flames. This is attributed to higher amounts of stable acetylene in ethylene flames. Acetylene is the first and major stable intermediate formed in ethylene flames. Thus, the formation of all the later intermediates directly involves acetylene. For instance, the species involved in benzene production, C4 species—butadiene, cyclobutadiene, diacetylene, and propargyl—are mainly formed from acetylene through various paths. Due to significantly high amounts of acetylene in ethylene flames than in propane flames at $\zeta_{st} = 0.064$, ethylene flames have higher amounts of benzene. Propene flames have much higher benzene due to multiple contributing routes.

$$\text{Allene} + \text{Allyl} \rightarrow C_6H_6 + H_2 + H \tag{R11}$$

$$\text{Propyne} + \text{Allyl} \rightarrow C_6H_6 + H_2 + H \tag{R12}$$

The increase in ζ_{st}, which is related to the increasingly higher oxygen content, has a significant impact on the decomposition of fuels and thus on the intermediates formed. For instance, in ethylene flames, for no dilution case ($\zeta_{st} = 0.064$) ethylene decomposes into vinyl accounting for 94% of vinyl formation, while for 100% dilution case ($\zeta_{st} = 0.782$), ethylene produces 85% vinyl, 10% ethyl, and 2% acetylene, with the rest being HCO and CH_3. This is one reason we see lower acetylene formation at higher ζ_{st} levels. The residence time also has some effect, as it decreases the fuel pyrolysis rate but increases oxidation reaction rates of PAHs with O and OH radicals, and thus decreases the amounts of soot precursors for all the three fuels, as

Table 2 Net rates (mol/cm^3 s) of important reactions contributing toward the consumption of propargyl in two propene flames for 0 and 100%N_2 dilution levels

Reaction#	Reactions	Net reaction rate (mol/cm^3 s)	
		$\zeta_{st} = 0.064$ (0%N_2)	$\zeta_{st} = 0.782$ (100%N_2)
R13	Propyne + M ↔ C_3H_3 + H + M	2.5×10^{-5}	-2.94×10^{-5}
R14	C_3H_3 + H ↔ C_3H_2 + H_2	3.16×10^{-5}	6.0×10^{-5}
R15	C_3H_3 + O ↔ CH_2O + C_2H	1.14×10^{-5}	4.67×10^{-5}
R16	C_3H_3 + OH ↔ C_3H_2 + H_2O	3.22×10^{-5}	6.77×10^{-5}
R17	C_3H_3 + OH ↔ C_2H_3 + HCO	8.51×10^{-5}	13.8×10^{-5}
R18	$2C_3H_3$ + M ↔ C_6H_6 + M	3.16×10^{-5}	1.09×10^{-5}
R19	$2C_3H_3$ ↔ C_6H_5 + H	5.22×10^{-5}	1.37×10^{-5}

Negative sign indicates that the reaction is reversed
From Ref. [22]

the oxygenation level is increased. The residence time effect is discussed further in this chapter.

Figure 7 clearly indicates that the amounts of benzene and pyrene formed decrease dramatically with increasing ζ_{st} or with increasing oxygenation level. This can be attributed to increased rates of reactions involving the oxidation of intermediates into smaller hydrocarbons. This reduction leads to lower amounts of intermediates that are responsible for benzene formation. In propene flames, allene, propyne, and propargyl are important intermediates involved in the formation of benzene. Increase in oxygenation level leads to increased concentrations of O and OH radicals that cause breakdown of these species into smaller hydrocarbons. Table 2 shows some representative peak net reaction rates in two propene flames with $\zeta_{st} = 0.064$ and 0.782, indicating the effect of oxygenation on intermediates. For instance, the peak net reaction rates of R15–R17 increase nearly by a factor of 2 or more, thus reducing propargyl into smaller intermediates, which have no role in benzene formation. Note that, C_3H_2 formed through R14 and R16 also decomposes into vinyl and HCO (not shown here). Another factor affecting propargyl formation in propene flames is due to reaction R13, which is reversed at higher ζ_{st}. For instance, without oxygenation, R13 contributes about 7% of total propargyl produced. As the level of oxygenation is increased, there is increased availability of H in the reaction zone, and this reaction is reversed and thus consumes propargyl.

3.1.3 Effect on Oxygenation and Unsaturation on Soot Formation

The sooting behavior is characterized in terms of soot number density and soot volume fraction. Since the soot number density is closely related to particle nucleation reaction rate, which is based on pyrene recombination reaction, the soot number density correlates closely with pyrene concentration. On the other hand, the soot

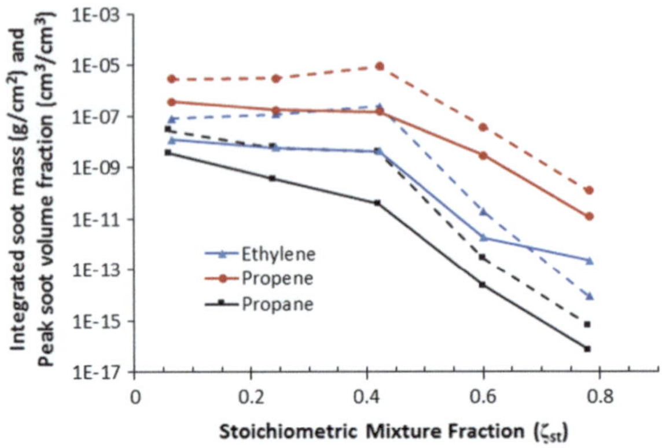

Fig. 8 Integrated soot mass (solid lines) and peak soot volume fraction (dotted) plotted versus ζ_{st} in ethylene, propane, and propene flames. From Ref. [22]

volume fraction also depends on the surface growth and soot oxidation rates, in addition to the nucleation rate. Figure 8 presents the variation of integrated soot mass (based on integrated soot volume fraction) and peak soot volume fraction (f_v) with respect to ζ_{st} for the three fuels. As ζ_{st} is increased, the total soot mass decreases monotonically for all three fuels. However, the peak f_v decreases monotonically with ζ_{st} in only propane flames. In ethylene and propene flames, the peak f_v first increases with ζ_{st} (for ζ_{st} < 0.423) and then decreases for ζ_{st} > 0.423. This is explained by considering reaction R5 or R20.

$$C_{soot}^{\cdot} + C_2H_2 \leftrightarrow C_{soot}{-}H + 2C + H \qquad \text{(R20)}$$

This reaction is part of the HACA mechanism, which describes the soot surface growth by converting the open spaces ($C_{soot}^{\cdot}$) into bulk carbon (soot). This reaction is highly temperature dependent, and the analysis indicates that for $\zeta_{st} < 0.423$, temperature in the soot formation region is increasingly higher as ζ_{st} is increased. For instance, for $\zeta_{st} = 0.064$, 0.243, and 0.423, the temperature values in soot-forming region are 1100, 1580, and 1990 K, respectively. These high temperatures coupled with high C_2H_2 concentration (Fig. 9) promote the R20 reaction rate. Consequently, the peak f_v increases with ζ_{st} for $\zeta_{st} < 0.423$, although the soot formation region narrows drastically. With further increase in ζ_{st}, the rate of surface growth decreases, mainly due to the sharply reduced C_2H_2 concentration. In addition, due to the increased availability of OH radicals at higher ζ_{st}, the soot oxidation rate increases. Consequently, for $\zeta_{st} > 0.423$, the soot volume fraction decreases as ζ_{st} is increased. In fact, for $\zeta_{st} =$ 0.602 and 0.782, flames for all three fuels can be considered as essentially soot-free since the total soot mass is negligibly small, of the order 10^{-13} g/cm^2 at $\zeta_{st} = 0.423$ compared to a value of the order of 10^{-8} g/cm^2 at $\zeta_{st} = 0.064$. Also, as mentioned earlier, the peak f_v decreases monotonically with ζ_{st} in propane flames. This is due to low C_2H_2 concentration in the soot-forming region of these flames.

Furthermore, as shown in Fig. 8, the integrated soot mass is higher in propene flames than that in ethylene and propane flames. This is due to the presence of

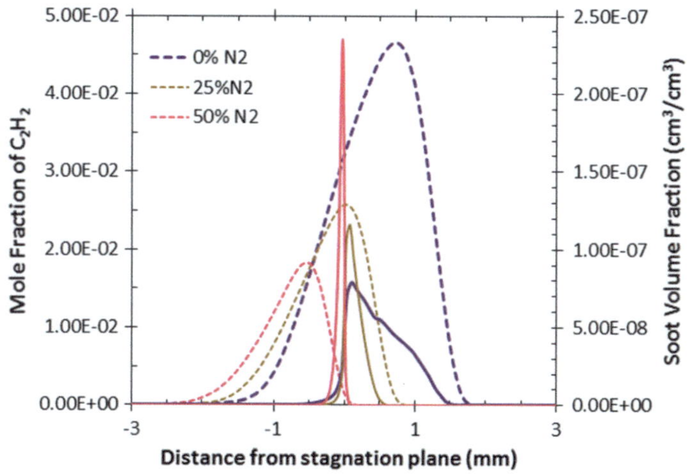

Fig. 9 Profiles of acetylene mole fraction (dotted lines) and soot volume fraction (solid) in ethylene/air flames for different levels of oxygenation. From Ref. [22]

double bond (C=C), which significantly affects the fuel pyrolysis chemistry and leads to higher concentrations of soot precursors (acetylene, benzene, and pyrene) in propene flames. Consequently, the propene flames are the most sooting, followed by ethylene flames, and then propane flames. Figure 10 plots the soot number density and volume fraction profiles in propane and propene flames for different oxygenation levels. As oxygenation is increased, the soot formation region shifts from oxidizer side to fuel side. Consequently, the soot formed in the fuel-rich region gets oxidized in the oxygen-rich region (hydrodynamic effect), and thus, the soot number density and total soot volume fraction decrease in propane and propene flames. However, the reduction is more pronounced in propane flames than in propene flames. This may be due to the higher concentrations of acetylene and PAH species in propene flames compared to those in propane flames, as discussed earlier.

3.1.4 Effect of Residence Time

Strain rate has been identified as an important parameter in establishing stable and soot-free flames. Fu et al. [34] and Wang and Chung [8] observed that in counterflow flames, as the strain rate is increased, the amounts of soot precursors and soot formed decrease noticeably due to the reduced residence time. An important observation from the current study at the fixed global strain rate, as the oxygenation level is increased, leads to a decrease in local residence time, and this can be attributed to the effect of oxygenation on the flame structure. Thus, with the increase in oxygenation level, the fuel decomposition and PAH formation occur in higher velocity region, implying lower residence time in this region [21]. Consequently, the fuel pyrolysis is expected

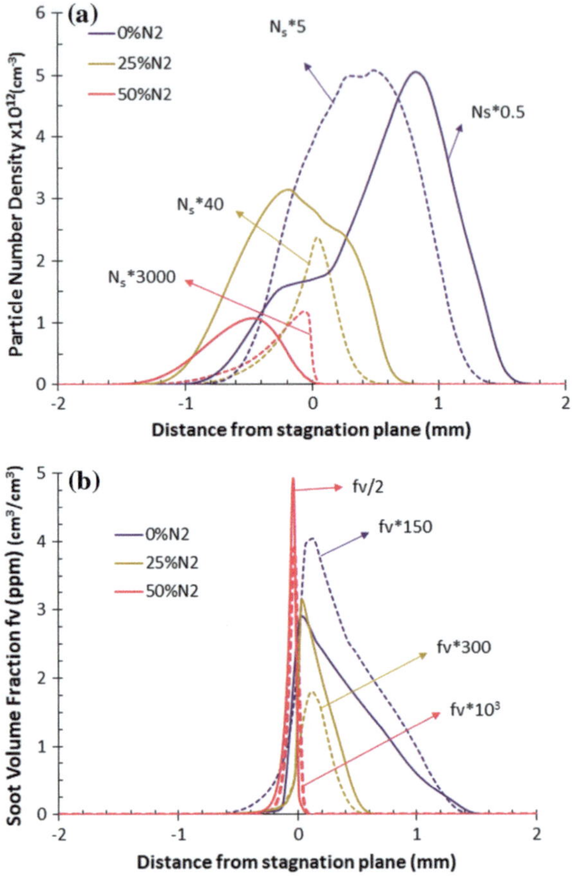

Fig. 10 Soot number density (**a**) and soot volume fraction profiles (**b**) in propane (dotted lines) and propene (solid) flames for different oxygenation levels. Note the values for propane are significantly lower, and thus, appropriate scaling factors are used for improved visibility. From Ref. [22]

to be incomplete resulting in more unburned hydrocarbons due to lower residence time. In contrast, as ζ_{st} is increased, the peak velocity decreases in the soot oxidation region (oxidizer side), implying increased residence time for soot oxidation in this region. Thus, the combined effect of increased O and OH radicals along with higher residence times on the oxidizer side leads to significantly reduced PAH and soot formation for all the three fuels.

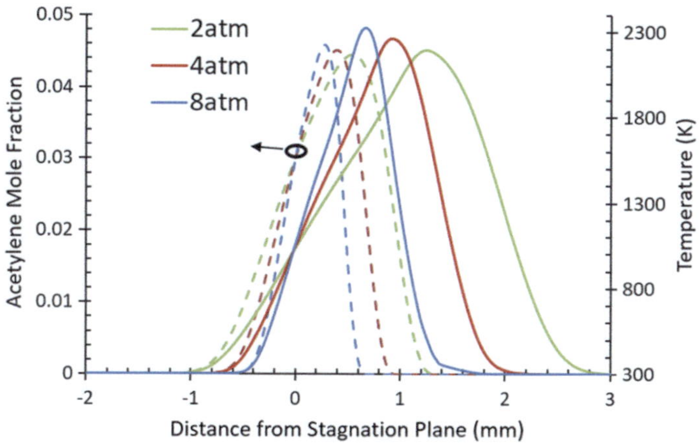

Fig. 11 Temperature (solid) and acetylene profiles in undiluted ethylene diffusion flames at 2, 4, 8 atm and global strain rate of 100 s^{-1}. From Ref. [28]

3.2 *Effect of Pressure*

3.2.1 Effect of Pressure on Flame Structure and Soot Precursors

Before examining the effect of oxygenation at elevated pressures, it is relevant to characterize the effect of pressure on the flame structure and soot precursors at reference conditions (no dilution). As indicated in Fig. 11, there is a significant effect of pressure on the flame structure. As pressure is increased, the flame thickness decreases and the flame location shifts toward the fuel side of the stagnation plane. Note that, an increase in pressure causes an increase in density, resulting in increased flow rates of the fuel and oxidizer streams, and a reduction in flame thickness. Also note that the mass diffusivity varies with temperature and pressure as $\mathcal{D} \propto T^{3/2} P^{-1}$. Thus, the reduced diffusivity at higher pressure reduces the flame thickness. Further, it is reported by Law [35] that flame thickness is proportional to $1/\surd(pa_G)$. In addition, the peak flame temperature increases from 2192 to 2320 K, as pressure increases from 2 to at 8 atm. The increase in peak flame temperature is related to the higher heat release rates at higher pressures [28]. This may be attributed to two factors. One is the reduction in flame thickness or reaction zone thickness. The other factor is related to the fact that major heat release is associated with the formation of CO_2 and H_2O species, and the peak rates of major CO_2 and H_2O formation reactions (R21 and R22) scale linearly with pressure. Peak CO_2 mole fractions increase with pressure (shown in Fig. 13).

$$CO + OH \Leftrightarrow CO_2 + H \tag{R21}$$

$$H_2 + OH \Leftrightarrow H_2O + H \tag{R22}$$

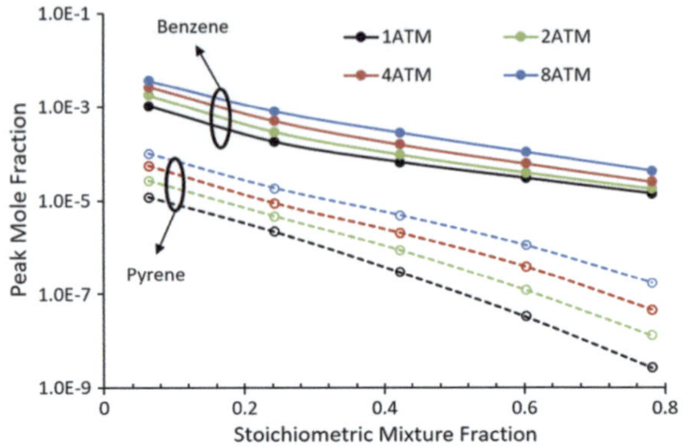

Fig. 12 Peak benzene (solid) and pyrene (dotted) mole fractions versus ζ_{st} in ethylene diffusion flames at 2, 4, and 8 atm. From Ref. [28]

Figure 11 also presents the acetylene mole fraction profiles in flames at 2, 4, and 8 atm. Note that acetylene plays an important role in soot surface growth rate, as discussed in a later section. We notice that the peak mole fraction of acetylene changes negligibly with pressure, ranging between 0.045 and 0.046 for pressures between 2 and 8 atm. As discussed in Refs. [16, 36], in atmospheric ethylene flames, acetylene is formed through H abstraction of vinyl, which is formed predominantly via ethylene decomposition (95%). The reaction path analysis indicated (Fig. 5) that with the increase in pressure, ethylene tends to decompose into acetylene directly while forming smaller amount of vinyl. For example, at $p = 1$ atm, the amounts of ethylene decomposing into vinyl and acetylene are 80 and 15%, respectively, while at $p = 8$ atm, these amounts are 52 and 42%, respectively. However, at higher pressures, due to increased participation of acetylene in the formation of butadiene (C_4H_6), diacetylene (C_4H_2), and cyclobutadiene (C_4H_4) and thus benzene, we notice no net change in the acetylene mole fraction with pressure. Figure 12 shows the peak benzene and pyrene mole fractions at various pressures. For these flames, there is a monotonic increase in peak benzene and pyrene mole fractions with pressure. This is due to the increase in C_4 intermediates formed from acetylene and ethylene. As discussed earlier, benzene in ethylene flames (Fig. 5) is formed via two routes, a C_2/C_4 path and a path through recombination of propargyl. As shown in Fig. 5, the increase in C_4 intermediates increase the contribution of C_2/C_4 path in benzene formation from 60 to 75% as pressure is raised from 1 to 8 atm. Thus, higher concentrations of C_4 intermediates lead to increased rates of reactions involving benzene formation and hence higher amounts of benzene. Further, the reaction path analysis indicated that benzene plays a significant role in the formation of pyrene, though the latter is formed through multiple paths. Hence, the increase in pyrene concentration correlates with the increase in benzene concentration mole fraction.

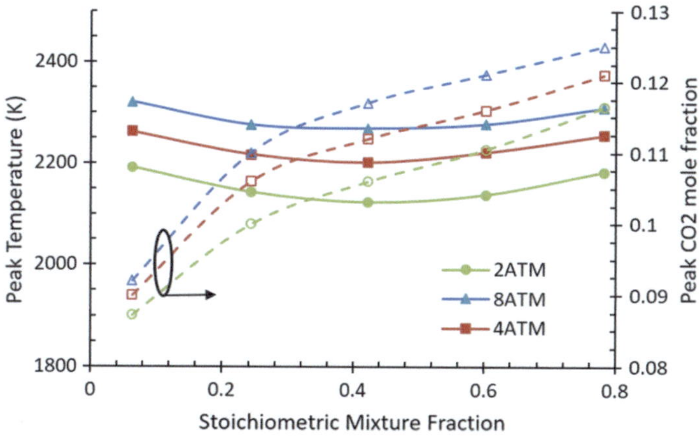

Fig. 13 Peak flame temperatures and peak CO_2 mole fractions versus ζ_{st} for flames at 2, 4, and 8 atm. From Ref. [28]

3.2.2 Effect of Oxygenation at Elevated Pressure

Figure 13 shows peak flame temperatures and CO_2 mole fractions plotted with respect to ζ_{st} for flames at 2, 4, and 8 atm. At any given pressure, we notice a relatively small variation in peak flame temperature with ζ_{st}. This is consistent with results for flames at 1 atm discussed earlier. The increase in peak CO_2 mole fraction with ζ_{st} at higher pressures is discussed later in this chapter. Figure 14a and b show the flame structures at 4 atm for $\zeta_{st} = 0.064$ and 0.423. It is interesting to note that the effect of ζ_{st} on flame structure qualitatively remains the same at higher pressures. Thus, an increase in ζ_{st} moves the flame from oxidizer side to the fuel side of the stagnation plane accompanied by a noticeable reduction in soot precursors. For example, benzene mole fraction decreases from 270 to 15 ppm as ζ_{st} is increased from 0.064 to 0.423. This result is consistent at any pressure considered in the study, as evident from Fig. 12.

Path analysis similar to that conducted at atmospheric pressure was also done at elevated pressures. Figure 15a and b present the benzene and pyrene mole fraction profiles in C/O coordinates for flames at 2, 4, 8 atm, and for different oxygenation levels ($\zeta_{st} = 0.064$, 0.423, and 0.782). The C/O ratio in the flame is defined as:

$$\frac{C}{O} = \frac{\sum_i x\chi_i C_x H_y O_z}{\sum_i z\chi_i C_x H_y O_z} \tag{5}$$

where x, y, z denote the number of carbon, hydrogen, and oxygen atoms, respectively, in species i, χ_i is species mole fraction, and the summation is over all species appearing in the mechanism.

An important observation from Fig. 15 is that at a given pressure, as ζ_{st} is increased, the regions of benzene and pyrene formation shift toward lower C/O ratios indicat-

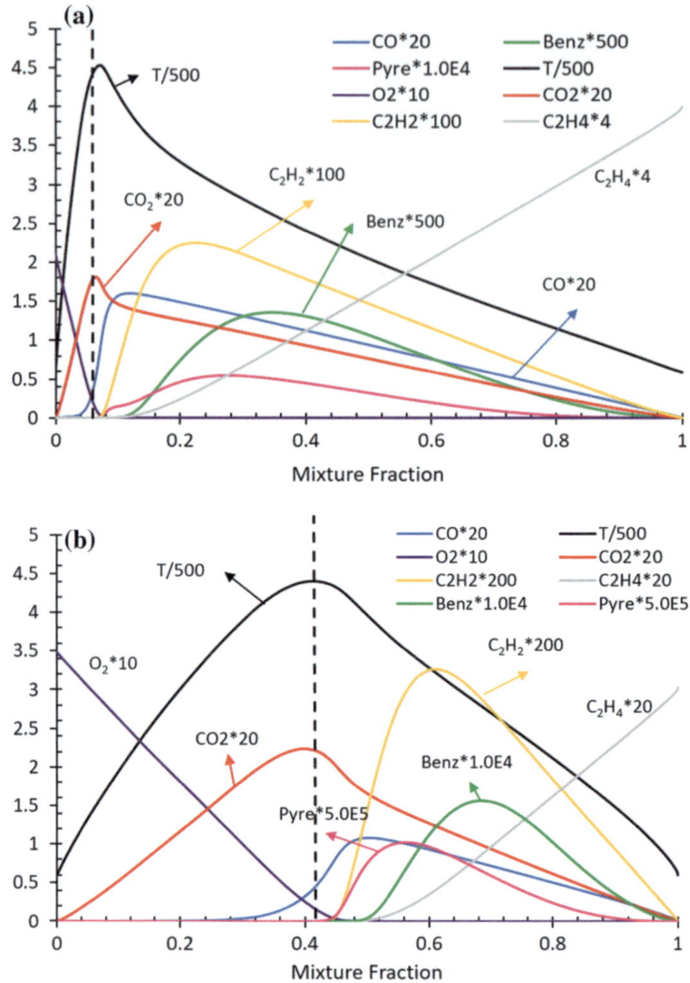

Fig. 14 Flame structures of two ethylene flames at 4 atm for $\zeta_{st} = 0.064$ (**a**) and 0.423 (**b**). Profiles of temperature and mole fractions of important species are plotted in mixture fraction coordinates. From Ref. [28]

ing increased availability of oxidizing species (O_2, OH, O). Skeen et al. [24] and Kalvakala et al. [22] have reported similar results for flames at atmospheric conditions. The reaction path analysis indicated that benzene formation routes are suppressed due to increased concentrations of oxidizing species (O_2, OH, O) in the benzene-forming regions, as key intermediates break down into small hydrocarbons or CO_2. For example, acetylene formed undergoes rapid oxidation through OH and O radicals, and reduces to CH_2, C_2H, and HCCO. This decreases the availability of acetylene (and similarly other hydrocarbons such as diacetylene and butadiene) for further reactions that involve soot precursor formation. Further, the effect of oxy-

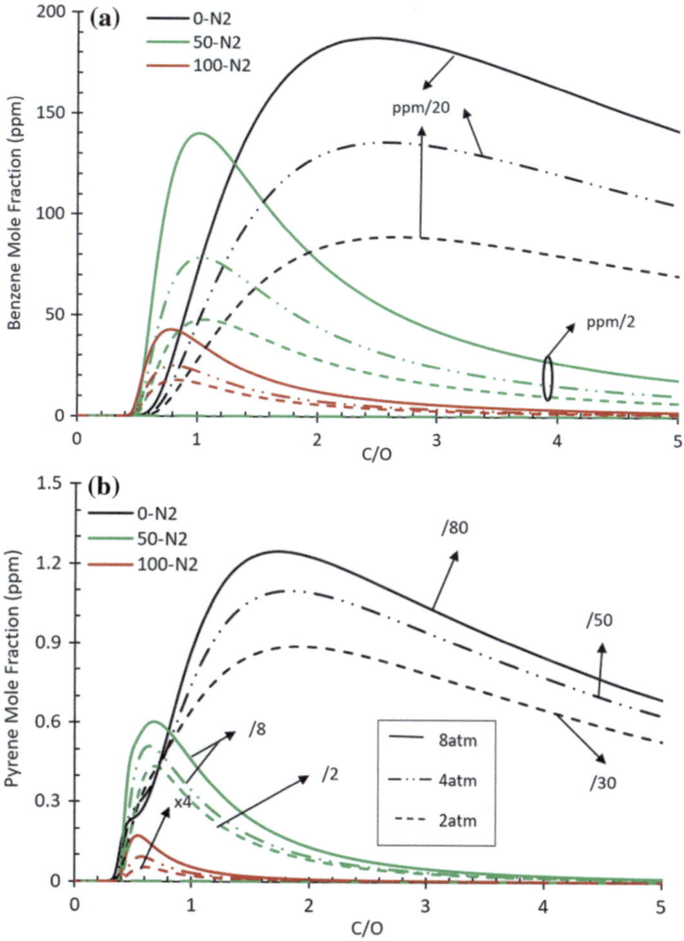

Fig. 15 Comparison of benzene (**a**) and pyrene (**b**) mole fraction profiles in C/O coordinates for flames at 2, 4 and 8 atm, and $\zeta_{st} = 0.064$ (0%N_2), 0.423 (50%N_2), and 0.782 (100%N_2). From Ref. [28]

genation is consistent with the increase in peak CO_2 mole fraction with ζ_{st}, as shown in Fig. 13. For example, the peak CO_2 mole fraction at 4 atm increases from 0.088 to 0.122 as ζ_{st} changes from 0.064 to 0.78. It is also interesting to note that our oxygenation strategy [16, 18] for significantly reducing soot precursors in atmospheric flames is also equally effective at higher pressures. For example, by changing ζ_{st} from 0.064 (0%N_2) to 0.781 (100%N_2) at 2 atm pressure, the peak benzene mole fraction decreases from 1780 to 20 ppm, while at 8 atm, it decreases from 3750 to 42 ppm, and thus maintaining a reduction factor close to 90. Thus, at any pressure, the effect of oxygenation is to increase the concentrations of oxidizing species, such as OH and O, in the reaction zone. As a result, the intermediate hydrocarbons, such

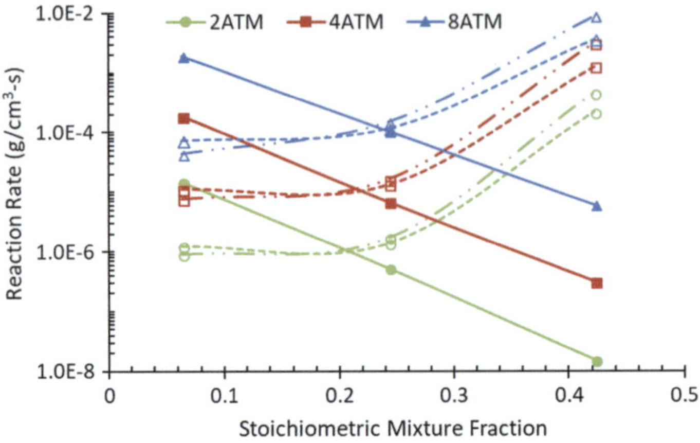

Fig. 16 Peak reaction rates for reactions R1 (solid), R6 (dashed), and R7 (dashed with dotted) plotted versus ζ_{st} for flames at 2, 4, 8 atm. From Ref. [28]

as acetylene, diacetylene, and butadiene, reduce into smaller hydrocarbons, such as CH_2O, CO, HCCO, and CH_2, causing significant reduction in the formation of soot precursors in oxygenated flames.

3.2.3 Effect of Pressure on Soot Formation in Oxygenated Flames

As discussed earlier, the nucleation rate is directly related to pyrene concentration. Consequently, at a given pressure, as pyrene concentration decreases due to oxygenation, the nucleation reaction rate also decreases significantly. Figure 16 presents the peak nucleation rate (based on reaction R1) versus ζ_{st} for flames at 2, 4, 8 atm. The peak rates of oxidation reactions (R6 and R7) are also plotted with respect to ζ_{st} at different pressures. Results show a dramatic decrease in the nucleation reaction rate with ζ_{st} at any pressure, but an increase with pressure at a given ζ_{st}. In addition, the peak soot oxidation rate increases with ζ_{st} and pressure, implying a reduction in soot formation rate.

Figure 17 presents the soot number density profiles in mixture fraction coordinates for flames established at $p = 2$ atm and different oxygenation levels ($\zeta_{st} = 0.064$ to 0.602 or 0 to 75%N_2 dilution). As indicated, the soot number density decreases sharply, nearly by a factor of about 2000, as ζ_{st} is increased from 0.064 (0%N_2) to 0.602 (75%N_2). In addition, as ζ_{st} is increased, it leads to a significant reduction in the soot formation region, and this region shifts toward the higher mixture values, i.e., from the oxidizer side to the fuel side. The sharp reduction in soot number density can be attributed to the change in flame structure, which causes a reduction in the amounts of intermediate species such as propargyl, acetylene, and diacetylene, which are responsible for the formation of benzene and pyrene, but are reduced to smaller hydrocarbons through reactions with O and OH radicals. This leads to a significant

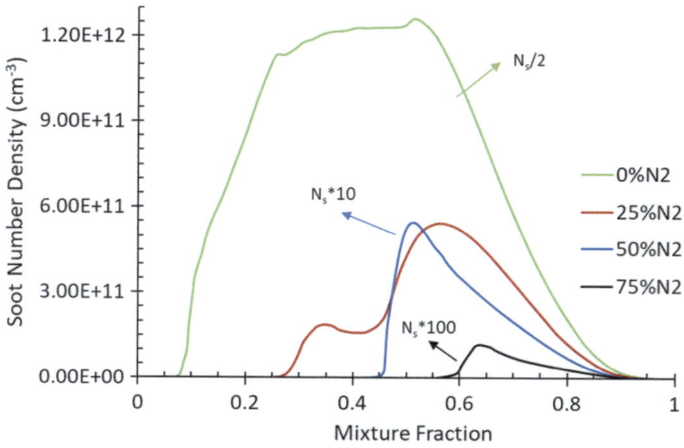

Fig. 17 Soot number density plotted in mixture fraction coordinates for ethylene flames at 2 atm and different levels of oxygenation; $\zeta_{st} = 0.064$ (0% N_2), 0.243 (25%N_2), 0.423 (50%N_2), and 0.602 (75%N_2). From Ref. [28]

reduction in the rate of nucleation reactions as ζ_{st} is increased at a given pressure, although the increase in pressure results in higher flame temperatures.

Figure 18 shows soot volume fraction profiles at 4 atm and for different oxygenation levels ($\zeta_{st} = 0.064, 0.243, 0.423$, and 0.602). As ζ_{st} is increased, the soot-forming region narrows drastically while the peak soot volume fraction increases sharply (for $\zeta_{st} \leq 0.423$). This implies competition between soot formation and soot oxidation reactions. Further analysis indicated that for flames with $\zeta_{st} \leq 0.423$, as ζ_{st} increases, the flame structure changes significantly, and the formation of soot precursors and soot is accompanied by higher local temperature. In addition, due to high acetylene concentration in these high-temperature zones, the soot surface growth rate (reaction R5) increases, promoting soot formation. However, as shown in Fig. 16, the peak soot oxidation rates also increase with ζ_{st}, due to the increased amounts of OH and O radicals. Also note that the peak nucleation rates, surface reaction rates, and oxidation reaction rates have similar values near $\zeta_{st} \approx 0.243$. Thus, due to the combined effect of narrower soot-forming regions and increased soot oxidation rate, the integrated soot volume fraction decreases with ζ_{st}, although the peak soot volume fraction increases. This is shown in Fig. 19, which plots the peak soot volume fraction and integrated (total) soot mass with respect to ζ_{st} at different pressures. Note that, the total soot mass is obtained by integrating soot volume fraction over the axial distance and multiplying by soot density. Results in Fig. 19 indicate that at any pressure, as ζ_{st} is increased, the total soot mass decreases, although the peak soot volume fraction increases for $\zeta_{st} \leq 0.423$. In contrast, for $\zeta_{st} > 0.423$, both the peak soot volume fraction and soot mass decrease, as ζ_{st} is increased. This may be attributed to two factors. One is due to a significant reduction in the amounts of soot precursors formed and thus reduction in the nucleation rate (reaction R1). The second factor is the reduction in total soot formation region as ζ_{st} is increased. Yet another

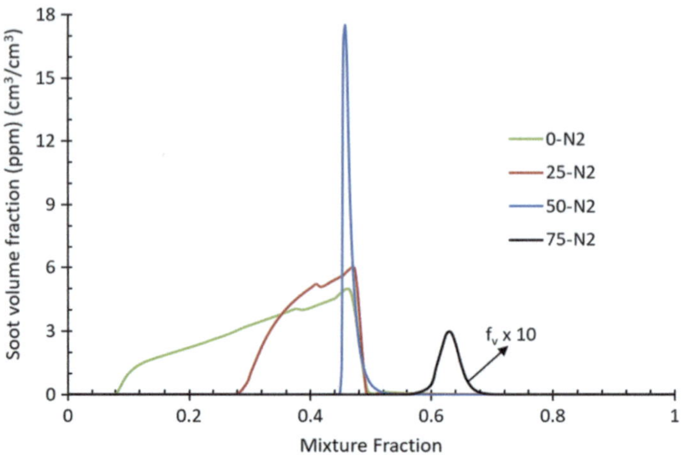

Fig. 18 Soot volume fraction plotted in mixture fraction coordinates for flames at 4 atm and $\zeta_{st} = 0.064$ (0%N_2), 0.243 (25%N_2), 0.423 (50%N_2), and 0.602 (75%N_2). From Ref. [28]

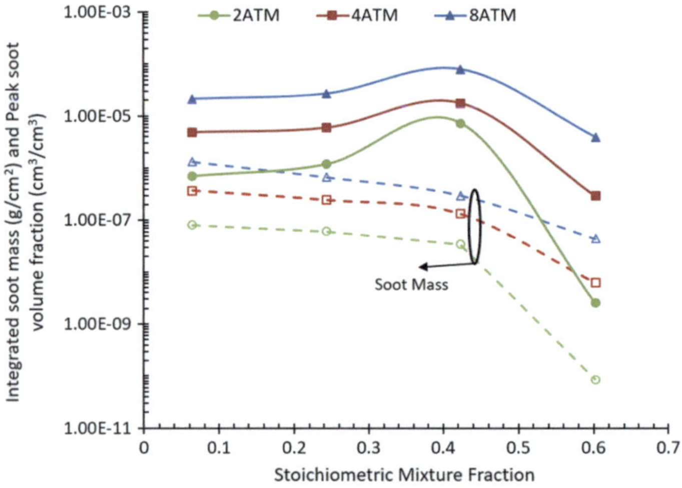

Fig. 19 Peak soot volume fraction (solid) and integrated soot mass (dash) versus ζ_{st} for flames at 2, 4, 8 atm. From Ref. [28]

factor is due to the fact that for $\zeta_{st} \leq 0.423$, the flow direction near the peak temperature location is from fuel to the oxidizer, while for $\zeta_{st} > 0.423$, the flow direction is reversed [27]. This change in flow direction helps in moving soot particles formed in fuel-rich zone into oxidizer zone and thus reducing the amount of soot formed. This is defined as a hydrodynamic effect, which plays an important role in reducing soot for flames at $\zeta_{st} > 0.423$ and at all pressures. Another important observation from Fig. 19 is that for a given ζ_{st}, both the peak soot volume fraction and total mass increase with pressure.

4 Conclusions

An important strategy of using oxygen-enhanced combustion for a dramatic reduction in soot in counterflow diffusion flames has been investigated. The strategy is based on modifying the stoichiometric mixture fraction by simultaneously using an oxygen-enriched oxidizer stream and a nitrogen-diluted fuel stream such that the adiabatic flame temperature is nearly constant. The effects of this strategy on the flame structure, soot precursors, and soot are investigated for three fuels in flames established at different pressures. Important observations are as follows:

1. The oxygen-enhanced combustion strategy is found to be very effective in significantly reducing the formation of soot precursors and soot in counterflow diffusion flames established at pressures 1–8 atm and for three fuels, namely ethylene, propane, and propene.
2. At a given pressure, as ζ_{st} is increased, the flame shifts toward the fuel side of the stagnation plane. Similarly, as pressure is increased at fixed ζ_{st}, the flame moves toward the fuel side of the stagnation plane. This change in flame structure has a direct impact on the formation and oxidation of soot precursors and soot. In addition, it modifies the local residence times for the formation of soot precursors and soot on the fuel side, and for their oxidation on the oxidizer side.
3. At any pressure, as ζ_{st} is increased, it causes a significant reduction in the formation of acetylene, benzene, and pyrene. This may be attributed to the change in flame structure that increases the concentrations of OH and O radicals in the peak flame temperature region. This promotes the oxidation of intermediate hydrocarbons, associated with the formation of soot precursors, into smaller species, and thus decreases the formation of benzene and pyrene for all three fuels. Consequently, the soot number density and total soot mass decrease significantly with the increase in ζ_{st}.
4. The fuel unsaturation causes a significant increase in the formation of soot precursors and soot. Consequently, the amount of soot formed is the highest in propene flames, followed by ethylene flames, and then propane flames.
5. The effect of pressure on soot formation is investigated in oxygenated ethylene flames. At any level of oxygenation, the amounts of PAH species and soot formed increase monotonically with pressure.
6. The oxy-combustion strategy examined here shows very promising results for a major soot reduction in laminar counterflow flames. A similar strategy therefore should be investigated for a broader class of flames including coflow laminar and turbulent flames.

References

1. Richter H, Howard JB (2000) Formation of polycyclic aromatic hydrocarbons and their growth to soot—a review of chemical reaction pathways. Prog Energy Combust 26:565–608
2. Wang H (2011) Formation of nascent soot and other condensed-phase materials in flames. Proc Combust Inst 33:41–67
3. Kennedy IM (2007) The health effects of combustion-generated aerosols. Proc Combust Inst 31(2):2557–2770
4. Oberdorster G, Sharp Z, Atudorei V, Elder A, Gelein R, Kreyling W, Cox C (2004) Translocation of inhaled ultrafine particles to the brain. Inhal Toxicol 16:437–445
5. Zheng M, Salmon LG, Schauer JJ, Zeng L, Kiang CS, Zhang Y, Cass GR (2005) Seasonal trends in PM2.5 source contributions in Beijing China. Atmospheric Environment. 39:3967–3976
6. Hansen J, Nazarenko L (2003) Soot climate forcing via snow and ice albedos. Proc Natl Acad Sci 101(2):423–428
7. Hiroyasu H, Kadota T (1976) Models for combustion and formation of nitric oxide and soot in direct injection diesel engines. SAE Paper 1976-760129
8. Wang Y, Chung SH (2016) Strain rate effect on sooting characteristics in laminar counterflow diffusion flames. Combust Flame 165:433–444
9. Yu Wang S, Park SM, Sarathy SH Chung (2018) A comparative study on the sooting tendancies of various 1-alkene fuels in counterflow diffusion flames. Combust Flame 192:71–85
10. Katsuki M, Hasegawa T (1998) The science and technology of combustion in highly preheated air. In: The combustion institute, pp 3135–3146
11. Gupta AK, Bolz S, Hasegawa T (1999) Effect of air preheat temperature and oxygen concentration on flame structure and emission. J Energy Res Technol 121:209–216
12. Gupta AK, Hasegawa T, Katsuki K, Kishimoto K, Morita M (2003) High temperature air combustion—from energy conservation to pollution reduction. In: Tsuji H (ed). CRC Press, Boca Raton, FL
13. Cavaliere A, de Joannon M (2004) Mild combustion. Prog Energy Combust Sci 30:329–366
14. Kukkadapu G, Kumar K, Sung CJ, Mehl M, Pitz WJ (2015) Autoignition of gasoline surrogates at low temperature combustion conditions. Combust Flame 162(5):2272–2285
15. Epping K, Aceves S, Bechtold R, Dec JE (2002) The potential of HCCI combustion for high efficiency and low emissions. In: SAE technical paper 2002-01-1923
16. Chang J, Kalghatgi G, Amer A, Viollet Y (2012) Enabling high efficiency direct injection engine with naphtha fuel through partially premixed charge compression ignition combustion. In: SAE technical paper 2012-01-0677
17. Akihama K, Takatori Y, Inagaki K, Sasaki S, Dean AM (2001) Mechanism of the smokeless rich diesel combustion by reducing temperature. In: SAE paper 2001-01-0655
18. Seepana S, Jayanti S (2012) Flame structure investigations of oxy-fuel combustion. Fuel 93:52–58
19. Desjardins P, Pitsch H, Malhotra R, Kirby SR, Boehman Andre L (2008) Structural group analysis for soot reduction tendency of oxygenated fuels. Combust Flame 154:191–205
20. Salamanca M, Sirignano M, Commodo M, Minutolo P, D'Anna A (2012) The effect of ethanol on the particle size distributions in ethylene premixed flames. Exp Thermal Fluid Sci 43:71–75
21. Du J, Axelbaum RL (1995) The effect of flame structure on soot-particle inception in diffusion flames. Combust Flame 100:367–375
22. Kalvakala K, Katta VR, Aggarwal SK (2018) Effects of oxygen-enrichment and fuel unsaturation on soot and NOx emissions in ethylene, propane and propene flames. Combust Flame 187:217–229
23. Chen R, Axelbaum RL (2005) Scalar dissipation rate at extinction and the effects of oxygen-enriched combustion. Combustion Flame 142(1–2):62–71
24. Skeen SA, Yablonsky G, Axelbaum RL (2010) Characteristics of non-premixed oxygen-enhanced combustion: II. Flame structure effects on soot precursor kinetics resulting in soot-free flames. Combust Flame 157:1745–1752

25. Sunderland PB, Axelbaum RL, Urban DL, Chao BH, Liu S (2003) Effects of structure and hydrodynamics on the sooting behavior of spherical microgravity diffusion flames. Combust Flame 132(1–2):25–33
26. Sugiyama G (1994) Non-luminous diffusion flame of diluted acetylene in oxygen enriched air. Symp (Int) Combust 25:601–608
27. Lin KC, Faeth GM (1996) Hydrodynamic suppression of soot emissions in laminar diffusion flames. J Propul Power 12(1):10–17
28. Kalvakala K, Aggarwal SK (2018) PAHs and soot emissions in oxygenated ethylene diffusion flames at elevated pressures. In: Proceedings of ASME turbo expo, turbomachinery technical conference and exposition
29. Design R (2015) CHEMKIN-PRO 15141, Reaction Design, San Diego
30. Puri IK, Seshadri K (1986) Extinction of diffusion flames burning diluted methane and diluted propane in diluted air. Combust Flame 65(2):137–150
31. Han X, Aggarwal SK, Brezinsky K (2013) Effect of unsaturated bond on NOx and PAH formation in n-heptane and 1-heptene triple flames. Energy Fuels 27:537–548
32. Appel J, Bockhorn H, Frenklach M (2000) Kinetic modeling of soot formation with detailed chemistry and physics: Laminar premixed flames of C2 hydrocarbons. Combust Flame 121:122–136
33. Colket MB, Hall RJ (1994) Successes and uncertainties in modeling soot formation in laminar, premixed flames, in soot formation in combustion: mechanisms and models. In: Bockhorn H (ed). Springer, New York
34. Fu X, Han X, Brezinsky K, Aggarwal SK (2013) Effect of fuel molecular structure and premixing on soot emissions from n-heptane and 1-heptene flames. Energy Fuels 27:6262–6272
35. Law CK (2006) Propagation, structure and limit phenomena of laminar flames at elevated pressures. Combust Sci Technol 178:335–360
36. Figura L, Gomez A (2014) Structure of incipiently sooting ethylene-nitrogen counterflow diffusion flames at high pressures. Combust Flame 161:1587–1603

CFD Methods in High-Speed Airbreathing Missile Propulsion Design

Debasis Chakraborty

Keywords Ramjet · Scramjet · Air-intake · Turbulence · Combustion · CFD

1 Introduction

Currently, CFD has matured into a rich and diverse subject for both basic and applied fluid dynamic research and provides a reliable and robust tool for aerospace propulsion design. Simultaneous developments of new computer architectures, robust numerical algorithms, advanced physical and chemical models of flow phenomena have made CFD an integral part of the propulsion design process. While experimental testing will always remain important for the design, CFD has reduced its dependence in the design exercise.

In DRDL, CFD methods have been used extensively in missile propulsion systems design. Commercial and open-source RANS solvers are used very effectively to analyse turbulent reacting and non-reacting flow fields inside different high-speed propulsion systems to suggest design improvements. Important user-defined functions (UDFs) are developed to augment the commercial CFD solvers for increased accuracy and range of applications. Open-source CFD software OpenFOAM [1] has been customized to solve many complex propulsion problems Through validations were carried out by comparing the simulation results with reliable experimental values before applying these CFD tools in the design exercises.

A compressible finite volume LES solver [2–4] is developed indigenously to analyse turbulent separated flows where RANS simulations are not adequate to provide design data. Hybrid approach is adopted by combining higher-order central difference

D. Chakraborty (✉)
Defence Research and Development Laboratory, P.O. Kanchanbagh, Hyderabad 58, India
e-mail: debasis_cfd@drdl.drdo.in

A. K. Gupta et al. (eds.), *Innovations in Sustainable Energy and Cleaner Environment*, Green Energy and Technology,
https://doi.org/10.1007/978-981-13-9012-8_12

scheme for better viscous terms resolution and a modified version of high-resolution upwind scheme (SLAU2) to obtain solution at discontinuity. Discontinuity sensor was adopted for smooth transition between the schemes and Gibbs phenomena sensors are used to separate out numerical and physical oscillations. Digital filtering method is implemented in LES solver to predict the inflow turbulence where spatiotemporal relations are correlated according to turbulent scaling laws. Developed LES solver is tested extensively for the number of canonical problems and is applied for mixing layer flows, subsonic/supersonic flows past backward-facing steps, shock-boundary layer interaction problems and solid rocket motor port flow analysis.

CFD methods facilitated the design of propulsion systems of various ongoing and future missile projects of DRDL. Extensive non-reacting and reacting simulations were carried out for the development of flightworthy scramjet propulsion system for hypersonic airbreathing cruise vehicle, and CFD simulations guided the experimental testing. Excellent match is obtained between experiment and pre-test prediction for various performance and flow parameters. Accurate estimation of heat flux obtained from high fidelity CFD simulations is used in thermo-structural design of the combustor. Aerodynamic and propulsion parameters obtained from end-to-end simulation (comprising of non-reacting flow in external surfaces and reacting flow in the combustor) provide vital input for mission design. CFD-based jet vane correlations are adopted in the flight computer for tactical missiles and form an integral part of missile control and guidance. The problem of high temperature in the base cavity of a long-range missile caused due to interaction of free stream and exhaust plume at high altitudes could be analysed only through CFD methods. Performance prediction of installed air intake of ramjet missiles, solid rocket motor (SRM) port flow field analysis, combustion instability prediction of SRM, plume-canister/plume-launcher interaction, etc. are some of the other notable applications of CFD methods in propulsion system design. Better understanding of complex reacting/non-reacting flows inside the propulsion system has made the design feasible and reduced the developmental cost and time of the system significantly. Maturity of simulations has enabled the designers to take stand-alone decisions without much experimental testing. Applications of CFD codes for the design of ramjet/scramjet propulsion systems are presented in the article.

2 Scramjet Combustor Design for Hypersonic Airbreathing Cruise Vehicle

The quest for high-speed airbreathing propulsion systems for both military and civil applications led the research on scramjet propulsion from the early 1960s [5–7]. Hydrogen fuel was considered for space applications; whereas, volume limited applications in lower hypersonic flow regime ($M < 8$) preferred liquid hydrocarbon fuels for higher energy density and ease in storability [8]. Better atomization, faster vaporization, deeper penetration into supersonic crossflow and faster reactions are the essential requirements for the realization of efficient hydrocarbon-fuelled scramjet combustor. A deeper penetration of fuel into a supersonic air stream is essential for faster mixing and sustained combustion. Liquid jet penetration into supersonic crossflow is studied for different dynamic pressure ratios [9] and observed 10–15 mm penetration depth into the crossflow pertaining to Mach 6–7 flight regime.

Different types of fuel injection systems, namely, wall injection, strut-based injection and pylon-based injection are investigated for hydrocarbon-fuelled scramjet combustor. For fuel injected from the wall, reaction occurs in the regions adjacent to the wall and cause excessive thermal loads on the combustor structure. Fuel injected from the strut/pylon in core supersonic flow can circumvent slow lateral fuel transport in the air stream. The oblique shocks generated from the struts also aid mixing for better combustion in high-speed propulsion devices.

Fuel injection from the struts has been studied for subscale scramjet engine including airframe integrated scramjet module [10, 11]. Reported experimental and numerical studies [12–15] on kerosene-fuelled scramjet combustor mostly address cavity-based flame holder and injection system in laboratory-scaled combustor. The studies on strut-based scramjet combustor with kerosene fuel are not adequately reported in open literature. Vinogradov et al. [16] carried out ignition, piloting and flame holding studies in a strut-based kerosene-fuelled scramjet combustor with fuel injected from the middle of the struts. Stable combustion of kerosene was achieved even after turning off pilot hydrogen. Bouchez et al. [17] also conducted experimental studies with water-cooled and liquid kerosene-cooled fuel injection struts and measured wall pressure, wall heat flux, total temperature at combustor exit, thrust, etc.

Powerful computers and robust numerical algorithms have enabled CFD to play a major role in understanding the key phenomenon in supersonic combustion. For accurately modelling of scramjet flow field, CFD must resolve three-dimensionality, shock-boundary layer interaction, turbulent mixing, atomization and combustion of liquid fuel. Limited numerical studies are available in the open literature on strut-based liquid-fuelled scramjet combustor. Dufour and Bounchez [18] have simulated Bouchez's scramjet experiment [17] using a 3D RANS solver and single-step chemical kinetics and reported good match between the computational and experimental wall static pressure. The effect of the combustor inlet Mach number and total pressure on the flow development in the scramjet combustor is numerically investigated by Manna et al. [19] and shown that higher combustor entry Mach number and distributed fuel injection are necessary to obtain predominant supersonic flow in the combustor.

Though the research had started the early 60s, successful flight demonstration of scramjet engine was done through Mach-7 flight test of X-43A (with hydrogen fuel) [20, 21] and X- 51A (with hydrocarbon fuel) [22] only recently. MBDA and ONERA developed a small-scale, 4.2 m long dual-mode scramjet-powered, experimental hypersonic vehicle [23]. Hypersonic Flight Experimental Vehicle, (Hyflex) was designed, developed and flight tested [24] in Japan. CFD tools were used extensively to study different physical phenomena of laminar/turbulent transition on forebody [25], aerothermodynamics [26], surface heating [27] and scramjet combustor [19] as well as in the design of various subsystems. Complete vehicle analysis integrating both external and internal flow together is scarce. Voland et al. [21] reported post-test CFD analysis for tip-to-tail for X-43 without adequate details.

The development of hypersonic airbreathing technology in India is presented by Paneerselvam et al. [28]. Demonstration of autonomous operation of kerosene-fuelled scramjet combustor at hypersonic flight speed (~6.0–6.5) for a flight duration of about 20 s is envisaged. Due to non-availability of powerful aircraft and advanced materials, ground launch option was adopted and vehicle integrated scramjet engine was put atop a solid booster. To protect the vehicle from severe aerodynamic heating during ascent phase, the vehicle is covered by heat shield fairing. The fairings and scramjet integrated cruise vehicle are separated from the booster at the desired altitude. Scramjet combustor configuration including the number of fuel injection struts, their positions and fuel injector locations, etc. were finalized through many connected pipe mode ground test [29, 30] and numerical simulations [31, 32]. The application of CFD in the development of scramjet combustor and the scramjet integrated hypersonic cruise vehicle is presented in the subsequent subsections.

2.1 Computational Methodology

Three-dimensional RANS equations are discretized using a commercial CFD software CFX [33]. The CFD code is fully implicit, finite volume method and is based on finite element geometry discretization. General non-orthogonal, structured, boundary-fitted grids are implemented. Convective terms are discretized through 2nd order scheme to capture high-speed flow features more accurately. $k - \varepsilon$ turbulence model with Launder and Spalding [34] wall-function is used. The logarithmic relation for the near wall velocity is given by:

$$u^{+} = \frac{U_{\mathrm{t}}}{u_{\tau}} = \frac{1}{\kappa}\ln\left(y^{+}\right) + C \quad \text{where} \quad y^{+} = \frac{\rho \Delta n u_{\tau}}{\mu} \quad \text{and} \quad u_{\tau} = \left(\frac{\tau_{\omega}}{\rho}\right)^{\frac{1}{2}} \tag{3}$$

u^{+}, u_{τ}, ρ, U_{t} and y^{+} *are the* near wall velocity, friction velocity, density, tangent velocity to the wall and dimensionless distance from the wall respectively. κ is the von Karman constant and C is a log-layer constant depending on wall roughness.

Thermal law-of-the-wall [35] is used to model wall heat flux distribution (q_{w}) as follows:

$$q_{\mathrm{w}} = \frac{\rho C_{\mathrm{p}} u_{\tau}}{T^{+}}(T_{\mathrm{w}} - T_{\mathrm{f}}) \tag{4}$$

where $\Gamma = \frac{0.01*(\mathrm{Pr} * y^{+})^{4}}{1+5*\mathrm{Pr}^{5} * y^{+}}$

$$\beta = \left(3.85\,\mathrm{Pr}^{1/3} - 1.3\right)^{2} + 2.12 * \ln(\mathrm{Pr}),\ \mathrm{Pr} = \frac{\mu * C_{\mathrm{p}}}{\lambda},\ u_{\tau} = \left(\frac{\tau_{\omega}}{\rho}\right)^{1/2}$$

$$\Gamma = \frac{0.01*(\mathrm{Pr} * y^{+})^{4}}{1+5*\mathrm{Pr}^{5} * y^{+}} \text{ and } y^{+} = \frac{\rho * \Delta n * u_{t}}{\mu}$$

Eddy dissipation combustion model [36] which is based on fast chemical reaction is employed to model combustion for its simplicity and robustness. It is assumed that the products are formed instantaneously when the fuel and oxidizer are mixed molecularly, and the reaction rate is proportional to ε/k, (k is the turbulent kinetic energy and ε is its rate of dissipation). Kerosene ($C_{12}H_{23}$)–air reaction is represented on a molar basis by,

$$C_{12}H_{23} + 17.75O_2 = 12CO_2 + 11.5H_2O$$

The mixing rate determined from the EDM is given as.

$$R_{k,\mathrm{EDM}} = -A_{\mathrm{ebu}}\rho\frac{\varepsilon}{k}\min\left\{Y_{\mathrm{f}}, \frac{Y_{\mathrm{o}}}{r_{\mathrm{k}}}, B_{\mathrm{ebu}}\frac{Y_{\mathrm{p}}}{1+r_{\mathrm{k}}}\right\}$$

where ρ and Y_{f}, Y_{o} and Y_{p} are the density and mass fractions of fuel, oxidizer and products, respectively, A_{ebu} and B_{ebu} are the model constants and r_k is the stoichiometric ratio.

Dispersed phase fluid (kerosene) is handled by Lagrangian particle tracking method (LPTM). Global time step is typically of the order 10^{-5} s and fourth order fall in log-normalized maximum residue is considered as convergence criteria. Thorough validation of the software is carried out by simulating the number of scramjet applications with hydrogen and kerosene fuel [37–41]. It is observed that computational and experimental results match well (within 5%) in the major thrust-producing zone of the combustor.

2.2 *Simulation of Isolated Scramjet Combustor*

The design of strut-based kerosene-fuelled scramjet combustor is finalized through the number of non-reacting and reacting CFD calculations. Initial scramjet combustor encountered severe thermal environment for the fuel injection struts. Use of exotic material like Tungsten and Niobium also could not alleviate the thermal problem. From the simulation, it was observed that the maximum gas temperature occurred at the locations where maximum erosion is observed in the experiment. The number of simulations is carried out with different positions of the struts and different fuel injection locations. The flow parameters are analysed in detail to find out the availability of fuel and oxidizer and heat release pattern to decide the thermally benign environment of the fuel injection struts. The strut position and fuel injection locations are further iterated to obtain the scramjet combustor performance. The combustor performance in terms of normalized thrust (the thrust has been normalized with maximum achievable thrust) and combustion efficiency for different struts and fuel injection locations are summarized in Table 1. It can be observed that the combustor performance is progressively improved with redistribution of fuel location and relocation of fuel injection struts. Important results of these simulations are available in

Refs. [31, 32, 42–45]. Summary and important results are briefly presented in this section.

2.2.1 Combustor Geometry and Inflow Details

Scramjet operating condition is achieved by passing high-pressure and high-temperature vitiated air through convergent and divergent nozzle. The schematic of combustor geometry and fuel injection strut is shown in Fig. 1. The details of the combustor geometry are available in Ref. [32]. Sweptback configuration of strut (similar to Marquardt Corporation, USA configuration [46]) is employed to increase the three dimensionality of flow for better mixing. All reacting flow calculations are carried out for equivalence ratio of 1.0 with Rosin Rammler particle distribution (diameter D = 37.32 μm and dispersion factor of $\gamma = 1.5$).

2.2.2 Computational Grid and Grid Independence of the Results

Block-structured grids of 250 × 62 × 96 (0.7 million) with fine grid near the leading and trailing edges of the struts, adjoining region between the nozzle and combustor, and near wall regions are made using ICEM CFD [47] for the entire domain. In the near wall region, minimum y^+ is about 13.6. Grid embedding is employed to capture

Table 1 Comparison of combustor performance with different struts and injection locations

Configuration	Combustion efficiency	Normalized thrust
Base line	74	0.88
Redistribution of fuel location	81	0.96
Relocation of fuel injection struts	82	1.0

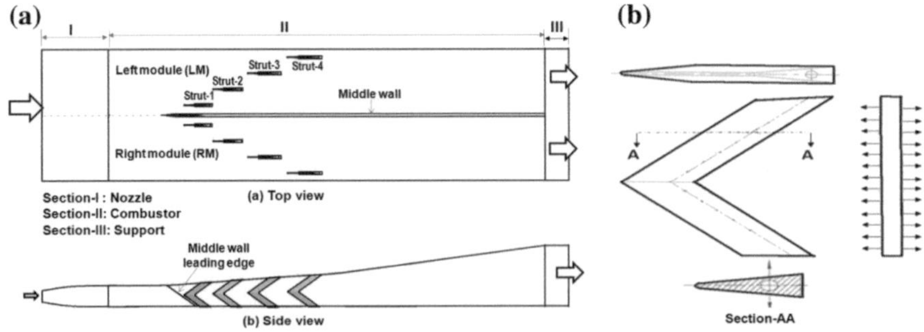

Fig. 1 Schematic of **a** scramjet combustor **b** fuel injection struts (Reproduced from Ref. [32], with permission from ASME)

very small injection holes. Computed surface pressures for two different grids (0.7 million and 1.4 million) are compared in Fig. 2 to demonstrate the grid independence of the results. Grid Convergence Index (GCI) [48–50] is also presented in the figure to show the error due to grid. Maximum GCI error of 3.33% indicates that the grid is adequate to capture most of features of the flow.

2.2.3 Reacting Flow Simulation Results

Transverse injection of liquid kerosene fuel into the supersonic airflow through the struts is simulated numerically. Mach number, static pressure and static temperature distribution at different axial stations [$X = -3.3$ h (nozzle entry), 0.0 (combustor entry), 2.9, 5.8, 8.7, 11.6, 14.5, 17.4 and 21.5 h (combustor exit)] are presented in Fig. 3a–c, respectively. As expected, heat release due to reaction brings down Mach number and increase static pressure and temperature in the reaction zone. Although the average Mach number has reached below unity at $X/h = 4.9 - 10.8$, flow is not fully subsonic in the region. Existence of few subsonic pockets due to reaction is responsible to bring the average Mach number below unity. Atmospheric pressure condition at the outflow boundary caused the flow compressed near the combustor exit. Mass fraction distributions of CO_2, O_2 and kerosene vapour at various axial locations are shown in Fig. 4a–c, respectively. Similar reaction pattern is visible in both the module with liquid kerosene droplets completely vaporize within the combustor. Computed top wall pressures match very well with experimental data as seen in Fig. 5. Experimental peak pressure is slightly more compared to the CFD values and also the experimental separation point is delayed slightly in right module compared to CFD.

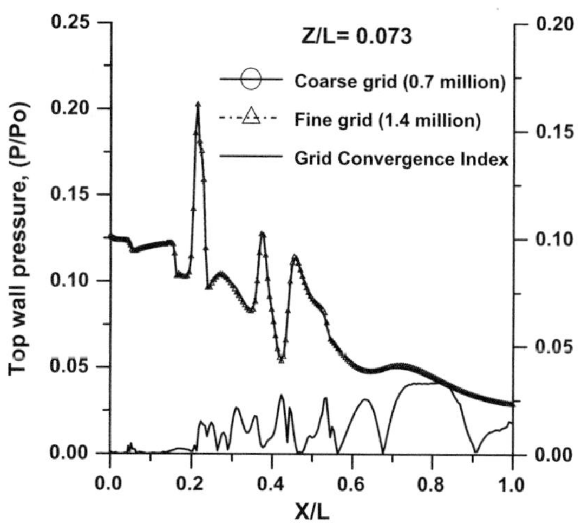

Fig. 2 Top wall surface pressure in comparison with two different grids (Reproduced from Ref. [31], with permission from Elsevier)

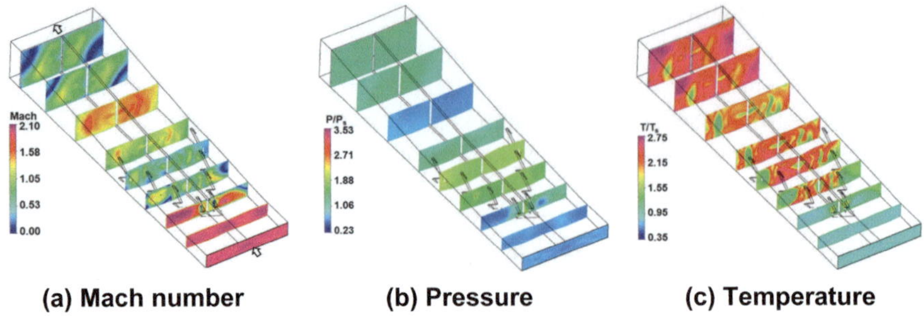

Fig. 3 Mach no, static pressure and static temperature distribution at various axial planes (Reproduced from Ref. [32], with permission from ASME)

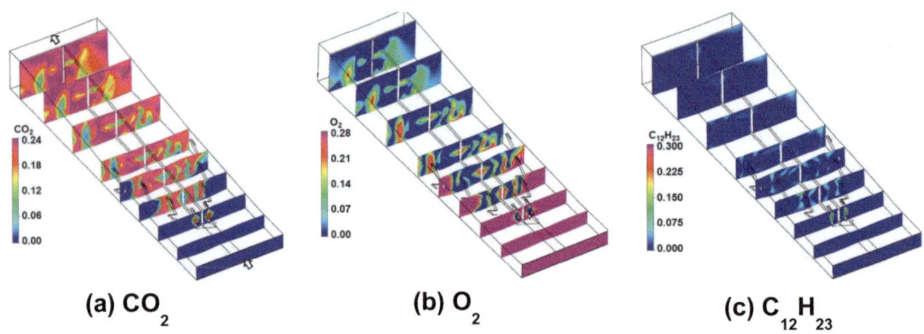

Fig. 4 Mass fraction distribution of CO_2, O_2 and kerosene ($C_{12}H_{23}$) vapour at various axial locations (Reproduced from Ref. [32], with permission from ASME)

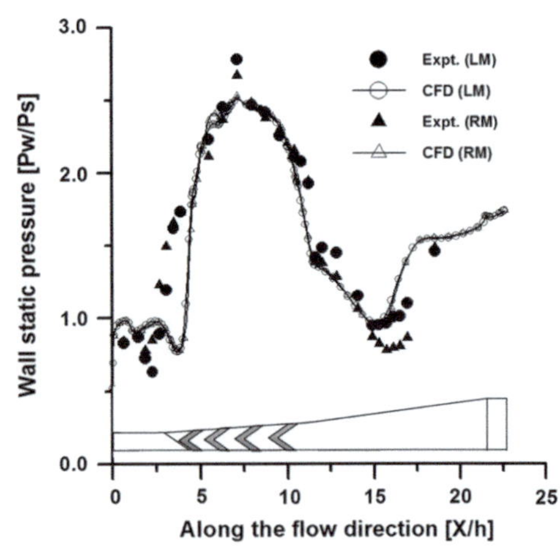

Fig. 5 Comparison of top wall pressure distribution (Z = 1.6 h) (Reproduced from Ref. [32], with permission from ASME)

2.2.4 Convective Heat Flux Estimation of Scramjet Combustor Wall

Although CFD methods are being employed to predict the overall performance of the scramjet engine in terms of thrust and combustion efficiency, the use of these methods in thermo-structural design of combustor walls is rather limited. High temperature inside the combustor due to reaction of fuel with incoming air stream and external aerodynamic heating by ambient hypersonic flow is the major source of heat load for combustor walls. Wall heat flux and resulting surface temperature are the key inputs for thermo-structural design of scramjet combustor. Convective heat flux contributes maximum (~85%) to the total heat flux in liquid rocket motors [51]. For scramjet combustor, since the flow field is convection dominated, convective heat flux will contribute even more (~90%) to the total heat flux (convective heat flux is 1.6 MW/m^2 out of total heat flux 1.8 MW/m^2). Measurements of these parameters are very difficult due to severe thermal and oxidizing environment of hypersonic flight Mach number.

For example, at $M_\infty > 6$, the combustion product total temperature and average wall heat flux would be around 2800 K and 5.0 MW/m^2 [52]. Wall heat flux and temperature in a supersonic model combustor are measured by Li et al. [53] with an integrated water-cooled sensor. Kennedy and Donbar [54] measured heat fluxes at four locations in a direct-connect gaseous hydrocarbon-fuelled cavity-based scramjet combustor operating at fuel equivalence ratios of 0.6–1.0. Measured heat fluxes are higher (0.6–2.0 MW/m^2) at reaction zone of the cavity flame holder; whereas, upstream and downstream regions of the flame holder experience comparatively lower heat fluxes. Zhang et al. [55] used a state observer-based method to estimate inner wall temperature from measured pressure and outer wall surface temperature of a scramjet combustor.

Well-resolved CFD tools can easily locate hot spot (region of extreme heat load) of the scramjet combustor wall which may be difficult to estimate through theoretical or experimental studies. Malsur et al. [42] validated the CFD software CFX [33] against experimentally measured heat flux data in supersonic convergent-divergent nozzle [56] and found out that 5 μm minimum grid spacing is required for accurate prediction of convective heat flux. The validated CFD tool was used to predict the convective heat flux (q_w) and heat transfer coefficient (h_c) for a flight sized kerosene-fuelled scramjet combustor. Important results are summarized in the section.

Computation of heat flux requires the resolution of thermal boundary layer. The 1st grid point adjacent to wall Δn of is varied from 5 to 80 μm. Clustering of grid is done in all the four walls and for Δn of 5 μm, y^+ value is less than 1. Comparison of non-dimensional convective heat flux ($q_\mathrm{w}/q_{\mathrm{w}_\mathrm{max}}$, q_{w_max} is maximum heat flux at 1st strut leading edge) and top wall surface pressure ($P_\mathrm{w}/P_{0_\mathrm{CI}}$, P_{0_CI} is the total pressure at combustor entry) along the flow direction near the mid-plane ($Z/h = 1.6$) for isothermal ($T_\mathrm{w} = 600$ K) reactive flow simulations is shown in Fig. 6a, b, respectively, for different grids. Computed pressures are seen to be invariant with grid size and computed heat fluxes vary very little for $\Delta n \leq 10$ μm, demonstrating the grid independence of the results.

Convective heat flux distributions at bottom, left, right and top walls are shown in Fig. 7a–d, respectively. Kerosene–air reaction caused higher heat flux behind the struts (Fig. 7a) top and bottom left wall corners. During scramjet operation, combustor wall temperature increases and causes the reduction of convective heat flux, and the combustor wall temperature is not known a priori. Isothermal reacting flow simulation corresponding to $T_w = 900$ K is carried out to determine the effect of wall temperature on heat transfer characteristics. Figure 8a shows the comparison of local heat flux distribution (at $Z/h = 1.6$) for top wall for two isothermal walls. Local heat transfer coefficient ($h_c = q_w/(T_{aw} - T_w)$) distribution (T_{aw} and T_w are the adiabatic and isothermal wall respectively) along the axial length (Fig. 8b) is shown to scale with the difference of the adiabatic wall temperature (T_{aw}) and the wall temperature (T_w). Hence, for getting the heat transfer characteristics of the scramjet combustor, it is not required to carry out computations with different wall temperature. One simulation for adiabatic condition and one isothermal condition is sufficient. Heat flux for any other wall temperatures can be obtained from heat transfer coefficient. The constancy of heat transfer coefficients with different wall temperature and requirement of one adiabatic and wall isothermal temperature in getting the heat flux values have been discussed in great detail with the number of validation cases in Ref. [57]. Computed heat fluxes are used in thermo-structural design of the scramjet combustor.

2.3 *Tip-to-Tail Simulation on Hypersonic Airbreathing Cruise Vehicle*

For high-speed airbreathing system, propulsion and aerodynamics are so strongly coupled that any demarcation of aerodynamics and propulsion surface is difficult. In fact, vehicle undersurfaces produce considerable thrust and act as propulsion device.

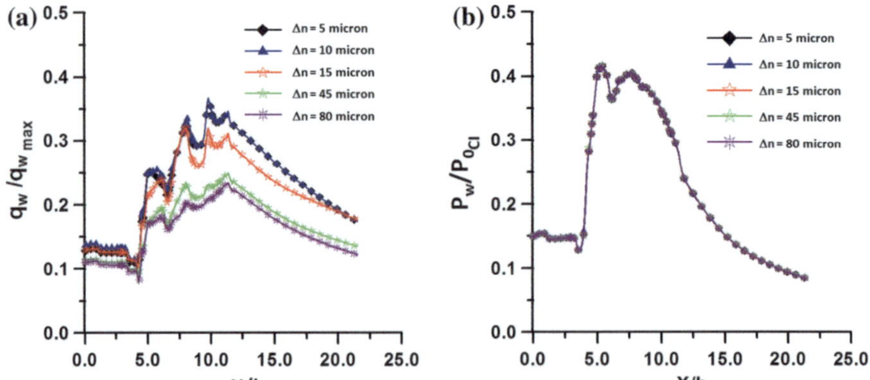

Fig. 6 Comparison of axial distribution of **a** heat flux and **b** top wall surface pressures for various grid sizes at $Z/h = 1.6$

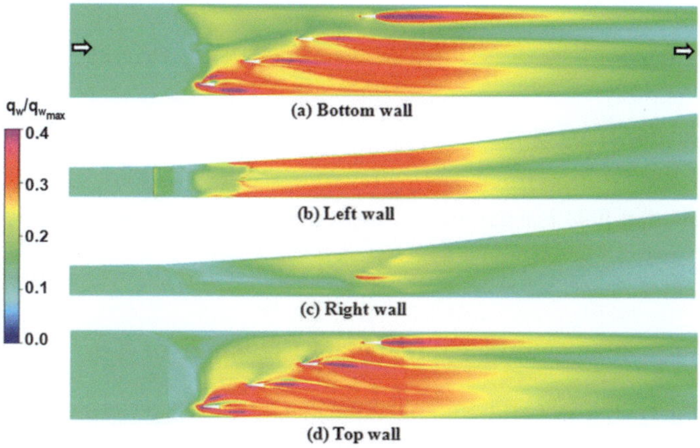

Fig. 7 Contour of heat flux on various walls

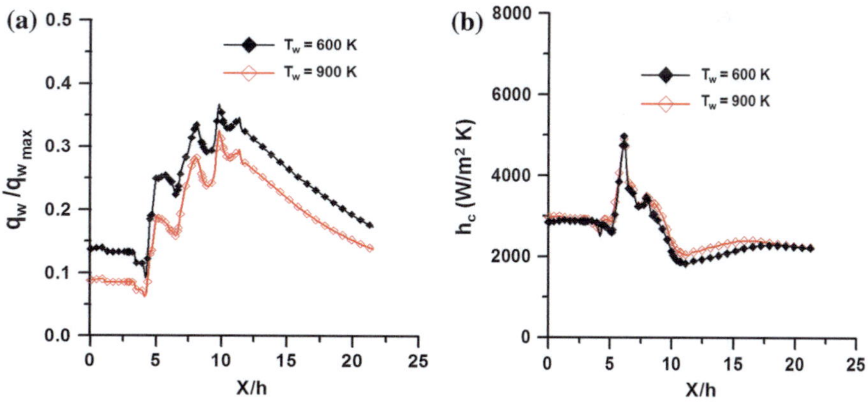

Fig. 8 Comparison of axial distribution of **a** heat flux and **b** heat transfer coefficient for two wall temperatures

Hence, a combined external-internal flow simulations including non-reacting flow in the vehicle forebody, fuselage, wing, tail, intake and reacting flow in the scramjet combustor have been carried out [58] to estimate all aeropropulsive parameters for mission design.

2.3.1 Geometrical Details of Cruise Vehicle with Intake, Combustor and SER Nozzle

The hypersonic vehicle schematic is shown in Fig. 9. The details of the geometry are available in Ref. [58]. The total length of the vehicle is 8.47*h*, *h* is the vehicle height. Kerosene fuel is injected through the number of injectors provided on either side of

Table 2 Mass flow rate and total pressure at the entry of the intake

AoA, α (deg.)	$\dot{m}_a/\dot{m}_{a_{\alpha=0}}$	$P_{rec} = P_o/P_{o\infty}$
0	1.0	0.47
6	1.26	0.31

the strut. The hot flow is exhausted through a SER nozzle with one-sided (upward) divergence angle.

2.3.2 Computational Grid and Boundary Condition

Good-quality hexahedral grids are generated in the computational domain containing 1560 blocks. Total grid of size 22.1 millions for half geometry [601 (X) × 224 (Y) × 164 (Z)] is used in the simulations. The grids are fine near the vehicle nose, ramps, intake entry, wings, fins, combustor strut regions, SER nozzle entry, vehicle base regions and near wall regions while relatively coarser grids are provided in remaining portion of the vehicle. Grid independence of the results is demonstrated by comparing the reacting flow surface pressure in Ref. [58]. Free-stream Mach number of 6.2, altitude 32.5 km, angle of attack 0° and 6°, pure air and liquid kerosene injection at equivalence ratio of 1.0 are considered for simulation.

2.3.3 Non-reacting Flow Simulation Results

The vehicle undersurface pressure and cowl surface pressures presented in Fig. 10 depict complex shock reflection process. Shock impingement on the cowl causes pressure rise at X/h of−1.75, whereas the under surface pressure remains constant till X/h of −0.75 before rising again due to cowl reflected shock impingement.

Air mass flow rate and total pressure recovery at the intake entry for two angles of attacks are tabulated in Table 2. Increase of 26% mass flow rate 34% less pressure recovery is observed for angle of attack $\alpha = 6°$ compared to $\alpha = 0°$.

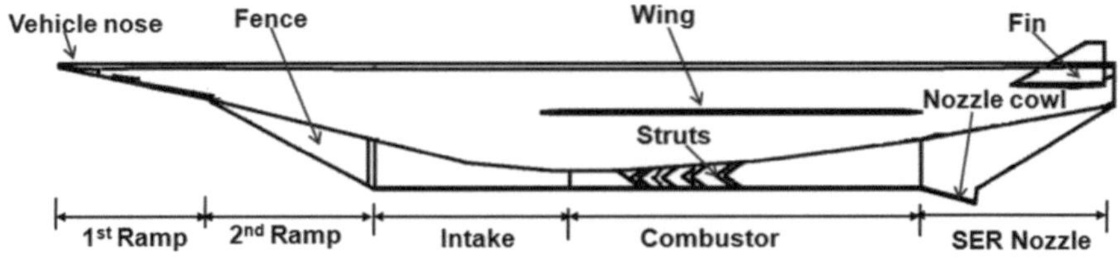

Fig. 9 Schematic of hypersonic airbreathing Cruise Vehicle (CV) (Reproduced from Ref. [58], with permission from AIAA)

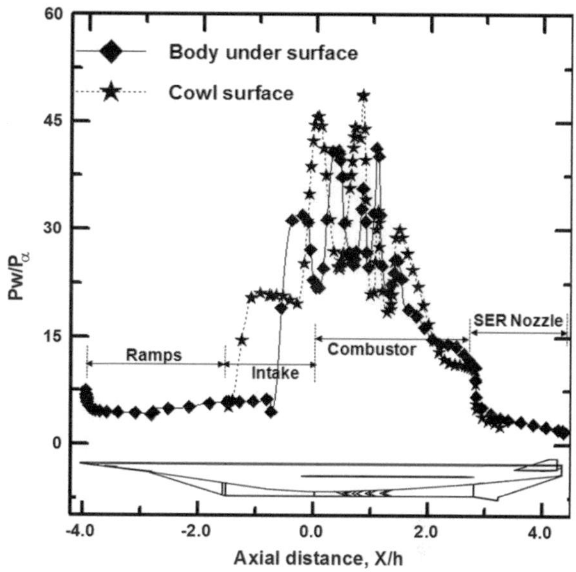

Fig. 10 Comparison of wall surface pressure (Reproduced from Ref. [58], with permission from AIAA)

2.3.4 Reacting Flow Simulation Results

Mach number distribution at combustor inlet, injection zone ($X/h = 1.0$), divergent region ($X/h = 2.0$) and combustor outlet ($X/h = 2.8$) is shown in Fig. 11 to depict the flow pattern in the combustor. The Mach number distributions at the combustor entry and exit are nearly uniform but few subsonic pockets are observed in the reaction zones in the strut regions. Minimum average Mach number of about 1.18 occurs at $X/h \sim 1.24$.

Computed wall pressure distribution at vehicle undersurface pertaining to non-reacting and reacting flow for two different grids (16.0 and 22.1 Millions) is compared in Fig. 12 for 0° angle of attack. Various shock reflections pertaining to non-reacting flow are crisply captured. Maximum error of 5% Grid Convergence Index (GCI) indicates that the grid is adequate to capture most of features of the flow and the solution in grid independent.

The cross-sectional distribution of CO_2 and O_2 mass fractions at various axial stations inside the combustor is shown in Fig. 13a, b, respectively. It is observed that the combustion of kerosene has occurred mostly in top wall, mid-wall and bottom wall regions. About 5.5% mass fraction of oxygen still left unburnt at the exit of the combustor. Various performance parameters, namely lift, drag, combustion efficiency and thrust from the combustor and the SER nozzle, for two angles of attack (0° and 6°) are tabulated in Table 3. Higher angle of attack produces more thrust due to increase in ingested mass flow rate.

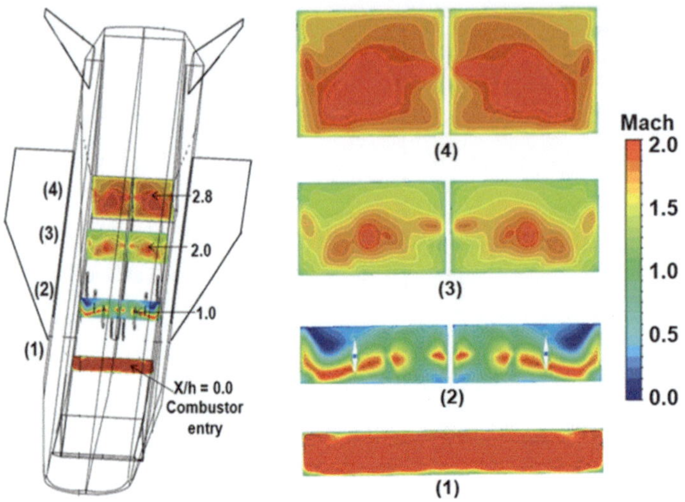

Fig. 11 Mach number distribution at different axial planes (Reproduced from Ref. [58], with permission from AIAA)

Fig. 12 Comparison of wall pressure at the vehicle bottom surface and GCI along the length of the vehicle from nose (Reproduced from Ref. [58], with permission from AIAA)

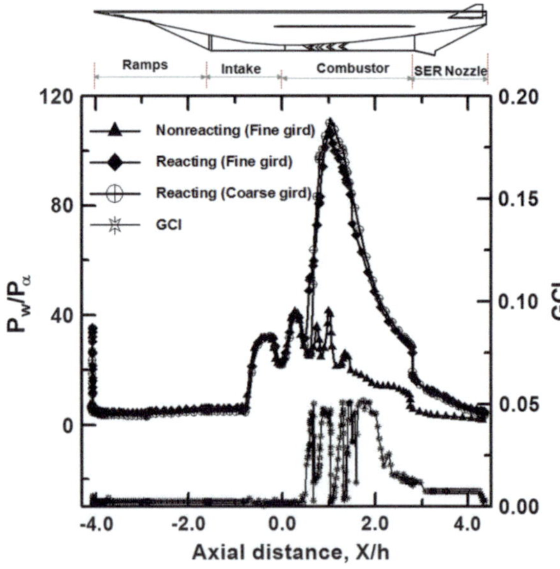

3 Simulation of Air Intake Flow Field for Ramjet/Scramjet Propulsion

Intake performance is very critical for ramjet and scramjet operations. The intake of a supersonic airbreathing engine captures and compresses free stream air. After heat addition, the flow is expanded in the nozzle to provide thrust. The design criteria

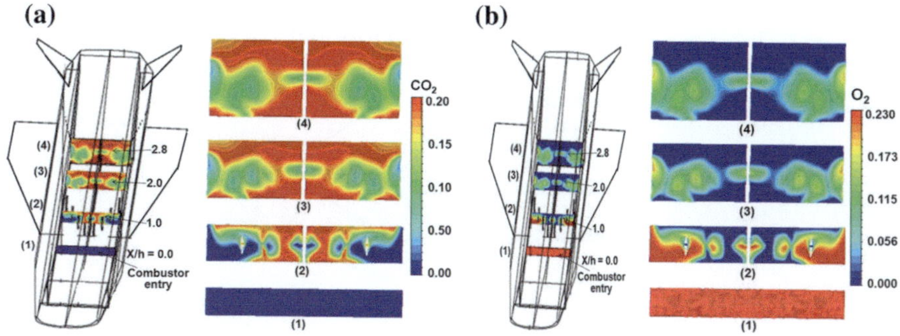

Fig. 13 **a** CO_2 mass fraction and **b** O_2 mass fraction at different axial stations (Reproduced from Ref. [58], with permission from AIAA)

Table 3 Comparison of performance parameters for two angles of attacks (Reproduced from Ref. [58], with permission from AIAA)

Parameters	Angle of attack	
	0°	6°
Combustion efficiency, %	76.0	81.8
Thrust (combustor), ($\Delta F/F$ci)	0.26	0.34
Vehicle drag (D_{CV}/Fci)	0.21	0.26
Thrust (SER nozzle), (F_{SERN}/Fci)	0.12	0.13
Net vehicle thrust (Total thrust-Drag) (F_{CV}/Fci)	0.17	0.21
Lift force, F_L/Fci	0.21	0.74
Specific impulse, I_{SP} (s)	940.5	1110.7

of supersonic air intake are well documented in the literature [59]. Intake flow is very sensitive to the interaction with the upstream external flow and downstream combustion process and hence exhibits complex flow phenomena over the expected range of operation. Detailed understanding of flow behaviour in the intake and the interaction with external flow is very important in the design exercise.

3.1 *Simulation of Installed Air Intake of a Ramjet Missile (Case 1)*

Supersonic intake for missile has to maintain good performance at manoeuvre (with different angles of attack and sideslip). If the air intakes are placed at the end of the nose cone, the boundary layer is not thick, but there is locally an overspeeding which makes the efficiency fall. On the contrary, if the intakes are located more downstream on the body, the overspeeding diminishes and yield increases. But in moving back the air intakes, the local boundary layer becomes thicker (in particular with incidence)

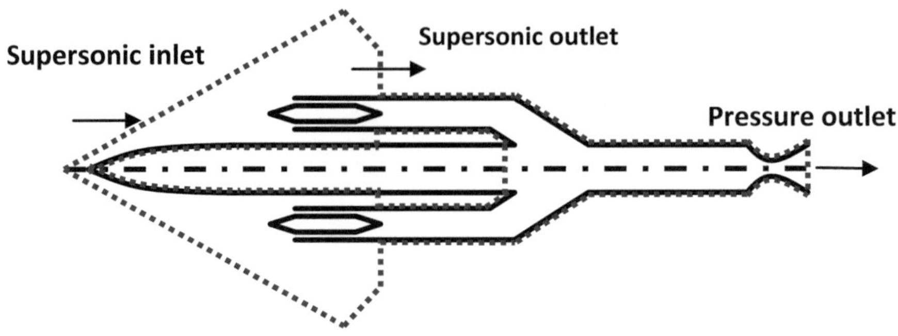

Fig. 14 Installed intake geometry and computational domain

and above all the zones affected by the vortices grow larger, which induces a decrease of performance. Influence of the distance of the air intakes from the nose of the engine is studied experimentally in Ref [60]. Herrmann and Gülhan [61] have shown from the experimental investigation that intake performance differed at different angles of attack ($-30° < \alpha < 30°$) when its axial position is varied along the missile body. Studies related to installed air intake have not appeared adequately in the literature.

The performance of individual intakes gets modified when it is installed to the core body at downstream location. The variation of the performance may be very significant when the vehicle is at angle of incidence. Saha et al. [62] carried out numerical simulation to study the flow field of ramjet installed intakes at angle of attack up to 6°. Pressure recovery versus mass flow characteristics of the intake at various angles of incidence are analysed and compared with the experimental result. The geometry which was investigated experimentally [63] consists of an ogive-cylinder core body and four integrated air intakes placed in a rear location. The schematic of the geometry along with the computational domain (marked with dotted line) is shown in Fig. 14. The computational domain of the problem includes the external flow field of the forebody and the internal flow path in the intakes and the dump chamber. As the interest of the study is to estimate the intake characteristic in installed mode, the external flow domain is terminated at certain downstream distance of the cowl lip. The four air intakes are connected to a dump chamber, and back pressure is simulated through a plug in the dump chamber. The air intake has centerbody with semicone angle of 27.5°. Different boundaries are indicated in the figure. At inflow boundary, uniform conditions of Mach number, static pressure and static temperature of 2.0, 0.28 atm and 261 K pertaining to wind tunnel condition are imposed.

Simulations are carried with different back pressure to find out the pressure recovery ($\pi = P_0/P_{0\alpha}$) versus mass capture ($\eta = m/m_c$) characteristics of the installed intake. Simulation started with zero angle of attack and back pressure of 0.8 atm. The back pressure is increased gradually to get the next operating point in the π-η curve. Mach number distribution in pitch and yaw plane of the air intake at 5° angle of incidence for three different back pressures ($p_b = 1.2$, 1.4 and 1.5 atm) are pre-

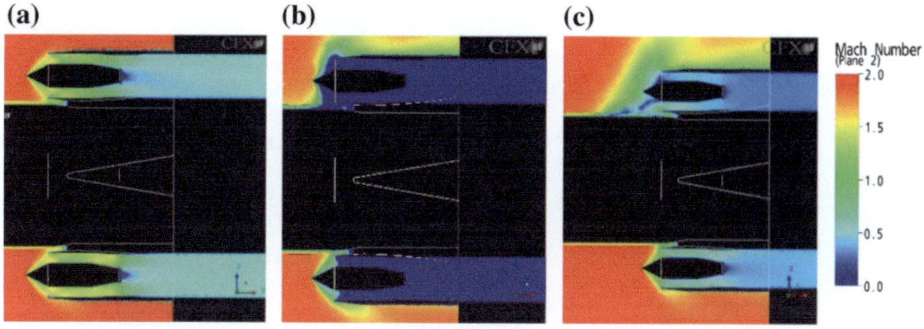

Fig. 15 Mach contour in pitch plane for 5° angle of incidence **a** $P_b = 1.2$ atm, **b** $P_b = 1.4$ atm and **c** $P_b = 1.5$ atm

sented in Fig. 15a–c, respectively. It can be observed that for lower back pressure, the flow structure at intake entry is same for both leeward and windward intakes (Fig. 15a) but at higher back pressure the flow structure are different (Fig. 15b, c). For highest backpressure of $P_b = 1.5$ atm (Fig. 15c), leeward side intake completely spills and the normal shock interacts with the corebody boundary layer. No spillage is observed for the intake in the yaw plane up to $P_b = 1.4$ atm (Fig. 16a, b) while spillage is seen to occur for $P_b = 1.5$ atm (Fig. 16c). The magnitude of the spillage in the yaw plane intake is much smaller for $P_b = 1.5$ atm compared to the spillage in leeward side intake in the pitch plane. The intake characteristics for angle of attack 5° have been plotted in Fig. 17. It can be seen that the mass capture is never full for the intakes in the supercritical range of operation (the vertical leg of the curve). It can be further seen that the East and West intakes perform similar, while the performance of the North and South intakes varies significantly in subcritical zone. This is due to the difference of flow pattern in the windward and leeward side intakes in the pitch plane for higher back pressure as explained previously while describing the qualitative features of the intake flow field at angle of incidence (Figs. 15 and 16). So the performance of the intakes at angle of attack is significantly different. The π-η characteristics of the intakes and dump plane at 6° angle of attack are compared with the experimental values in Fig. 18. A good overall agreement has been obtained. In this case also, top intake performance gets deteriorated fast compared to other intakes.

3.2 *Simulation of Installed Air Intake of a Ramjet Missile (Case 2)*

Many practical variable ducted ramjet missiles mount intakes asymmetrically (90° apart) in the windward side to obtain ideal performance. Intakes positioned close to combustor are favourable because of better internal and less weight. Asymmetrically

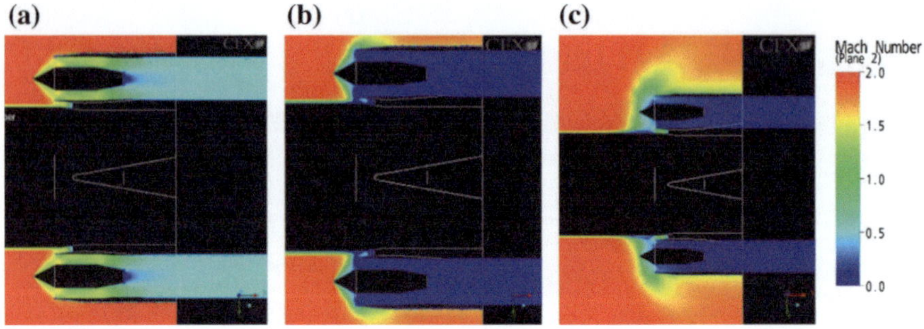

Fig. 16 Mach contour in Yaw plane for 5° angle of incidence **a** $P_b = 1.2$ atm, **b** $P_b = 1.4$ atm and **c** $P_b = 1.5$ atm

Fig. 17 Intake characteristics for AOA = 5°

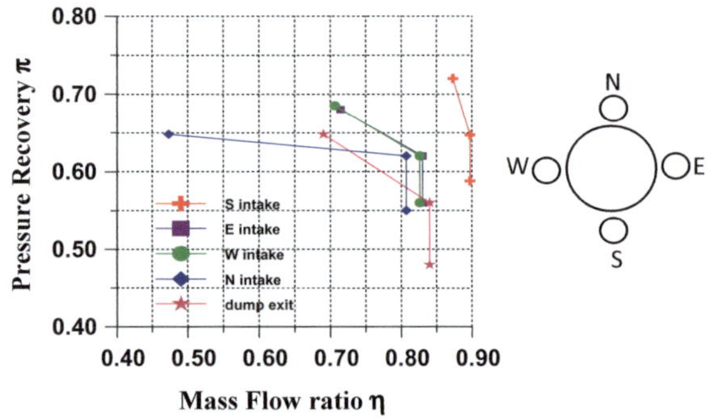

placed twin intake mounted on an ogive cylinder ($L/D = 14$) X-tail configuration is experimentally investigated Heyes [64] in Mach no range of 2.5–3.95 and Reynolds number of 6.56×10^6 per metre. The details of the geometry are available in Ref. [64] and are shown schematically in Fig. 19. Bhandarkar et al. [65] carried out combined external and internal flow simulation of Heyes experimental configuration (cheek mounted rectangular twin intake (B1I2T2 configuration) of Ref. [64]) using RANS methodology and evaluated intake performance parameters in terms of mass capture ratio and pressure recovery. The computational domain is shown in Fig. 20.

All the flow characteristics including oblique shock generated at the nose, flow spillage are clearly captured in the Mach number distributions at symmetry plane presented in Fig. 21 for three different angles of attacks. Mass capture ratio and pressure recovery coefficients of the intake for different angles of attacks are tabulated in Table 4. Increase in angle of attack decreases mass capture ratio by 9.8% at $\alpha = 7°$ compared to $\alpha = 0°$. Pressure recovery also decreases with increase in angle of attack. The cross-flow vorticity contours at different axial locations up to the intake plane for 7° angle of attack presented in Fig. 22 depict the growth of the boundary layer along the missile length. Flow quality at the intake entrance is clearly visible in the figure. The comparisons of computed aerodynamic coefficients (Drag (C_D), Normal force (C_N) and pitching moment (C_m) with experimental results [64] for different

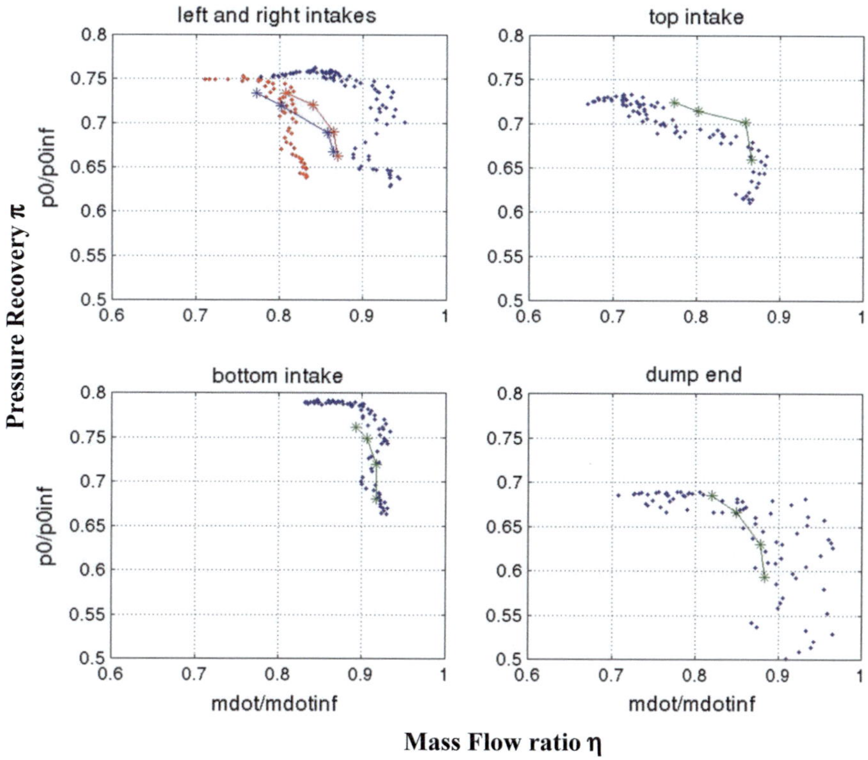

Fig. 18 Installed intake characteristics—Mach = 2.0, α = 6°

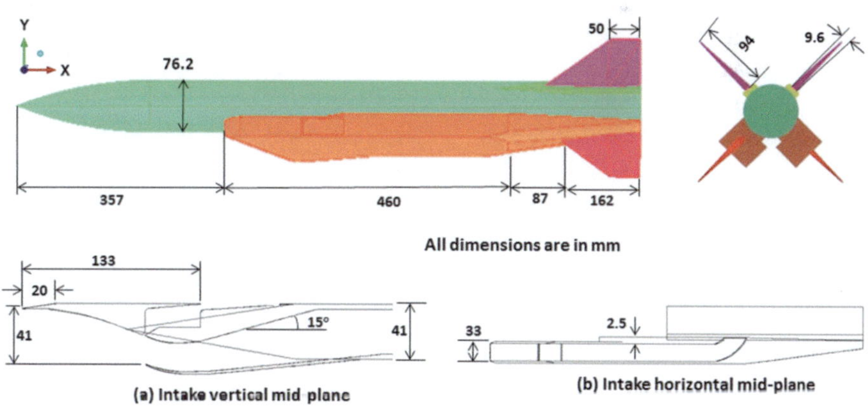

Fig. 19 Model configuration for which the computations are carried out (Reproduced from Ref. [65], with permission from DESIDOC)

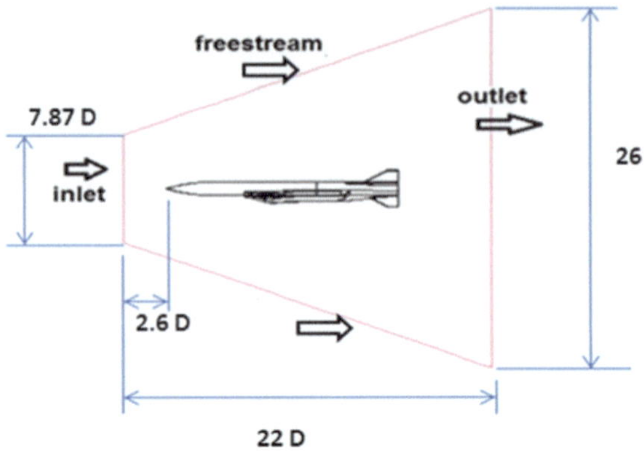

Fig. 20 Computational domain along with boundary conditions (Reproduced from Ref. [65], with permission from DESIDOC)

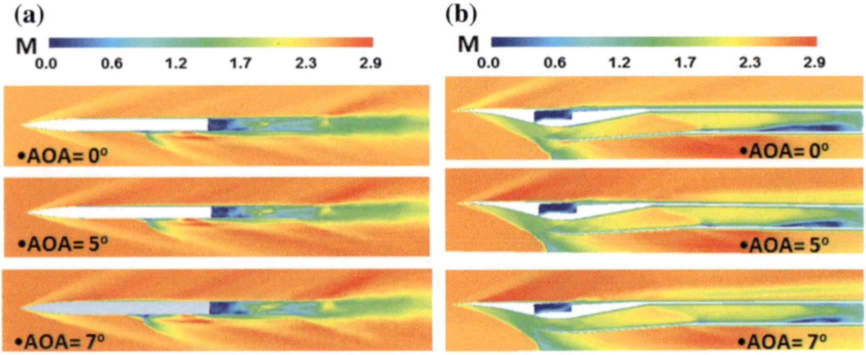

Fig. 21 Mach no contour at symmetry plane **a** full configuration, **b** with blown-up view near the intake (Reproduced from Ref. [65], with permission from DESIDOC)

Table 4 Mass capture ratio and pressure recovery at various angles of attack (Reproduced from Ref. [65], with permission from DESIDOC)

Angle of attack	Mass capture ratio	Pressure recovery	Angle of attack
0°	0.8108	0.75	0°
5°	0.7455	0.73	5°
7°	0.7314	0.73	7°

angles of attack are presented in Fig. 23a, b. A 4.2–8.5% deviation is observed in drag coefficient for $M_\infty = 2.5$ as α is changed from and 0° to 7°. The computed normal force coefficient has shown very good match (~0.5%) with experimental results at α = 5° and 7°; however, it differs by 6.5% for α = 0°.

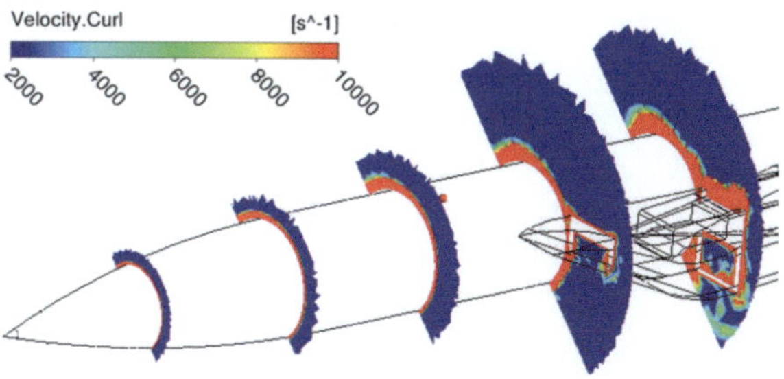

Fig. 22 Vorticity contours at different axial planes (Reproduced from Ref. [65], with permission from DESIDOC)

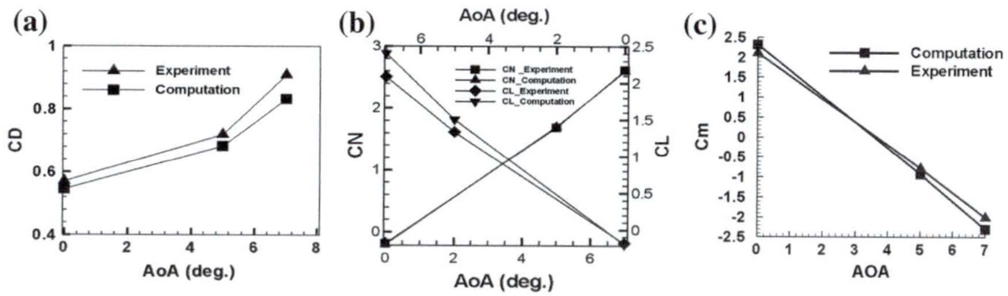

Fig. 23 Comparison of Aerodynamic characteristics **a** C_D, **b** C_N, **c** C_m (Reproduced from Ref. [65], with permission from DESIDOC)

3.3 Simulation of Starting/Unstarting Flow in Hypersonic Air Intake

Mixed compression intake (combination of internal and external compression) is generally considered for hypersonic airbreathing propulsion system. The schematic flow pattern in mixed compression intake is shown in Fig. 24. The incoming airflow is first compressed by the forebody bow shock and undergoes the number of compressions at vehicle undersurface compressions before entering into the intake. The external shock system coalesces at the intake cowl and turns inward into the intake by cowl geometry. The flow undergoes further shock reflections inside the intake duct. The interaction of the centerbody and cowl shocks with forebody boundary layer may cause flow separation

Mixed compression hypersonic intakes often plagued by "unstart problem" caused due to over-contraction, variation of flight conditions, increase of back pressure in combustor, etc., or due to a combined effect of these factors. Timofeev et al. [66] defined theoretical maximum permissible area ratio based on the theory of oblique shocks for the starting of internal compression intake. This, however, does not account for the interaction of shock and boundary layer in the intake which I very predominant in hypersonic flow. Kantrovich [67] proposed various methods, namely (1) variable intake geometry, (2) bleed bypass, (3) overboard spillage and (4) starting

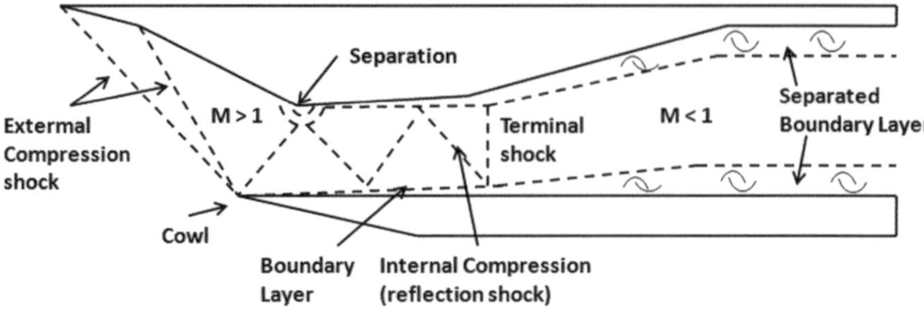

Fig. 24 Schematic of the flow field in mixed compression intake (Reproduced from Ref. [68], with permission from DESIDOC)

using unsteady effects, etc. to start supersonic intakes at any flight condition. In hypersonic flow, any complex mechanical control system may cause severe structural and cooling problems. The predictions of intake unstart and its mitigation plan are very much essential for hypersonic intake design. Experimental and numerical studies for hypersonic intakes continue to be very active research topics in recent literature. Important researches in this area are presented in Ref. [68].

The effect of geometrical variation on the starting characteristics of the isolated intake at free-stream Mach number 3.6 is investigated experimentally by Thiagarajan and Satyanarayana [69]. Saha and Chakraborty [68] carried out unsteady CFD simulation of the intake geometry of Ref. [69] at test condition and compared computed results with experimental data to obtain insights into the complex starting process through analysis of various flow variables.

A 1:8 scale intake model is tested in 340 mm diameter axisymmetric supersonic wind tunnel at DRDL at Mach 3.6 and 8×10^6 Reynolds number to simulate the non-reacting flow path of the intake of the integrated scramjet vehicle up to the combustor exit. The schematic of the test model is shown in Fig. 25 which includes 8 fuel injection struts similar to the flight configuration. Further, details of experimental condition are available in Ref. [69]. The cowl was set at 0° position at the beginning to start the intake and is opened up gradually to ingest more mass flow of air. It was observed that increasing the cowl angle from 7.5° to 8.5° the flow through the intake gets unstarted, resulting in high spillage.

Unsteady simulations are conducted out for free-stream Mach number 3.6, total pressure of 14 bar and total temperature of 300 K similar to the test condition. Two different simulations corresponding to cowl opening angle of 0° and 8.5° are carried out to study the effect of cowl opening angle on starting characteristics of isolated intake. Mach number plots in the symmetry plane corresponding to 0° cowl opening angle (cowl is parallel to ramp surface) at three different time instants (t_0, $t_0 +$ 1.5 ms, and $t_0 + 2.9$ ms) are shown in Fig. 26 to depict the unsteady flow field for free-stream Mach number 3.6. No shock structure ahead of intake cowl indicates that the intake is in the started condition in spite of the presence of large separation regions inside the intake. The oscillating separation bubbles (changing position with

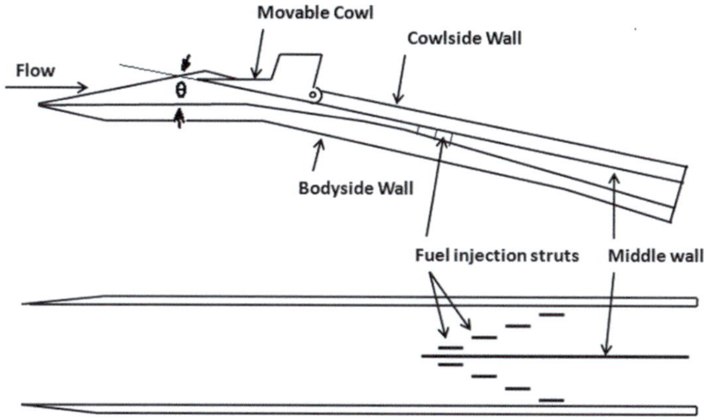

Fig. 25 Test model of hypersonic intake (Reproduced from Ref. [68], with permission from DESIDOC)

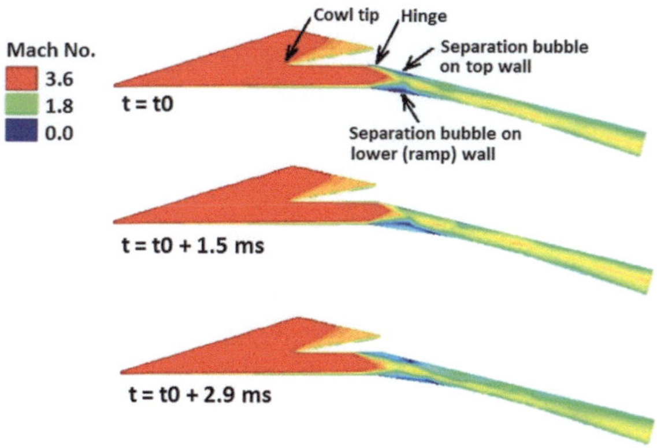

Fig. 26 Mach number contour at various instants of time showing unsteadiness of flow in a started intake (Reproduced from Ref. [68], with permission from DESIDOC)

time) are due to shock boundary layer interaction indicate unsteadiness of the flow in the intake. Figure 27 presents the comparison of computed ramp wall pressures with test data [69] at four different time instants. (In the absence of any unsteady pressure measurement data, time-averaged experimental pressure distribution on the ramp wall is compared with simulation results.) A reasonable match between experiment and computation is obtained. Ramp surface pressure is seen to fluctuating in time. The amplitude of 0.2 bar fluctuation of in static pressure (over an average pressure of 0.5 bar) and time period of 5 ms is observed at a location of the separation bubble.

Mach number plots in the symmetry plane for 8.5° cowl opening angle presented in Fig. 28 shows a large bubble of separated flow ahead of intake entrance causing intake unstarting. The changing position of the separation bubbles both at cowl and corebody surfaces depicts the unsteadiness of the flow. Computed pressures at different time instants at bottom ramp surface are compared with experimental data

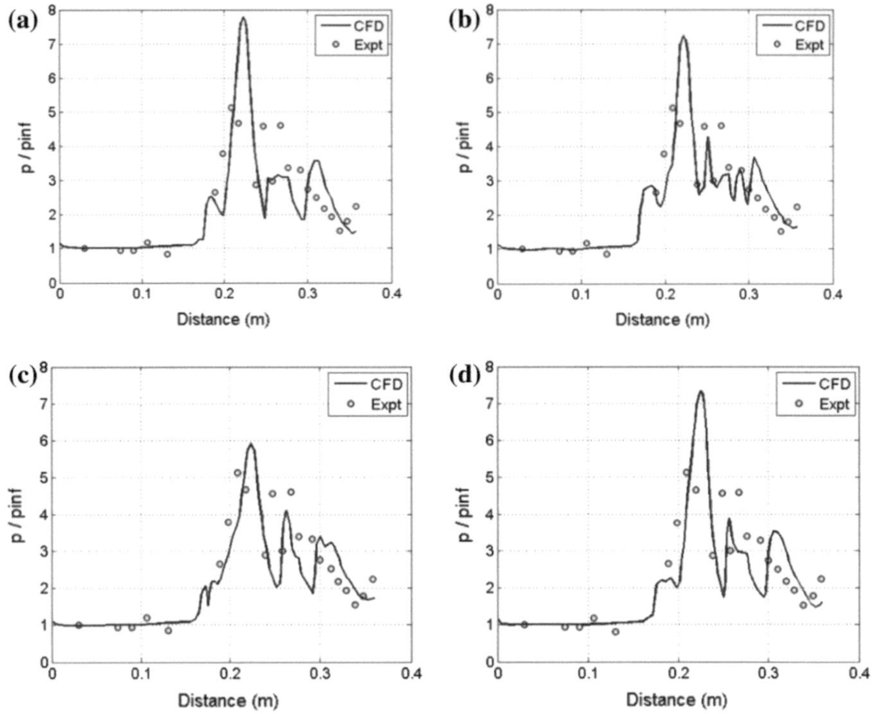

Fig. 27 Ramp wall pressure comparison at various time instants for 0° cowl opening **a** $t = t_0$, **b** $t = t_0 + 1.9$ ms, **c** $t = t_0 + 2.5$ ms, **d** $t = t_0 + 4.5$ ms (Reproduced from Ref. [68], with permission from DESIDOC)

in Fig. 29. The development of flow as the intake gets unstarted is shown in Fig. 30. The intake unstarting Mach number for 8.5° cowl opening angle is in between 3.0 and 3.2, in comparison to the experimental value of Mach 3.6.

4 Concluding Remarks

The role of CFD in the design of high-speed airbreathing systems is presented. 3D RANS equations are solved along with SST-k-ω turbulence model. Infinitely, fast single-step chemistry and Lagrangian particle tracking are employed to describe the liquid hydrocarbon combustion. All numeric are well resolved, and grid independence of the simulations is demonstrated by not only comparing the flow parameters with different grids but also analysing the Grid Convergence Index. CFD simulations guided the development of flight-sized isolated kerosene-fuelled scramjet combustor for hypersonic airbreathing mission. Insights of mixing and combustion process insights the combustor obtained from numerical simulations has helped to make design improvements by relocating fuel injection struts and injector locations to

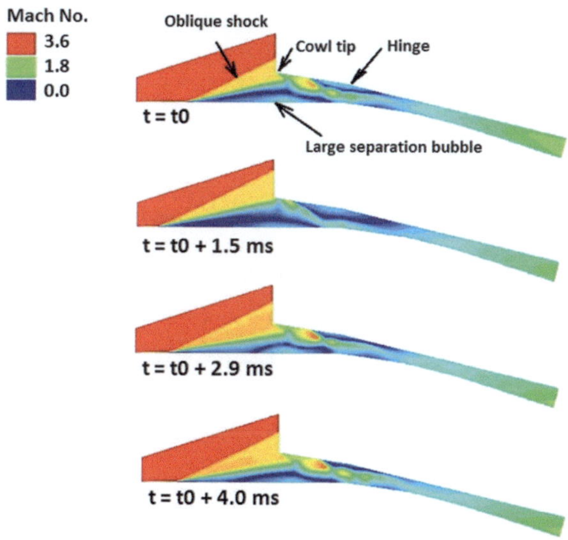

Fig. 28 Mach number contour at various instants of time showing unsteadiness of flow in an unstarted intake (Reproduced from Ref. [68], with permission from DESIDOC)

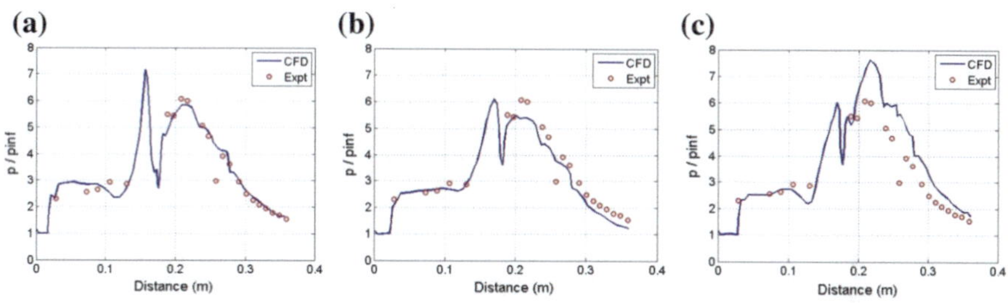

Fig. 29 Ramp wall pressure at different instants of time showing flow unsteadiness in an unstarted intake. **a** $t = t_0$, **b** $t = t_0 + 0.75$ ms, **c** $t = t_0 + 2.5$ ms (Reproduced from Ref. [68], with permission from DESIDOC)

achieve optimized performance of the combustor. Very good match of the wall pressure has been obtained with experimental data for both non-reacting and reacting flows. Convective heat fluxes in the scramjet combustor are estimated through well-resolved thermal boundary layer simulations. It is found that a minimum spacing of 10 microns adjacent to the wall is necessary for accurate prediction of wall heat flux. Tip-to-tail simulations with coupled non-reacting flow in the vehicle external surface and intake and reacting flow in the scramjet combustor for a complete hypersonic airbreathing vehicle demonstrate positive thrust minus drag and evaluated aerodynamics and propulsion parameters are used in mission design.

The performance parameters of installed air intake of ramjet powered vehicles are estimated through coupled simulation of the external and internal flow field. The

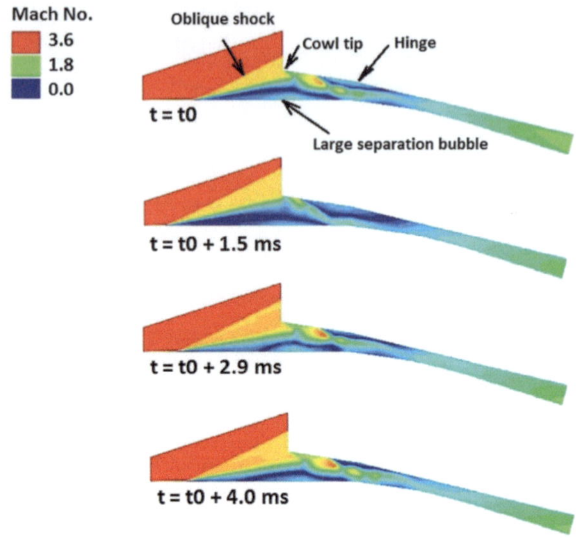

Fig. 30 Mach number contours showing unstarting of intake at inlet $M = 3.0$ (Reproduced from Ref. [68], with permission from DESIDOC)

computed pressure recovery and mass flow rate characteristics match well with experimental data for different angles of attacks. It was found that the operation regime for the intake in the leeward side in the pitch plane moves towards the subcritical regime faster than the other intakes at higher angle of attack and causes significant flow spillage. The starting and unstarting characteristics of a hypersonic intake are evaluated through unsteady RANS simulations for different Mach number and cowl opening angles. The computed pressures match well with experimental data for both starting and unstarting conditions. The flow is found to be unsteady in the intake even in the started condition and oscillate with amplitude of 0.2 bar and time period of about 5 ms. Both free-stream Mach number and cowl opening angle are seen to affect the starting characteristics of hypersonic air intakes. The reliability of CFD methods as an efficient and reliable design tool for high-speed airbreathing vehicle is demonstrated.

Acknowledgements The works presented in the article are carried out by the members of Computational Combustion Dynamics (CCD) Division of Directorate of Computational Dynamics (DOCD) of DRDL. The author greatly acknowledges the contributions of Dr. P. Manna, Sri Soumyajit Saha, Ms. Souraseni Basu, Sri Malsur Dharvath and Sri Anand Bhandarkar in preparing the articles. Thanks are due to the scientists of DRDL for providing the geometrical configurations and experimental data for the simulation and comparisons.

References

1. Weller HG, Tabor G, Jasak H, Fureby C (1998) A tensorial approach to computational continuum mechanics using object-oriented techniques. Comput Phys 12(6):620. https://doi.org/10.1063/1.168744
2. Chakravarthy K, Chakraborty D (2014) Modified SLAU2 scheme with advanced shock stability. Comput Fluids 100:176–184
3. Chakravarthy K, Arora K, Chakraborty D (2014) A simple hybrid finite volume solver for compressible turbulence. Int J Numer Methods Fluids 77:707–731. https://doi.org/10.1002/fld.4000
4. Chakravarthy K, Chakraborty D (2017) Damping numerical oscillations in hybrid solvers through detection of Gibbs phenomenon. Int J Numer Methods Fluids 84:699–714. https://doi.org/10.1002/fld.4367
5. Feri A (1964) Review of problems in application of supersonic combustion. J R Aeronaut Soc 68:575–597
6. Northam GB, Anderson GY (1986) Supersonic combustion ramjet research at Langley. AIAA paper no. 86-0159
7. Waltrup PJ (1987) Liquid-fueled supersonic combustion ramjet: a research perspective. J Propul Power 3(6):515–524. https://doi.org/10.2514/3.23019
8. Curran ET (2001) Scramjet engines: the first forty years. J Propul Power 17(6):1138–1148. https://doi.org/10.2514/4.866609
9. Edward AK, Joseph AS (1973) Liquid jet injection into a supersonic flow. AIAA J 11(9):1223–1224. https://doi.org/10.2514/3.50567
10. Miyajima H, Chinzei N, Mitani T, Wakamatser Y, Maita M (1992) Development status of the NAL scramjet engine test facility and subscale scramjet engine. AIAA paper no. 92-5094
11. Sunami T, Sakuranaka N, Tani K, Kiraiwa T, Shimura T (1997) Mach 4 test of a scramjet engine—effect of isolator. In: Proceeding of 13th international symposium of air breathing engine. AIAA, Washington D.C., pp 615–625
12. Mathur T, Gruber M, Jackson K, Donbar J, Donaldson W, Jackson T, Billig F (2001) Supersonic combustion experiments with a cavity-based fuel injector. J Propul Power 17(6):1305–1312. https://doi.org/10.2514/2.5879
13. Yu KH, Wilson KJ, Schadow KC (2001) Effect of flame holding on supersonic—combustion performance. J Propul Power 17(6):1287–1295. https://doi.org/10.2514/2.5877
14. Gruber MR, Donbar JM, Carter CD, Hsu KY (2004) Mixing and combustion studies using cavity-based flame holders in a supersonic flow. J Propul Power 20(5):769–778. https://doi.org/10.2514/1.5360
15. Yu G, Li JG, Chang XY, Chen LH, Sung CJ (2003) Fuel injection and flame stabilization in liquid-kerosene fueled supersonic combustor. J Propul Power 19(5):885–893. https://doi.org/10.2514/2.6179
16. Vinogradov VA, Kobigsky SA, Petrov MD (1995) Experimental investigation of kerosene fuel combustion in supersonic flow. J Propul Power 11(1):130–134
17. Bouchez M, Dufour E, Montazel X (1998) Hydrocarbon fueled scramjet for hypersonic vehicles. AIAA paper no. 1998–1589
18. Dufour E, Bouchez M (2001) Computational analysis of a kerosene fueled scramjet. AIAA paper no. 2001–1817
19. Manna P, Ramesh B, Chakraborty D (2008) Liquid fueled strut based scramjet combustor design—a computational fluid dynamics approach. J Propul Power 24(2):274–281. https://doi.org/10.2514/1.28333
20. McClinton CR, Rausch VL, Shaw RJ, Metha U, Naftel C (2005) Hyper-X: foundation for future hypersonic launch vehicles. Acta Astronaut 57:614–622. https://doi.org/10.1016/J.actaastro.2005.03.061
21. Voland RT, Huebner LD, McClinton CR (2006) X-43A hypersonic vehicle technology development. Acta Astronaut 59:181–191. https://doi.org/10.1016/J.actaastro.2006.02.021

22. Rondeau MC, Jorris RT (2013) X-51A scramjet demonstrator program: waverider ground and flight test. In: SETE 44th international/SETP southwest flight test symposium, Ft Worth, TX, 28 Oct–1 Nov 2013
23. Fedioun I, Orlik E (2012) Boundary layer transition on the LEA hypersonic vehicle forebody. AIAA 2012-5864
24. Sakurai H, Kobayasi M, Yamazaki I, Shirouzu M (1997) Shirouzu M (1997) Development of the hypersonic flight experimental vehicle. Acta Astronaut 40(2–8):105–112
25. Lockwood MK, Petley DH, Martin JG, Hunt JL (1999) Airbreathing hypersonic vehicle design and analysis methods and interactions. Prog Aerosp Sci 35:1–32
26. Gnoffo PA, Weilmuenster KJ, Hamilton HH II (1999) Computational aerothermo-dynamic design issues for hypersonic vehicles. J. Spacecraft Rockets 36(1):21–43
27. Fuhrmann HD, Hildebrand J, Lalicata T (1999) Aerothermodynamic overview, X-34. J. Spacecraft Rockets 36(2):153–159
28. Pannerselvam S, Thiagarajan V, Ganesh TKA, Geetha JJ, Ramanuchari V, Prahlada (2005) Airframe integrated scramjet design and performance analysis. ISABE Paper 2005-1280
29. Chandrasekhar C, Ramanujachari V, Reddy TKK (2014) Evaluation of kerosene fuelled scramjet combustor using a combination of cooled and uncooled strut. Defense Sci J 64(1):5–12 (2014). https://doi.org/10.14429/dsj.64.2733
30. Ramanujachari V, Chandrasekhar C, Satya V, Panneerselvam S (2009) Experimental investigations of a strut based scramjet combustor using kerosene fuel. In: 7th Asia-Pacific conference on combustion. National Taiwan University, Taipei, Taiwan. Paper ID 10161
31. Manna P, Dharavath M, Sinha PK (2013) Chakraborty D (2013) Optimization of a flight-worthy scramjet combustor through CFD. Aerosp Sci Technol 27:138–146
32. Malsur D, Manna P, Sinha PK, Chakraborty D (2016) Numerical analysis of a kerosene fuelled scramjet combustor. ASME J Therm Sci Eng Appl 8(8):0111003
33. CFX Computational Fluid Dynamics Software (2001) Version 2.11.1. AEA Technology Engineering Software Ltd
34. Launder BE, Spalding DB (1974) The numerical computation of turbulent flows. Comput Methods Appl Mech Eng 3:269–289
35. Kader BA (1981) Temperature and concentration profiles in fully turbulent boundary layers. Int J Heat Mass Transf 24(9):1541–1544
36. Magnussen BF, Hjertager BH (1976) On mathematical modeling of turbulent combustion with special emphasis on soot formation and combustion. In: Sixteenth symposium (international) on combustion. The Combustion Institute, pp 719–725
37. Manna P, Chakraborty D (2005) Numerical simulation of transverse H_2 combustion in supersonic airstream in a constant area duct. J. Inst. Eng. 86:47–53
38. Saha S, Chakraborty D (2006) Reacting flow computation of staged supersonic combustor with strut injection. In: Proceedings of 36th AIAA fluid dynamics conference, San Francisco, USA. AIAA paper no. 2006–3895
39. Afroz J, Chakraborty D (2006) Numerical simulation of supersonic combustion of pylon injected hydrogen fuel in scramjet combustor. J Inst Eng 87:1–6
40. Manna P, Ramesh B, Chakraborty D (2007) Thermochemical exploration of a cavity based supersonic combustor with liquid kerosene fuel. J Aerosp Sci Technol 59(4):246–258
41. Ramesh B, Chakraborty D (2006) Numerical simulation of kerosene fueled ramp cavity based scramjet combustor. J Aerosp Sci Technol 58(2):104–112
42. Malsur D, Manna P, Chakraborty D (2018) Prediction of heat flux in a scramjet combustor with kerosene fuel through CFD. J Aerosp Sci Technol 70(2):85–97
43. Manna P, Malsur D, Chakraborty D (2009) Performance improvement of kerosene fuelled scramjet combustor through modified fuel injection—a CFD study. J Aerosp Sci Technol 61(2):496–503
44. Malsur D, Manna P, Chakraborty D (2015) Effect of turbulence models and spray parameters on kerosene fuelled scramjet combustor. J Aerosp Sci Technol 67(3):369–383
45. Malsur D, Manna P, Chakraborty D (2015) Study of mixing and combustion in scramjet combustor with ethylene fuel through CFD. Acta Astronaut 117:305–318. https://doi.org/10.1016/J.actaastro.2015.08.014

46. Marquardt Corporation (1964) Supersonic combustion tests in an 8000 FPS air stream. Report No. 6064, Van Nuys, California, USA
47. ANSYS ICEM-CFD-11, Installation and Overview, Jan 2007
48. Cutler AD, Danehy PM, O'Byrne S, Rodrigues CG, Drummond JP (2014) Supersonic combustion experiment for CFD model development and validation. AIAA paper no. 2004-266
49. Roache PJ (2003) Error base for CFD. AIAA paper no. 2003-0408
50. Roache PJ (1998) Verification and Validation in computational science and engineering. Hermon Publishers, New Mexico
51. Lefebvre AH (1989) Atomization and sprays, 1st edn. Hemisphere Publishing Corporation, Bristol
52. Song KD, Choi SH, Scotti SJ (2006) Transpiration cooling experiment for scramjet engine combustion chamber by high heat fluxes. J Propul Power 22(1):96–102. https://doi.org/10.2514/1.11300
53. Li L, Fan X, Wang J (2011) Measurements of wall heat flux and temperature in a supersonic model combustor. AIAA paper, 2011-5916
54. Kennedy PJ, Donbar M (2011) Heat flux measurement in a scramjet combustor using direct write technology. AIAA Paper, 2011-2330
55. Zhang C, Qin J, Yang Q, Zhang S, Chang J, Bao W (2015) Indirect measurement method of inner wall temperature of scramjet with a state observer. Acta Astronaut 115:330–337. https://doi.org/10.1016/J.actaastro.2015.05.030
56. Back LH, Massier PF, Gier HL (1965) Convective heat transfer in a convergent–divergent nozzle. Technical report 32-415, Jet Propulsion Laboratory, California Institute of Technology, Pasadena, CA
57. Bhandarkar A, Malsur D, Murty MSRC, Manna P, Chakraborty D (2016) A novel CFD method to estimate heat transfer coefficient for high speed flows. Defense Sci J 66(3):203–209. https://doi.org/10.14429/dsj/66.8873
58. Malsur D, Manna P, Chakraborty D (2015) Tip to tail simulation of a hypersonic airbreathing cruise vehicle. J Propul Power 31(5):1370–1379. https://doi.org/10.2514/1.B35686
59. Seddon J, Goldsmith EL (1985) Intake Aerodynamics. AIAA Education Series
60. Laruelle G (1984) Air intakes for a probative missile of rocket ramjet. NASA TM-77407
61. Herrmann D, Gülhan A (2010) Influence of intake orientation on Ramjet performance. J Propul Power 26(4):848–857. https://doi.org/10.2514/1.47140
62. Soumyajit S, Sinha PK, Chakraborty D (2010) CFD prediction of Ramjet intake characteristics at angle of attack. J Aerosp Sci Technol 62(3):159–165
63. Theerthamalai P, Sekharan M, Umakant J, Selvarajan S (2003) Experimental investigation of an installed air intake model. In: National conference on air breathing engine, Bangalore
64. Hayes C (1983) Aerodynamic characteristics of a series of twin-Inlet airbreathing missile configurations. In: NASA Technical Memorandum 84559, Langley Research Center, Virginia
65. Bhandarkar A, Basu S, Manna P, Chakraborty D (2016) Aerodynamic characterization of ramjet missile through combined external-internal CFD simulation. Defense Sci J 66(6):624–629. https://doi.org/10.14429/dsj/66.9677
66. Timofeev EV, Tahir RB, Sannu M (2008) On recent developments related to flow starting in hypersonic air intakes. AIAA paper 2008-2512
67. Kantrowitz A, Donaldson C (1945) Preliminary investigation of supersonic diffusers. NACA WR L-713
68. Saha S, Chakraborty D (2015) Numerical simulation of a hypersonic air intake. Defense Sci J 65(3):189–195. https://doi.org/10.14429/dsj.65.8254
69. Thiagarajan V, Stayanarayana A (2013) Effect of variations of geometry parameters on air-intake starting characteristics. In: Proceedings of 8th symposium on applied aerodynamics and design of aerospace vehicles, Hyderabad, India, pp 1–7

Role of Analysis Led Design Approach in Diesel Engine-Based After-Treatment System

Ambarish Khot and Nitin Tripathi

Keywords ALD · After-treatment system · Adblue · Spray · FEA

Abbreviations

ANR	Ammonia NO_x Ratio
ALD	Analysis Lead Design
CES	Cummins Emission Solutions
CFD	Computational Fluid Dynamics
CTR	Cummins Technical Report
DEF	Diesel Exhaust Fluid
ESC	European Stationary Cycle
FEA	Finite Element Analysis
LES	Large Eddy Simulations
OBD	On-board Diagnostics
PDF	Probability Density Function
PID	Proportional Integral Derivative
SDI	Species Distribution Index
UI	Uniformity Index

A. Khot (✉) · N. Tripathi
Cummins Technical Centre India, Pune, India
e-mail: ambrish.d.khot@cummins.com

N. Tripathi
e-mail: nitin.tripathi@cummins.com

A. K. Gupta et al. (eds.), *Innovations in Sustainable Energy and Cleaner Environment*, Green Energy and Technology,
https://doi.org/10.1007/978-981-13-9012-8_13

ANR Factor—Ratio of local ANR to commanded ANR. Set of ANR factors calculated from probability density function are given as an input for running parallel channel 1D model using AVL boost.
Transient Temperature Behavior—In ESC cycle, temperature takes longer time to stabilize and sometimes even after 2 min duration, temperature values are not stabilized and shows the transition. So, modeling ESC cycle as steady-state runs in boost may give higher deviation from the actual test, since temperature may or may not be stabilized.

1 Introduction

1.1 Role and Growth of ALD in Industry

In Industries as we are moving ahead, there is a great push to deliver product faster and cheaper. ALD, as a tool, has been proven beneficial in terms of speed and cost, since it reduces the need for physical experimentation, shortens the time to market, and improves design reliability.

ALD is used in a wide variety of industries, including automotive, power generation, aerospace, and even in medical research and its impact is increasing day by day, since it can cross the boundary/limitation of the physical experiments such as analyzing things at microlevels or to find optimized design among a large number of available samples. Its impact is enormous as the equation of motion provides us with everything that is meaningful to know about the domain. For example, an after-treatment engineer not using ALD will make many assumptions about the fluid mechanics while designing after-treatment system with lower pressure drop and that can lead to significant assumptions about the fluid motions and chemical kinetics. ALD allows one to simulate the after-treatment system without making an assumption about the flow patterns and chemical reactions to design the system correctly in the first attempt. With the advancement of computational power, ALD has become more powerful and can solve ambitious numerical problems in a very short time and hence is the lifeline for the research departments of industries.

After-treatment system is an additional part of an engine and comes up with additional problems like backpressure, which directly impact the fuel economy of the engine. Also, UWS spray unit, which is needed to reduce engine-out NO_x has issues like urea freezing in supply lines at cold conditions, solid deposit formation of urea and lesser evaporation at lower temperature and many more. Addressing all these issues through test is expensive and is difficult to understand in detail and that is why ALD is important in order to deal with those issues.

Based on the above phenomenon and many others, there are various applications of CFD in after-treatment industries which involves pressure drop prediction, thermal management, reaction modeling, optimization, spray modeling, freezing, melting, and boiling, etc.

1.2 After-Treatment System

History of the diesel engine is very old and has gone through numerous changes with time. With technology advancement, diesel engine performance has changed drastically. There have been changes that are driven by government regulations and norms to reduce emissions from engines. Restrictive emission norms forced industries to move toward sophisticated after-treatment systems, which later became an integral part of the engine. Figure 1 depicts the evolution of technologies with time.

Pollution and health-related issues are well documented. Extensive research has been done to find the root cause of health issues due to emissions as reported in Muzio 2001 [1]. Increased health hazard forced the government to impose norms on the engine exhaust gas emission which includes restriction on hydrocarbons and nitrous oxides. Initially for less restrictive norms, tweaking within an engine or exhaust gas reactor was sufficient to control the emission. But as norms become stringent, after-treatment system starts playing an important role which has started its journey from the catalytic converter to more advanced technologies. Latest after-treatment systems have various controllers, sensors, and PID logics to ensure its performance at every operating condition.

After-treatment system has evolved over a period of time and there has been a significant reduction in its weight and size due to market demand. Compactness of after-treatment system brings several challenges and issues like increased backpressure, low NO_x and hydrocarbon conversion, thermal issues, sensors sensitivity, and many others.

Compactness also brings trade-off between various parameters, for e.g., improving exhaust gas and urea uniformity at catalyst face with the help of perforation tubes and mixers also contributes to increased backpressure which directly affects the fuel economy.

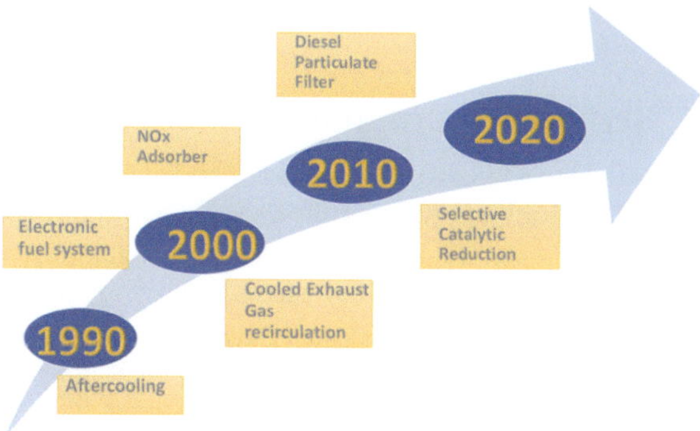

Fig. 1 Evolution of diesel engine technology

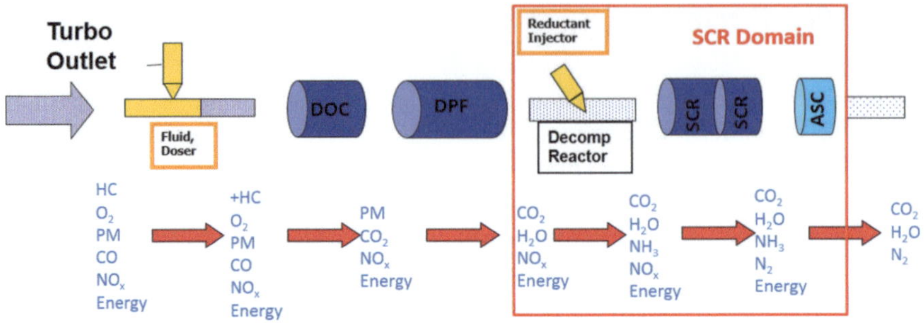

Fig. 2 Schematic of an after-treatment system

2 Basics of After-Treatment System

Basic schematic of after-treatment system is provided in Fig. 2. Once coming from the engine exhaust, gases first enter an oxidation catalytic converter positioned near the engine where hydrocarbon and carbon monoxide are converted to carbon dioxide and water, and after that, the gases reach diesel particulate filter. Here, particles are removed from the gas stream and collected in the filter structure. Filter is regenerated in regular intervals and the carbon particles are removed by combustion. To reduce nitrous oxides to a minimum level, when exhaust gases come out of the diesel particulate filter, 32.5% aqueous urea solution named "Adblue" is sprayed into the gas stream. This releases ammonia into the exhaust system. Subsequently, De-NO_x catalyzer or selective catalytic reduction (SCR) reduces nitrogen oxides to nitrogen and water. Urea spray undergoes primary and secondary atomization to form fine droplets which decompose to form ammonia and urea vapor. Also, at lower temperatures, some urea due to lack of heat energy forms solid deposits at the walls of the after-treatment system.

3 Technology Evolution and ALD

Any technology or product follows three phases when entering in market, i.e., birth, growth, and maturity. Among all the parameters, usually, profitability remains positive at growth and maturity since technology/product at these phases has been accepted by the market and it starts generating revenue. While on the other hand, it involves significant initial investment at birth phase because the technology needs improvements and seen as a risk by a consumer, due to which revenue is minimal. So, investment at birth phase becomes very critical and a systematic approach is needed to optimize the investment at this phase (Fig. 3).

Leveraging ALD at concept and designing phase helps us to optimize the design quickly and reduce effort/cost for testing and support.

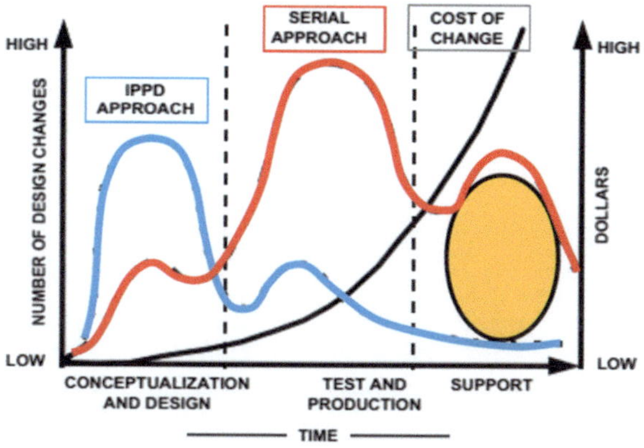

Fig. 3 Serial approach versus IPPD approach [2]

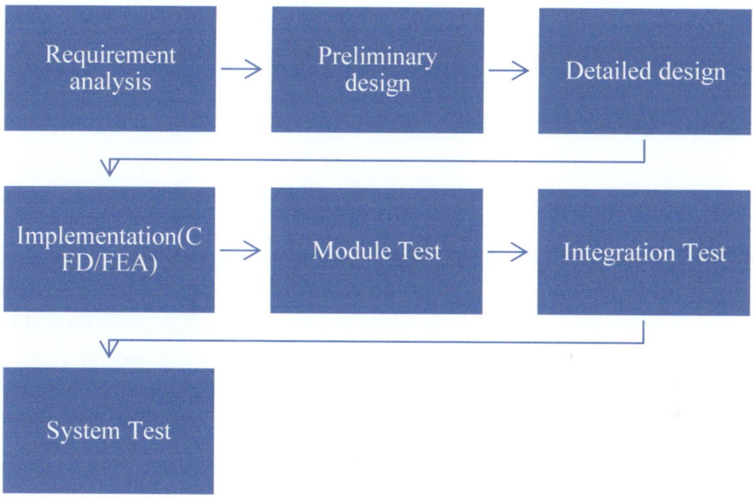

Fig. 4 Product life cycle

Product development goes through different phases. Figure 4 shows us the product life cycle of an after-treatment system which starts from gathering the requirements followed by designing and once the initial design is ready, it is analyzed through CFD/FEA and this process of design and analysis goes back and forth till the optimized designed obtained as per the requirements. Once the design gets optimized, it moves forward to physical component testing and engine testing with A/T integrated to it.

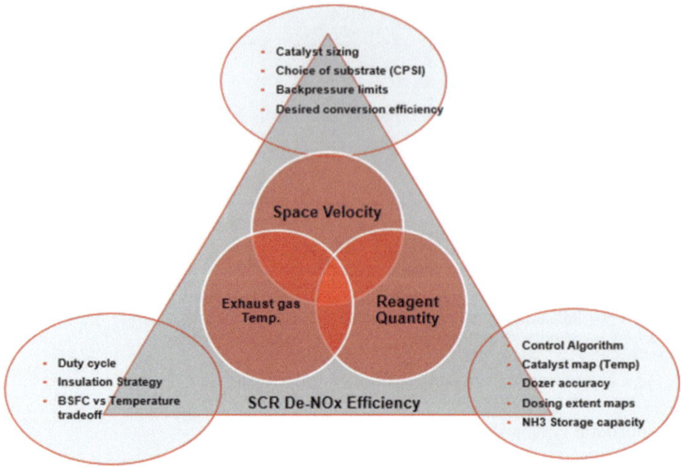

Fig. 5 Product optimization triangle

4 Product Optimization

The after-treatment system has been proven a very efficient device in controlling emission but is an additional burden to the engine performance and sizing. So, optimization of the after-treatment system also becomes essential for better engine performance along with its own performance and there are different parameters associated with that such as backpressure, skin temperatures, and catalyst utilization (Fig. 5).

ALD has made a strong contribution in predicting the performance of these parameters at a very early stage and in a later section, we will be giving details about all these parameters, failure modes in the after-treatment system, and also some case studies are provided which will explain various performance predicting capabilities and issue resolution through ALD.

4.1 Optimization Through ALD

4.1.1 Optimization Through CFD

CFD provides wide range of performance parameters for the after-treatment system. Figure 6 shows the various parameters which CFD can predict for different components of the after-treatment system. Explaining some of them below.

- Uniformity Index or UI tells the distribution of velocities or some species across a plane, which is very useful to check the flow distribution at catalyst inlet to assess the catalyst utilization. Also, it is helpful in optimizing location for different sensors, since we need uniform flow at the sensor plane for its good performance. *This index is* popularly known as flow distribution index for gas velocity distribution and if its value is 1, flow is perfectly uniform across that surface/plane. UI of ANR is one of the parameters in 3D-CFD, which defines the performance of the system. It is proportional to NO_x conversion efficiency which is either measured in a test or calculated through 1D tools like AVL boost.

- Backpressure prediction or pressure drop across after-treatment system is the most critical measure against which performance of every A/T is evaluated since it is directly related to fuel economy. With an increase in backpressure, fuel consumption goes up.
- Urea deposits are also one of the critical issues which hamper system performance and sometimes stop its functioning. ALD provides a proactive approach to predict deposit formation and helps to prepare a design with lower probability of deposit formation.
- Other component issues which are critical for system optimization and can be predicted through CFD are skin temperature, % thaw volume prediction in DEF supply unit and Dozers and many more mentioned in Fig. 6.

4.1.2 Optimization Through FEA

After-treatment Systems are developed to deliver minimum noise and emissions with maximum durability and performance. FEA simulates different structural and acoustic failure modes of after-treatment system. Figure 7 shows the various FE analyses

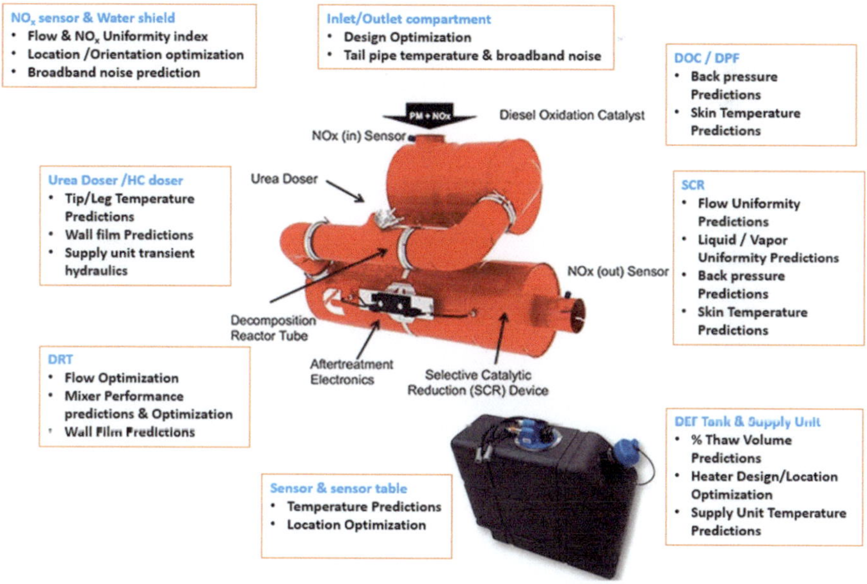

Fig. 6 Different areas of after-treatment system where CFD can predict the performance

used to predict different structural and acoustic failure mode of after-treatment system and optimizing a design at the very initial level and resolving issues in test cell or field.

For simulating structural failure mode like road vibration is simulated through random vibration analysis (PSD). To validate under static loading "static analysis" is used. For exhaust gas, pressure pulsation can be simulated through harmonic analysis. Urea dozer mount interface sealing requirement can be simulated using sealing analysis (nonlinear analysis). Checking for thermal performance, thermal analysis is performed, to check expansion and contraction of the exhaust system, and stresses developed because of these phenomena.

For validating acoustic performance, transmission loss and insertion loss (IL) analysis are done. Transmission loss is the acoustic power level difference between the power incidents on the muffler and transmitted downstream into an anechoic termination so it is used to compare different design against critical frequency. Insertion loss is the difference between the radiated sound pressure level without a muffler (straight pipe replaces muffler) and that with the muffler. IL can be changed by different engine running-condition, environment, and terminal conditions. Thus, in the engine test, it will be best to use the same engine/running-condition as that used by the customer. Final evaluation of muffler performance is based on the insertion loss analysis.

FEA is very useful for proactively identifying failure mode and to mitigate failure mode at the early phase of the design cycle which in turn would help to reduce cost, design cycle time, no. of prototype testing, and to get optimized design.

In the later section, some case studies have been provided which will show the application of some analysis and ALD methods developed in-house along with validation.

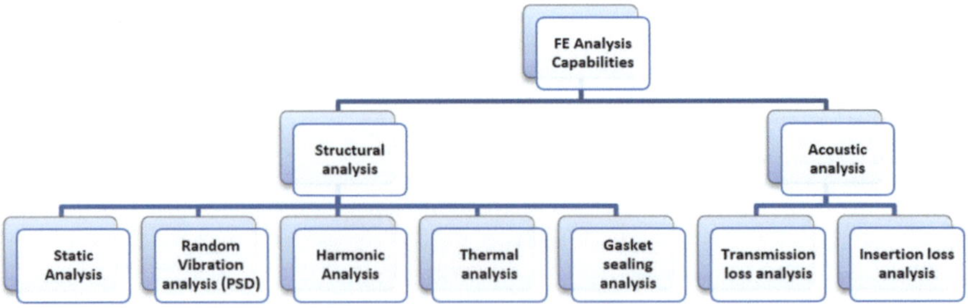

Fig. 7 FEA analysis capabilities (above figure represents the different areas where FEA plays a major role in optimizing a design at the very initial level and resolving issues in test cell or field)

NH3(liquid)- NH3(Gas)
HNCO(liquid)- HNCO(Gas)
HNCO(Gas)+H2O)Gas)- NH3(Gas)+CO2(Gas)

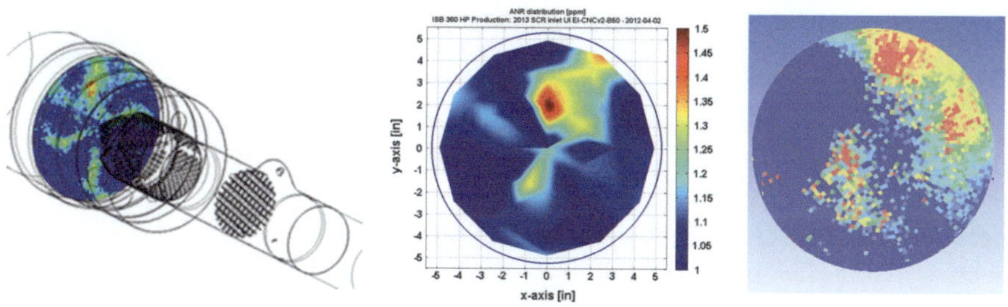

Fig. 8 2010 US EPA AFT geometry (left), ANR uniformity contour predicted experimentally (middle), and ANR contour predicted through CFD (right) using in-house decomposition model [3] in fluent 15

5 Case Studies of Typical After-Treatment System Analysis

5.1 Urea Spray Analysis

Uniformity of NH_3 and ANR in front of catalyst is very essential for after-treatment performance and for the proper catalyst utilization. Significant improvement has been made in modeling urea spray over the years. In Cummins Emissions Solutions group, Achuth Munnannur developed a multicomponent model [3] which was an improvement over a simple water spray modeling. This model includes improved heat and mass transfer model along with modeling of thermolysis and hydrolysis processes. Figure 8 shows 2010 geometry with ANR contours plotted through experimental methods at catalyst face and showing CFD results using fluent 15.0.7 with RNG turbulence model and standard wall functions. Although the correlation between test and CFD looks fine but still need some improvements.

Later versions of fluent show further improvement in spray modeling. The latest approach also involves urea reaction to form ammonia and isocyanic acid and with fluent 18.0, it is showing better correlation with test for ANR distribution (Fig. 9).

Left most ANR contour of Fig. 9 shows the results of fluent 16.2, which was under predicting urea decomposition as compared to experiment. Middle contour shows the results of fluent 16.2 with in-house decomposition model [3] for urea decomposition and over prediction of urea decomposition has been observed as compared to experiment. Rightmost contour shows the results of fluent 18.0 and is matching well with the experiment.

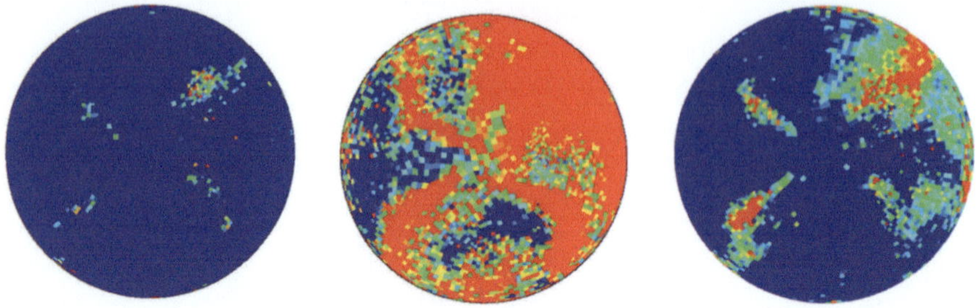

Fig. 9 Left (fluent 16.2), middle (fluent 16.2 + in-house decomposition model), and right (fluent 18.0)

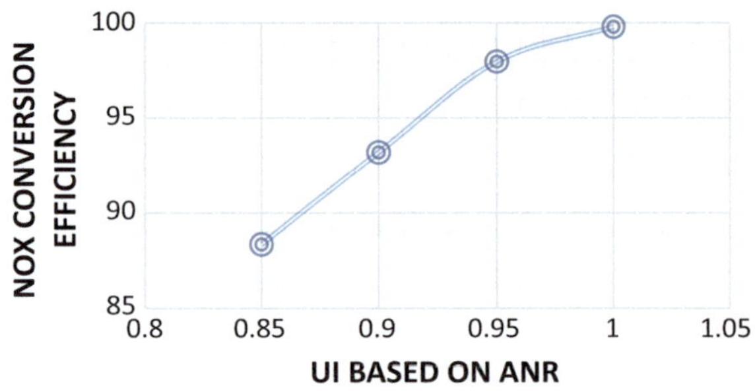

Fig. 10 NO_x conversion efficiency versus Uniformity Index

5.2 *Coupling 3D CFD with 1D (AVL Boost) to Predict the Impact of Non-uniformity of ANR Distribution on NO_x Conversion Efficiency*

Uniformity Index (UI) which is one of the important deliverables of spray analysis is not directly comparable with testing deliverables like ammonia slip and NO_x conversion efficiency. So, it is important to have UI acceptance criteria for a design which is targeting ongoing emission norms. 1D performance modeling (AVL boost) is playing an important role for CES because of its nature of capturing catalytic reactions within SCR and it can model the reaction for NO_x reduction within catalyst and predict the amount of NO_x which is coming out of catalyst and hence evaluate the NO_x conversion efficiency. NO_x conversion efficiency is a major parameter to select the SCR volume. Figure 10 shows the relation between ANR UI and NO_x conversion efficiency for one of the catalysts.

It is always needed a correlation between Uniformity Index (UI) and NO_x conversion efficiency to define UI target values in a different program. Since NO_x conversion target is based on the emissions norms, UI target values should be chosen corresponding to NO_x conversion target to avoid over or under designing of a product, while analyzing in CFD. Also, considering non-uniformity of ANR while calculating NO_x conversion efficiency through ALD will give more accurate results.

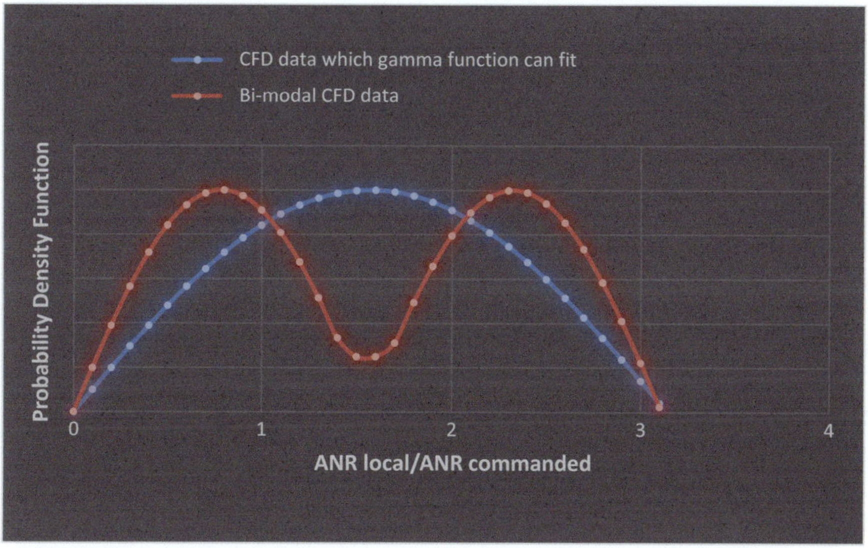

Fig. 11 Sample ANR distribution from CFD and gamma distribution fit [6]

Work has been done by McKinley [5] to understand the impact of ANR UI at SCR inlet on NO_x conversion efficiency. In McKinley [5], ANR distribution is represented by probability density function (beta function) and is used to calculate ANR factors.

Later, similar approach has been used by Apoorv [6] with gamma function, where actual CFD data were taken and fit into the gamma distribution to calculate ANR factors, which represents the ANR UI and the ANR factors are used as inputs for running boost simulation.

CFD data can be bimodal (Fig. 11) sometimes and gamma distribution may not exactly fit the data. To address this, Apoorv Kalyankar developed a MATLAB script to incorporate empirical PDF-based approach to generate ANR factors from CFD data.

In the next section, results for one of the case using "Empirical PDF-based approach" at ESC cycle have been shown.

5.2.1 Results and Findings

It was found that with 3D-1D coupling, conversion efficiency predicted was very closer to the test data (Fig. 12). Also, it was found that, at lower temperatures, exact simulation of ESC cycle gives good correlation with the test as compared to running steady-state analysis for each mode of ESC cycle. In ESC, temperature does not reach steady state in the allotted time for each mode (as shown in Fig. 13), so results of steady-state analysis may deviate from the actual NO_x conversion due to "Transient Temperature Behavior."

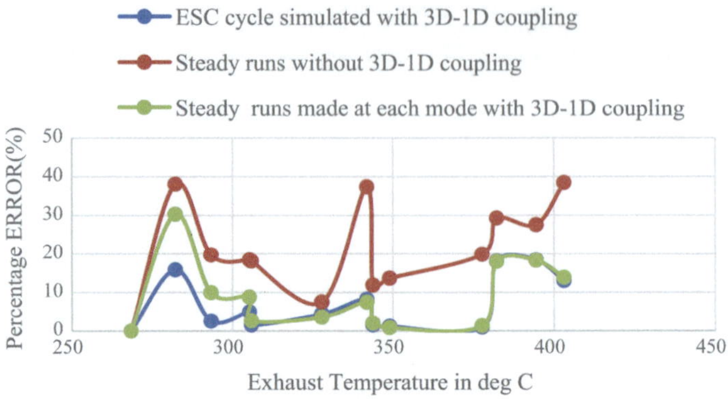

Fig. 12 Plot is showing NO_x conversion efficiency percentage error with test

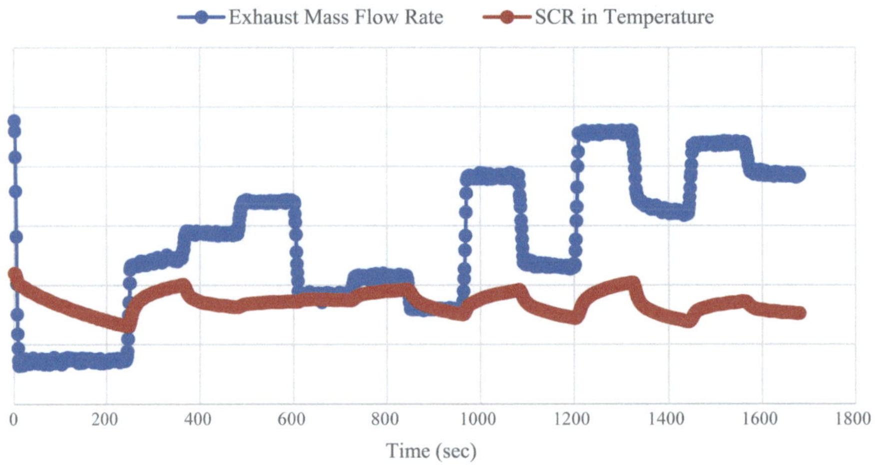

Fig. 13 Mass flow rate and temperature in vertical axis

5.3 NO_x Sensor Location Optimization Analysis

With the compulsion of OBD, NO_x sensors play a very crucial role in meeting emission targets. NO_x sensors either shaped as a probe or a wheel and its location is very critical to calculate the amount of NO_x in the domain. ALD becomes very important in this case because there are NO_x sensor probes which sense ammonia as NO_x, so 3D CFD along with 1D (AVL boost) provides us the capability to model NO_x reduction reaction within catalyst and NO_x/NH_3 distribution at downstream of the catalyst. And we can identify the location which is rich in NO_x and lean in ammonia.

Procedure—Apoorv Kalyankar developed this approach to predict NO_x and NH_3 distribution downstream of the catalyst. Transfer function generated with the help of

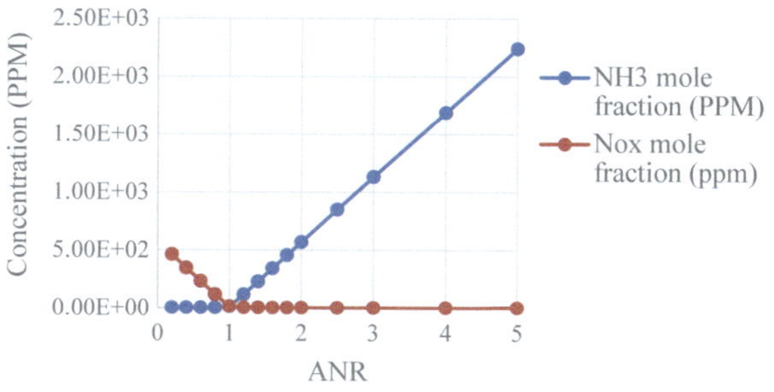

Fig. 14 Transfer function (NO_x and NH_3 mol fraction vs. ANR)

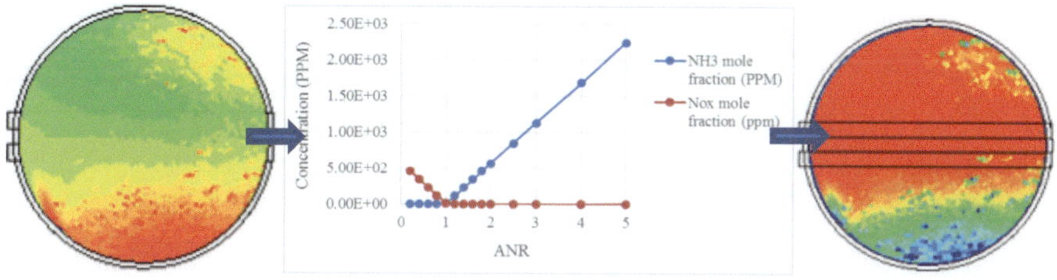

Fig. 15 Transfer function (middle) generated from the SCR inlet data (left) and give NO_x and NH_3 distribution at outlet (right)

AVL boost gives catalyst out NO_x and ammonia values as a function of catalyst inlet ANR. Figure 14 show some samples of the transfer function.

ANR distribution at inlet comes from 3D spray analysis and this transfer function is used to calculate NO_x and NH_3 at the outlet of catalyst (Fig. 15).

Once obtained the NO_x and NH_3 distribution at catalyst outlet, steady-state species analysis needs to be run on a reduced domain to get the NO_x and NH_3 distribution at the downstream of the after-treatment system.

For NS5, NS6, and BS6 emission norms, EGP will be having closed-loop control system for exhaust NO_x measurement. Testing conducted at the initial stage shows that normalized error in NO_x measurement changes significantly with the location of NO_x sensor as compared to actual system out NO_x ppm. Inconsistent trends correspond to design changes were observed in NO_x testing, which leads to conduct CFD analysis for better understanding of critical parameter for NO_x sensor measurement accuracy.

- Qualitative ALD prediction of NO_x and NH_3 ppm (including catalyst kinetics effect are within ±20% agreement with test results.
- ALD guidelines for sensor position selection are suggested in terms of SDI, % deviation in ppm, and sensor mass flow rates.
- Design modification has been done based on CFD results to improve sensor measurement accuracy at particular location (Fig. 16).

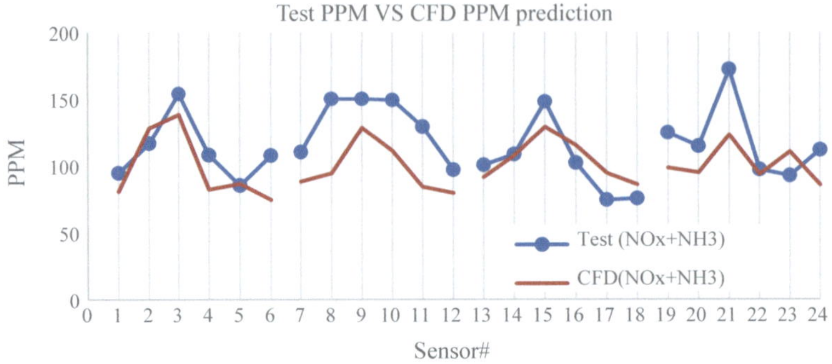

Fig. 16 Above plot showing test versus CFD predicted $NO_x + NH_3$ PPM

5.4 Case Study on Shaker Test Simulation Using PSD Analysis

The emission control devices like diesel oxidation catalyst [DOC], diesel particulate filter [DPF], and selective catalyst reduction [SCR] system are collectively known as After Treatment (AT) Emission Solution System used in off-road heavy-duty vehicles and mounted on chassis. The off-road heavy-duty vehicles are subjected to uncertain forces through vibrations, which are highly random in nature. A current PSD analysis process using ANSYS has been used to calculate the fatigue life of AT system component [DOC/DPF/SCR]. Figure 17 (left) is a picture for actual experimental shaker test setup block diagram and Fig. 17 (right) shows FE model setup block diagram.

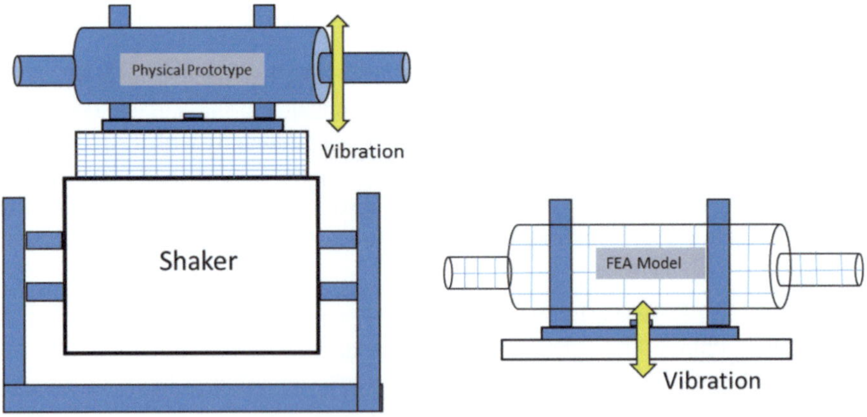

Fig. 17 Block diagram Shaker test setup (left) and FE model setup (right)

The correlation of the acceleration spectrum obtained from the test and the FE analysis RPSD (acceleration response PSD) are compared and are shown in Fig. 18. Variation of RPSD (acceleration response PSD) total gRMS is less than 10% in FEA and testing. The analysis results can be used to give meaningful judgements on whether a component will survive random vibration testing.

Damage is the fraction of life that is consumed during a test or simulation. Therefore, if an analysis computes a damage value of 0.5 at a location, then the part could survive for twice the duration of the analysis. The FE-based fatigue analysis accurately identified the failure locations and estimated fatigue damage within a factor of 2 of observed test results means it could survive for half of the duration. Fatigue damage from the test for all three prototypes noted in the range of 1.25–2 means part could survive for 0.75–0.5 the duration. Fe-based calculation well-correlated compared to three prototype testing (Figs. 19 and 20).

5.5 *LES of Backward-Facing Step Flow in Open FOAM*

Flow over a backward-facing step (BFS) is one of the most commonly studied flows for turbulent analysis due to the physical significance of separated flows and yet, simplicity in geometry. Extensive research work involving experimental as well as computational studies using large-eddy simulation (LES), and Reynolds-Averaged Navier-Stokes (RANS) has been carried out on the backward-facing step geometry. DNS in most cases is not affordable due to the huge computational costs involved and RANS gives us the average quantities of flow fields rather than concentrating on the detailed but random turbulent phenomena. In order to realize the smaller-scale

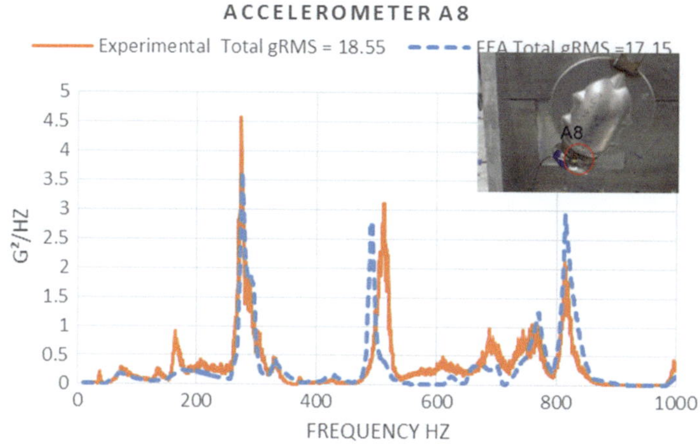

Fig. 18 Acceleration response comparison of PSD FE analysis with PSD testing

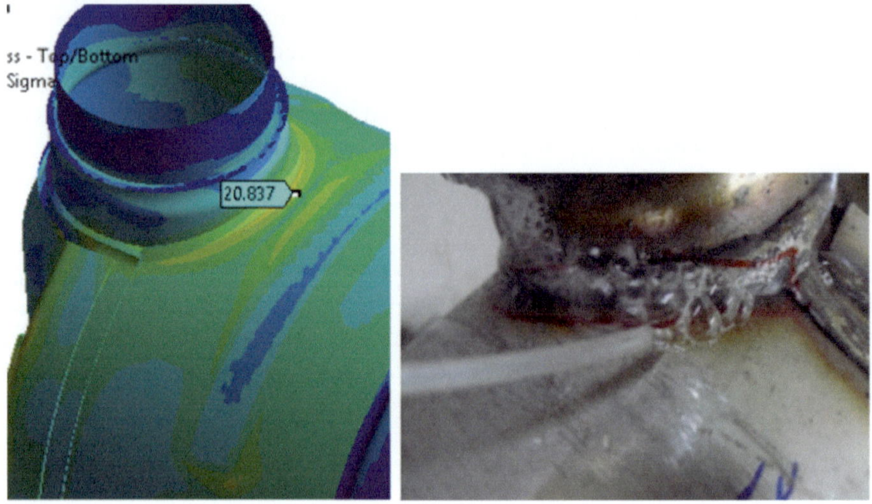

Fig. 19 Stress plot and damage location in testing of outlet joint weld

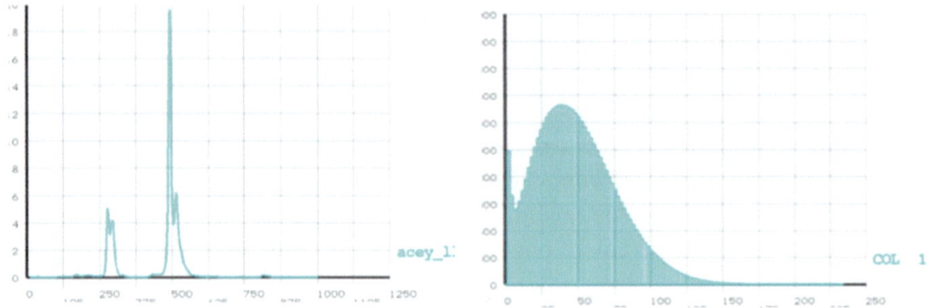

Fig. 20 Fatigue calculation of inlet joint weld

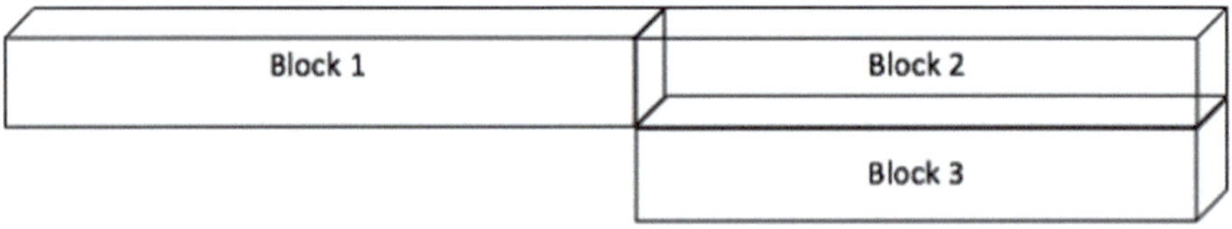

Fig. 21 Schematic of the geometry showing blocks

kinematic features and at the same time save on the computational resources (as compared to DNS), LES has emerged as a middle ground with tremendous potential.

Computational Domain and Mesh Generation. Figures 21 and 22 provide a description of the computational domain used for simulations.

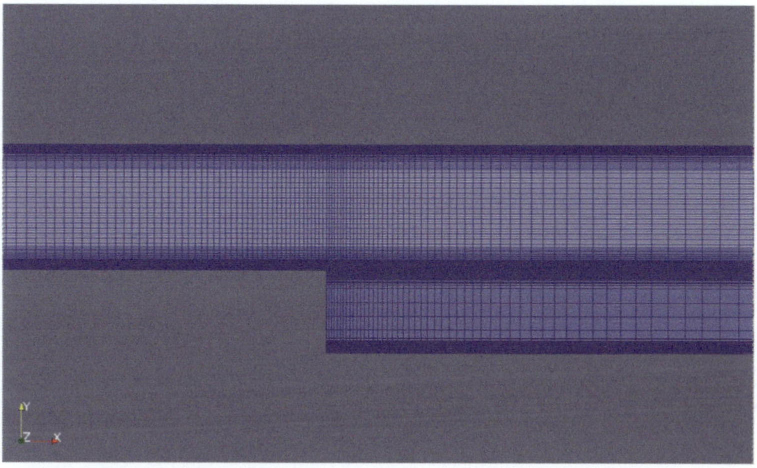

Fig. 22 View of the structured mesh

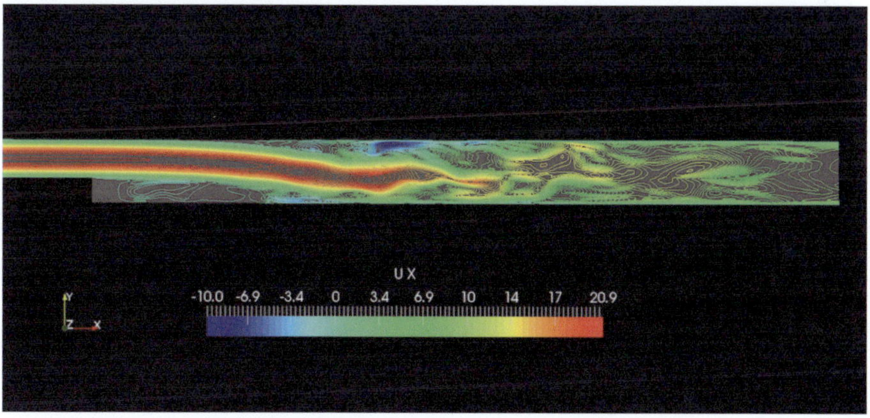

Fig. 23 Velocity contours at the mid-plane of LES simulations in OpenFOAM

5.5.1 Results and Discussions

The success of the simulations on BFS flow mainly depends on accurately predicting the recirculation length. However, one of the major reasons for using LES for the simulations is to adequately capture the turbulent vortical structures in the flow which otherwise cannot be captured using the RANS approach. Presented in this section are some of the results obtained from the LES simulation on Mesh 3 after 27,000 iterations (Figs. 23, 24 and 25).

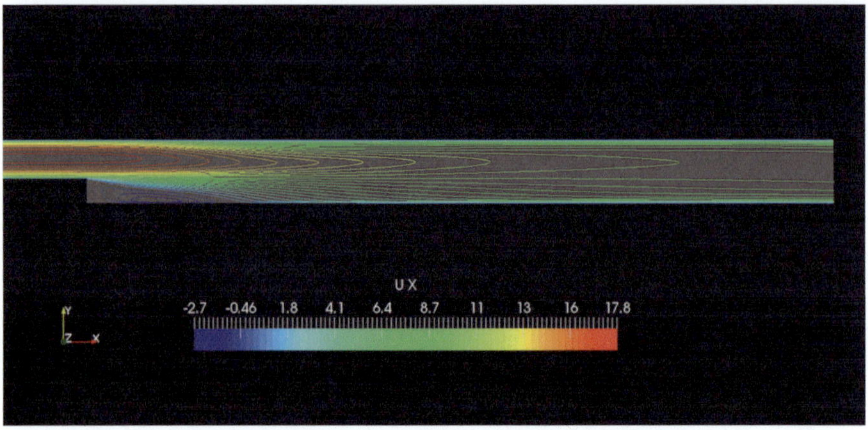

Fig. 24 Velocity contours at the mid-plane of RANS simulation in OpenFOAM

Re	Tool	Model	Reattachment Length Ratio	%Deviation from Experimental Results	%Deviation of Open FOAM Results (Mesh1)
3615	Experiment		6.45	-	
	ANSYS	SST k-ω	6.38	-1.09	33.23
		LES	8.32	28.99	2.16
		DDES	13.93	115.97	-38.98
		ELES (Zonal LES)	6.13	-4.96	38.66
	Open FOAM	LES	8.5	31.78	-

Fig. 25 Percentage deviation of experimental, ANSYS, and OpenFOAM results

5.5.2 Major Findings from This Study

LES simulations were carried out on a BFS case with an expansion ratio of 1.6 and a rather low Reynolds number of 3615. Initially, simulations were carried out in ANSYS on various models including LES, DDES, SST k-ω and ELES (Zonal LES). A study was then carried out in Open FOAM with a view of comparison. Different Sub-grid scale models including Smagorinsky, one equation eddy (k-equation) and dynamic k-equation model were used in the Open FOAM study. These models seemed to have minimal effect on the value of recirculation length. The most influential parameters were the grid size and inflow conditions. Although the recirculation length was overpredicted in most of the LES cases, it was found that the case with synthetic turbulence conditions at the inlet (with random fluctuations added in order to trigger turbulence) performed better than the cases with a uniform velocity profile at the inlet. SST k-ω model and ELES were successful in accurately predicting the recirculation length.

The grid size, particularly in the near-wall regions, is a significant factor for wall-bounded flows like in the current case. In LES, scales up to the inertial range are resolved and the once in the dissipation range are modeled. Improper resolution in the near-wall regions leads to the unresolved or small-scale eddies to falsely add to the effective viscosity and hence the SGS stress tensor. This leads to damping of turbulence. At a certain distance from the wall, all large-scale eddies in the flow reduce to Kolmogorov eddies (smallest eddies in turbulence). All-important turbulence parameters are defined by Kolmogorov microscales in this region. The turbulent length scales become anisotropic close to the wall with the wall-normal scale tending to zero while the other two scales remain finite. This causes streaky structures to be developed in the region close to the wall. Moreover, the small eddies carry a great amount of kinetic energy. Therefore, most of the production of kinetic energy takes place in the near-wall region. It is very important to have a proper mesh resolution (sometimes 90% of the total mesh size) in the near-wall regions where this phenomenon occurs since there is a need to resolve or filter all small-scale turbulence in this region. This requirement is at times as high as the mesh requirement of DNS in near-wall flows and heavily adds to the computational efforts. The total computational effort for a wall-resolved LES scales from $Re^{1.8}$ to $Re^{2.4}$ according to studies performed in the past. For DNS, a scaling of Re^3 is expected. This is seen as one of the major limitations of LES.

This was also observed in the simulations performed in the current study. Improper mesh resolution in the near-wall regions had a damping effect on turbulence in the inlet section. The adverse velocity gradients in the near-wall region expected in a fully developed turbulent flow could not be captured. In contrast to this, RANS simulations with the use of wall functions were able to do this according to the law-of-the-wall. Owing to the above reasons, a hybrid RANS-LES method, such as detached eddy simulation (DES), is preferred in modern times. In such a method, RANS methodology is used in the near-wall regions (to model near-wall effects) where the LES grid requirements are too high, and LES is performed for regions away from the wall. This helps in capturing the large energetic eddies in the flow more accurately using LES while not being damped by inaccurate sub-grid stresses in the near-wall regions. However, attaining the correct grid requirements for hybrid methods like DES or delayed detached eddy simulation (DDES) is a challenging task. DDES performed on the current case in ANSYS over predicted the reattachment length ratio by 115%. The log layer is not accurately resolved in this hybrid method and there is a known over prediction of kinetic energy and under prediction of wall shear stress by these models if a proper grid resolution is not used. It has been observed in some past studies that DES and DDES are not suitable wall modeling techniques particularly in case of low Reynolds number flows.

SST k-ω model, with its known advantages in separated flows and low Reynolds number applications, performed better than any of the models in the study (1.09%

deviation from experimental results). The model accurately predicts the turbulent kinetic energy in near-wall regions where it uses the transport equations for k and ω.

A Zonal LES study with SST model used for the initial length of domain and pure LES used in the region close to and after the step was performed in ANSYS. It performed better than the pure LES method in predicting the recirculation length with only 4.9% deviation from experimental results. This could be attributed to the ability of the SST k-ω model to simulate developing turbulent flow accurately in the inlet section even with the use of a coarse grid. During the switch between SST model and LES, the LES simulation already has an input resembling "actual" turbulence. This can be seen as an emphasis on the significance of accurate inlet conditions for LES simulations.

In conclusion, it can be said that LES could not capture the Kolmogorov eddies in near-wall region due to inadequate grid resolution and application of inaccurate inflow boundary conditions for a fully developed turbulence to be attained upstream of the backward-facing step.

6 Conclusion

With more stringent norms, after-treatment system has become an integral part of automotive industries and ALD playing an important role in after-treatment product life cycle. CFD and FEA together ensure the performance and durability of the product.

Product optimization at early phase found to be very beneficial in this competitive environment and ALD prove itself as a useful tool to optimize after-treatment system. ALD capability has been enhanced to a level, where it can predict performance very close to experiments and hence providing capability to deliver a product right first time.

Various capabilities validation has been introduced in this chapter:

1. Predicting urea distribution at catalyst inlet with advanced decomposition models and validated with test results.
2. Coupling with AVL boost gives us advantage to consider catalytic reactions and hence gives improvement in predicting NO_x conversion efficiency and predicting NO_x distribution at catalyst outlet to optimize NO_x sensor location.
3. Shaker test simulation through PSD analysis identifies failure location and fatigue damage estimation.
4. Also, high fidelity model like LES has been shown which are advantageous for separated flows.

Acknowledgements Special thanks to all folks mentioned below for continuous improvement of methods and processes in area of ALD for After-treatment Industry. Excerpt of their contribution has been selected for building this chapter.
Achuth Munnannur—Tech Advisor, CES US

Apoorv Kalyankar—Sr. Engineer, CES US
Kishor Deshmukh—Technical specialist, CES CTCI
Suraj Khalate—Technical specialist, CES CTCI
PrachiChirputkar—MS Intern (Friedrich-Alexander University, Erlangen-Nuremberg).

References

1. Muzio JE (2001) Twenty-five years of SCR evolution: implications for US application and operation. EPA, DOE, EPRI and AWMA Combined Power, Chicago
2. DoD guide to Integrated product and process development (version 1) (1995)
3. Munnannur A, Liu Z (2010) Development and validation of a predictive model for DEF injection and urea decomposition in mobile SCR De-NO_x systems. SAE Technical Paper 2010-01-0889
4. FLUENT A (n.d.) ANSYS fluent theory guide
5. McKinley T, Alleyne A, Lee C (2010) Mixture non-uniformity in SCR: modeling and uniformity index requirements for steady-state and transient operation. SAE Int J Fuels Lubr 3(1):486–499
6. Kalyankar A, Munnannur A, Liu Z (2015) Predictive modeling of impact of ANR non-uniformity on transient SCR system DeNO_x performance. SAE Technical Paper 2015-01-1055

Part IV
Combustion Systems

Design Philosophy for a Laboratory Scale Gas Turbine Combustor

Dinesh Kumar Roshan, Ramana Sreenivas Burela and Abhijit Kushari

Keywords Gas turbine combustor · Design · Aviation turbine fuel · Pressure loss

Nomenclature

I.S.A	International Standard Atmosphere
$\dot{m}_3$	Inlet air mass flow rate (unit: kg/s)
$\dot{m}_{SW}$	Mass flow rate through swirler (unit: kg/s)
$\dot{m}_{ZP}$	Mass flow rate in primary zone (unit: kg/s)
$\dot{m}_f$	Mass flow rate of fuel (unit: kg/s)
$\Delta P_{3-4}/P_3$	Combustor pressure loss
$\Delta P_{3-4}/q_{ref}$	Combustor pressure loss factor
$\frac{\Delta P_{SW}}{q_{ref}}$	Pressure loss across swirler
$\frac{\Delta P_h}{P_3}$	Pressure loss across holes
A_{ref}	Reference area (unit: m^2)
A_{ft}	Cross sectional combustor area (unit: m^2)
A_{sw}	Swirler area (unit: m^2)
A_h	Area of the hole (unit: m^2)
A_{an}	Area of the annulus (unit: m^2)
V_c	Volume of the combustor (unit: m^3)
C_f	Flow coefficient
C_d	Coefficient of discharge of the flow meter
C_v	Coefficient of velocity due to vena-contracta

D. K. Roshan · R. S. Burela · A. Kushari (✉)
Department of Aerospace Engineering, Indian Institute of Technology Kanpur, Kanpur 208016, India
e-mail: akushari@iitk.ac.in

A. K. Gupta et al. (eds.), *Innovations in Sustainable Energy and Cleaner Environment*, Green Energy and Technology,
https://doi.org/10.1007/978-981-13-9012-8_14

C_{dh}	Hole discharge coefficient
D_{ref}	Reference diameter (unit: m)
D_{ft}	Combustor diameter (unit: m)
d_h	Hole diameter (unit: m)
K_{SW}	Swirler concordance factor
L_{Dz}	Secondary zone length (unit: m)
P_3	Inlet pressure (unit: P)
R_a	Specific gas constant (unit: J/Kg^{-1} K^{-1})
R	Outer radius of the swirler (unit: m)
r	Inner radius of the swirler (unit: m)
T_3	Inlet temperature (unit: K)
T_4	Exit temperature (unit: K)
T_{max}	Max exit temperature (unit: K)
K	Factor of pressure loss
K_1, K_2, K_3	Empirical constants
q_{ref}	Reference dynamic pressure (unit: kg/(m s^2)
TQ	Temperature quality factor
M_c	Air velocity loading parameter
LHV	Lower heating value (KJ/Kg)
N_h	No. of holes
$(F/A)_{act}$	Fuel to air ratio actual
$(F/A)_{St}$	Fuel to air ratio stoichiometric

Greek letters

β_{SW}	Turning angle of the airflow (unit: °)
Ω_c	Air loading parameter
Ω	Non-dimensional efficiency parameter
ϕ_{PZ}	Primary zone equivalence ratio
ϕ_{global}	Global equivalence ratio
α	Orifice area ratio
μ	Bleed to orifice area ratio

1 Introduction

Substantial progress have been made in recent years in gas turbine combustor design process, but still much of the design methodology relies upon empirically derived design rules. As combustion chamber is one of the most important part of the aircraft engine, its development should be along with rest of engine components such as intakes, compressor, turbine and nozzle.

An aircraft engine operate over a wide range of inlet pressure and temperature within its flight envelope. A typical airliner flies at a cruise altitude of 11,000 m, where the ambient pressure and temperature are 22.7 kPa and 216.8 K, compared to the sea level values of 100 kPa and 288.15 K [1]. Therefore, at the cruising condition, the combustor has to operate at much lower air density and mass flow rate. But it needs to maintain approximately the same fuel/air ratio as the sea level to maintain the required value of turbine inlet temperature [2]. Therefore, the designer has to account for this large variation in choosing a combustor configuration. An annular combustion chamber, used in air borne gas turbine engines, was designed by Mark and Selwyn [3]. They considered a constant pressure enthalpy addition process to design this combustor and used Siemens NX 8.0 modeling software for optimizing the design. A valuation of the design was made in [4] and the design methodology for the combustion chamber operated with producer gas was presented.

Typically, in gas turbine combustors, the overall air/fuel ratio is of the order of 100:1, while the stoichiometric ratio for the combustion of Aviation Turbine Fuel (ATF) and air is approximately 15:1. Such a lean mixture is beyond the flammability limit of this fuel and, hence, in order to have sustained combustion, the air is added in stages into the combustor. About 20% of the overall air is added in the primary zone to burn the fuel within its flammability limit. About 40% air is added in the secondary zone to burn the unburned products of the primary zone while avoiding local cooling of the flame and thereby reducing the reaction rate in that neighborhood. Finally, in the tertiary or dilution zone, the remaining 40% air is added that mixes with the products of combustion to cool them down to the required turbine entry temperature dictated by the turbine material [5].

The actual operation of gas turbine combustors is well understood. However, a methodology for the systematic design of such combustors in a consolidated form is not available in open literature. Therefore, in this chapter we present a methodical procedure of how an airborne gas turbine combustor should be designed in order to be tested experimentally in a laboratory at atmospheric conditions. In-order to achieve this design, various empirical relations are used from several literatures to reduce the design time. In the proposed design and the combustor developed based on this design, air is provided in four stages, viz., primary zone, secondary zone, atomizing zone and quenching zone. The requirement of air staging for the design are estimated as per the standard practice [5] and the air supply holes are distributed accordingly at various locations in the combustor. It should be noted that this design methodology is for an atmospheric test rig, but in order to operate it at higher pressures modifications are required.

2 Design Methodology

The design of a gas turbine combustor should ensure varied operational requirements along with complete combustion of the fuel inside a short combustor, as well as substantially low levels of emissions per unit flow rate of fuel with respect to other engines. This chapter presents the design of a 170 kW swirl stabilized can combustor, which operates at atmospheric pressure. For most part of the design process, design criteria is followed from the book by Melconian and Modak [6] which is described in Fig. 1. It should be noted that the steps in dashed lines, although essential for the design of a practical combustor, are not required for preliminary design of a laboratory scale combustor presented in this chapter.

2.1 Combustor Type and Fuel Selection

The choice of a particular combustor type is dependent on the overall engine design and the optimum use of the available space to achieve desired compactness. There are two basic types of combustor, tubular and annular. "Tubo-annular" or "can-annular" combustor are also in operation [7]. In airborne gas turbine engines, annular combustor is normally used [8]. However, scaled-down test of an annular combustor is complex because of the difficulty in manufacturing and requirement of very high

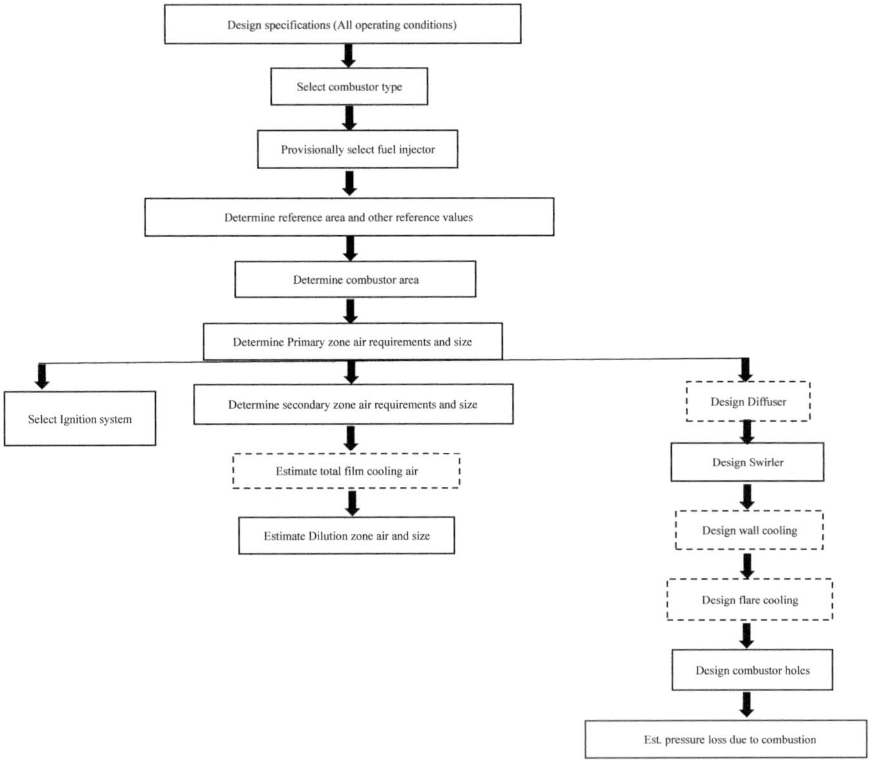

Fig. 1 Preliminary design procedure [6]

air flow rates. However, a section of annular combustor could be used for testing, but it would not resemble the complete annular combustor. Therefore, a can combustor rig is the natural choice for scaled-down experiments.

Two parameters are used to define the can combustor rig air flow which simulate actual conditions of a full engine. The 'air loading parameter' represents the rate of reaction and the residence time of the combustible mixture at low power and the 'air velocity parameter' represents the characteristics of the flow at high power conditions [9].

$$\text{Air loading parameter, } \Omega_c = \frac{K_1 \dot{m}_3}{P_3^{1.8} e^{\frac{T_3}{K_2}} V_c} \tag{1}$$

$$\text{Air velocity parameter, } M_c = \frac{K_3 \dot{m}_3 \sqrt{T_3}}{P_3 A_{\text{ft}}} \tag{2}$$

For the design discussed in this chapter, the value of air loading parameter is taken to be 0.68, which ranges between 0.6 and 0.8 for the idle condition i.e., the least power condition [10]. In order to estimate the empirical constants, the inlet conditions for the air mass flow rate, upstream temperature and pressure, the fuel-air ratio and the volume of the combustor are taken from [11] and corresponding design parameters for the scaled combustor for laboratory tests are estimated. The values of empirical constants K_1 and K_2, used for the present design, are shown in Table 1. It should be noted that the present combustor is designed only for low power condition, therefore the air velocity parameter is not relevant.

Based on the specified design conditions, the mass flow rate for this design is estimated to be 0.133 kg/s. In this design, the inlet pressure and temperature are not the same as the actual engine but corresponds to the available conditions in the laboratory. However, the overall fuel-air ratios are kept constant at 0.015 and 0.03 respectively for the test cases for the design validation. The designed combustor is to be tested with liquid ATF (whose lower heating value (LHV) is 43.8 MJ/Kg). Therefore, a swirling jet air-blast atomizer is used to atomize the liquid fuel [12].

2.2 *Combustor Area and Volume*

After the type of combustor and the kind of fuel is identified, the next step would be determining the length and the diameter of the combustors. In a can combustor the length (L)/diameter (D) should be in the range of $1 < L/D < 4$ [13].

Table 1 Values of the empirical constants used in the design

Empirical constants	K_1	K_2
Value	2122×10^4	3500

$$\text{Combustion intensity} = \frac{\dot{m}_{\text{f}} * LHV}{V_{\text{c}} * \text{Pressure}} \, \text{kW/m}^3\text{atm} \quad (3)$$

Two different fuel flow rates ($\dot{m}_f$) i.e., 0.002 and 0.004 kg/s, were used as the initial design parameter for the estimation of combustor volume. The combustion intensity, given in Eq. 3, for air borne gas turbine combustors are in the range of 20–50 MW/m^3 kPa [7]. The volume of the combustion chamber is a function of area and length. The best possible values for combustor length, *L*, and combustor diameter, *D*, were ascertained to be 0.3 and 0.12 m iteratively with corresponding combustion intensities of 25.4 and 50.9 MW/m^3 atm for the two fuel flow rates respectively. Thus, the desired value of L/D of this combustor is 2.5, which is in the range of conventional gas turbine combustors. Corresponding to this geometry, the volume and the area of the combustor are calculated as $3.39 \times 10^{-3} \text{m}^3$ and 0.14 m respectively.

The reference area (A_{ref}) of a combustor is defined as the maximum cross-sectional area of the casing without a liner and the corresponding diameter in the reference diameter of a circular combustor as shown in Eq. 4. The desired reference area is to be chosen considering either chemical or pressure loss limitations.

$$A_{\text{ref}} = \pi \frac{D_{\text{ref}}^2}{4} \quad (4)$$

2.3 Aerodynamic Considerations

The representative non-dimensional values of pressure loss terms for can combustors are shown in Table 2 [6]. ΔP_{3-4} is the combustor over all pressure loss, P_3 is the total pressure at the combustor inlet, q_{ref} is the dynamic pressure at the maximum cross-sectional area, $\dot{m}_3$ is the total air mass flow rate and T_3 is total temperature at the combustor inlet. Equation 5 [6] can be used to estimate the reference area for the non-dimensional parameters listed in Table 2.

$$A_{\text{ref}} = \left[\frac{R_a}{2} \left(\frac{\dot{m}_3 \sqrt{T_3}}{P_3} \right)^2 \left(\frac{\Delta P_{3-4}}{q_{\text{ref}}} \Big/ \frac{\Delta P_{3-4}}{P_3} \right) \right]^{0.5} \quad (5)$$

In Eq. (5) the value of R_a is 143.5 J kg^{-1} K^{-1}. By assuming the values of P_3 and T_3 to be 105,325 pa and 320 K respectively, which are the actual values available in the laboratory for an atmospheric test, the value of the reference area is estimated

Table 2 Pressure loss terms for can combustor [6]

Type of combustor	$\Delta P_{3-4}/P_3$	$\Delta P_{3-4}/q_{\text{ref}}$	$\dot{m}_3\sqrt{T_3}/P_3$
Can-type	5.3	40	3×10^{-3}

to be 5.41×10^{-3} m^2 using Eq. 5. However, the desired diameter of the combustor is 0.12 m for the desired combustion intensity as discussed earlier and hence the actual cross-sectional area for a cylindrical combustor is 0.011 m^2. The reference area corresponding to the actual area, estimated using Eq. 6 [6] is 0.016 m^2. The difference between the desired value (estimated using Eq. 5) and the actual value from Eq. 6, is quite large due to the absence of casing in the present combustor.

$$A_{\mathrm{ft}} = 0.7 A_{\mathrm{ref}} \tag{6}$$

2.4 Chemical Considerations

For any operating condition within the envelope of flammability and stability limits, the combustion efficiency is a function of non-dimensional efficiency parameter Ω, defined by Lefebvre [14], as shown in Eq. 7. Lefebvre and Ballal [7] have presented the empirical relation between combustion efficiency and the parameter Ω. Typically when Ω is 73×10^{-6}, the combustion efficiency close to 100% [7].

$$\Omega = \frac{P_3^{1.75} A_{\mathrm{ref}} D_{\mathrm{ref}}^{0.75} \exp(T_3/b)}{\dot{m}_3} \tag{7}$$

The parameter b in Eq. 7 is the temperature corrector factor, which can be estimated using the empirical Eqs. 8 or 9 [15], depending on the equivalence ratio in primary zone represented in Eq. 10.

$$b = 245(1.39 + \ln \phi_{\mathrm{PZ}})\ 0.6 < \phi_{\mathrm{PZ}} < 1.0 \tag{8}$$

$$b = 170(2.00 + \ln \phi_{\mathrm{PZ}})\ 1.0 < \phi_{\mathrm{PZ}} < 1.4 \tag{9}$$

$$\phi_{\mathrm{PZ}} = \frac{\phi_{\mathrm{global}}}{\dot{m}_{\mathrm{ZP}}/\dot{m}_3} \tag{10}$$

Sometimes a modified expression (Eq. 11) is used for the estimation of combustion efficiency parameter that includes the aerodynamic consideration. However, in the present study, since we are not using the casing, Eq. 7 was used.

$$\Omega = \frac{P_3^{1.75} A_{\mathrm{ref}} D_{\mathrm{ref}}^{0.75} \exp(T_3/b)}{\dot{m}_3} \left(\frac{\Delta \mathrm{P}_{\mathrm{ft}}}{q_{\mathrm{ref}}}\right)^{0.4} \tag{11}$$

$$\phi_{\mathrm{global}} = \frac{(F/A)_{\mathrm{act}}}{(F/A)_{\mathrm{St}}} \tag{12}$$

In Eq. 10, $\dot{m}_{\mathrm{ZP}}$ is the air flowrate in primary zone of the combustor and ϕ_{global} is the overall equivalence ratio estimated using Eq. 12. For 0.002 kg/s of fuel, with air as oxidizer, the design fuel to air ratio $(F/A)_{act}$ is 0.018, whereas stoichiometric fuel air ratio $(F/A)_{st}$ is 0.066 for ATF. The ratio of these two parameters would give ϕ_{global}, which is 0.27. It should be noted that for 0.004 g/s of fuel flow, ϕ_{global} is 0.56.

Considering around 40% of the total air to pass through the primary zone, which is comprising of primary air, quenching air and atomizing air, the mass flow rate in the primary zone $\dot{m}_{ZP}$ is 0.053 kg/s. Corresponding to this mass flow rate, the equivalence ratio in the primary zone is $\phi_{PZ} = 0.675$. Hence for the parameter 'b' Eq. 8 should be considered. By substituting all the parameters in Eq. 7, we get $\Omega = 61 \times 10^6$, with an estimated combustion efficiency around 85–90% for the combustor reported here.

2.5 Length of Secondary Zone

In order to estimate the length of the secondary zone (or) dilution zone, the temperature traverse quality (TQ) in the combustor exit should be calculated. Temperature traverse quality (TQ) is defined as the difference between the peak exit temperature and the mean exit temperature divided by the mean temperature rise across the combustor, as shown in Eq. 13. The desired range of TQ is 0.05 and 0.3 [6].

$$\text{Temperature Traverse Quality (TQ)} = \frac{T_{\max} - T_4}{T_4 - T_3} \tag{13}$$

Table 3 shows the dilution zone length to combustor diameter ratio (L_{DZ}/D_{ft}) as function of TQ for different values of pressure loss factor [6]. The length of the secondary zone (or) the dilution zone is estimated by taking various values of TQ ranging from 0.05 to 0.03 keeping in mind the overall combustor length estimated earlier. For the present design, considering TQ = 0.175 for $\Delta P_{3-4}/q_{ref} = 50$ and $D_{ft} = 0.12$ m, the estimated length of the dilution zone is 0.165 m. However, the overall length of the combustor is 0.3 m. Hence, at 0.135 m from the dump plane, secondary holes should be present.

Table 3 Dilution zone length to combustor diameter ratio as a function of TQ for different values of pressure loss factor [6]

$\Delta P_{3-4}/q_{ref}$	L_{Dz}/D_{ft}
30	$2.96 - 9.86\ TQ + 13.3\ TQ^2$
50	$2.718 - 12.64\ TQ + 28.51\ TQ^2$

2.6 Swirler Design

The air-fuel mixture should have enough residence time in the primary zone to achieve desired flame stability. The axial velocity of the flow can be reduced effectively by generating a recirculating flow in the primary zone. Such recirculating flows can be effectively induced in the primary zone by using a swirler around the fuel injector [7]. An axial swirler is simpler than radial swirler. For desired flame stability, the turning angle of the air flow, β_{SW}, should lie between 45° and 70° [6]. The effective flow area for an axial swirler can be estimated by the relationship proposed by Knight and Walker [16] as shown in Eqs. 14 and 15.

$$A_{sw} = \pi\left(R^2 - r_o^2\right) \quad (14)$$

$$\frac{\Delta P_{SW}}{q_{ref}} = K_{SW}\left[\left(\frac{A_{ref}}{A_{SW}}\right)^2 \sec^2\beta_{SW} - \left(\frac{A_{ref}}{A_{ft}}\right)^2\right]\left(\frac{\dot{m}_{SW}}{\dot{m}_3}\right)^2 \quad (15)$$

The concordance factor K_{SW} is 1.30 for straight radial swirler blades and 1.15 for thin curved blades [16]. For the present design, an axial swirler with straight radial blades aligned at 45° angle to the axial direction was fitted in the inlet pipe. The outer radius (R) of the swirler is 0.025 m and the inner radius (r_o) is 0.0141 m as shown in Fig. 2. So, the area of the swirler is 6.25E−4 m^2. By considering a 45° swirler and 25% of the total air to pass through the swirler, the pressure loss across the swirler ($\Delta P_{SW}/q_{ref}$) is estimated (from Eq. 15) to be 102.6.

$$\Delta P_{SW} = \Delta P_{snout} - \Delta P_{diffuser} \quad (16)$$

It should be noted that for practical combustors, the total pressure loss across the swirler is difference between the total pressure loss inside the snout and the total pressure loss in the diffuser as shown in Eq. 16.

Fig. 2 3-D view of a 45° swirler

2.7 Air Flow Holes

The air should enter the combustor in multiple stages through well designed holes in order to have the desired residence time, proper fuel-air mixing and good pattern factor without local chilling of the flow. The design of these holes is not trivial as the flow through these holes depend on their sizes, the pressure drop across them, the duct geometry and the flow conditions in the vicinity of the hole. It is also necessary to verify proper distribution of air through these holes. This is an iterative process that follows the method described below:

(a) Choose of the bleed ratio (B) that represents the ratio between the hole mass flow rate ($\dot{m}_h$) and the annulus mass flow rate ($\dot{m}_{\text{an}}$).
(b) Choose the discharge coefficient ($C_{\text{d,h}}$) for the hole (which is chosen to be 0.5 at the beginning of the iterations)

$$\frac{\Delta P_h}{P_3} = \frac{143.5\, \dot{m}_h^2 T_3}{P_3^2 C_{\text{dh}}^2 A_h^2} \tag{17}$$

(c) From Eq. 17 estimate the hole total area (A_{h}) for each zone. $\Delta P_{\text{h}}/P_3$ is approximately taken as 6%.
(d) Determine the orifice area ratio ($\alpha = A_{\text{h}}/A_{\text{an}}$) and the bleed/orifice area ratio ($\mu = B/\alpha$).

$$K = 1 + \delta^2 \left\{ 2\mu^2 + \left[4\mu^4 + \left(\mu^2/\delta^2 \right)\left(4B - B^2 \right) \right]^{0.5} \right\} \tag{18}$$

(e) The factor of pressure loss, K, is calculated using Eq. 18. The factor of momentum loss, δ, varies between 0.75 and 0.9 [17].

$$C_{\text{dh}} = \frac{(K - 1)}{\delta \left[4K^2 - K(2 - B)^2 \right]^{0.5}} \tag{19}$$

(f) The value of discharge coefficient (C_{dh}) is obtained using Eq. 19 [17].

The calculations are repeated till the value of C_{dh} converges. The value of total area of holes (A_{h}) is then estimated using Eq. 17 and the corresponding hole diameter is estimated using Eq. 20.

$$d_{\text{h}} = 2\sqrt{\frac{A_{\text{h}}}{\pi N_{\text{h}}}} \tag{20}$$

The number of holes (N_{h}) in each zone is specified in this iterative process and the corresponding hole diameter is estimated. However, it should be ensured that the sum of the hole's diameters is less than the combustion chamber diameter. If this condition is violated then it is necessary to redefine the number of holes (or) to reconsider some other parameter such as $C_{dh.}$

In the present combustor design, there are two specific locations where the hole design is necessary, i.e., at the quenching and secondary air admission holes. For the quenching air holes, the factor of pressure loss (K) is 12,231 with the coefficient of discharge to be 0.62. Thus, there are 25 holes with 0.002 m diameter. These quenching holes were located at the base plate of the combustor. For the secondary air admission holes, the pressure loss factor (K) is 2689 and the discharge coefficient C_{dh} is 0.62, thus having 12 holes of 0.0127 mm diameter, placed at 0.135 m from the dump plane.

2.8 Air Staging Design for Combustor

In order to have high combustion efficiency, and to avoid local chilling of the flame, combustion air should be carefully introduced into the combustor. The entry of combustion air should be in such a way that it should not drastically reduce the reaction rate surrounding the air admission holes. The effective areas for all the admission holes are calculated through various iterative and empirical relations. These areas are tabulated in Table 4. For the atomization of liquid fuel, a swirling jet air blast atomizer is used. In this atomizer a small amount of atomization air is required. In the present design, corresponding to fuel flow rate of 0.002 kg/s, the atomizing air mass flow rate was also chosen to be 0.002 kg/s for maintaining the atomizing air to liquid fuel ratio to be 1.0.

By assuming the velocities at all the sections are same and using the continuity equation, dividing individual area by total effective area gives the individual staging requirements. This lead to the air staging requirements for the present combustor which 18.3% for Primary air, 44.4% for Secondary air, 2.29% for Quenching air, and 0.198% for Atomizing air.

Table 4 Effective area of air admission holes

Address of air admission holes	Effective area (m^2)
Primary air (or) Swirling air	6.25×10^{-4}
Quenching air	7.85×10^{-5}
Secondary air	1.520×10^{-3}
Atomizing air	6.785×10^{-6}
Total effective area	3.423×10^{-3}

3 Experimental Setup for Design Validation

The test setup used to validate the design is schematically represented in Fig. 3. As per the design, the combustion chamber diameter is 0.12 m and the length of the combustor is 0.3 m. The combustor is fabricated using Stainless steel (Type SS 306). Dry compressed air from a laboratory compressor is supplied to the realized combustor through different ports for primary, secondary, quenching and atomizing air. For every inlet port of air supply to the combustor, there is a settling or plenum chamber, where the air enters, settles down and then gets uniformly distributed across all the ports after being metered and controlled using Mass flow meter/controllers. For example, quenching settling chamber is located just before the base plate. The corresponding air flowrate is controlled using an Alicat® mass flow controller (MFC) having a capacity of 1000 SLPM. The base plate of the combustor is a circular disc made of stainless steel. This disc is 10 mm thick, having 25 holes of 2 mm diameter located at 40 mm from the center. Similarly, primary and atomizing air are also controlled using separate mass flow controllers (MFC). The secondary air is metered using an orifice meter. A differential pressure transducer is placed across its two ports to measure the pressure difference and a stand-alone pressure sensor is used for density correction. The schematic representation of the test setup is shown in Fig. 3 and a photograph of the combustor is given in Fig. 4.

The mass flow rate of combustion air at secondary zone is calculated using Eq. (21).

$$Q = c_f A \sqrt{2\rho(P_1 - P_2)} \tag{21}$$

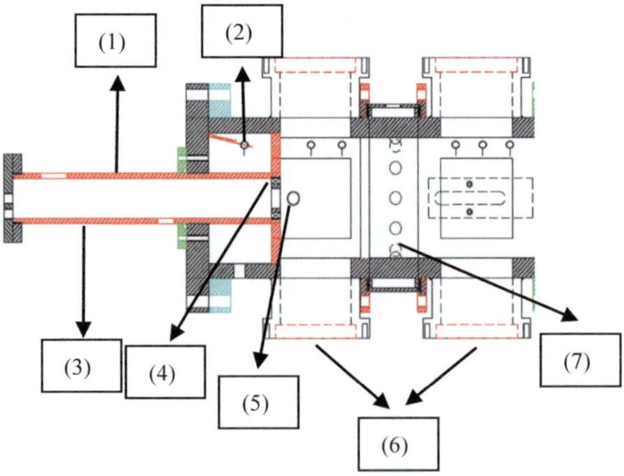

Fig. 3 Schematic of the combustor: (1) Location of pressure measurement on inlet pipe, (2) Location of pressure measurement at quenching air settling zone, (3) Inlet pipe, (4) Swirler, (5) Location of spark plug, (6) Optical windows, (7) Secondary air holes

Fig. 4 Photograph of the can-combustor test rig

Where c_f is flow coefficient, which is the product of C_d (coefficient of discharge of the flow meter) and C_v (the coefficient of velocity due to the formation of vena-contracta). ρ is the density of air corresponding to the upstream pressure at 150 mm ahead of the orifice plate. This secondary air enters radially into the combustion chamber through 12 holes of 1.27 cm diameter at a distance of 13.75 cm from the base plate.

Two additional pressure sensors, one installed at the center of the quenching air settling chamber and another installed at the middle of the inlet pipe were used to estimate the pressure loss across the combustor. The pressure reading are acquired through NI 9205®, 32-Channel, 16 Bit Analog Input Module that is capable of measuring at 250 k Samples/s. All the pressure measurements are taken simultaneously.

The primary and secondary air flowrates are kept constant at 23.7 and 63 g/s respectively. The atomizing air is maintained at 3 g/s and the fuel flowrate is kept constant at 2 g/s so that the atomizing air to liquid ratio (ALR) remains at 1.5. However, the quenching air flow rate is varied between 0.98 and 9.88 g/s to quantify it's impact on the pressure drop so as to validate the design. It should be noted that only a meager change in the global fuel to air ratio (FAR) i.e., from 0.02 to 0.022 occurred for this variation in quenching air flowrate. The validation experiment was started with the maximum amount of quenching air mass flow rate (9.88 g/s) and then reducing this flow rate in steps of 0.49 g/s till the minimum flow rate (0.98 g/s) was attained. This set of experiments is termed as forward path. Similarly, from the minimum amount of quenching air, the flow rate was gradually increased up to the maximum value and that set is termed as the return path.

4 Results and Discussions

The main idea of this experimental work is to validate the design, which means that the pressure loss across the combustor should be about 5%. However, in this study the main focus is presented on how the pressure loss is affected by an increase and decrease of quenching air mass flow rate. To estimate the pressure loss, the pressure upstream of the combustion chamber i.e., in the settling chamber pressure is subtracted from the atmospheric pressure (101,325 Pa) and divided by the settling chamber pressure.

In airborne gas turbine combustors, the pressure loss across the combustor should be less than 5%, because increase in pressure loss directly affects the thrust of the engine. Pressure loss between the primary inlet pipe (upstream of the swirler representing the primary air) and the test rig, shown in Fig. 5, conveys that the pressure loss between the swirler and the combustion chamber is 0.06. There is a drift between the forward path (decreasing quenching air) and the return path (increasing quenching air) due to the heating of the combustor with time. But, in general, the this pressure loss is almost independent of the variation of the quenching air suggesting the primary and quenching zones are decoupled, which was one of the aims of this design so that individual air staging effects can be elucidated.

In practical combustors, quenching air is introduced through the base plate so that high temperature hot gases in the primary zone do not affect the base plate. Figure 6 shows the variation of the pressure loss between the quenching zone and combustor, which is measured to be less than 3.5% and gradually increasing with the increase in the quenching air as expected. The pressure loss shown in Fig. 6 is almost at par with an efficient gas turbine combustor. There is also a drift in the return path to the forward path mainly due to the preheating of existing air by the hot combustor neighborhood.

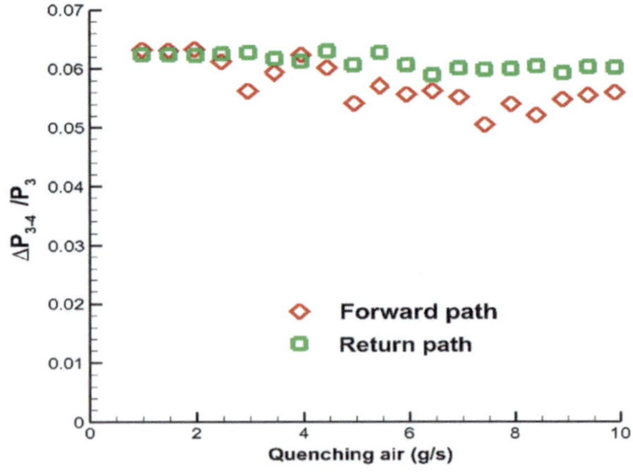

Fig. 5 Estimation of pressure loss between primary inlet pipe and combustor

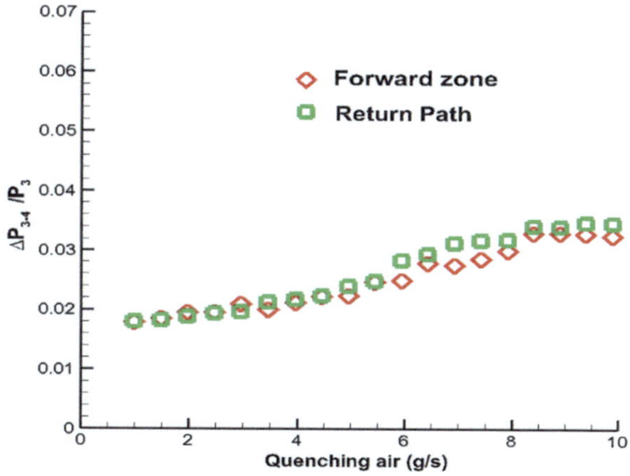

Fig. 6 Estimation of pressure loss between quenching zone and combustor

5 Conclusions

A can combustor has been designed and the methodology followed is explained by using similarity parameters between an annular combustor to a can combustor. Based on the fuel flow rate and the combustion intensity, the diameter and the length of the combustor are calculated. By aerodynamic and reaction rate considerations area of the combustor was obtained using empirical relationships. Using similar empirical relations, the length of the secondary zone (or) dilution zone was estimated, along with the swirler geometry and its area was calculated. To determine the number of holes in the quenching zone and the secondary zone, an iterative process is used assuming the values of coefficient of discharge through holes (C_{dh}) and pressure loss. Air staging is done for different sections and using the continuity equation. The measurement of pressure loss across the combustor was conducted to validate the design and the values obtained were closely matching with the estimated values.

References

1. Cavcar M (2000) The international standard atmosphere (ISA). Anadolu University Turkey
2. Talay TA (1975) Introduction to the aerodynamics of flight. National Aeronautics and Space Administration, vol NASA SP-36, pp 6–9, Washington, D.C.
3. Mark CP, Selwyn A (2016) Design and analysis of annular combustion chamber of a low bypass turbofan engine in a jet trainer aircraft. Propuls Power Res 5(2):97–107
4. Srinivasa SG, Murali Krishna MVS, Reddy DN (2013) Design and analysis of gas turbine combustion chamber for producer gas as working fuel. Int J Comput Eng Res 3(1):444–447
5. Saravanamuttoo H, Rogers G, Cohen H (1996) Gas turbine theory, 4th edn. LONGMAN group, p 455
6. Melconian JW, Modak (1985) Combustor's design, Sawyer's gas turbine engineering handbook: theory and design, vol 1. Turbomachinery International Publications, Connecticut
7. Lefebvre AH, Ballal DR (2010) Gas turbine combustion: alternative fuels and emissions

8. Kiameh P (2002) Gas turbine combustors, Power generation handbook, Chapter 12 McGraw Hill Professional, Technology & Engineering
9. Melconian JO (1980) The design and development of gas turbine combustors, Northern Research and Engineering Corporation, Massachusetts, Woburn, USA
10. Sampath P, Shum F (1985) Combustion performance of hydrogen in a small gas turbine combustor. Int J Hydrogen Energy 10(12):829–837
11. Ramraj SH, Sekher C, Srihari Dinesh Kumar J, Raut A, Kushari A (2017) Sensitivity analysis of inlet conditions on emissions for different biofuel blend. In: Proceeding of 1st national aerospace propulsion conference
12. Srihari Dinesh Kumar J, Mariappan S, Kushari A (2016) Spray dynamics in a swirling jet air-blast atomizer. In: ILASS—Asia, Department of Aerospace Engineering Indian Institute of Technology Kanpur
13. Beran M, Koranek M, Axelsson AL-UE (2014) US. Patent, vol 2, no 12
14. Lefebvre AH (1966) Theoretical aspects of gas turbine combustion performance, vol CoA NOTE N, no 163
15. Herbert MV (1961) A theoretical analysis of reaction rate controlled systems : part II. In: Eighth symposium on combustion, vol 8, no 1, pp 970–982
16. Knight HA, Walker RB (1957) The component pressure losses in combustion chambers
17. Kaddah KS (1964) Discharge coefficients and jet deflection angles for combustor liner air entry holes. Ph.D. Thesis, p 63, June

Auto-Ignition of Hydrogen-Rich Syngas-Related Fuels in a Turbulent Shear Layer

Panagiotis Simatos, Fabian Hampp and Rune Peter Lindstedt

Keywords Auto-ignition · Syngas · Premixed turbulent flames · Turbulent shear layer · Experimental and computational study

1 Introduction

The development of low carbon footprint and clean energy technologies is essential to resolve the concerns of climate change and diminishing fossil fuel resources [1]. State-of-the-art gas turbines for power generation are typically optimised for natural gas-based operation [2] and offer single-digit nitrogen oxide emissions [3] by reducing peak temperatures [4]. Related burning modes exhibit chemical timescales of the same order as the flow timescales [5, 6]. Hydrogen-enriched fuel blends provide a route to decarbonise existing technologies. Such fuel blends have the potential to expand the lean operation limit [7–9], improve engine efficiency and reduce CO and NO_x emissions [10, 11]. Wang et al. [12] have shown that hydrogen addition enhances the initial combustion process under fuel-lean conditions. The substitution of kerosene with hydrogen in gas turbines has shown similar improvements [13]. Hydrogen-rich fuel blends are frequently related to syngas or biomass-derived gas. However, the variability of available mixtures can lead to fuel flexibility concerns for engine manufacturers [14]. Moreover, the elevated hydrogen concentration results in safety concerns due to the higher risk of explosions or transition to detonation [15–18]. For example, the syngas mixture reactivity is strongly dependent upon hydrogen content, often its primary component. This has direct implications for flame propa-

P. Simatos · F. Hampp · R. P. Lindstedt (✉)
Department of Mechanical Engineering, Imperial College, London SW7 2AZ, UK
e-mail: p.lindstedt@imperial.ac.uk

P. Simatos
e-mail: p.simatos15@imperial.ac.uk

F. Hampp
e-mail: f.hampp11@imperial.ac.uk

A. K. Gupta et al. (eds.), *Innovations in Sustainable Energy and Cleaner Environment*, Green Energy and Technology,
https://doi.org/10.1007/978-981-13-9012-8_15

gation speeds, explosion overpressures, auto-ignition and turbulence–chemistry interactions [19]. The latter can lead to combustion instabilities via differences in flame stabilisation [1, 14]. Lieuwen et al. [20] showed significantly reduced auto-ignition delay times with increasing H_2 content for various H_2/CH_4 and H_2/CO mixtures at $\Phi = 0.4$ and $P = 15$ atm. Li et al. [21] measured flame speeds and explosion overpressures generated in an obstructed flame tube for a wide range of binary and ternary hydrogen-enriched CH_4 and CO blends. Methane exhibited a distinctly stronger inhibiting impact on the reactivity compared to CO. A scaling based on the amount of air required to fully oxidise the fuel mixture was found to correlate the impact of fuel composition on the explosion overpressures [21] and turbulent flow field [22] very well. The impact of hydrogen content on laminar burning properties in binary fuel mixtures has been studied extensively, e.g. [19, 23, 24]. Generally, CH_4 was found to exhibit a significantly stronger inhibiting effect on the hydrogen reactivity compared to CO [20, 21]. However, despite the recent progress in hydrogen safety and fundamental combustion behaviour, a lack of experimental data was identified by Messaoudani et al. [25].

The auto-ignition in turbulent flows is of significant importance for a variety of practical applications such as flame stabilisation, high altitude re-light, diesel and homogeneous charge compression ignition engines, and lean premixed combustors [26, 27]. The auto-ignition process is strongly dependent upon the chain branching reactions and turbulent transport [28]. Mastorakos et al. [26] used direct numerical simulation (DNS) to investigate auto-ignition in turbulent flows and differences in auto-ignition trends were correlated with the scales of turbulence and partial premixing. Cabra et al. [29, 30] introduced a vitiated co-flow burner with well-defined boundary conditions that renders it ideal to advance the fundamental understanding of flame stabilisation and auto-ignition. This has yielded a variety of experimental and computational studies [29–32]. O'Loughlin and Masri [33] developed a new burner configuration to investigate auto-ignition of spray flames experimentally [34] and computationally [35, 36]. The transported probability density function (PDF) approach has been used extensively in computational studies and has shown to be able to predict auto- and re-ignition and extinction phenomena for a wide range of hydrogen and hydrocarbon flames [27, 31, 37–40]. Transported PDF methods solve the highly non-linear chemical source term in closed form and without approximation. The method is thus ideal for the investigation of flows where the chemistry interacts with turbulence and, in particular, where turbulence–chemistry interactions exert a dominant influence.

The current work presents a systematic experimental and numerical study on the impact of fuel reactivity changes by gradual enhancement of methane or carbon monoxide/air mixtures with hydrogen on the auto-ignition in a turbulent shear layer. Flame stabilisation in turbulent shear layers is a key consideration for the operation of gas turbines where the injected fuel interacts with (re-circulating) hot combustion products. The study was conducted using a variant of a vitiated co-flow lifted flame burner [29, 30]. Binary fuel mixtures of H_2/CH_4 have been studied by a significant number of researchers [41–45], though typically not under conditions of relevance to the current study. Such fuel mixtures are of interest for the decarbonisation of installed power generating capacity. Investigations covering fuel blends of H_2/CO, on the other hand, are uncommon and mainly related to research on syngas utilisation and fuel flexibility [20, 46, 47]. The current study covers a total of 34 lean (stoichiometry of 0.80) premixed gas mixtures over a wide range of H_2/CH_4 and H_2/CO fuel blends. The effect of additional inert gas dilution is also investigated. The data provide a consistent and unique database detailing the reactivity of hydrogen-rich syngas-related fuel blends in a turbulent flow.

2 Experimental Configuration

2.1 *Burner Configuration*

The vitiated co-flow burner, schematically depicted in Fig. 1, was based on the design of Cabra et al. [29]. The gas injection system was modified to accommodate multi-component premixed gas mixtures and ensure safe operation under hydrogen-rich conditions as detailed below. The reactants to generate the hot pilot in the vitiated co-flow were first injected into a primary mixing chamber. The premixed gas mixture was subsequently injected into a secondary mixing chamber through a circumferential ring featuring 32 radial nozzles. A sintered plate with a thickness of 10 mm and a maximum pore size of 76 µm acted as flashback arrestor and separated the secondary mixing chamber from the reservoir leading up to a perforated plate. The perforated plate exhibited a diameter of 210 mm, prepared with 2200 holes of diameter 1.58 mm that resulted in a blockage ratio of 87%. The perforated plate was used to anchor lean premixed flames that provided a well-controlled pilot stream surrounding the centre nozzle.

The gas components for the central jet mixture were injected into a separate gas mixer. The mixing chamber was divided by a sintered disk (maximum pore size of 76 µm) that acted as a flashback arrestor. The premixed gas mixtures were subsequently passed directly to the jet nozzle with an inner diameter d = 4.2 mm. The nozzle exit was located 70 mm above the perforated plate to ensure a uniform co-flowing pilot. Dry and filtered air from Howden compressors and other reactants were supplied at a pressure of 4.0 bar(g). The purities of the cylinder gases were: H_2 (99%), CH_4 (99%), CO (99%) and N_2 (99%) [48]. Gases were metered via digital Bronkhorst mass flow controllers featuring a flow uncertainty $<0.8\%$ [49] and operated using a purpose written LabView [50] interface.

2.2 *Jet Conditions and Syngas Mixtures*

The current study investigated 34 binary fuel blends of H_2/CH_4 and H_2/CO at a constant equivalence ratio of $\Phi = 0.8$. The H_2/CH_4 mixtures covered the full range

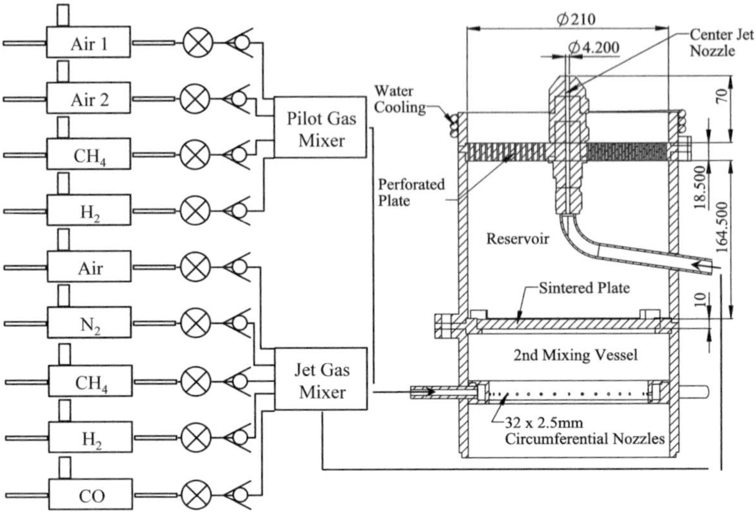

Fig. 1 Schematic of the lifted flame burner test facility

between the pure components and the intermediate steps of 90/10, 80/20, 70/30, 60/40, 50/50, 40/60 and 25/75 in mole per cent of hydrogen and methane, respectively. The corresponding H_2/CO blends covered mixtures of 90/10, 80/20, 70/30, 60/40, 50/50, 40/60 and 30/70 mol per cent. Furthermore, dilution of the fuel blend with nitrogen was also investigated by doubling the amount of N_2 in the mixture. The investigated mixtures are summarised in Table 1. The wide range of gas mixtures resulted in differences in gas properties, e.g. the unburned gas density (ρ_j) and dynamic viscosity (μ_j). The intrusion of a free jet into an essentially stagnant medium is primarily governed by the jet momentum ($M_j = \rho_j \cdot u_j^2$), which accordingly was maintained constant. The value of the jet momentum was defined such that the jet velocity of the mixture with the highest reactivity (i.e. pure H_2) had an exit jet velocity of $u_j \approx 100$ m/s. The density changes in the reactants and the required conservation of momentum resulted in a jet velocity range from $89.2 < u_j$ (m s^{-1}) < 100 for the case from pure CH_4/air to H_2/air mixtures, respectively. The corresponding changes in the reactant properties lead to an approximately constant Reynolds number ($Re_j = \rho_j u_j \cdot d/\mu_j$) in the range $20.5 \times 10^3 \leq Re_j \leq 23.2 \times 10^3$.

2.3 *Co-flowing Hot Combustion Products*

The temperature of the pilot was varied to accommodate the wide range of mixture reactivities. Therefore, the equivalence ratio of the lean premixed pilot flames was adjusted ($0.28 < \Phi_p < 0.54$). To ensure stable flame anchoring on the perforated plate for all pilot conditions, small quantities of CH_4 were added to the hydrogen

Table 1 Overview of all investigated fuel mixtures, where u_j is the jet velocity and X the mole fraction each component k. The first column specifies the fuel blend percentage with the residual being H_2, e.g. 20% CH_4 represents the 80% H_2/20% CH_4 mixture. Inert gas dilution is indicated by + N_2. Density values at 298 K

Mixture	u_j (m/s)	X_{H_2}	X_{CH_4}	X_{CO}	X_{N_2}	X_{Air}	ρ_j (kg/m^3)
0% CH_4	100	0.252	0	0	0	0.748	0.898
10% CH_4	96.8	0.185	0.021	0	0	0.795	0.960
20% CH_4	94.7	0.139	0.035	0	0	0.826	1.00
30% CH_4	93.3	0.105	0.045	0	0	0.850	1.03
40% CH_4	92.2	0.080	0.053	0	0	0.868	1.06
50% CH_4	91.4	0.059	0.059	0	0	0.882	1.08
60% CH_4	90.8	0.043	0.064	0	0	0.893	1.09
75% CH_4	90.1	0.024	0.071	0	0	0.905	1.11
100% CH_4	89.2	0	0.078	0	0	0.923	1.14
0% CO	100	0.252	0	0	0	0.748	0.898
10% CO	98.7	0.226	0	0.025	0	0.748	0.924
20% CO	97.3	0.201	0	0.050	0	0.748	0.951
30% CO	96.0	0.176	0	0.076	0	0.748	0.981
40% CO	94.7	0.151	0	0.101	0	0.748	1.00
50% CO	93.5	0.126	0	0.126	0	0.748	1.03
60% CO	92.3	0.101	0	0.151	0	0.748	1.06
70% CO	91.1	0.076	0	0.176	0	0.748	1.09
0% CH_4 + N_2	95.5	0.158	0	0	0.372	0.470	1.00
10% CH_4 + N_2	93.6	0.114	0.013	0	0.386	0.488	1.03
20% CH_4 + N_2	92.4	0.084	0.021	0	0.395	0.500	1.06
30% CH_4 + N_2	91.5	0.063	0.027	0	0.402	0.508	1.08
40% CH_4 + N_2	90.9	0.047	0.032	0	0.407	0.515	1.08
50% CH_4 + N_2	90.4	0.035	0.035	0	0.411	0.520	1.10
60% CH_4 + N_2	90.1	0.025	0.038	0	0.414	0.524	1.11
100% CH_4 + N_2	89.1	0	0.045	0	0.422	0.534	1.12
0% CO + N_2	95.5	0.158	0	0	0.372	0.470	1.00
10% CO + N_2	94.8	0.142	0	0.016	0.372	0.470	1.00
20% CO + N_2	94.0	0.127	0	0.032	0.372	0.470	1.02
30% CO + N_2	93.2	0.111	0	0.047	0.372	0.470	1.04
40% CO + N_2	92.5	0.095	0	0.063	0.372	0.470	1.06
50% CO + N_2	91.7	0.079	0	0.079	0.372	0.470	1.07
60% CO + N_2	91.0	0.063	0	0.095	0.372	0.470	1.09
70% CO + N_2	90.4	0.047	0	0.111	0.372	0.470	1.12

flames [4]. The realised pilot temperatures were in the range from $1040 < T_p$ (K) < 1490. The lower temperature limit was restricted by blow-off and the upper limit by the material properties of the burner. The pilot temperature was measured 40 mm above the perforated plate with a 50 μm R-type thermocouple throughout the entire experiment to monitor any drift. A cold gas bulk flow velocity of $u_p = 1.0$ m s^{-1} and total volumetric pilot flow rate of 3.41×10^{-2} m^3 s^{-1} was used.

2.4 Safety Precautions

The explosion (e.g. hydrogen) and toxicity (e.g. carbon monoxide) hazards associated with the reactant gases used in the current study demand a reliable safety control system that prohibited the release of unburned reactants. Thus, the temperature of the pilot was also monitored with a second, robust, thermocouple (shielded K-type). A shut-off signal was sent to the mass flow controllers if the pilot temperature fell below 950 K. The later was defined as high as feasible for a fast response time while allowing for minor temperature variations during operation.

2.5 Measurement and Analysis

The current mixture matrix has a strong focus on hydrogen-rich fuel blends—including pure H_2. Thus, the flame lift-off height was measured by means of flame chemiluminescence using an interline-transfer CCD-camera (LaVision Intense Camera) connected to an intensified relay optics (IRO) unit (LaVision, photocathode S20, phosphor P43) to include the chemiluminescence of hydroxyl (OH^*). The measured chemiluminescence signal is comprised of the entire spectral range and is limited by the quantum efficiency of the IRO and the transmission efficiency of the detection system. Species that contribute to the signal are, amongst others, formaldehyde (CH_2O^*), OH^*, methylidyne (CH^*) and carbon dioxide (CO_2^*), e.g. Yu et al. [51]. García-Armingol and Ballester [52, 53] have shown a shift of the chemiluminescence spectra away from a OH^* dominated signal to a broadband spectral range with increasing hydrogen concentration in binary H_2/CH_4 and ternary $H_2/CH_4/CO$ fuel mixtures. Hwang et al. [54] utilised CH_2O^* chemiluminescence to investigate ignition and low temperature heat release and the comprised spectra of CO^*, CH^*, OH^*, amongst others, to identify high temperature heat release in homogeneous charge compression ignition engines. Sheinson and Williams [55] evaluated the contribution of various species to the chemiluminescence signal for blue and cool acetaldehyde flames. Formaldehyde was identified as the main chemiluminescence source, while the contribution of other species such as CO_2^* was minor. The current detection system was calibrated for an image interrogation region of 250×190 mm (axial $\times$ radial) that was resolved by 1376×1040 pixels. The intensifier gain was adjusted to compensate for the weaker signal obtained with increasingly stretched reaction zones. The exposure time was set to 50 μs.

To achieve statistically independent data, 1000 images were recorded for each set of conditions with a repetition rate of 10 Hz. An average background image was compiled from 100 images for each pilot condition and was subtracted from the raw images. All subsequent calculations, i.e. determination of the average, root mean square and instantaneous flame lift-off height, were performed on the basis of the corrected images. Entrainment of ambient air into the hot pilot stream restricted the maximum lift-off height that can be determined reliably to $x/d \simeq 50$ (blow-off limit). All data points exceeding this limit were excluded from further analysis.

3 Transported Probability Density Function Modelling

A two-dimensional parabolic flow solver was used in combination with the joint-scalar transported probability density function (PDF) approach of Lindstedt and co-workers [38, 39, 56, 57]. The turbulence properties were obtained at second moment level using the Speziale, Sarkar and Gatski (SSG) closure by Speziale et al. [58]. The corresponding joint-scalar mass density function (MDF) can be written as,

$$\mathscr{F}_\phi \left(\phi_\alpha,\ H; \underline{x}\right) \tag{1}$$

where $\phi_\alpha, \alpha = 1, \ldots, a$ are the species mass fraction, H is the enthalpy, and $\underline{x}$ is the physical position. The transport equation for $\mathscr{F}_\phi$, given in Eq. (2), was solved using a Lagrangian particle-based Monte Carlo method [59] and coupled with the CFD solver, which provides the velocity field and timescale information.

$$\frac{\partial \mathscr{F}_\phi}{\partial t} + \frac{\partial}{\partial x_l}[\widetilde{u_l}\mathscr{F}_\phi] + \frac{\partial}{\partial \Psi_\alpha}[S_\alpha(\underline{\Psi})\mathscr{F}_\phi] = -\frac{\partial}{\partial x_l}[\langle u_l'' \mid \underline{\Psi}\rangle \mathscr{F}_\phi] + \frac{\partial}{\partial \Psi_\alpha}\left[\left\langle \frac{1}{\rho}\frac{\partial J_l^\alpha}{\partial x_l} \mid \underline{\Psi}\right\rangle \mathscr{F}_\phi\right] \tag{2}$$

The terms on left-hand side of Eq. (2) are closed and denote transport in (i) physical space and (ii) scalar space (chemical reaction) via the chemical source term ($S_\alpha(\underline{\Psi})$). The terms on the right-hand side were modelled. The first represents transport in physical space due to turbulent fluctuations and was closed via a gradient diffusion approximation with the turbulent Prandtl/Schmidt number set to 0.70. The second term stands for transport in scalar space due to molecular mixing and was closed using the modified Curl's model [60], where the calculation of the scalar turbulent mixing timescale τ_ϕ was provided by the second moment closure.

$$\tau_\phi^{-1} = \frac{\widetilde{\epsilon_\phi}}{\widetilde{\phi''^2}} = \frac{C_\phi}{2}\frac{\widetilde{\epsilon}}{\widetilde{k}} = \frac{C_\phi}{2}\tau_T^{-1} \tag{3}$$

In the above, $\widetilde{\epsilon_\phi}$ is the scalar dissipation rate, $\widetilde{\phi''^2}$ is the scalar variance, and C_ϕ represents the turbulent to mixing time ratio constant with a common value of 2.3 [38]. The radiative losses were calculated based on the optically thin assumption [61] accounting for the radiation of H_2O, CH_4, CO and CO_2 [57]. The chemical mechanism featured 44 chemical species and 271 reactions [62] resulting in a 45-dimensional joint PDF (including enthalpy). Following Gkagkas et al. [40], the rate by Sutherland et al. [63] was used for the reaction $O + H_2 \rightleftharpoons H + OH$. The 44 species are: $H_2, O_2, H, OH, O, H_2O, HO_2, H_2O_2, C_2H_6, C_2H_5, C_2H_4, C_2H_2, CH_4, CH_3, {}^1CH_2, {}^3CH_2, CH, C_1$, CO, CO_2, CH_2O, CHO, CH_3OH, CH_3O, CH_2OH, C_2O, C_2H, C_2HO, C_2H_2O, C_2H_3, CH_3CHO, CH_3CO, C_2, CH_2CHO, H_2C_2, CH_3OO CH_3OOH, $CHCH_2O$, C_2H_4O, C_2H_4OOH, C_2H_5O, C_2H_5OO, C_2H_5OOH and N_2.

3.1 Computational Details

The experimental reactant temperature was set to $T_j = 295$ K. However, heat transfer from the vitiated co-flow to the centre nozzle may result in preheating. Garcia [64] measured a nozzle exit temperature of 350 K at a lower jet velocity, similar pilot temperature and thinner nozzle wall thickness. A sensitivity analysis on the jet reactant temperature was conducted in the range $295 < T_j$ (K) < 350 for the 50% H_2/50% CH_4 mixture at different pilot temperatures. No significant differences were observed in the computationally determined flame lift-off height as illustrated in Fig. 2. Consequently, the jet reactant temperature was set to 320 K consistent with previous studies by Cabra et al. [30] and Gkagkas [27, 40]. The mixture composition of the hot pilot was determined based on chemical equilibrium calculations using the experimental reactant flow rates and the measured temperature. The effect of a pilot stream containing incomplete combustion products (i.e. approaching equilibrium) was also investigated with no significant differences in the flame lift-off height determined. A mesh sensitivity analysis was performed using 92, 142 and 284 control volumes (N_c) along with 400 and 800 stochastic particles (N_p) per cell. No significant differences were observed between the cases. Consequently, the computational domain was resolved by 142 control volumes in the radial direction combined with 400 stochastic particles per cell. The axial (downstream) step was determined using a restrictive CFL criterion. The computational domain started at the nozzle exit of the fuel jet and extended axially to $x/d = 100$ and radially to $r/d = 9$.

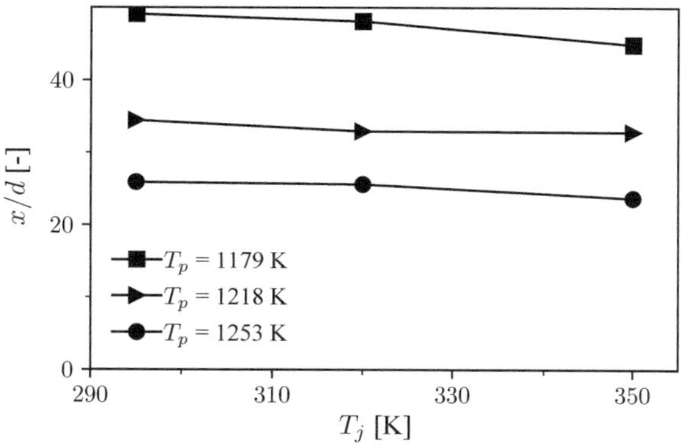

Fig. 2 Fuel jet temperature (T_j) sensitivity on the flame lift-off height (x/d) for three different pilot temperatures (T_p) for the 50% H_2/50% CH_4 mixture

4 Lift-Off Height Definition

4.1 Analysis of Experimental Data

The large number of different mixtures, the use of an average of five pilot temperatures per mixture and the required 1000 images (N) per data point resulted in a total of approximately 165,000 instantaneous flame images. Therefore, a purpose written MATLAB [65] algorithm was used to automatically detect the flame lift-off height. The instantaneous chemiluminescence signal was radially integrated over the entire domain, and the resulting profile was subsequently smoothed using a moving average filter with a filter size of ~1 mm. The instantaneous flame lift-off height (x_n) was determined based on the location of the steepest axial gradient of the smoothed chemiluminescence profile ($\bar{\mathrm{I}}_s$) as defined by Eq. (4). Subsequently, the mean flame lift-off height ($\overline{x}$) and root mean square (x') were calculated to evaluate the experimental variation. The method was found to provide an unambiguous detection of the instantaneous lift-off height and was independent of the background intensity. However, the determination of the steepest gradient from the computational results was found to be sensitive to the spatial extent of the interrogation frame. Therefore, a second approach was developed and used for the comparison with the computational results. Accordingly, the mean chemiluminescence intensity images were integrated radially over the entire domain, and the resulting profile was subjected to the same smoothing filter ($\widetilde{\mathrm{I}}$). The alternative flame lift-off height ($\widetilde{x}$) was based on the location of the 50% value of the peak of $\widetilde{\mathrm{I}}$ as shown by Eq. (5). The second method was found to enable robust comparisons of experimental and computational data. The average difference between $\overline{x}$ and $\widetilde{x}$ was found to be 3.6 x/d ($\simeq$ 15 mm) with uncertainties for individual cases outlined below (Fig. 3).

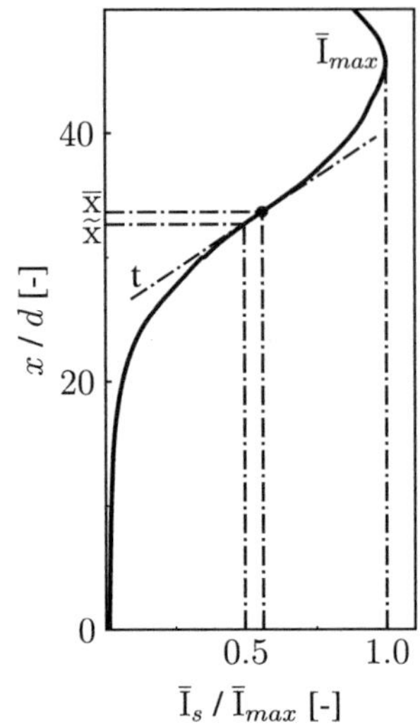

$$x_n = x\left|\left(\nabla \bar{\mathrm{I}}_s\right)_{max}\right.$$

$$\bar{x} = \frac{1}{N} \textstyle\sum_{n=1}^{N} x_n \qquad (4)$$

$$x' = \left(\frac{1}{N} \textstyle\sum_{n=1}^{N} (\bar{x} - x_n)^2\right)^{1/2}$$

$$\widetilde{x} = x\left|\left(\frac{1}{2}\, \widetilde{\mathrm{I}}_{max}\right)\right. \qquad (5)$$

Fig. 3 Data analysis schematic, where I_p is the peak intensity, $\bar{x}$ and $\widetilde{x}$ is the lift-off height based on Eqs. (4) and (5), and t the tangent on the steepest gradient of $\bar{\mathrm{I}}_s$

4.2 Analysis of Computational Data

The chemiluminescence spectra depend on the fuel blend composition [52, 53] and the reaction conditions (e.g. temperature) [54]. The formation of CH_2O is characteristic of the low temperature reaction onset for hydrocarbon fuels, and the distinct low temperature chemistry is particularly important during auto-ignition [27]. The build up of HO_2 and H_2O_2 is also pivotal in the low temperature chemistry [66] as the decomposition of these species leads to the onset of high temperature chain branching and the formation of OH in the heat release layer. The latter is accordingly approximately identified by the cross-correlation of OH and CH_2O [4]. The OH chemiluminescence signal is a strong contributor to the flame spectra of $H_2/CH_4/CO$ fuel blends [52, 53]. However, chemical mechanisms that reliably predict the formation of electronically excited species, e.g. OH* [67, 68], have not been validated for the wide range of conditions investigated in the current study. Consequently, it is assumed that concentrations of electronically excited species concentrations are statistically correlated with the corresponding ground state species. Accordingly, lower and upper limit lift-off heights based on the species concentration of CH_2O and OH were defined. Both species contribute to the experimentally detected chemiluminescence signal with CH_2O representative of cool flames and the reaction onset [55], while OH is used for the identification of high temperature reactions [54]. An example of computationally determined distributions of CH_2O, H_2O_2 and OH are shown along with temperature in Fig. 4 for a 50% H_2/50% CH_4 mixture with a $T_p = 1218$ K. The computationally determined lift-off height was defined based on Eq. (5) using the CH_2O and OH concentration profiles (instead of I). A moving average filter (on

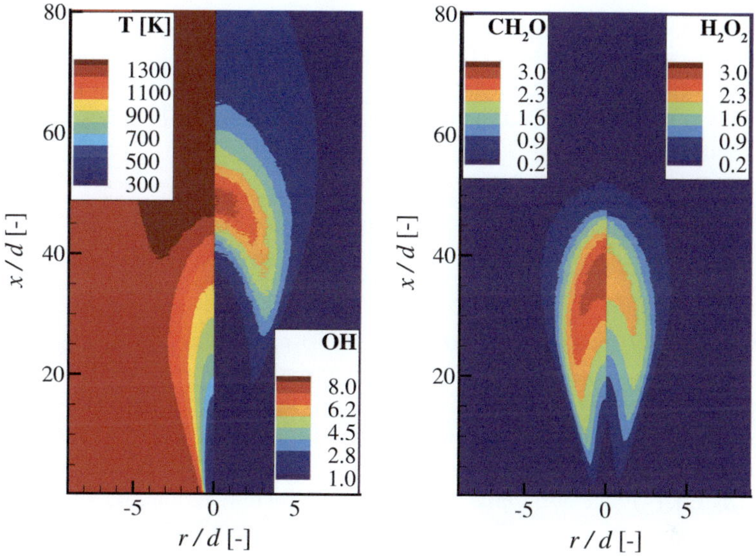

Fig. 4 Computed temperature, OH, CH_2O and H_2O_2 molar concentrations [μ mol/m^3] for the 50% H_2/50% CH_4 mixture at a pilot temperature of T_p = 1218 K

average 3 *d*) was used to remove noise resulting from the stochastic Monte Carlo approach. The smoothing resulted in continuous concentration profiles and an unambiguous detection of the flame lift-off heights.

5 Results and Discussion

The main objective of the current investigation was to determine the impact of fuel reactivity on flame stabilisation in a turbulent shear layer by gradually substituting H_2 by either CH_4 or CO. Therefore, the jet equivalence ratio ($\Phi = 0.8$), jet momentum and Reynolds number ($\simeq 22 \times 10^3$) were maintained as close to constant as practically possible. Chemistry-related effects were further highlighted by varying the pilot flame temperature, responsible for the auto-ignition of the fuel blend, for each mixture within the experimental operating limits.

The experimental mean flame chemiluminescence of the 60% H_2/40% CH_4 flame series is shown as an example in Fig. 5 for the pilot temperature range $1112 < T_p$ (K) < 1227. The corresponding computational results are illustrated in terms of the CH_2O and OH concentrations for T_p = 1112, 1153 and 1227 K in Fig. 6. The latter figure clearly shows the separation of the low and high temperature reaction zones as characterised by CH_2O and OH, respectively. By contrast, as outlined above, the experimental images contain contributions from both combustion regimes. The flame lift-off height (stabilisation point) reduced significantly with increasing T_p due to the shorter auto-ignition delay time. This trend is qualitatively consistent for all mixtures. However, the quantitative reduction of the flame lift-off height with increasing T_p was found to be dependent on the mixture reactivity and fuel blending components as discussed below.

5.1 Experimental Results

The experimentally determined dimensionless lift-off heights (x/d) using Eq. (4) are depicted in Fig. 7. The pilot temperature was varied between blow-off (lower limit) and flame attachment to the nozzle (upper limit) for each mixture. The former was defined as $x/d > 50$. The absolute minimum and maximum temperatures were 1040 $< T_p$ (K) $<$ 1490. The vertical bars illustrate the rms (x') of the lift-off height as defined by Eq. (4). The horizontal bars show the pilot temperature fluctuations (rms), determined using the R-type thermocouple, as mentioned in Sect. 2.3.

The measured x/d of the H_2/CH_4 fuel blends are shown in the top left of Fig. 7. The flame lift-off height increased significantly with successive H_2 substitution by CH_4. This is directly related to the decrease in mixture reactivity and the corresponding increase in auto-ignition delay times. For example, values of x/d of mixtures with 0, 10 and 20% CH_4 at $T_p \approx 1040\ K$ increased from $x/d = 6.8$, 11 to 21, respectively, and illustrate the sharp reduction in fuel reactivity caused by the introduction of

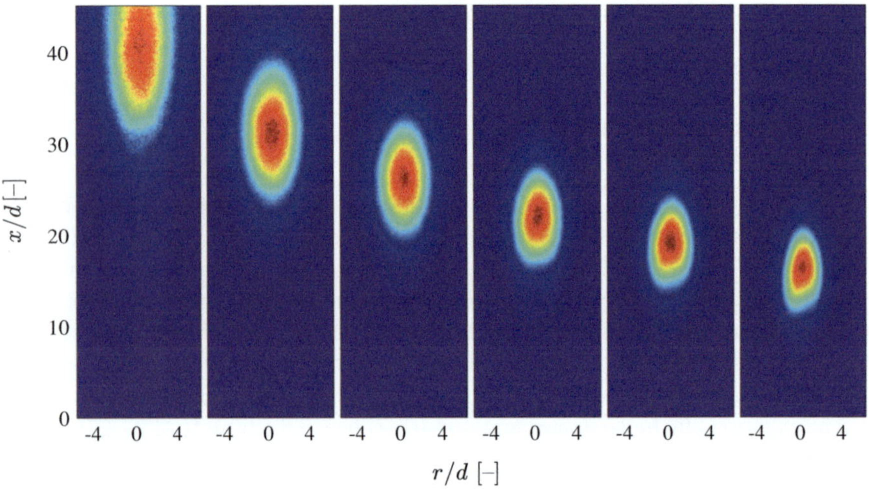

Fig. 5 Experimentally determined mean flame chemiluminescence images for the 60% H_2/40% CH_4 case. The pilot temperatures (T_p) increase from (left to right) with T_p = 1112, 1137, 1153, 1173 1193 and 1227 K. The colour code represents the detected mean flame chemiluminescence in arbitrary units

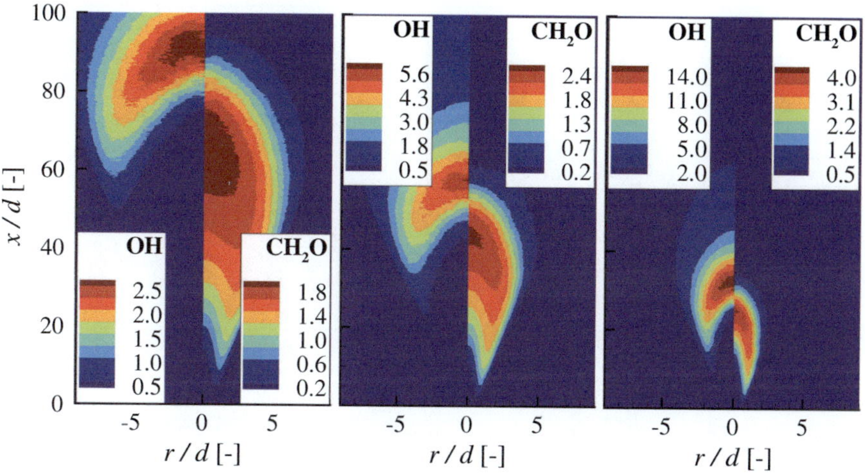

Fig. 6 Computationally determined CH_2O and OH concentrations [μ mol/m^3] for the 60% H_2/40% CH_4 case. The pilot temperatures (T_p) increase from (left to right) with $T_p = 1112$, 1153 and 1227 K

methane. The required change in T_p also indicates a reduced mixture reactivity. For example, cases with pure H_2 and with a 10% CH_4 substitution were restricted to a maximum $T_p \approx 1095\ K$ due to the risk of flashback. By contrast, the maximum T_p for the 20% CH_4 fuel blend was increased to 1160 K. Furthermore, mixtures with > 30% CH_4 exhibited a higher temperature sensitivity as indicated by the steeper gradients in Fig. 7. Additional substitution of H_2 with CH_4 continues the same trend with an increase in the required minimum and the possible maximum pilot temperatures. Such a sharp impact of H_2 substitution with CH_4 was also reported by Lieuwen et al. [20].

The dimensionless flame lift-off heights for H_2/CO blends see top right panel of Fig. 7, increased with CO content. However, compared to the CH_4 mixtures, the trend is much less pronounced, particularly for mixtures with a CO content below 50%. The current results correspond qualitatively very well to the findings by Lieuwen et al. [20] and can be related to the higher intrinsic reactivity of CO compared to CH_4. In this context, it may be noted that H_2/CO fuel blends with less than 50% CO were stabilised at the lowest possible $T_p \approx 1040$ K. Further CO addition resulted in a more significant increase in flame lift-off height at a given temperature and a more pronounced non-linearity. This suggests that the auto-ignition process of H_2/CO mixtures is substantially governed by H_2 up to high volumetric mixture fractions of CO. A qualitatively similar behaviour was observed by Fotache et al. [46] for H_2/CO fuel blends in a study featuring ignition against a hot air stream in a counterflow arrangement. Three ignition regimes were defined: (1) a hydrogen-dominated regime for $100\% < X_{H_2} < 17\%$, (2) a transition regime spanning from $17\% < X_{H_2} < 7\%$ and (3) a hydrogen catalysed regime for $X_{H_2} < 7\%$. According to this classification, all mixtures of the current study are within the hydrogen-dominated regime.

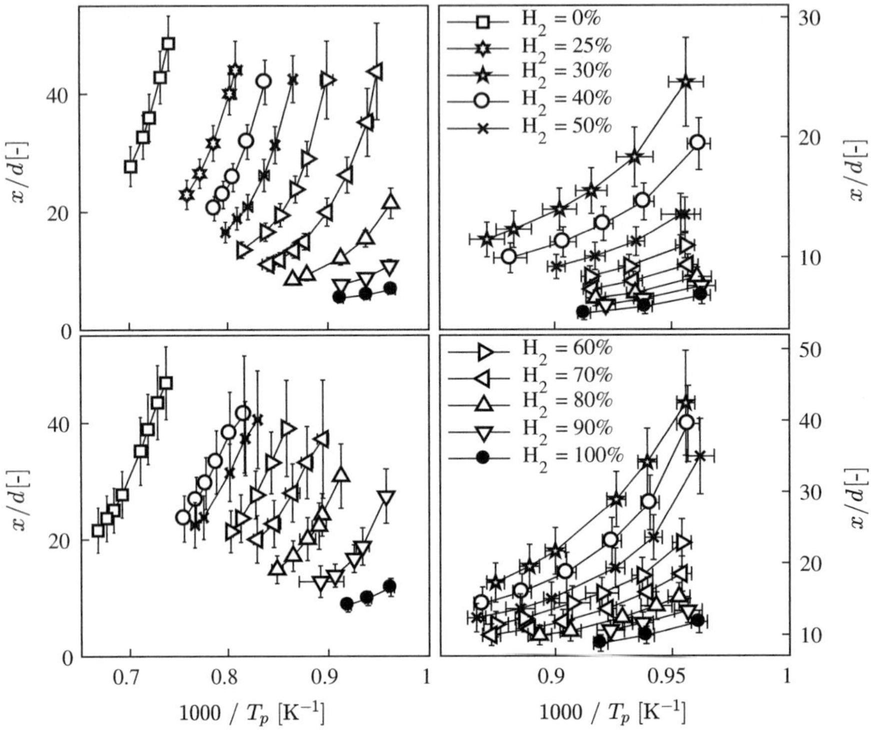

Fig. 7 Experimentally determined flame lift-off height/stabilisation points normalised by the nozzle diameter (x/d) as a function of pilot temperature (T_p). Top left: H_2/CH_4; Right: H_2/CO; Bottom left: $H_2/CH_4/N_2$; Right: $H_2/CO/N_2$. Note that due to the number of conditions covered, the legend outlining the amount of H_2 in the fuel blends was split over the two right-hand side panels

However, a noticeable increase in the ignition temperature for volumetric hydrogen concentrations of $X_{H_2} \leq 50\%$ was observed in the present experiments. The ignition temperature remained approximately constant for mixtures with a higher hydrogen content. The current findings suggest that the hydrogen-dominated regime is narrower, with the transition regime starting with $X_{H_2} < 50\%$ as also suggested by Sung et al. [47].

The frequent practical availability of highly diluted syngas motivated the introduction of additional inert gas as listed in Table 1. This reduces the fuel component concentration levels, adds a heat sink, affects third body reactions and increases the auto-ignition delay time [69]. The measured flame stabilisation point (x/d) for $H_2/CH_4/N_2$ and $H_2/CO/N_2$ fuel blends are shown in the bottom of Fig. 7. The substitution of H_2 with CH_4 combined with the added thermal ballast has a pronounced impact on the mixture reactivity with a blending factor of only 10% of CH_4 exhibited a strong influence on the diluted mixture. For higher CH_4 blending factors, the influence of N_2 dilution resulted in the need for higher T_p. For the extreme example of 100% CH_4, the investigated temperature range was $1355 < T_p$ (K) < 1425 for

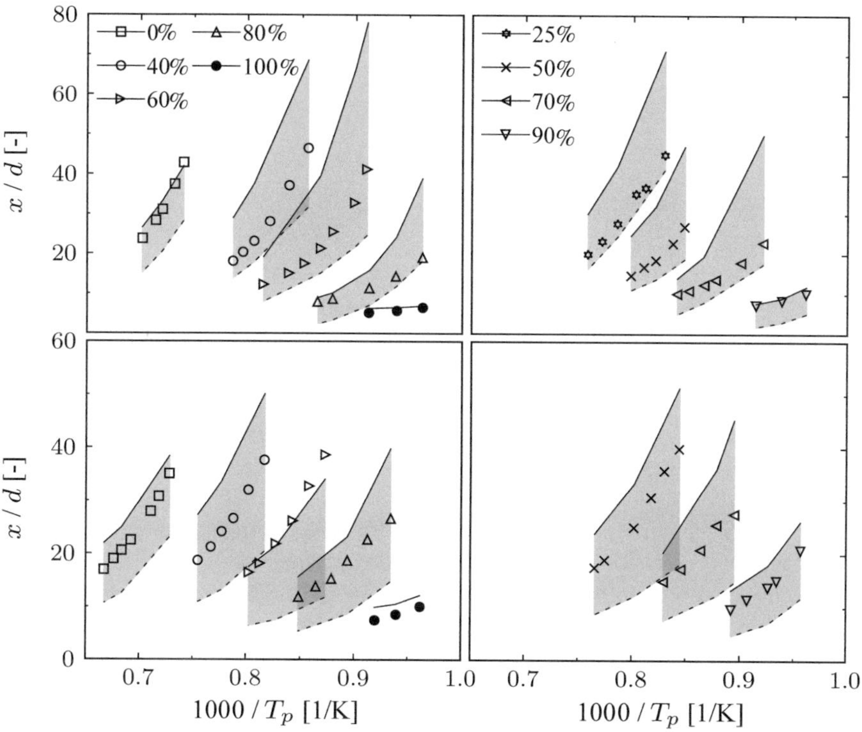

Fig. 8 Computationally determined flame lift-off heights/stabilisation points (x/d) for H_2/CH_4 fuel blends as a function of the reciprocal pilot temperature (top: without N_2 dilution; bottom: with N_2 dilution). The solid (dashed) line is the lift-off height based on the calculated OH (CH_2O) concentration. The shaded grey area indicates the area in between. The markers show the experimental data based on Eq. (5). The legend indicate the H_2 percentage of the fuel blend

the non-diluted mixture, while the diluted mixture required a temperature range of $1380 < T_p$ (K) < 1480. Thus, for a given T_p and fuel component ratio, the additional inert gas dilution extracts heat from the reaction zone and reduces the radical pool, which consequently leads to a substantially increased ignition delay and a correspondingly higher lift-off height.

The additional N_2 dilution resulted in a larger separation between the H_2/CO mixtures compared to the non-diluted counterparts. Small quantities of CO (e.g. 10%) yield a distinct increase in x/d, and the ignition characteristics of CO become more pronounced. The transition regime [47] for the present N_2-dilution was expanded to mixtures up to 80% H_2 as evident in the bottom right of Fig. 7. The slope and linearity indicate a predominant H_2 ignition regime at high temperatures, while the impact of CO became apparent at lower T_p. Thus, the mixture transferred into the transition regime with decreasing T_p. This behaviour can be attributed to the difference in auto-ignition temperatures of the separate components or a failure by the hydrogen component to release sufficient energy on a suitable timescale to trigger ignition of the carbon containing fuel mixture component. This observation is consistent with the noted differences in behaviour between CO and CH_4 and should ideally be investigated further.

5.2 Computational Results

The computational results for H_2/CH_4 and $H_2/CH_4/N_2$ blends are shown in Fig. 8. The solid lines show the computationally determined lift-off height based on the OH concentration. The dashed lines show the equivalent based on the CH_2O concentration. The experimental data, for consistency determined using Eq. (5), are shown for comparison. The OH-based lift-off height accurately predicts the experimental x/d for pure CH_4 and for blends with H_2 concentrations $\geq$90%. The latter cases are dominated high temperature chain branching, and the experimentally determined chemiluminescence signal is accordingly determined by OH* [70]. The high pilot temperature of the pure methane case also results in high temperature reaction zones with strong OH* emission [54]. However, for the intermediated cases (i.e. 20% $\leq H_2 \leq$ 80%) the experimentally determined lift-off height is closer to the values based on CH_2O due to the increased importance of low temperature chemistry. The discrepancy between experimental and computational data was evaluated using Eq. (4), where N is the number of data points per mixture and $\widetilde{x}_{i,e}$ and $\widetilde{x}_{i,c}$ are the experimental and computational lift-off height, respectively.

$$\epsilon = \sqrt{\frac{\sum_{i=1}^{N} \left(\widetilde{x}_{i,e} - \widetilde{x}_{i,c}\right)^2}{N}} \tag{4}$$

The differences as measured by ϵ in computed and experimental flame stabilisation points (x/d) based on the OH concentration for pure H_2 and 90% H_2/10% CH_4 were $\epsilon = 0.82$ and 1.1 x/d compared to measured flame stabilisation points of 8.85 and 11.3 x/d. The prediction accuracy for all H_2/CH_4 mixtures with 20–80% H_2 based on the OH lift-off height resulted in $\epsilon = 17 \pm 5.4\ x/d$. The corresponding CH_2O-based lift-off height predictions improved to $\epsilon = 6.5 \pm 2.9\ x/d$. The difference between the flame lift-off height based on the OH and the CH_2O concentration increases with decreasing T_p. This can be attributed to a thermochemical state that favours persistent CH_2O concentrations in absence of high temperature chain branching reactions. For the N_2 diluted cases, the predicted lift-off height based on the OH concentration shows a better agreement with the experimental data compared to the CH_2O based x/d due to the increased pilot flame temperatures. The respective ϵ are 6.8 ± 3.7 and $11 \pm 3.7\ x/d$.

The OH-based lift-off heights for the CO (H_2/CO and $H_2/CO/N_2$) blends agree comparatively well with the experimental data as shown in Fig. 9. This can, at least partly, be expected as the reaction path through CH_2O is not dominant for H_2/CO fuel blends, and, accordingly, the experimental chemiluminescence signal is dominated by OH*. The corresponding lift-off height can consequently be determined without ambiguity. Overall, the computationally determined lift-off heights are modestly over-predicted. The deviations with experiments were $\epsilon = 0.75 \pm 0.17\, x/d$ and 2.6 $\pm\, 1.0\, x/d$ for all H_2/CO and H_2/CO/N_2 mixtures, respectively. Thus, the data suggest that H_2 acts as an initiator of reaction progress and results in prompt high temperature chain branching for the current CO-based data series.

Considering the results for all mixtures, it appears clear that for fuel blends containing methane, there is a comparatively wide range of conditions where the low and high temperature reaction zones, characterised by formaldehyde and the hydroxyl radical, are spatially distributed. The delineation of the two zones can potentially be done experimentally by determining the cross-correlation of the two species as an indicator of the heat release layer [4]. The issue is clearly important when comparing experimental and computational data for the stabilisation point of lifted flames – in particular if simplified (e.g. topological) calculation methods are used that do

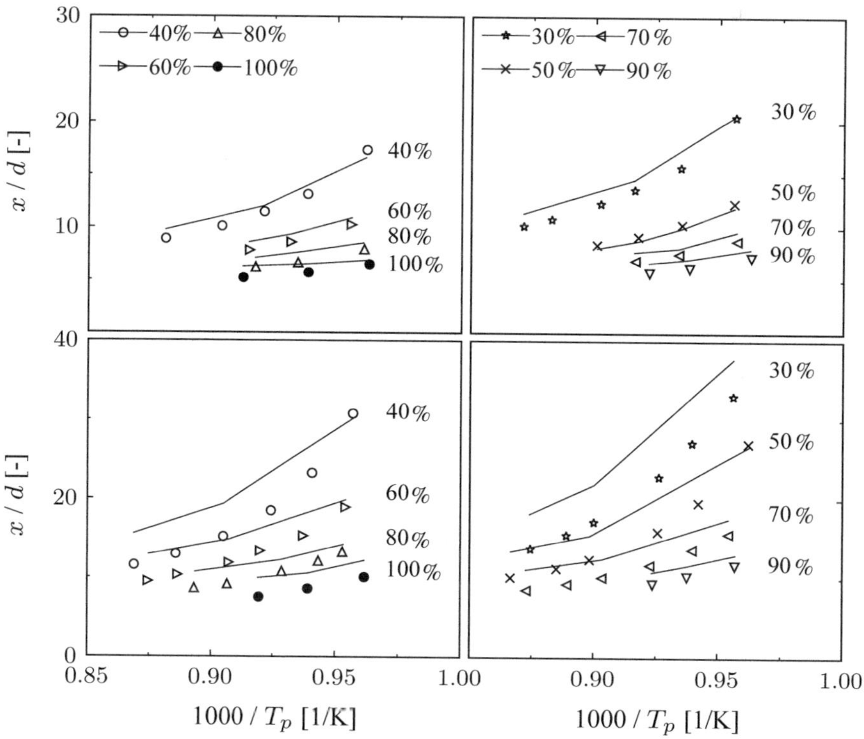

Fig. 9 Computationally determined flame lift-off heights (x/d) for H_2/CO fuel blends as a function of the reciprocal pilot temperature (top: without N_2 dilution; bottom: with N_2 dilution). The solid line is the lift-off height based on the calculated OH concentration. The markers show the experimental data based on Eq. (5). The legend indicate the H_2 percentage of the fuel blend

not provide comprehensive information on turbulence–chemistry interactions via the joint PDF.

5.3 Fuel Blend Comparison

Li et al. [21] introduced a scaling factor (β), see Eq. (5), to compare binary and ternary $H_2/CH_4/CO$ mixtures based on the amount air required to fully oxidise the fuel.

$$\beta = \frac{\frac{X_{H_2}}{(X_{H_2}/X_A)_{st}}}{\frac{X_{H_2}}{(X_{H_2}/X_A)_{st}} + \frac{X_F}{(X_F/X_A)_{st}}} \tag{5}$$

The scaling successfully correlated the explosion overpressure and turbulent flow field for a wide range mixtures. Consequently, the β scaling was used here to highlight the different inhibiting effects of CH_4 and CO blending on the auto-ignition in a turbulent shear layer. The 90% H_2/10% CH_4 and the 70% H_2/30% CO cases share the same $\beta \simeq 0.7$. The two cases showed a very similar flame lift-off height and were stabilised using the same pilot temperature range as depicted in Fig. 10. Evidently, the β factor provides a successful scaling for x/d for mixtures with high reactivity. However, the success is diminished for mixtures with reduced reactivity, i.e. 80% H_2/20% CH_4 and 40% H_2/60% CO with $\beta = 0.5$. The increasing discrepancy can be expected as the auto-ignition process in the current turbulent shear layer is strongly governed by the chemistry. The trend is further emphasised by the spread between the CH_4 and CO fuel blends with an identical β value for mixtures with additional N_2 dilution.

6 Conclusions

The current work presents an experimental and computational study on the impact of mixture reactivity on the auto-ignition/flame stabilisation in a turbulent shear layer using a vitiated co-flow burner. The mixture reactivity was varied by gradual substitution of hydrogen with methane or carbon monoxide. A total of 34 lean (stoichiometry of 0.80) premixed gas mixtures were investigated over a wide range of H_2/CH_4 and H_2/CO fuel blends. The effect of additional inert gas dilution was also investigated and found to accentuate the trends observed with the undiluted mixtures. To accommodate the wide range of mixture reactivities, the pilot temperature was varied to assure ignition while avoiding flame blow-off and flame attachment to the fuel jet nozzle. The experimental flame lift-off height was measured using flame chemiluminescence (including CH_2O^* and OH^*) imaging. A transported PDF approach, closed at the joint-scalar level, was used for the computational prediction

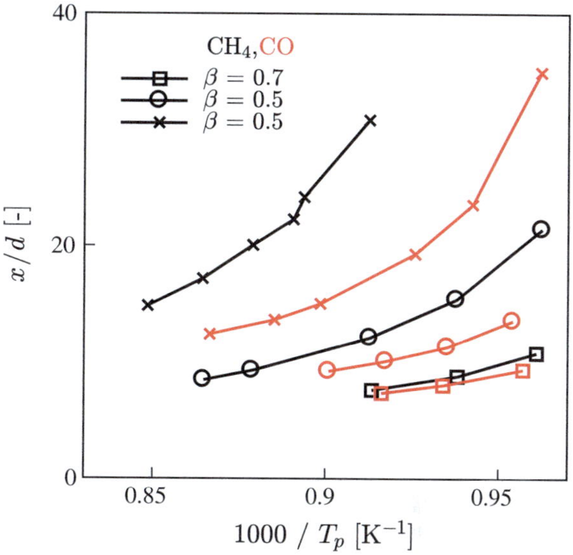

Fig. 10 Experimentally determined flame lift-off heights/stabilisation points (x/d) for mixtures with constant β values. Black lines show CH_4 and red lines CO based mixtures. The □ markers correspond to 10% CH_4 and 30% CO. The ○ markers correspond to 20% CH_4 and 30% CO. The × markers show N_2 diluted mixtures with 10% CH_4 and 30% CO

of the experimental results. The experimentally determined flame lift-off height was strongly dependent upon mixture reactivity (i.e. H_2 content), fuel blending component (i.e. CH_4 or CO) and the pilot temperature. The flame lift-off height, and thus auto-ignition delay, decreased substantially with increasing H_2 fuel concentration. The same general trend was evident for increasing pilot temperatures. The blending of CH_4 showed a significantly stronger inhibiting effect on the H_2 chemistry compared to CO. Moreover, the auto-ignition process of H_2/CH_4 mixtures exhibited a stronger temperature dependency. The additional inert gas dilution resulted in an increased auto-ignition delay and higher flame lift-off for a given pilot temperature due to the reduced precursor radical pool concentration and reduced flame temperature. The computational flame lift-off heights were calculated based on the CH_2O and OH concentration levels, representing low and high temperature reaction zones. Predictions of the flame lift-off height/stabilisation point based on the OH concentration were found to be good for H_2/CO and $H_2/CO/N_2$ mixtures due to the much-reduced importance of the formaldehyde chemistry. Good agreement was also found for pure CH_4, due to the use of high temperature pilot flames, and H_2/CH_4 mixtures with $H_2 \geq 90\%$ due to the intrinsic mixture reactivity. However, the lift-off height of H_2/CH_4 mixtures with an intermediate H_2 concentration could not be predicted accurately based on the computed OH concentration due to the importance of CH_2O chemiluminescence. The predictive capability was improved by a factor of three when taking into account the CH_2O concentration levels. This was attributed to a thermochemical state that favours persistent CH_2O concentrations with delayed high temperature chain branching reactions due to a separation of the low and high temperature reaction zones. The latter observation is important when comparing flame stabilisation point more generally and, overall, the current study provides a unique and consistent data set that furthers understanding of the impact of hydrogen enrichment of syngas related fuel streams.

Acknowledgements The authors gratefully acknowledge the support of the ETI under the High Hydrogen project (PE02162) for the construction of some aspects of the experimental facility and the procurement of the experimental data. The contributions by Prof. H. J. Michels are also gratefully acknowledged. Support for the computational work was derived with the permission of Toyota Motor Europe NV/SA, and the authors wish to express their gratitude to Dr. K. Gkagkas.

References

1. Lieuwen T, Yang V, Yetter R (eds) (2009) Synthesis gas combustion: fundamentals and applications. CRC Press, Boca Raton. https://doi.org/10.1201/9781420085358
2. Daniele S, Jansohn P, Mantzaras J, Boulouchos K (2011) Turbulent flame speed for syngas at gas turbine relevant conditions. Proc Comb Inst 33(2):2937–2944. https://doi.org/10.1016/j.proci.2010.05.057
3. Milosavljevic VD, Lindstedt RP, Cornwell MD, Gutmark EJ, Váos EM (2006) Combustion instabilities near the lean extinction limit. In: Yu KH, Whitelaw JH, Witton JJ, Roy G (eds) Advances in combustion and noise control. Cranfield University Press, pp 149–165
4. Hampp F, Shariatmadar S, Lindstedt RP (2018) Quantification of low Damköhler number turbulent premixed flames. Proc Comb Inst. https://doi.org/10.1016/j.proci.2018.06.079 (in press)
5. Hampp F, Lindstedt RP (2017) Quantification of combustion regime transitions in premixed turbulent DME flames. Combust Flame 182:248–268. https://doi.org/10.1016/j.combustflame.2017.04.006
6. Hampp F, Lindstedt RP (2017) Strain distribution on material surfaces during combustion regime transitions. Proc Comb Inst 36(2):1911–1918. https://doi.org/10.1016/j.proci.2016.07.018
7. Korb B, Kawauchi S, Wachtmeister G (2016) Influence of hydrogen addition on the operating range, emissions and efficiency in lean burn natural gas engines at high specific loads. Fuel 164:410–418. https://doi.org/10.1016/j.fuel.2015.09.080
8. Zhang B, Ji C, Wang S (2016) Performance of a hydrogen-enriched ethanol engine at unthrottled and lean conditions. Energy Convers Manage 114:68–74. https://doi.org/10.1016/j.enconman.2016.01.073
9. Amrouche F, Erickson PA, Park JW, Varnhagen S (2016) An experimental evaluation of ultra-lean burn capability of a hydrogen-enriched ethanol-fuelled Wankel engine at full load condition. Int J Hydrogen Energy 41(42):19231–19242. https://doi.org/10.1016/j.ijhydene.2016.07.267
10. Bell SR, Gupta M (1997) Extension of the lean operating limit for natural gas fueling of a spark ignited engine using hydrogen blending. Combust Sci Technol 123(1–6):23–48. https://doi.org/10.1080/00102209708935620
11. Ma F, Wang Y, Liu H, Li Y, Wang J, Zhao S (2007) Experimental study on thermal efficiency and emission characteristics of a lean burn hydrogen enriched natural gas engine. Int J Hydrogen Energy 32(18):5067–5075. https://doi.org/10.1016/j.ijhydene.2007.07.048
12. Wang J, Huang Z, Tang C, Zheng J (2010) Effect of hydrogen addition on early flame growth of lean burn natural gas–air mixtures. Int J Hydrogen Energy 35(13):7246–7252. https://doi.org/10.1016/j.ijhydene.2010.01.004
13. Frenillot JP, Cabot G, Cazalens M, Renou B, Boukhalfa MA (2009) Impact of H_2 addition on flame stability and pollutant emissions for an atmospheric kerosene/air swirled flame of laboratory scaled gas turbine. Int J Hydrogen Energy 34(9):3930–3944. https://doi.org/10.1016/j.ijhydene.2009.02.059

14. Lieuwen T, McDonell V, Santavicca D, Sattelmayer T (2008) Burner development and operability issues associated with steady flowing syngas fired combustors. Combust Sci Technol 180(6):1169–1192. https://doi.org/10.1080/00102200801963375
15. Lowesmith BJ, Mumby C, Hankinson G, Puttock JS (2011) Vented confined explosions involving methane/hydrogen mixtures. Int J Hydrogen Energy 36(3):2337–2343. https://doi.org/10.1016/j.ijhydene.2010.02.084
16. Salzano E, Cammarota F, Di Benedetto A, Di Sarli V (2012) Explosion behavior of hydrogen–methane/air mixtures. J Loss Prev Process Ind 25(3):443–447. https://doi.org/10.1016/j.jlp.2011.11.010
17. Ma Q, Zhang Q, Chen J, Huang Y, Shi Y (2014) Effects of hydrogen on combustion characteristics of methane in air. Int J Hydrogen Energy 39(21):11291–11298. https://doi.org/10.1016/j.ijhydene.2014.05.030
18. Hampp F, Lindstedt RP (2016) Fractal grid generated turbulence–a bridge to practical combustion applications. In: Sakai Y, Vassilicos C (eds) Fractal flow design: how to design bespoke turbulence and why. Springer International Publishing, Cham, pp 75–102. https://doi.org/10.1007/978-3-319-33310-6_3
19. Tang C, Zhang Y, Huang Z (2014) Progress in combustion investigations of hydrogen enriched hydrocarbons. Renew Sustain Energy Rev 30:195–216. https://doi.org/10.1016/j.rser.2013.10.005
20. Lieuwen T, McDonell V, Petersen E, Santavicca D (2008) Fuel flexibility influences on premixed combustor blowout, flashback, autoignition, and stability. J Eng Gas Turbines Power 130(1):011506–011506–10. https://doi.org/10.1115/1.2771243
21. Li T, Hampp F, Lindstedt RP (2018) Experimental study of turbulent explosions in hydrogen enriched syngas related fuels. Process Saf Environ Prot 116:663–676. https://doi.org/10.1016/j.psep.2018.03.032
22. Li T, Hampp F, Lindstedt RP (2018) The impact of hydrogen enrichment on the flow field evolution in turbulent explosions. Submitted to Comb Flame
23. Zhang Y, Shen W, Fan M, Zhang H, Li S (2014) Laminar flame speed studies of lean premixed H_2/CO/air flames. Combust Flame 161(10):2492–2495. https://doi.org/10.1016/j.combustflame.2014.03.016
24. Donohoe N, Heufer A, Metcalfe WK, Curran HJ, Davis ML, Mathieu O, Plichta D, Morones A, Petersen EL, Güthe F (2014) Ignition delay times, laminar flame speeds, and mechanism validation for natural gas/hydrogen blends at elevated pressures. Combust Flame 161(6):1432–1443. https://doi.org/10.1016/j.combustflame.2013.12.005
25. Messaoudani Zl, Rigas F, Binti Hamid MD, Che Hassan CR (2016) Hazards, safety and knowledge gaps on hydrogen transmission via natural gas grid: a critical review. Int J Hydrogen Energy 41(39):17511–17525. https://doi.org/10.1016/j.ijhydene.2016.07.171
26. Mastorakos E, Baritaud TA, Poinsot TJ (1997) Numerical simulations of autoignition in turbulent mixing flows. Combust Flame 109(1):198–223. https://doi.org/10.1016/S0010-2180(96)00149-6
27. Gkagkas K, Lindstedt RP (2007) Transported PDF modelling with detailed chemistry of pre- and auto-ignition in CH_4/air mixtures. Proc Comb Inst 31(1):1559–1566. https://doi.org/10.1016/j.proci.2006.08.078
28. Hilbert R, Thévenin D (2002) Autoignition of turbulent non-premixed flames investigated using direct numerical simulations. Combust Flame 128(1):22–37. https://doi.org/10.1016/S0010-2180(01)00330-3
29. Cabra R, Myhrvold T, Chen JY, Dibble RW, Karpetis AN, Barlow RS (2002) Simultaneous laser raman-rayleigh-lif measurements and numerical modeling results of a lifted turbulent H_2/N_2 jet flame in a vitiated coflow. Proc Comb Inst 29(2):1881–1888. https://doi.org/10.1016/S1540-7489(02)80228-0
30. Cabra R (2004) Turbulent jet flames into a vitiated coflow. Technical report, California University, Berkeley. https://ntrs.nasa.gov/search.jsp?R=20040042477
31. Masri AR, Cao R, Pope SB, Goldin GM (2004) Pdf calculations of turbulent lifted flames of H_2/N_2 fuel issuing into a vitiated co-flow. Combust Theory Modell 8(1):1–22. https://doi.org/10.1088/1364-7830/8/1/001

32. Cao RR, Pope SB, Masri AR (2005) Turbulent lifted flames in a vitiated coflow investigated using joint PDF calculations. Combust Flame 142(4):438–453. https://doi.org/10.1016/j.combustflame.2005.04.005
33. O'Loughlin W, Masri AR (2011) A new burner for studying auto-ignition in turbulent dilute sprays. Combust Flame 158(8):1577–1590. https://doi.org/10.1016/j.combustflame.2010.12.021
34. Medwell PR, Masri AR, Pham PX, Dally BB, Nathan GJ (2014) Temperature imaging of turbulent dilute spray flames using two-line atomic fluorescence. Exp Fluids 55(11):1840. https://doi.org/10.1007/s00348-014-1840-3
35. Williams A, Shcherbik D, Bibik O, Lubarsky E, Zinn BT (2015) Autoignition of a jet-a fuel spray in a high temperature vitiated air flow. ASME Turbo Expo: Power for Land, Sea, and Air, Volume 4A: Combustion, Fuels and EmissionsASME Turbo Expo: Power for Land, Sea, and Air, Volume 4A: Combustion, Fuels and Emissions p V04AT04A021 https://doi.org/10.1115/GT2015-42199
36. Mastorakos E (2017) Forced ignition of turbulent spray flames. Proc Comb Inst 36(2):2367–2383. https://doi.org/10.1016/j.proci.2016.08.044
37. Tang Q, Xu J, Pope SB (2000) Probability density function calculations of local extinction and no production in piloted-jet turbulent methane/air flames. Proc Comb Inst 28(1):133–139. https://doi.org/10.1016/S0082-0784(00)80204-0
38. Lindstedt RP, Louloudi SA, Váos EM (2000) Joint scalar probability density function modeling of pollutant formation in piloted turbulent jet diffusion flames with comprehensive chemistry. Proc Comb Inst 28(1):149–156. https://doi.org/10.1016/S0082-0784(00)80206-4
39. Lindstedt RP, Louloudi SA (2002) Joint scalar transported probability density function modeling of turbulent methanol jet diffusion flames. Proc Comb Inst 29(2):2147–2154. https://doi.org/10.1016/S1540-7489(02)80261-9
40. Gkagkas K, Lindstedt RP (2009) The impact of reduced chemistry on auto-ignition of H_2 in turbulent flows. Combust Theory Modell 13(4):607–643. https://doi.org/10.1080/13647830902928524
41. Scholte TG, Vaags PB (1959) Burning velocities of mixtures of hydrogen, carbon monoxide and methane with air. Combust Flame 3:511–524. https://doi.org/10.1016/0010-2180(59)90057-4
42. Cheng RK, Oppenheim AK (1984) Autoignition in methane–hydrogen mixtures. Combust Flame 58(2):125–139. https://doi.org/10.1016/0010-2180(84)90088-9
43. Ilbas M, Crayford AP, Yılmaz I, Bowen PJ, Syred N (2006) Laminar-burning velocities of hydrogen–air and hydrogen–methane–air mixtures: an experimental study. Int J Hydrogen Energy 31(12):1768–1779. https://doi.org/10.1016/j.ijhydene.2005.12.007
44. Gersen S, Anikin NB, Mokhov AV, Levinsky HB (2008) Ignition properties of methane/hydrogen mixtures in a rapid compression machine. Int J Hydrogen Energy 33(7):1957–1964. https://doi.org/10.1016/j.ijhydene.2008.01.017
45. Di Sarli V, Benedetto AD (2007) Laminar burning velocity of hydrogen–methane/air premixed flames. Int J Hydrogen Energy 32(5):637–646. https://doi.org/10.1016/j.ijhydene.2006.05.016
46. Fotache CG, Tan Y, Sung CJ, Law CK (2000) Ignition of $CO/H_2/N_2$ versus heated air in counterflow: experimental and modeling results. Combust Flame 120(4):417–426. https://doi.org/10.1016/S0010-2180(99)00098-X
47. Sung CJ, Law CK (2008) Fundamental combustion properties of H_2/CO mixtures: ignition and flame propagation at elevated pressures. Combust Sci Technol 180(6):1097–1116. https://doi.org/10.1080/00102200801963169
48. Hampp F, Lindstedt RP (2018) Quantification of external enthalpy controlled combustion at unity Damköhler number. In: Runchal AK, Gupta AK, Kushari A, De A, Aggarwal SK (eds) Energy for propulsion : a sustainable technologies approach, green energy and technology. Springer, Singapore, pp 189–215. https://doi.org/10.1007/978-981-10-7473-8_8
49. Geipel P, Goh KHH, Lindstedt RP (2010) Fractal-generated turbulence in opposed jet flows. Flow Turbulence Combust 85(3):397–419. https://doi.org/10.1007/s10494-010-9288-x
50. Elliott C, Vijayakumar V, Zink W, Hansen R (2007) National instruments LabVIEW: a programming environment for laboratory automation and measurement. J Assoc for Laboratory Automation 12(1):17–24. https://doi.org/10.1016/j.jala.2006.07.012

51. Yu X, Zha K, Luo X, Taraza D, Jansons M (2013) Simulation and experimental measurement of CO_2*, OH* and CH_2O* chemiluminescence from an optical diesel engine fueled with n-Heptane. In: 11th international conference on engines & vehicles, SAE International. https://doi.org/10.4271/2013-24-0010
52. García-Armingol T, Ballester J (2014) Flame chemiluminescence in premixed combustion of hydrogen-enriched fuels. Int J Hydrogen Energy 39(21):11299–11307. https://doi.org/10.1016/j.ijhydene.2014.05.109
53. García-Armingol T, Ballester J (2014) Influence of fuel composition on chemiluminescence emission in premixed flames of $CH_4/CO_2/H_2/CO$ blends. Int J Hydrogen Energy 39(35):20255–20265. https://doi.org/10.1016/j.ijhydene.2014.10.039
54. Hwang W, Dec J, Sjöberg M (2008) Spectroscopic and chemical-kinetic analysis of the phases of HCCI autoignition and combustion for single- and two-stage ignition fuels. Combust Flame 154(3):387–409. https://doi.org/10.1016/j.combustflame.2008.03.019
55. Sheinson RS, Williams FW (1973) Chemiluminescence spectra from cool and blue flames: electronically excited formaldehyde. Combust Flame 21(2):221–230. https://doi.org/10.1016/S0010-2180(73)80026-4
56. Lindstedt RP, Ozarovsky HC (2005) Joint scalar transported PDF modeling of non-piloted turbulent diffusion flames. Combust Flame 143(4):471–490. https://doi.org/10.1016/j.combustflame.2005.08.030
57. Lindstedt RP, Louloudi SA (2005) Joint-scalar transported PDF modeling of soot formation and oxidation. Proc Comb Inst 30(1):775–783. https://doi.org/10.1016/j.proci.2004.08.080
58. Speziale CG, Sarkar S, Gatski TB (1991) Modelling the pressure–strain correlation of turbulence: an invariant dynamical systems approach. J Fluid Mech 227:245. https://doi.org/10.1017/S0022112091000101
59. Pope SB (1985) PDF methods for turbulent reactive flows. Prog Energy Combust Sci 11(2):119–192. https://doi.org/10.1016/0360-1285(85)90002-4
60. Janicka J, Kolbe W, Kollmann W (1979) Closure of the transport equation for the probability density function of turbulent scalar fields. J Non-Equilib Thermodyn 4(1):47–66. https://doi.org/10.1515/jnet.1979.4.1.47
61. Grosshandler WL (1993) RADCAL – a narrow-band model for radiation calculations in a combustion environment. Technical Report NBS TN 1402, National Bureau of Standards, Gaithersburg, MD. https://doi.org/10.6028/NIST.TN.1402
62. Lindstedt RP, Waldheim BBO (2013) Modeling of soot particle size distributions in premixed stagnation flow flames. Proc Comb Inst 34(1):1861–1868. https://doi.org/10.1016/j.proci.2012.05.047
63. Sutherland JW, Michael JV, Pirraglia AN, Nesbitt FL, Klemm RB (1988) Rate constant for the reaction of $O(^3P)$ with H_2 by the flash photolysis–shock tube and flash photolysis–resonance fluorescence techniques; 504 $K \leq T \leq 2495$ K. Symp (Int) Combust 21(1):929–941. https://doi.org/10.1016/S0082-0784(88)80325-4
64. Garcia CE (2018) Characterization of nanoparticles generated in reacting flows. PhD thesis, Imperial College, London
65. MATLAB and Statistics Toolbox 8.3 (2014) The MathWorks Inc. Natick, Massachusetts, United States
66. Bahrini C, Herbinet O, Glaude PA, Schoemaecker C, Fittschen C, Battin-Leclerc F (2012) Quantification of hydrogen peroxide during the low-temperature oxidation of alkanes. J Am Chem Soc 134(29):11944–11947. https://doi.org/10.1021/ja305200h
67. Hall JM, Petersen EL (2006) An optimized kinetics model for OH chemiluminescence at high temperatures and atmospheric pressures. Int J Chem Kinet 38(12):714–724. https://doi.org/10.1002/kin.20196
68. Nori V, Seitzman J (2008) Evaluation of Chemiluminescence as a Combustion Diagnostic Under Varying Operating Conditions. In: 46th AIAA aerospace sciences meeting and exhibit, American Institute of Aeronautics and Astronautics, Aerospace Sciences Meetings https://arc.aiaa.org/doi/10.2514/6.2008-953

69. Spadaccini LJ, Colket MB (1994) Ignition delay characteristics of methane fuels. Prog Energy Combust Sci 20(5):431–460. https://doi.org/10.1016/0360-1285(94)90011-6
70. Yoo SW, Law CK, Tse SD (2002) Chemiluminescent OH* and CH* flame structure and aerodynamic scaling of weakly buoyant, nearly spherical diffusion flames. Proc Comb Inst 29(2):1663–1670. https://doi.org/10.1016/S1540-7489(02)80204-8

Emissions from HEFA Fuelled Gas Turbine Combustors

H. Fujiwara, S. Nakaya, M. Tsue and K. Okai

Keywords Hydro-treated ester and fatty acid (HEFA) · Fuel property · O-ring testing · Mode analysis · Combustion test · Particulate matter

Nomenclature

ANA	All Nippon Airways
ASTM	American Society for Testing and Materials
CCS	Carbon dioxide capture and storage
CLD	Chemiluminescence detector
CO	Carbon monoxide
CO_2	Carbon dioxide
EI	Emission Index
FAME	Fatty and methyl ester fuel
FT-SPK	Fischer-Tropsch Synthetic Paraffin Kerosene
HEFA	Hydro-treated Ester and Fatty Acid
IATA	International Air Transport Association
ICA	Independent Component Analysis
IRHD	International Rubber Hardness Degrees

H. Fujiwara (✉) · K. Okai
Propulsion Research Unit, Aeronautical Research Directorate,
Japan Aerospace Exploration Agency, 7-44-1 Jindaiji-Higashi,
Chofu, Tokyo 182-8522, Japan
e-mail: fujiwara.hitoshi@jaxa.jp

S. Nakaya · M. Tsue
Department of Aeronautics and Astronautics, Graduate School of Engineering,
The University of Tokyo, 7-3-1 Hongo, Bunkyo, Tokyo 113-8656, Japan

A. K. Gupta et al. (eds.), *Innovations in Sustainable Energy and Cleaner Environment*, Green Energy and Technology,
https://doi.org/10.1007/978-981-13-9012-8_16

JAXA	Japan Aerospace Exploration Agency
KAKENHI	Grants-in-Aid for Scientific Research (Name from Japanese)
MHPS	Mitsubishi-Hitachi Power Systems
NCA	Nippon Cargo Airlines
NDIR	Nondispersive infrared detectors
NEDO	New Energy Industrial Technology Development Organization
NO_x	Nitrogen oxides
nvPM	Non-volatile particulate matter
PAH	Polycyclic aromatic hydrocarbon
PASS	Photo acoustic soot sensor
PetroOxy	Rapid Small-Scale Oxidation Tester made by Anton Paar ProveTec
POD	Proper Orthogonal Decomposition
PVC	Processing Vortex Core
Re	Reynolds number
RQL	Rich burn Quick quench and Lean burn
SOFC	Solid Oxide Fuel Cell
SPK	Synthetic Paraffin Kerosene
TEC	Toyo Engineering Cooperation
THC	Total hydrocarbons
UOP	Universal Oil Products

1 Introduction

Global climate change due to the rapid increase in CO_2 emissions especially caused by aviation is one of the critical issues that can be solved through international collaborations. Though the amount of CO_2 emission from aviation is only around two per cent of the total CO_2 emission, it is important to start any possible measures now to suppress aviation CO_2 emission. Recent rapid growth of aviation transportation has caused many activities that could curtail CO_2 reduction form aviation. Much efforts are in progress in other fields, such as automotive and power generation, wherein electric and hybrid vehicles are on the rise and carbon dioxide capture and storage (CCS) has already been installed in some plants. It should be noted that more than 90% of CO_2 emissions from commercial aircraft operations are generated by large aircraft, which indicates to pursue research to reduce commercial aircraft emissions with a focus on technology applicable to large commercial aircraft [1].

In aviation, alternative fuels are considered one of the important options to suppress CO_2 emission, while their specifications are strictly defined in ASTM D7566 "Standard Specification for Aviation Turbine Fuel Containing Synthesized Hydrocarbons" [2]. Its Annexes define not only the chemical and physical properties but also the manufacturing process of those fuels, which is of prime importance from the aviation safety point of view.

In Japan, R&D towards sustainable energy have been conducted quite actively due to the limited amount of available petroleum resources and existing potential of using biomass and algae. However, up to now, there is no plant which can supply enough bio-derived alternative fuel for aviation. There are only three examples of actual utilization of certified alternative aviation fuel for use in flights of large commercial aircraft. The first one is Japan Airline's flight in 2008, which was used as a part of the examination to certify HEFA (Bio-derived synthetic paraffinic kerosene) [3]. The second and third flights are both ferry flights of B787 for All Nippon Airways (ANA) [4] and B747-8F for Nippon Cargo Airlines (NCA) [5] in 2012. The fuel for the first flight test by JAL was shipped from the USA to Japan, while fuels for the second and third ferry flights were refuelled in the USA.

In Japan, discussions and investigations of bio-derived aviation alternative fuels started in 2010. At JAXA, the impact of introducing biofuels has been investigated [6], and combustion testing begun with available (not certified and general) bio-fuel (Fatty acid methyl ester, FAME fuel) in 2011 [7]. As expected, the fuel had higher CO emissions at low-load conditions due to low flame temperature [8]. In 2012, the University of Tokyo, Airlines and Airports in Japan and related institutes formed the initiatives for Next-generation Aviation Fuels (INAF) and published a report on a roadmap for establishing a supply chain for the next-generation aviation fuels in 2015 [9]. In the same year, the Ministry of Economy, Trade and Industry of Japan established a process study committee for clarifying problems to be considered and planning the future process in realizing flights using biojet fuel by 2020 to commemorate the Summer Olympic Games and Paralympic Games in Tokyo [10]. Inspired from these discussions and proposals, three major active projects embarked that had a focus on the production of bio-derived aviation fuels. In 2015, Euglena and their partners announced their "Made-in-Japan Biofuels Project" to produce and supply bio jet/diesel fuels by 2020 [11]. In 2017, IHI and Kobe University started a project to commercialize bio jet fuel production by developing an integrated production process of microalgae-based biofuel supported by New Energy Industrial Technology Development Organization (NEDO) [12]. Also in 2017, Mitsubishi-Hitachi Power System (MHPS) with partners (Toyo Engineering Cooperation (TEC), Chubu Electric Power and JAXA) started a project to conduct a pilot-scale plant testing on a Fischer-Tropsch Synthetic Paraffin Kerosene (FT-SPK) fuel production derived from lignocellulosic biomass supported by NEDO [13]. For this project, JAXA participated to conduct the final combustion tests of the product. In addition, the University of Tokyo and JAXA conducted a fundamental research on biofuels combustion supported by KAKENHI.

This article presents the results of investigations on aviation certified biofuels to understand the limitations and potentials of bio-derived aviation fuels as a contribution towards more environmentally friendly aviation.

The alternative turbine fuel used in this study was hydro-treated ester and fatty acid (HEFA) made from tallow fat, provided by Honeywell UOP/Nikki Universal. The manufacturing process is specified in Annex 2 of ASTM D7566. The HEFA fuel was available for commercial flights, as long as the blending ratio did not exceed 50 vol% as specified in Annex 2.

2 Property Measurement and Atmospheric Pressure Testing

In this chapter, different methods are presented which lead to the properties of the analysed fuels. For that reason, a series of basic measurement and testing were conducted, which were different in pressure and temperature compared to actual conditions in combustors.

2.1 Chemical Analysis of Hydro-Treated Ester and Fatty Acid (HEFA)

Specifications for HEFA are provided in ASTM D7566 Annex 2 [14]. To understand the similarities and differences of HEFA fuel compared to the baseline Jet A1 fuel, the properties of HEFA fuel were analysed.

2.2 O-Ring Sink Test

ASTM D7566 specification limits the blending ratio of HEFA to Jet A1 to the max. 50% by volume. There are some technical rationales for these limitations, among which, the most remarkable one being that certain amount of aromatic compounds are necessary to eliminate the risk of fuel leakage through sealing rings in piping. ASTM D7566 specifies that the resultant blended fuel should contain at least 8% of aromatics compound by volume, though it includes a note that as yet no technical rationale prevails for limiting the value of 8% and that it should be studied in the future. The impact of HEFA on O-rings on the fuel passage was examined through a 70-h sink test. Fluorocarbon O-rings, for example, Viton, and a Nitrile rubber O-ring were selected tests wherein they were both sunk in HEFA and Jet A1 fuel for 70 h.

2.3 Oxidation Test

Compared to certified fuels for aviation, bio-derived alternative fuels such as fatty acid methyl ester (FAME) were much affected by the oxidation without adequate treatment [14]. A long-time oxidation test for both Jet A1 and HEFA was also conducted. The test method was a typical Rancimat method with a temperature of 100 °C and a duration time of 12 h each.

2.4 Basic Combustor Testing for Mode Analysis

For investigating the basic swirling flow of a simple combustor test, mode analysis with proper orthogonal decomposition (POD) and independent component analysis (ICA) was performed [15]. These analyses can separate out important structures of combustion with swirl flows.

Figure 1 shows a schematic of the experimental apparatus for the mode analysis. Spray combustion testing with the swirled flow was realized using a Parker-Hannifin type double-swirl burner [8]. CH* Chemiluminescence of the flame was taken using high-speed cameras (Vision Research Inc., Phantom Micro LC) at 2000 fps; 2000 images of the test data were used to analyse the combustion behaviour with POD and ICA. Test conditions are shown in Table 1.

3 Combustor Testing and Conditions

Two types of single-injector combustor were used for testing at JAXA AP7 medium-pressure test rig, with the maximum inlet temperature, pressure and air-mass flow of 1000 K, 10 bar and 2 kg/s, respectively. Figure 2 shows the schematic overview of the combustion test rig while Fig. 3 shows a photograph of the combustion test rig.

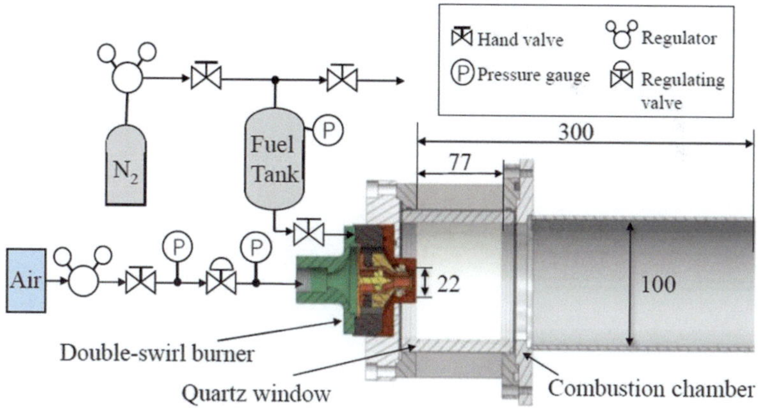

Fig. 1 Schematic of the experimental apparatus for modal analysis

Table 1 Combustion test condition for mode analysis

Case	Air flow rate (g/s)	Swirl number	Fuel
1	15	0.86	Jet A1
2	15	0.86	HEFA

This test rig is usually used for the demonstration of new combustor concepts and for the development of innovative measurement technologies. Crystal glass windows were installed in both the facility pressure casing and the combustor liner so that the high-pressure combustion phenomena can directly be seen from the outside of the casing (see Figs. 4 and 5). An exhaust gas sample probe with eight ψ0.8 mm sampling holes was located at the exit of the combustor liner (Fig. 6).

The samples from the exhaust gas were led to the measurement instruments through a stainless steel tube connected to a valve to control the mass flow and temperature of the sample gas (see Figs. 7 and 8). NO_x concentration was measured using a chemiluminescence detector (CLD); CO and CO_2 concentrations were measured through nondispersive infrared detectors (NDIR), and total hydro carbon (THC) concentration was measured using a flame ionization detector (FID), respectively, Horiba MEXA 7000. Non-volatile particle matter (nvPM) mass concentration was measured using a photo acoustic soot sensor (PASS), AVL MSS 483.

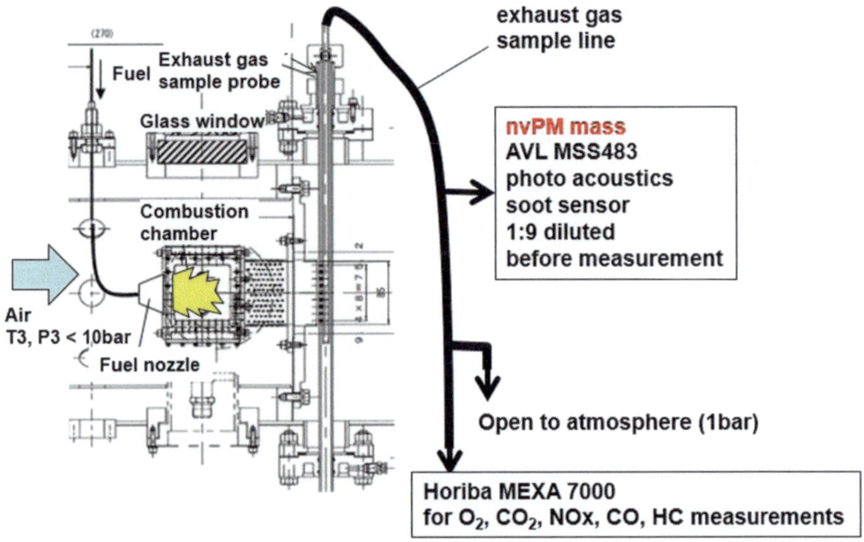

Fig. 2 Overview of the combustion test rig

Fig. 3 Combustion test rig at JAXA AP7

Fig. 4 Photograph of pressure casing with windows

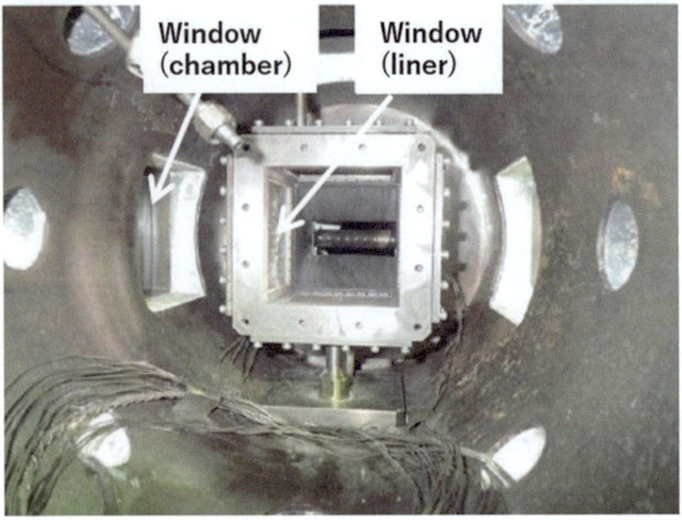

Fig. 5 View inside the pressure casing (viewed from upstream)

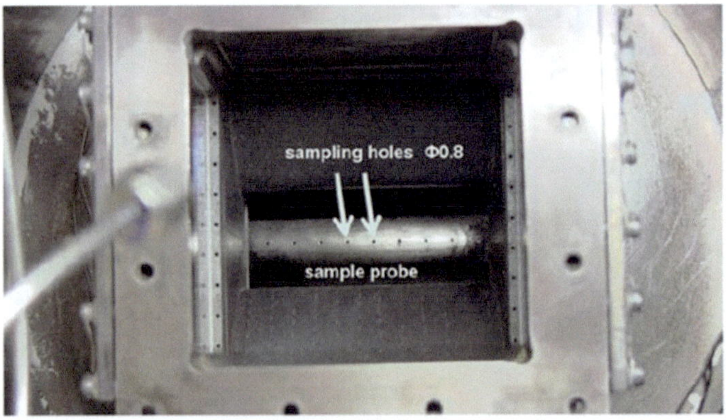

Fig. 6 Exhaust gas sample probe (viewed from upstream)

Fig. 7 Exhaust gas sample line

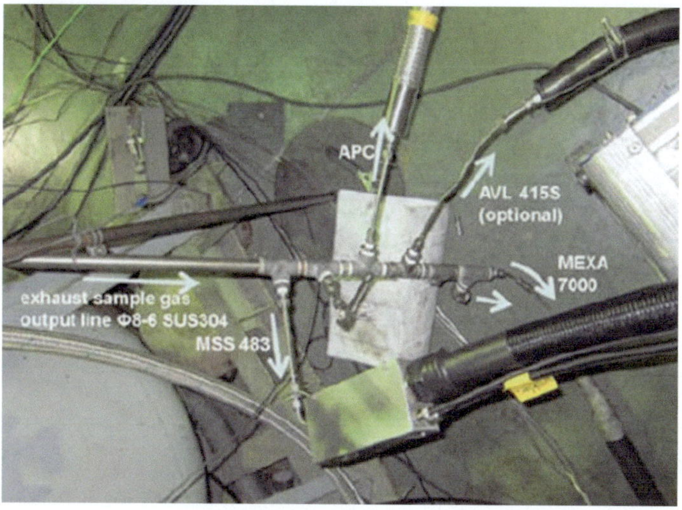

Fig. 8 Exhaust gas distribution connected to measuring instruments

Fig. 9 Test chamber for RQL combustor

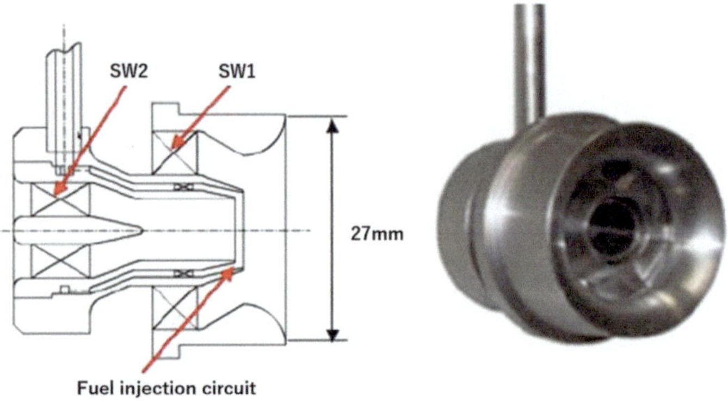

Fig. 10 Parker-Hannifin type air-blast fuel nozzle

One of the tested combustors (case 1) was a RQL combustor [16] as shown in Fig. 9, wherein 10% of total air flow entered through the upstream Parker-Hannifin type air-blast fuel nozzle (Fig. 10), while the remaining 90% of total air entered through the air holes located on the combustor liner (seen in Fig. 9 as combustion/dilution air holes). The other combustor (case 2) was a concentric lean burn burner [17] as seen in Fig. 11, which consisted of a pilot diffusion burner located at the centre and a lean premix main burner surrounding it. Only the pilot burner was fuelled at low-load inlet air condition, while at high load conditions both pilot and main burner were fuelled.

Combustion tests were performed for the conditions shown in Table 2.

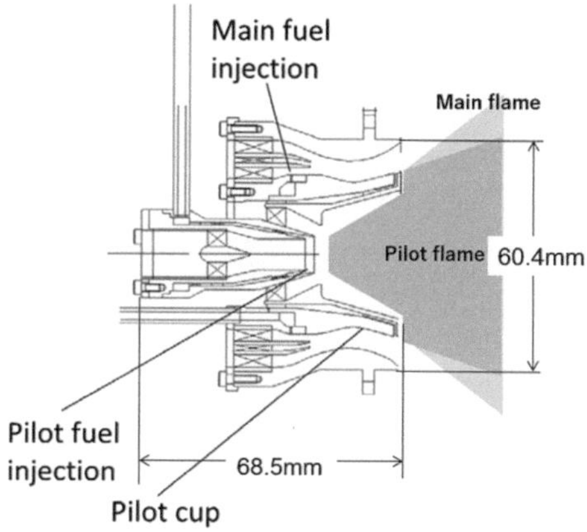

Fig. 11 Concentric lean burn burner

Table 2 Combustion rig test conditions

Case	Combustor	Inlet temperature (K)	Inlet pressure (kPa)	Pressure loss ratio (%)	Fuel injection type
1-1	RQL	500	500	4.0	N/A
1-2		550	800	4.0	N/A
2-1	Lean burn	450	360	4.0	Pilot only
2-2		760	700	4.0	Pilot + main

4 Results on Property Measurements and Their Considerations

Table 3 shows a comparison of the major components of D7566 specifications and the results of the analysis for HEFA, Jet A1 and their 50:50 blend. Numerals with bold type indicate that the analysed fuel did not satisfy the D7566 specification. Several considerations on the analysis are summarized below:

1. HEFA showed a very low freezing point, considered to be caused by branched paraffin.
2. Dynamic viscosity and surface tension of HEFA and Jet A1 differ only little from each other. Though, the slight differences might affect atomization behaviour.
3. The density of HEFA was lower than that of Jet A1. The density of Jet A1 refined in Japan tends to be lower than those from other countries that could possibly cause in the density of fuel blend to be outside of the required specification. To meet the specification, a slightly lower ratio of HEFA (e.g. 30%) would be acceptable in Japan.
4. The lubricity of HEFA is lower than of Jet A1 due to lack of sulphur content. The refined Jet A1 fuel in Japan tends to have smaller amount of sulphur content,

Table 3 Chemical and physical analysis of the fuels

Properties	HEFA	D7566 Table A2.1-2	Jet A1	Jet A1: HEFA = 50:50	D7566 Table 1	Test method
Freezing point (°C)	−58.5	<−40	−48.0	−51.5	<−47	JIS K 2276
Dynamic viscosity (mm^2/s) @15 °C	1.963	N/A	1.655	1.786	<8 (−20 °C)	JIS K 2283
Surface tension (mN/m)	22.5	N/A	23.6	23.1	N/A	JIS K 2241
Density (g/cm^3) @15 °C	0.7554	0.73–0.77	0.7886	**0.7732**	0.775–0.840	JIS K 2249-1
Net heat of combustion (MJ/kg)	44.14	N/A	43.38	43.73	>42.8	JIS K 2279
Physical distillation						JIS K 2254
Initial boiling point (°C)	146.5	N/A	148.5	148.5	N/A	
10% recovered temp. (°C)	165.0	<205	164.5	165.0	<205	
50% recovered temp. (°C)	208.0	Report	193.5	198.5	Report	
90% recovered temp. (°C)	253.5	Report	237.5	247.0	Report	
Final boiling point (°C)	269.0	<300	259.0	262.0	<300	
Lubricity (mm)	1.04	N/A	0.83	**0.87**	<0.85	ASTM D5001
Carbon (mass%)	84.7	N/A	86.1	85.4	N/A	JPI-5S-65
Hydrogen (mass%)	15.2	N/A	13.8	14.5	N/A	

(continued)

Table 3 (continued)

Properties	HEFA	D7566 Table A2.1-2	Jet A1	Jet A1: HEFA = 50:50	D7566 Table 1	Test method
Sulphur (mass%)	<0.0001	<0.0015	0.0006	0.0003	<0.3	JIS K 2541-6
Composition						ASTM D1319
Paraffin (vol%)	98.4	Report	79.1	89.1	N/A	
Olefin (vol%)	0.9	N/A	3.0	1.6	N/A	
Total aromatics (vol%)	0.7	N/A	17.9	9.3	8–25	
Aromatics						ASTM D6379
Benzenes (vol%)	<0.1	N/A	19.1	10.2	N/A	
(mass%)	<0.1	N/A	21.3	11.6	N/A	
Naphthalene (vol%)	<0.1	N/A	0.1	0.1	N/A	
(mass%)	<0.1	N/A	0.2	0.1	N/A	
Total aromatics (vol%)	<0.1	N/A	19.2	10.3	8.4–26.5	
(mass%)	0.1	N/A	21.5	11.8	N/A	

which might cause insufficient lubricity of the 50:50 blend fuel specified in D7566.

5. The most remarkable difference between the two fuels (Jet A1 and HEFA) is its aromatics content. HEFA contains little aromatic compounds while Jet A1 contains about 20% of aromatic compounds on volume which turns out to mainly consist of benzenes and its derivatives with little polycyclic aromatic hydrocarbons (PAHs).

As stated before, due to lack of aromatics content in HEFA, sealing characteristics with O-rings are of concern. Table 4 shows a comparison of hardness, tensile strength and elongation of the two O-rings before and after 70 h sink test for the investigated fuels. The result shows that the effect of the fuels on the fluorocarbon O-ring is small. While it was fairly large on the Nitrile rubber O-ring, wherein volumetric expansion with a loss of both tensile strength and hardness could be observed. On the other hand, the effect of HEFA on Nitrile rubber O-ring was smaller than that of Jet A1. The difference causes the limited amount (ratio) of HEFA, which can be approved by using conventional fuel feeding lines.

Table 4 Effect of fuel change on O-rings

Specification		Fluorocarbon	Nitrile rubber
		M83248/1	MS29513
Initial value	Hardness IRHD-M	71	62
	Tensile strength (MPa)	14.7	14.1
	Elongation (%)	199	329
Jet A1 70 h sink test result (relative change)	Hardness (%)	+0	−12
	Tensile strength (%)	+2	−49
	Elongation (%)	+4	−26
	Volume expansion (%)	+0.3	+24.8
HEFA 70 h sink test result (relative change)	Hardness (%)	+0	−5
	Tensile strength (%)	−4	−16
	Elongation (%)	−4	−6
	Volume expansion (%)	+0.4	+6.5

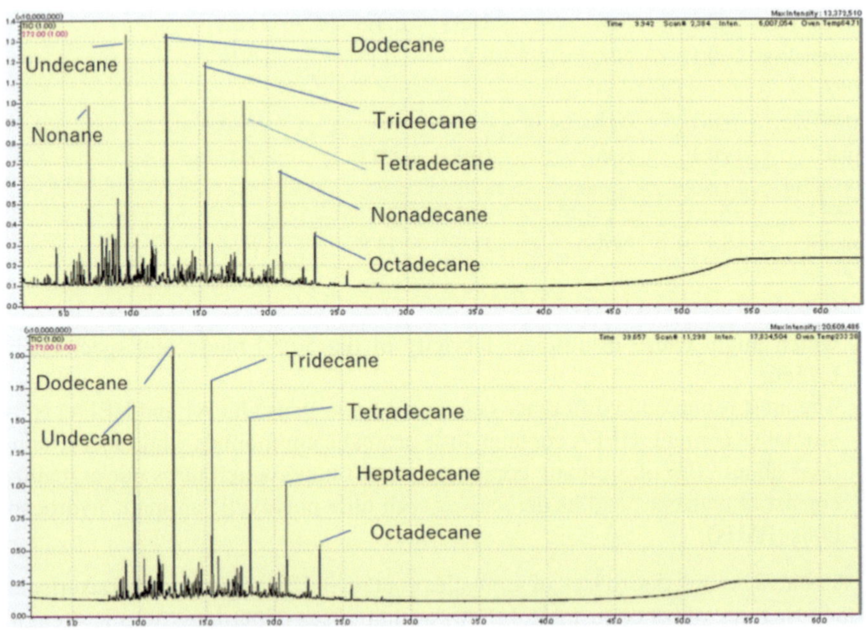

Fig. 12 Oxidation test for Jet A1, before testing (top) and after testing (bottom)

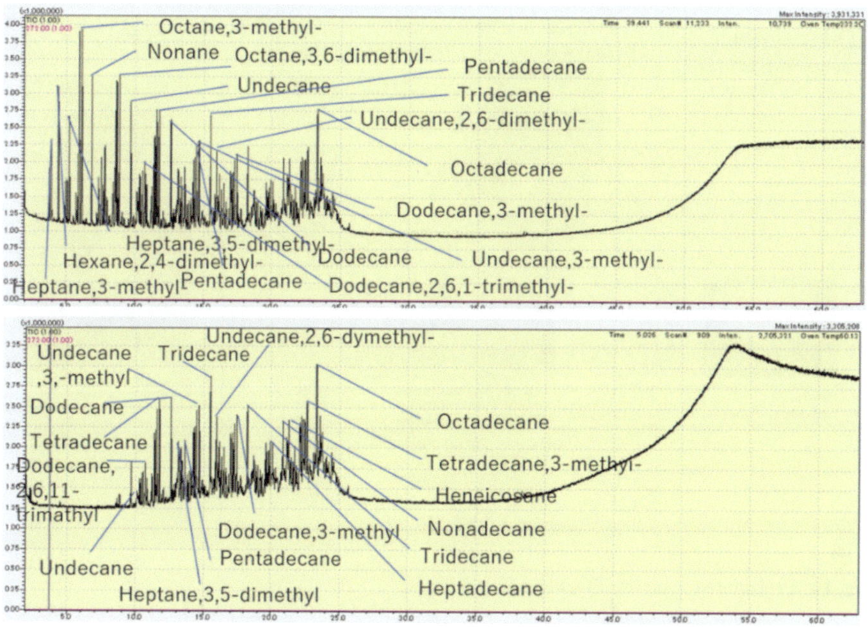

Fig. 13 Oxidation test for HEFA, before testing (top) and after testing (bottom)

Oxidation stability tests were conducted both for Jet A1 and HEFA fuel. Figures 12 and 13 show the results of the component analysis for Jet A1 and HEFA, respectively. Both fuels were primarily composed of similar paraffin and with additives, thereby there was almost no composition change due to oxidation. Amara et al. [18] have demonstrated oxidation stability testing using PetroOxy for both Jet A1 and HEFA (without additives) and concluded that HEFA was less stable than Jet A1 due to the difference in chemical composition (HEFA does not have cyclic compounds). The present test result showed almost no difference in stability between the two fuels, whereas Amara et al. [18] displayed clear differences by showing less stability for HEFA than for Jet A1. The test condition for the latter case was at higher pressure (7 bar) and without additives in the case of HEFA fuel. The present results are both certified and obtained from the market therefore had anti-oxidant additives. Among the test condition, the storage stability (oxidation stability) for Jet A1 and HEFA was quite similar.

To understand the effect of fuel on the atomization process in a combustor, a combustion test and mode analysis were conducted. Figure 14 illustrates the procedure of mode analysis (POD and ICA) with the data taken in the present study. Our previous report showed that the effect of the change of swirl number (swirl intensity) and airflow rate can be well illustrated with these mode analyses for the Jet A1 fuel case. As seen from Fig. 14, there were four representative modes from the present ICA (three modes from processing vortex core (PVC) [20] and one from

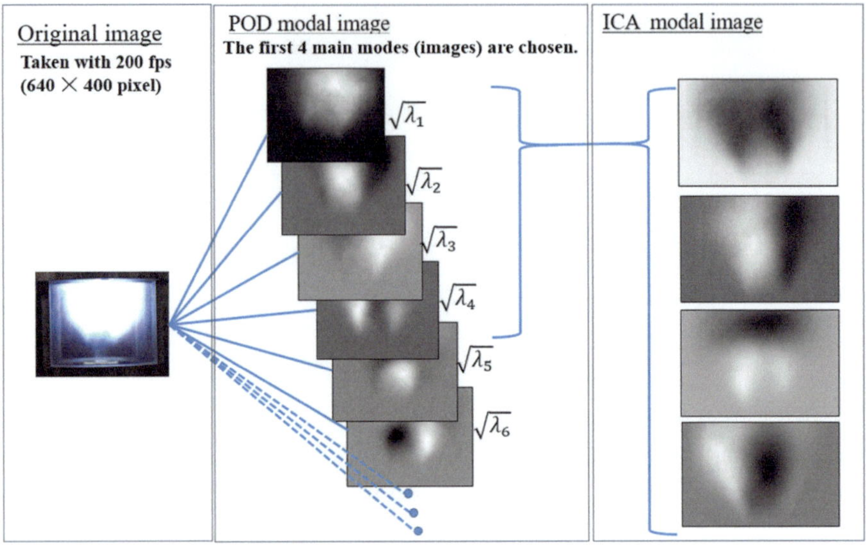

Fig. 14 Procedure to acquire POD and ICA images

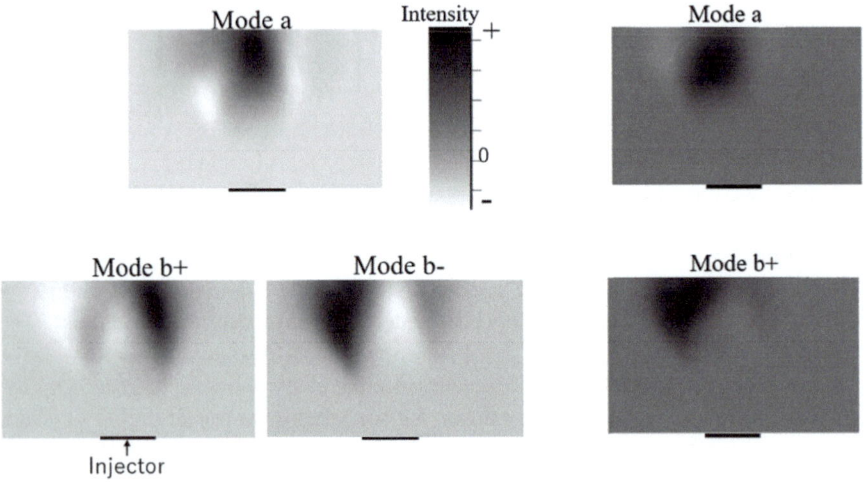

Fig. 15 Comparison of ICA images for Jet A1(left) and HEFA (right)

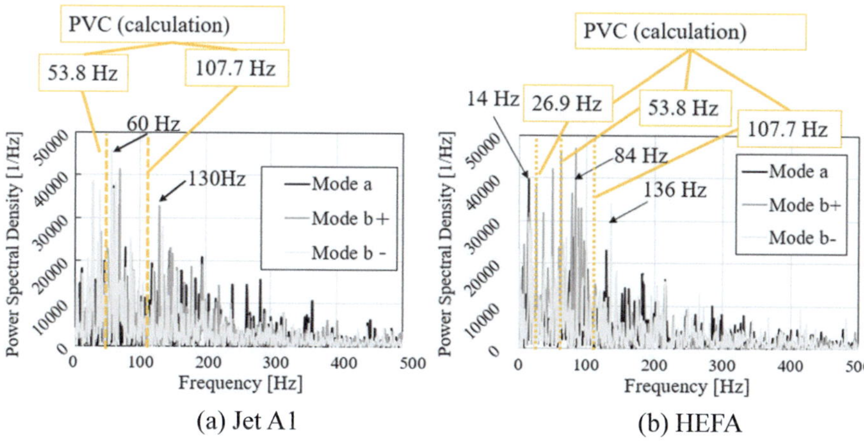

Fig. 16 Power spectral density

longitudinal acoustic instability). In the present study, the condition of the air flow rate and swirl number remained the same, and only the fuel was changed and mode analysis data compared. Figure 15 shows the representative ICA images obtained (mode a, b+, b−) for both fuels (Jet A1 and HEFA). In the present case, three modes from PVC were observed. Mode a is a quasi-uniform flow structure, and modes b+ and b− are coupled PVC conditions. For both of the fuel cases, similar modes were observed. To compare the two fuel cases, power spectral density with time variation for the observed modes is presented in Fig. 16. From the Jet A1 (Fig. 16a) data, mode a and mode b have strong peak points with frequencies around 60 and 130 Hz, respectively. Simple PVC calculation revealed the primary frequency of 53.8 Hz (Re = 40,000 in the present experiment). The frequency and its double frequency represent well the observed peak frequencies, indicating that the swirl-flow spray combustion case had a dominance of PVC. For the HEFA case (Fig. 16b) contrary to the Jet A1 case (Fig. 16a) peaks of each mode were not so clear, still the peak frequency was almost identical to that of the Jet A1 case. To understand the cause of this difference, time-averaged CH* chemiluminescence images for the two fuel cases were compared and are shown in Fig. 17. A comparison of the images showed HEFA to reveal a slightly weaker intensity, and the distance from the flame rib to the nozzle exit was longer than for the Jet A1 case. As seen in Table 2, Jet A1 and HEFA resemble in physical property in general, but there are several differences. HEFA has a smaller density and slightly less surface tension as seen from Table 2. These values affect the atomization characteristics. For example, classical representation of the disintegration of liquid sheets was given by the average droplet diameter of conical liquid sheets, which reveals a function of a characteristic wavelength on those liquid sheets and a function of density and surface tension of the fuel [21]. Smaller density and surface tension of HEFA compared with Jet A1 resulted in larger droplets (less enhanced atomization). The differences in density and surface tension were small,

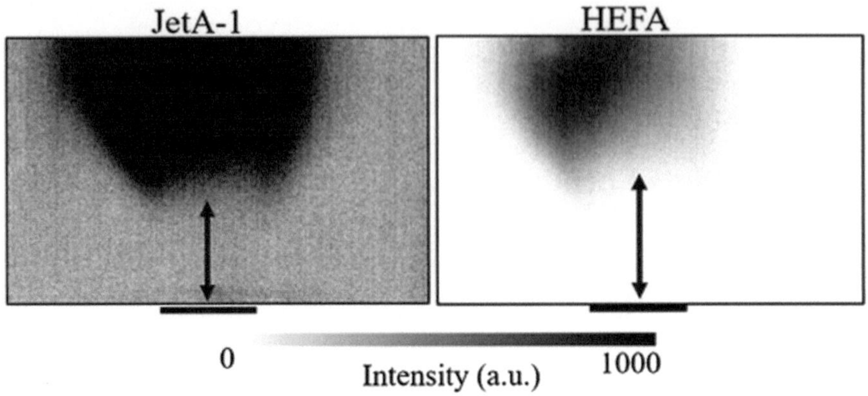

Fig. 17 Time-averaged image of CH* chemiluminescence

but the resultant numbers seen in Fig. 16 suggest that there would be cases with a high content of HEFA in HEFA/Jet A1 mixture, which tends to deteriorate atomization behaviour when using the same atomizer. Although the atomization for the same experiment looked different for the two fuels, the measured combustion efficiency was almost the same.

5 Results of Combustor Testing at Real Pressure and Temperature Conditions and Their Considerations

Combustion rig tests were conducted using two different combustors (case 1: rich burn quick quench lean burn, RQL, Combustor and case 2: Concentric Lean Burn Combustor). In this chapter, the results obtained from each combustor are presented.

5.1 Case 1: Combustion Testing with RQL Combustor[1]

Two inlet air conditions (case 1-1 and 1-2) shown in Table 2 were tested with total air/fuel ratios ranged from about 50 to 150. Jet A1 and HEFA were blended before supplying to the fuel nozzle through a T-junction.

Combustion tests with air flow condition case 1-1, with each having 100% HEFA and 100% Jet A1 supplied separately were conducted to investigate the effect of the fuel on the combustion behaviour. Figure 18 shows the non-volatile PM (nvPM), NO_x, CO and THC emission index (in g/kg fuel) as a function of the air/fuel ratio.

[1]From AIAA 2016-4953 [22]; reprinted by permission of the American Institute of Aeronautics and Astronautics, Inc.

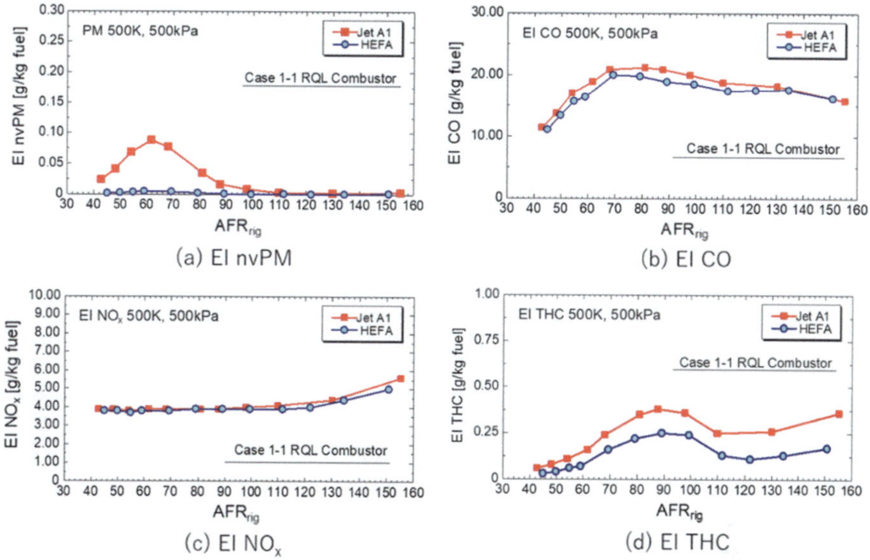

Fig. 18 Combustion test results (case 1-1)

The result indicates that the nvPM emission index was greatly reduced with HEFA while other exhaust gas components were similar to each other.

Figure 18 also shows that the maximum value of nvPM emission index appeared at the condition of air–fuel ratio of approximately 65, wherein 95% reduction of nvPM emission index was observed when using HEFA instead of Jet A1. Figure 19 compares the direct flame images at air–fuel ratio of 65 with HEFA and Jet A1 images captured from outside of the pressure chamber by a usual digital camera without changing any conditions of the camera including the aperture. The images show that HEFA had less flame brightness than that from Jet A1 combustion; this is consistent with the above results of the reduced nvPM emission when using HEFA.

In order to further examine the effect of the blending ratio of the two fuels to the exhaust gas emission index, the blending ratio of HEFA was changed in increments of 10%, keeping the air–fuel ratio at 65. The result is shown in Fig. 20. The results show that the nvPM mass emission index was almost linearly reduced with increase in HEFA blend ratio while NO_x, CO and THC remained fairly constant.

Combustion tests inlet air condition for cases 1-2 at higher air temperature and pressure were also performed in a similar way to case 1-1, and the results are summarized in Figs. 21 and 22. The combustion tests of case 1-2 also showed a similar trend as that shown in case 1-1.

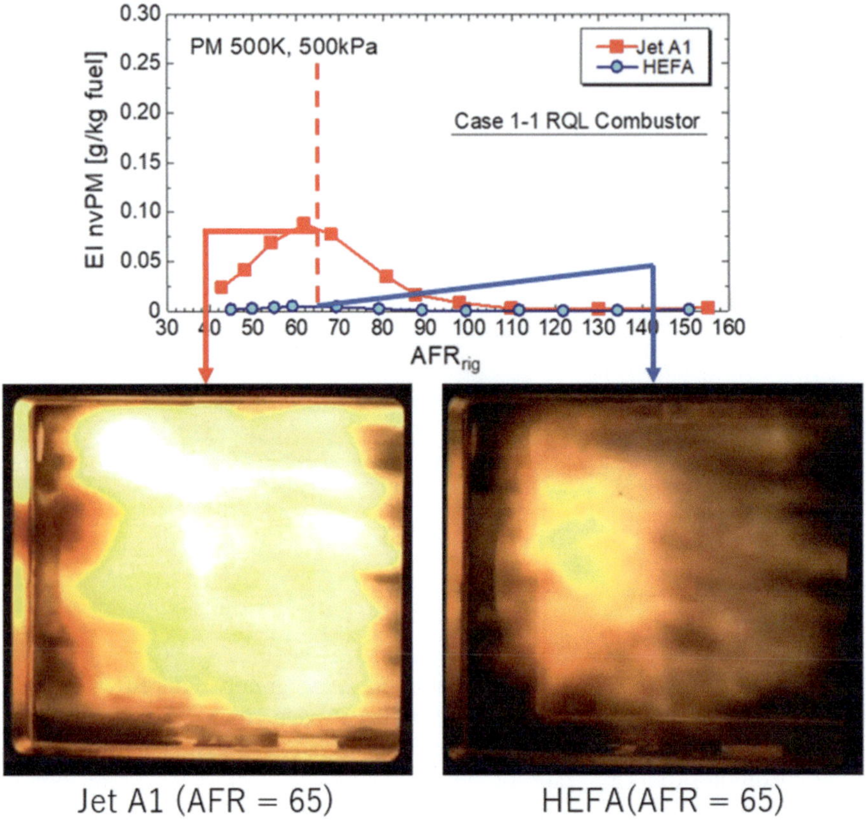

Fig. 19 Comparison of the global flame images (case 1-1, AFR = 65)

5.2 *Case 2: Combustion Testing with Concentric Lean Burn Combustor*[2]

Cases 2-1 and 2-2 were different in fuel injection schedule. Case 2-1 had only pilot fuel injected, which represents low-load condition. Case 2-2 had both pilot and main fuel injected and represents high load condition.

At first, the results of case 2-1 (pilot only) are shown. In this case, air fuel ratio was 100 and direct flame images of the condition with pure fuel (both Jet A1 and HEFA) were compared as shown in Fig. 23. A blue flame was dominant with HEFA, while a yellow flame was dominant with Jet A1. NO_x, THC, CO emission indices and nvPM mass concentration of the exhaust gas were compared and the results shown in Fig. 24. The NO_x, THC and CO showed similarities between Jet A1 and HEFA

[2]From AIAA 2018-1474 [23]; reprinted by permission of the American Institute of Aeronautics and Astronautics, Inc.

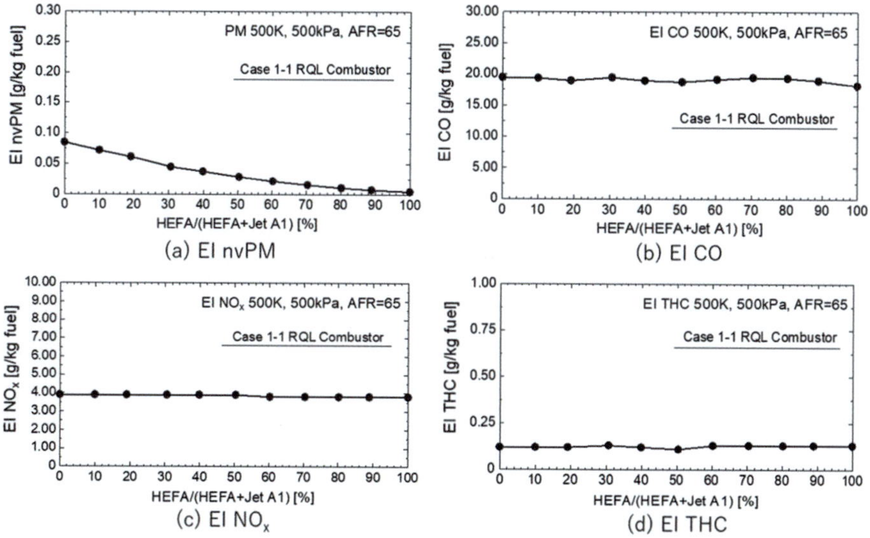

Fig. 20 Combustion test results (case 1-1, AFR = 65, blended fuel)

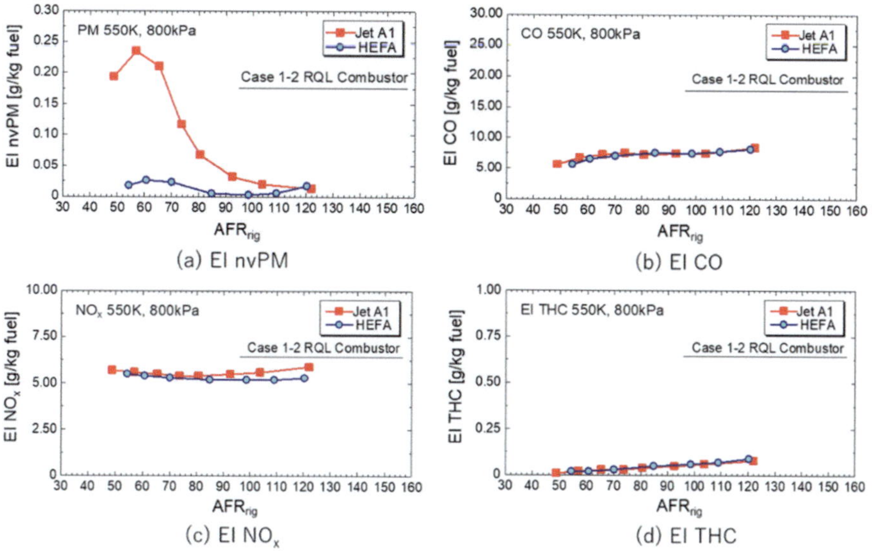

Fig. 21 Combustion test results (case 1-2)

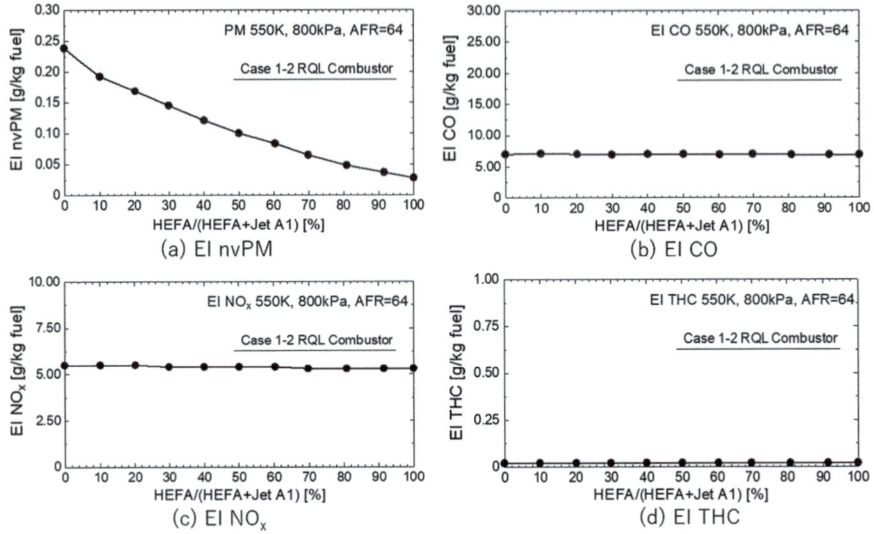

Fig. 22 Combustion test results (case 1-2, AFR = 64, blended fuel)

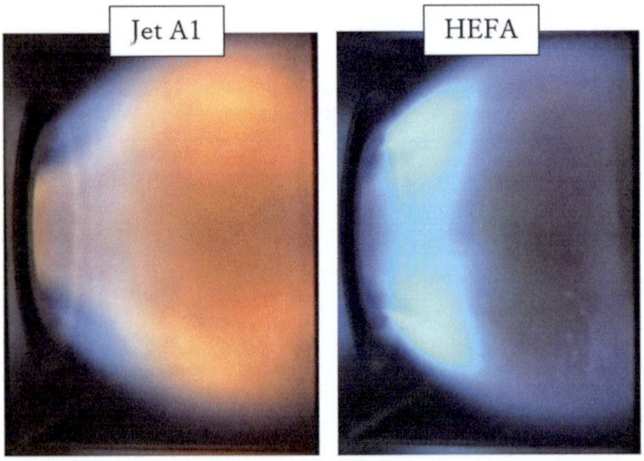

Fig. 23 Comparison of the flame visualization (case 2-1, pilot burner only, AFR = 100)

while nvPM was lower with HEFA. A similar tendency was also observed for the cases 1-1 and 1-2 with RQL combustor given earlier in this chapter.

The result of case 2-2 was different from the result shown earlier. In case 2-2, both pilot and main burner were fuelled.

In the previous study, such a combustion instability was observed and its phenomena were reported, when the same combustor and injector configuration were used [22]. As expected, in the present study with Jet A1 fuel, combustion instability was observed at a pilot fuel ratio of up to 15%.

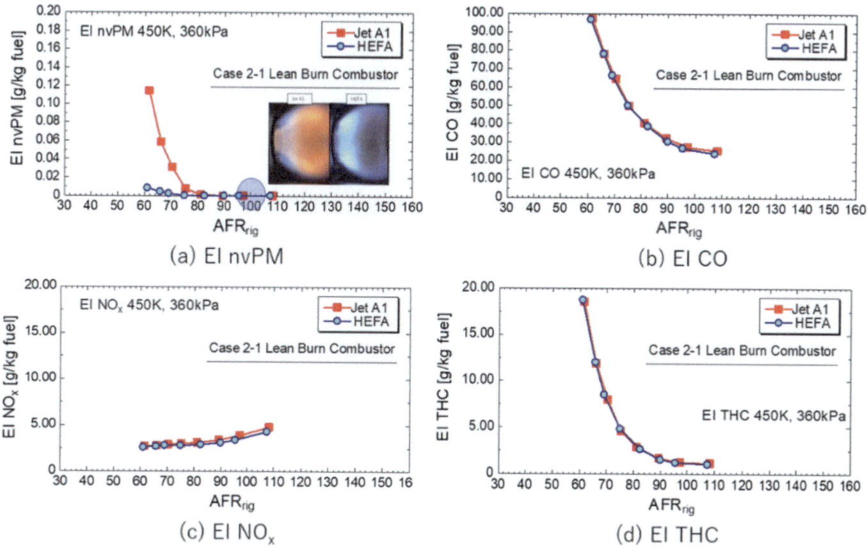

Fig. 24 Combustion test results (case 2-1)

Initially, the fuel injection condition without inducing instability was examined. No combustion instability was observed when the pilot fuel ratio was at 20% for both pure fuel cases (Jet A1 and HEFA). Figure 25 shows a direct comparison of the direct flame images when the pilot fuel ratio was 20% and air–fuel ratio was 35. The results showed that both main and pilot flames are blue when HEFA was used, while the main flame was blue, and the pilot flame was yellow when Jet A1 was used as fuel. In the case of no instability (pilot fuel ratio was 20%), the emission indices were compared in Fig. 26, indicating that NO_x, THC and CO emission were similar to each other. Non-volatile PM was much smaller than in case 2-1, with only the pilot burner operational.

Figure 27 shows a comparison of NO_x emission between the two fuels under the condition of instability occurrence in the Jet A1 case (pilot fuel ratio is 15%). Other emission indices were not properly obtained due to oscillation. Figure 27 shows the stated pressure oscillation, as measured using a pressure transducer located on the liner wall at the condition where air–fuel ratio was 40.7, pilot fuel ratio was 15% and Jet A1 was used as the fuel (top right side of the figure). The upper figure shows time-dependent pressure oscillation while the lower shows a spectrum of oscillation. The figure also shows pressure oscillation of around 550 Hz using Jet A1 fuel. Bottom right of Fig. 27 shows the same pressure oscillation as given above (top right side with Jet A1) at the condition when air–fuel ratio was 38.5, pilot fuel ratio was 15% using HEFA as the fuel. In the HEFA test, no oscillation was observed under any condition as seen from the figure. Although the number of test conditions was limited, the results indicate that the fuel change from Jet A1 to HEFA might affect the combustion instability. In the previous report using the same combustor, a simple

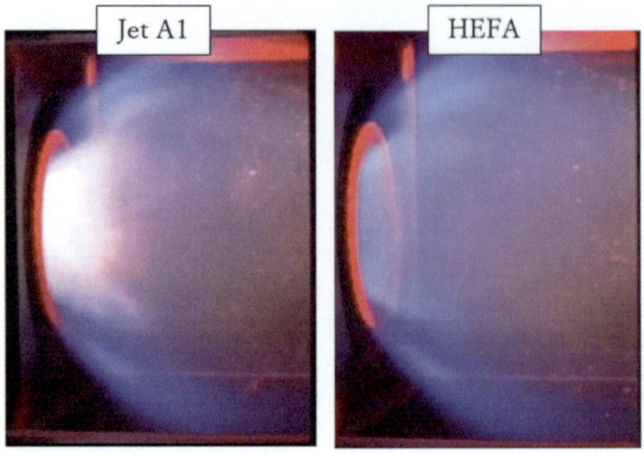

Fig. 25 Comparison of the flame visualization (case 2-2, pilot and main burner, AFR ~ 35, pilot fuel ratio = 20%)

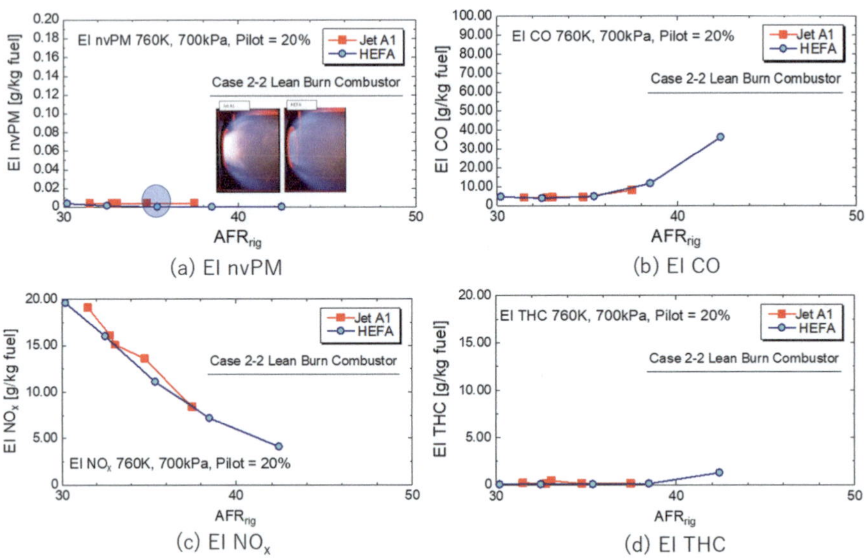

Fig. 26 Combustion test results (case 2-2, pilot fuel ratio = 20%)

acoustic velocity coupling mechanism was proposed as the driving mechanism to generate heat release oscillations which was used to explain the observed oscillation. In the proposed mechanism, there were coupled features inducing the oscillation, including the path of "fuel atomization, evaporation and mixing". With the mode analysis, it was suggested that the slightly different physical properties (density and surface tension) might affect the liquid atomization characteristics for the two fuels.

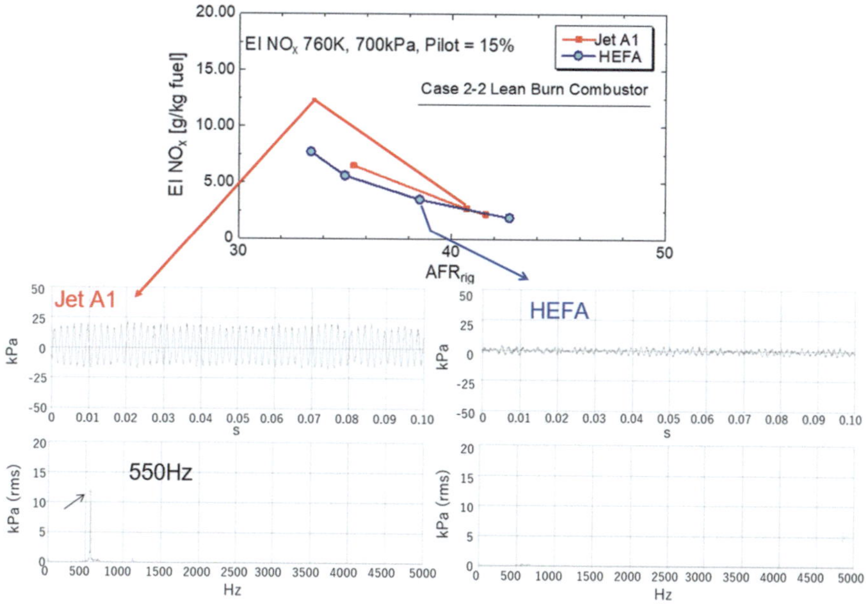

Fig. 27 Pressure oscillation measured (case 2-2, pilot fuel ratio = 15%)

5.3 *Discussion and Future Perspectives*

Combustion test cases in which spray and diffusion flame region are dominant; the effect of fuel change from Jet A1 to HEFA is remarkable especially in soot formation phenomena. The results showed a clear reduction of emission index (EI) nvPM for HEFA and mixture fuels compared to pure Jet A1, and this is attributed largely to the reduced aromatic concentration [8].

The other combustion properties, EI of CO, NO_x and THC were almost unaffected by the fuel change. In case 2-2, an unstable condition exhibited when using Jet A1, but no visible instability occurred when using HEFA at the same condition. This might be from the slight differences in atomization behaviour due to different properties, such as density and surface tension. This suggests that the usage of pure HEFA might require a change of fuel injection design.

By introducing HEFA and bio-derived fuels to aviation, the results showed carbon emission reduction that shifted towards carbon neutrality. If the upper limit of alternative fuels in mixtures for aviation propulsion is increased, the contribution to reduce carbon emission would be enhanced. From these investigations, major issues associated with increasing the ratio of e.g. HEFA in the fuel mixture above 50% would be to cause in swelling of O-rings, a different energy content due to different density. These drawbacks seem technically solvable in the medium-term perspective.

Various conceptual and experimental approaches for reduced fuel consumption are in progress that are partly due to the existing target for reducing CO_2 emissions as per the International Air Transport Association (IATA) roadmap [23]. According to the IATA technology roadmap, CO_2 emissions must reach half of 2005s baseline by 2050. To accomplish this task, the introduction of low-carbon fuels and new technology advancements are necessary and expected. Any certified alternative turbine fuel for aviation will be expected to have an aggregate composition, which is essentially equivalent to petroleum-based jet fuels [1]. The carbon emissions from an engine burning alternate fuel will be equivalent to those from the combustion of petroleum-based jet fuel. Biomass-derived alternative fuels have the potential to reduce life-cycle emissions compared to conventional petroleum-based jet fuels, since biomass-based hydrocarbons absorb CO_2 from the atmosphere when they grow, and the CO_2 emitted during fuel combustion is equivalent to the absorbed amount during its growth. These biomass-based fuels can be used in a mixture with petroleum-based jet fuel of up to 50% by volume due to the reasons given earlier. The reduction of CO_2 emissions from aircraft should be done by integrating possible and useful measures. The use of alternative fuel is one such measure. The committee on propulsion and energy systems, which focuses on the reduction of commercial aviation carbon emissions, recommended the following four approaches to be considered as the highest priority for execution: advances in aircraft/propulsion integration, improvements in gas turbine engines, development of turboelectric propulsion systems and advances in sustainable alternative jet fuels [1]. One of the highest priority approaches is turboelectric propulsion systems. A question arises whether the advances in sustainable alternative jet fuels have positive meanings with the advancements of electric aircraft technologies. Electric aircraft and aircraft electric propulsion technologies are highly expected as a promising measure to dramatically reduce CO_2 emission from aircraft, and in approximately five years small size electric air vehicles might be entering into the market. These small-scale vehicles would have totally battery powered propulsion systems with no need for jet fuel. However, as mentioned above, more than 90% of global emissions in the aviation industry are from large-scale commercial aircraft [1]. For these large-scale aircraft, aircraft propulsion systems generate and consume electricity during flight. Proposed aircraft propulsion systems for large-scale aircraft have turboelectric systems that use gas turbines to drive electrical generators, which power electric motors, which in turn drive the propulsion fans [1]. In these cases, electric aircraft for commercial flights are expected to use jet fuels since gas turbines are still a primary power source. Furthermore, many of such systems have a more independent gas turbine system than conventional aircraft propulsion systems (mostly high-bypass turbofan engines), because electric motors for propulsion fans and generators connected to gas turbines are connected via an electrical transmission line and are designed for both gas turbines and electric motors to drive the propulsion fans. This architecture makes them fairly wider than the components in conventional turbofan engines. Gas turbine systems necessitate jet fuel, but it is expected that alternative SPK fuels are more compatible than petroleum-based jet fuels. One strong point of alternative SPK fuels is the lack of sulphur and aromatics. Such a lack of sulphur and aromatics would result in a longer lifetime of gas turbines, and some of the design constraints of high-temperature parts can be alleviated. Very low nitrogen, aromatics and sulphur will fundamentally also conclude in a cleaner burning fuel [8]. Low aromatic concentration in the fuel results in lower concentration of "seed" molecules, which contribute to the formation of soot nuclei [8]. This further results in smaller probability to produce contrails during flight. But presently, sulphur and aromatics are taking important roles in lubrication and fuel supply, therefore more care should be taken into the design of the whole fuel composition.

Future aircraft power systems are expected to become more and more electrical, and gas turbine might rather take a main role as the power production system. The introduction of a high-temperature fuel cell [e.g. solid oxide fuel cell (SOFC)] in combination with a gas turbine is a highly promising solution for both primary and secondary power sources in aircraft power systems [24]. SOFC can use hydrocarbon fuel with an internal and an external reformer. Very low sulphur and aromatics contribute to a better system design for the SOFC system in terms of desulphurization and reforming.

From these points of view, among the jet fuels, alternative aviation SPK fuels (including FT-SPK and HEFA) are very much promising medium-term measures to reduce CO_2 emission used for conventional turbofan engine and for further compatible with electrical power generation systems (gas turbine and fuel cell/gas turbine hybrid). The reduction of CO_2 by introducing these biomass-derived fuels can be achieved without impacting the time frame or suitability of other potential carbon lowering approaches. These fuels have additional advantages of consideration in turboelectric systems.

6 Conclusion

In this chapter, several important aspects of an original aviation fuel (Jet A1) and an alternative aviation fuel (HEFA) were compared and investigated.

Chemical and physical property analysis, O-ring sink test, oxidation test, modal analysis and combustion rig test with two different combustor types Jet A1 were performed and investigated using both the fuels and their blends.

The exhaust gas measurement data showed that the non-volatile PM mass reduced when using HEFA and its blends with Jet A1, which was also supported from the flame visualization studies.

Non-volatile particulate matter (nvPM) mass was reduced with HEFA only at pilot mode for the concentric lean burn burner, while NO_x, THC and CO emissions were similar between the two fuels.

Combustion instabilities were observed with Jet A1 under certain condition of pilot and main fuel injection mode for the concentric lean burn burner, while those were not observed with HEFA at any conditions within the present test conditions examined. Fuel changes may affect combustion instability characteristics.

Acknowledgements This work was in part supported by JSPS KAKENHI Grant Number 16H04586.

The authors thank Mr. Patrick Salman of the University of Tokyo and Professor Ashwani K. Gupta of the University of Maryland for their support and comments and revisions.

References

1. National Academies of Sciences, Engineering, and Medicine (2016) Commercial aircraft propulsion and energy systems research: reducing global carbon emissions. The National Academies Press, Washington DC. https://doi.org/10.17226/23490
2. ASTM D7566-18 (2018) Standard specification for aviation turbine fuels containing synthesized hydrocarbons. ASTM International, West Conshohocken, PA
3. Rahmes TF, Kinder JD, Henry TM, Crenfeldt G, Leduc GF, Zombanakis GP, Abe Y, Lambert DM, Lewis C, Juenger JA, Andac MG, Reilly KR, Holmgren JM, McCall MJ, Bozzano AG (2009) Sustainable bio-derived paraffinic kerosenes (Bio-SPK) jet fuel flights and engine tests results. In: 9th AIAA aviation technology, integration, and operations conference (ATIO), Hilton Head, South Carolina, USA (AIAA 2009-7002)
4. ANA press release (2012) World-first ferry flight of B787 using biofuel, 17 Apr 2012 (in Japanese). https://www.ana.co.jp/pr/12_0406/12-ana-boeing0417.html. Accessed 3 Sept 2018
5. NCA press release (2012) About arrival of the first-made B747-8F, 25 July 2012 (in Japanese). http://www.nca.aero/news/2012/documents/RELEASEJ.pdf. Accessed 3 Sept 2018
6. Fukuyama Y, Fujiwara H, Okai K (2009) Alternative fuels and their impact on the turbofan engine design and performance under the realistic flight conditions. In: ACGT paper No. 51, Asian joint congress on gas turbines 2009 (ACGT 2009), Tokyo, Japan (ACGT2009-051)
7. Okai K, Fujiwara H, Hongoh M, Shimodaira K (2012) Application of a bio-fuel to a single sector combustor for an experimental small aero-engine. In: Asian joint congress on gas turbines 2012 (ACGT 2012), Shanghai, China (ACGT 2012-2121)
8. Lefebvre AH, Ballal DR (2010) Gas turbine combustion: alternative fuels and emissions, 3rd edn. CRC press, Taylor & Francis
9. Initiatives for Next-generation Aviation Fuels (2015) Report of the initiatives for next-generation Aviation fuels (English translation of the original Japanese version)
10. Ministry of Economy, Trade and Energy press release, July 2 2015, Agency for Natural Resources and Energy, Establishment of a Committee for the Study of a Process Leading to Introduction of Bio Jet Fuel for the 2020 Summer Olympic Games and Paralympic Games in Tokyo. http://www.meti.go.jp/english/press/2015/0702_01.html. Accessed 3 Sept 2018
11. Terasaki NH (2017) Road to Tokyo 2020 and beyond; Japan's initiatives. In: IATA alternative fuel symposium, 16–17 Nov 2017
12. IHI press release (2017) IHI to implement Pilot scale experiment of algae mass cultivation for biofuel in Thailand, 6 Nov 2017. https://www.ihi.co.jp/en/all_news/2017/other/2017-11-06/index.html. Accessed 3 Sept 2018
13. Mitsubishi-Hitachi Power Systems (2017) An activity of R&D and verification of bio fuel production-a pilot-scaled testing of throughflow production process (in Japanese). http://www.nedo.go.jp/content/100870869.pdf. Accessed 3 Sept 2018
14. Abe Y, Toba M, Mochizuki T, Yoshimura Y (2009) Oxidative degradation behaviour of fatty acid methyl ester in fish oil biodiesel and improvement of oxidation stability by partial hydrogeneration. J Jpn Pet Inst 52(6):307–315 (in Japanese)
15. Hyvärinen A, Karhunen J, Oja E (2001) Independent component analysis. Wiley-Interscience Publication, John Wiley & Sons, Inc. https://doi.org/10.1002/0471221317
16. Makida M, Yamada H, Shimodaira K (2014) Detailed research on rich-lean type single sector and full annular combustor for small aircraft engine. In: 29th Congress of the international council of the aeronautical sciences (ICAS 2014), St. Petersburg, Russia (ICAS2014-0628)
17. Yamamoto T, Shimodaira K, Yoshida S, Kurosawa Y (2013) J Eng Gas Turb Power 135:031502

18. Amara AB, Kaoubi S, Starck L (2016) Toward an optimal formulation of alternative jet fuels: enhanced oxidation and thermal stability by the addition of cyclic molecules. Fuel, Elsevier, pp. 98–105. https://doi.org/10.1016/j.fuel.2016.01.040
19. Iida H, Andou S, Ishikawa H, Fujiwara H, Nakaya S, Tsue M (2017) Mode analysis of Jet A1/Bio fuel spray combustion behaviour in swirl flow under atmospheric pressure. In: 55th Symposium (Japanese) on Combustion, Toyama, Japan (in Japanese) (Paper number E342)
20. Chanaud RC (1965) Observation of oscillatory motion in certain swirling flows. J Fluid Mech 21(1):111. See also, Gupta, AK, Lilley, DG, Syred N (1984) Swirl flows. Abacus Press/Gordon and Breach publishers, UK
21. York JL, Stubbs HF, Tek MR (1953) The mechanism of disintegration of liquid sheets. Trans ASME 75:1279–1286
22. Okai K, Fujiwara H, Makida M, Shimodaira K, Yamada H, Nakamura M (2016) The effect of the fuel change from petroleum kerosene to HEFA alternative jet fuel on the emission of an RQL type gas turbine combustor. In: 14th International energy conversion engineering conference, AIAA Propulsion and Energy Forum, (AIAA 2016-4953)
23. Okai K, Fujiwara H, Yoshida S, Yamamoto T, Shimodaira K (2018) The effect of the fuel change from petroleum kerosene to HEFA alternative jet fuel on the emission of a concentric lean-burn burner for gas turbine. AIAA Aerospace Sciences Meeting, AIAA SciTech Forum, (AIAA 2018-1474)
24. Tachibana A, Saito K, Yamamoto T, Makida M, Kitano T, Kurose R (2015) Experimental and numerical investigation of thermo-acoustic instability in a liquid-fuel aero-engine combustor at elevated pressure: validity of large-eddy simulation of spray combustion. Combust Flame 162:2621–2637
25. IATA Sustainable Aviation Fuel Roadmap 1st edn. ISBN 978-92-9252-704-4, 2015, International Air Transport Association, Montreal-Geneva. http://www.iata.org/whatwedo/environment/Pages/alternative-fuels.aspx. Accessed 3 Sept 2018
26. Okai K, Himeno T, Watanabe T, Nomura, H, Tagashira T (2015) Investigation of FC/GT hybrid core in electrical propulsion fan aircraft. In: 51st AIAA/ASME/SAE/ASEE joint propulsion conference, Orlando, Florida, (AIAA 2015-3888)

Lean-Dome Pilot Mixers' Operability Fundamentals

Xiao Ren, Xin Xue, Kyle B. Brady, Chih-Jen Sung and Hukam C. Mongia

Keywords Lean direct injection · Airblast · Lean blowout · NO_x emissions

1 Introduction

To meet increasingly stringent aviation emissions standards, NASA and industry have been conducting research and technology activities aimed at developing low emissions aviation engines, including rich burn-quick mix-lean burn (RQL) [1, 2], lean premixed prevaporized (LPP) [3], and lean direct injection (LDI) [4]. While an industry-wide push to higher system efficiencies has driven engine manufactures to higher combustor inlet temperatures and pressures, these conditions have led to some difficulties in terms of emissions and operability. For instance, RQL-concept combustors often have difficulty controlling smoke emissions and avoiding thermal NO_x-producing near-stoichiometric conditions, particularly when balanced against the need for optimized dome cooling [5]. Likewise, LPP concepts have been found to be prone to problems such as auto-ignition, instability, and flashback at high pressure and temperature conditions [6], thereby limiting their operating ranges. The LDI concept—research on which began as early as 1990 [7]—has shown promise

X. Ren · X. Xue · C.-J. Sung
University of Connecticut, Storrs, CT, USA

X. Xue
Beihang University, Beijing, China

K. B. Brady
National Research Council Research Associateship Program, WPAFB, Dayton, OH, USA

H. C. Mongia (✉)
CSTI Associates, LLC, Yardley, PA, USA
e-mail: hmongia43@hotmail.com

A. K. Gupta et al. (eds.), *Innovations in Sustainable Energy and Cleaner Environment*, Green Energy and Technology,
https://doi.org/10.1007/978-981-13-9012-8_17

for further technology development and demonstration as a method to achieve sufficiently premixed conditions without the need for separate premixing chambers and hence holds potential for mitigating limitations associated with LPP-based designs while retaining a globally fuel-lean architecture. The new generation of LDI has the potential to further reduce NO_x by 50% compared to the current engines [8]. Meanwhile, the iteration on the specific design features of the LDI air swirler/fuel injector design is ongoing, as such features are critical to overall LDI combustion performance. In particular, the swirler vane angle has been intensively studied to improve both emissions and operability of the LDI concept. In this chapter, the influences of swirler vane angle on the reacting flow field, flame structure, lean blowout limit, and NO_x emissions are illustrated by various non-intrusive diagnostics results on single element LDI mixers. In addition, the performance comparisons between a representative swirl-venturi LDI configuration and a conventional airblast injector configuration have been provided.

2 Multipoint LDI

The swirler vane angle, which determines the swirl strength, has been shown to significantly influence NO_x emissions and operability in the first-generation LDI concept (LDI-1), consisting of nine identical simplex element injectors [4]. In this swirl-venturi (SV) concept, each fuel injector had an axial air swirler and a simplex fuel nozzle, followed by a converging-diverging venturi section for quick mixing of the fuel and air before combustion, as shown in Fig. 1. Although the NO_x emissions of LDI-1 were shown to be substantially reduced from the then-current state of the art, its emission levels were not low enough to meet program goals. The impact of swirler vane angle (swirl strength) was investigated in [4] as shown in Fig. 2, demonstrating that overall NO_x emissions lowered as swirl angle decreased and that the drop-off in NO_x emissions index (NO_xEI) significantly increased for reduced swirl angles. However, while the LDI-1 configuration with a swirler vane angle of

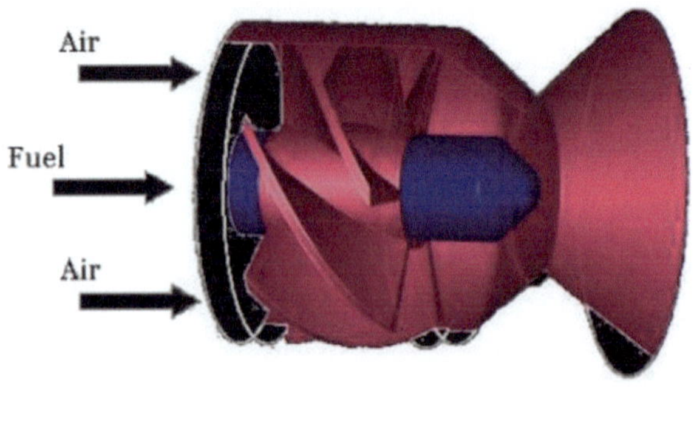

Fig. 1 An example of single element first-generation LDI (LDI-1) injector, consisting of an air swirler and a simplex fuel injector [4]

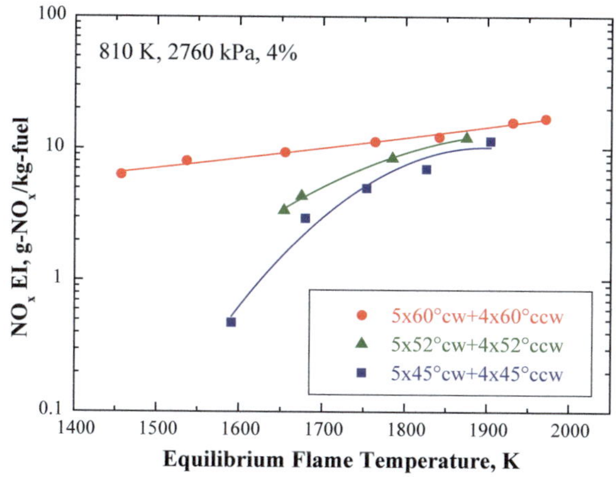

Fig. 2 Typical LDI-1 NO_xEI versus combustor exit temperature ($T4$) at combustor inlet temperature ($T3$) and pressure ($P3$) of 810 K and 2760 kPa, respectively. The data plotted were taken from [4]

45° had the lowest NO_x emissions, its operability was also reduced to unacceptable levels. Moreover, fuel staging did not extend low power operability.

The second generation of LDI (LDI-2) was developed to improve the low power operational range and further reduce NO_x emissions. New features of the single element injector included airblast injectors in place of the original simplex injectors, as shown in Fig. 3. In this fuel injection method, the inner swirling air flow assists fuel atomization and facilitates mixing of the air and fuel. In addition, in the LDI-2 designs, the size of each fuel-air mixer varies and may be in-plane with the dome or recessed. Three second-generation SV-LDI variations tested by NASA included 13 element injectors [9] that were called the flat dome configuration, the 5-recess configuration, and the 9-recess configuration, as shown in Fig. 4. The specifications of the mixers (injectors and swirlers) used in the three LDI-2 configurations are listed in Table 1. The pilot swirler for all three configurations used relatively larger vane

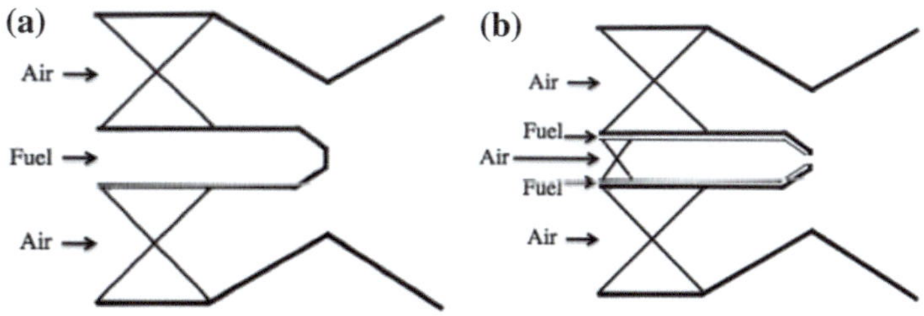

Fig. 3 Illustration of a single second-generation LDI (LDI-2) schematic [9]: **a** simplex fuel injector and **b** airblast fuel injector

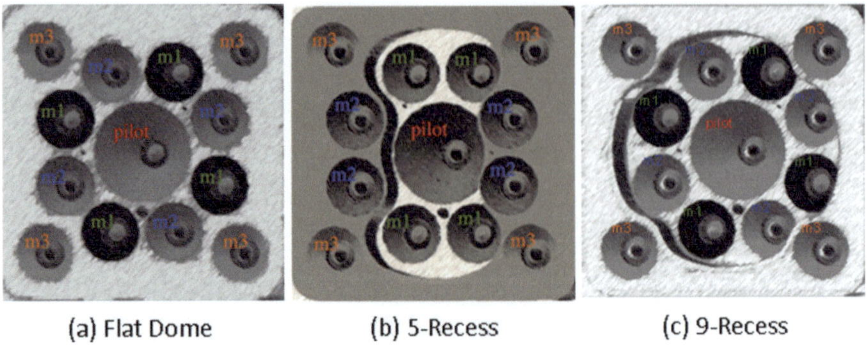

Fig. 4 Second-generation 13-element lean direct injection schematic [9]

angles (57°) in an effort to improve low power operation. All main stages for the three configurations employed a swirler vane angle of 45° to reduce the emissions. The injectors for Main 1 and Main 2/Main 3 were simplex and airblast, respectively, which was thought to maintain the operability at low power while reducing emissions at high power.

By using fuel staging strategy that fuel/air mixers were configured differently for each stage, the LDI-2 configurations were shown to increase the operating range and combustion efficiency at low power conditions and improve upon overall NO_x emissions relative to the LDI-1 designs. Among these three variations of LDI-2, the 5-recess configuration generally yielded slightly lower NO_x emissions than the flat dome or 9-recess configurations, as displayed in Fig. 5. Specifically, ~85% landing/takeoff (LTO) NO_xEI margin from the CAEP/6 standards was achieved with the 5-recess LDI-2 configuration. This is a significant improvement over two premier large engines, Trent1000 and GEnx, as shown in Fig. 6. Furthermore, the combustion dynamics characteristics of these three LDI-2 configurations were also studied in [10], demonstrating that while such dynamics were typically small, all three LDI-2 configurations had high combustion dynamics when inlet temperatures were low. With identical overall equivalence ratio, it was found that shifting half of the main-stage fuel flow from the simplex Main 1 stage to the airblast Main 2 stage could appreciably reduce the combustion acoustics, as well as decrease NO_x emissions concurrently.

Table 1 Definitions of injectors and swirlers used in LDI-2 configurations [9]

Configuration	Pilot Injector	Pilot Swirler	Main 1 Injector	Main 1 Swirler	Main 2 Injector	Main 2 Swirler	Main 3 Injector	Main 3 Swirler
Flat Dome	Simplex	55°ccw	Simplex	45°ccw	Airblast	IAS:45°cw OAS:45°cw	Airblast	IAS:45°cw OAS:45°cw
5-recess	Airblast	IAS:57°cw OAS:57°ccw	Simplex	45°cw	Airblast	IAS:45°cw OAS:45°ccw	Airblast	IAS:45°cw OAS:45°ccw
9-recess	Airblast	IAS:57°cw OAS:57°ccw	Simplex	45°ccw	Airblast	IAS:45°cw OAS:45°cw	Airblast	IAS:45°cw OAS:45°cw

For each stage, the type of fuel injector and the air swirler angle(s) are given
OAS outer air swirler; *IAS* inner air swirler; *cw* clockwise; *ccw* counterclockwise

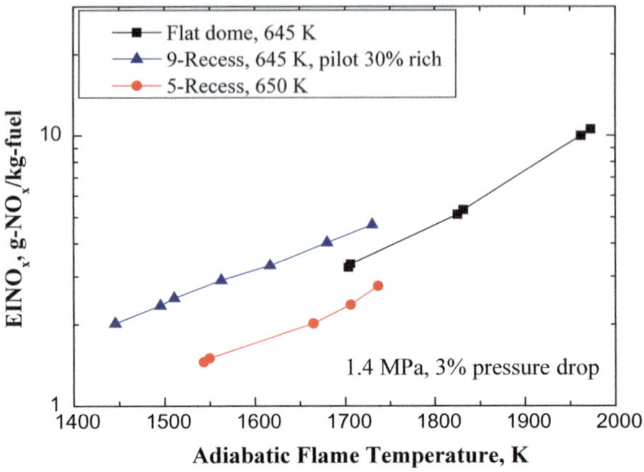

Fig. 5 NO_xEI comparison between flat dome (no-recess), 5- and 9-element recess dome LDI-2 configurations. The data plotted were taken from [9]

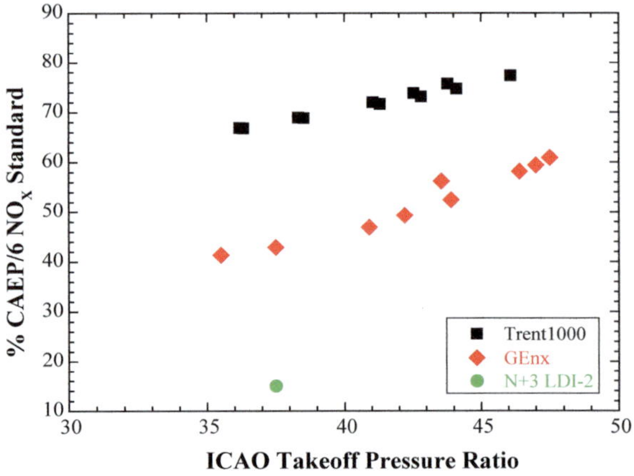

Fig. 6 Comparison of LDI-2 technology potential with two premier large engines' LTO NO_x emissions. The data of Trent1000 and GEnx were taken from the ICAO Aircraft Engine Emissions Databank [11]

3 Single Element LDI Research

Since the LDI performance has been shown to be sensitive to the swirler vane angle and the injector type, the fundamental behavior of the individual swirl injector is critical to understanding and predicting the performance of the whole system. A parametric study of the axial swirler vane angle effect on the non-reacting flow field was conducted in [12] for LDI-1. In the swirler vane angle range of 40°–65°, the strength of swirl demonstrated a clear impact on the mean and turbulent flow fields, as illustrated in Fig. 7. With small swirler vane angles (40° and 45°) weakly swirling flow resulted in no central recirculation zone, though small corner recirculation zones were present, with their size decreasing with increasing swirl angle. Large swirler vane angles (60° and 65°) produced strongly swirling flows, generating a central recirculation zone by virtue of vortex breakdown and the resulting adverse pressure gradient. Overall size of this zone increased with larger swirler vane angles, and

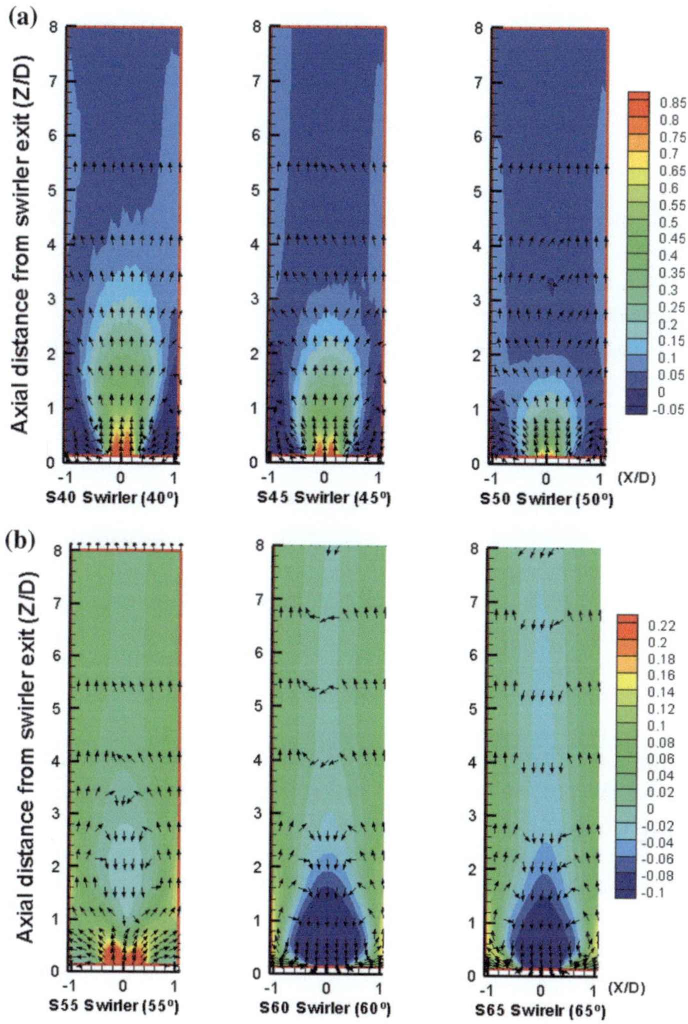

Fig. 7 Mean normalized axial velocity distributions for simplex injector of LDI-1 with varying air swirler vane angles of **a** 40°, 45°, and 50° and **b** 55°, 60°, and 65°. Taken from [12] with the permission of Professor San-Mou Jeng

almost no corner recirculation zone was observed for these cases. For the intermediate swirler vane angles (50° and 55°), they exhibited similar velocity profiles as the weak swirl cases near the swirler exit plane, with the 55° case representing transitional conditions where a small, weak recirculation zone was generated 1–3 D downstream of the dump plane, where $D = 22.5$ mm as the swirler exit diameter.

It is also noted from [12] that the critical swirl number (SN) for the development of central recirculation zone was approximately unity. Swirl number—routinely used to characterize swirl strength—represents a non-dimensional ratio between the axial flux of the angular momentum and the axial flux of the axial momentum, as shown below [13].

$$\mathrm{SN} = \frac{\int \rho W U r \mathrm{d}A}{R \int \rho U^2 \mathrm{d}A} \tag{1}$$

Here, W is the mean tangential velocity, U is mean axial velocity, and R is the radius of the swirler outer boundary. Thus, the aforementioned results indicate that when the axial flux of the angular momentum exceeds the axial flux of the axial momentum, a central recirculation zone would be generated.

In addition to the non-reacting flow field, the effects of swirler vane angle on spray angle and droplet size distribution have also been studied for the LDI-1 configurations by examining cold flow mixing using air and water [14]. A contour plot of the axial velocity of the droplets overlaid with the vector field is shown in Fig. 8. The results show that the spray angle increases dramatically with larger swirler vane angles, commensurate with the non-reacting flow fields. Among the tested configurations, the 60° configuration was determined to have the most even spray distribution. The Sauter mean diameter (SMD) profiles 11 mm downstream of the venturi exit for each swirler vane angle are shown in Fig. 9. While the SMD distributions were similar for swirler vane angles of 45° and 52°, namely smaller droplets at the center of the spray cone and larger droplets near the confinement wall, the SMD distribution range was narrower for the swirler vane angle of 60°, largely between 40 and 60 μm.

Other studies have also explored the effects of varying swirler vane angles, finding their significant impacts on the velocity field, spray characteristics, ignition and extinction performance, and emissions. Alkabie and Andrews [15] showed that for a radial swirler configuration, varying the vane angle in the range of 0–60° dramatically altered extinction, combustion efficiency, and NO_x emissions. Consistent with the previously discussed results of [12], it was found that reducing the vane angle from 60° to 20° significantly reduced NO_x emissions, but at the cost of worse lean blowout performance. When the vane angle was reduced from 60° to 45°, lean blowout equivalence ratio of natural gas/air flame increased by 0.068 with 400 K inlet temperature. Comparing among swirlers with the vane angles of 15°, 30°, 45°, and 60°, Raj and Ganesan [16] found that the 45° vane swirler produced what they

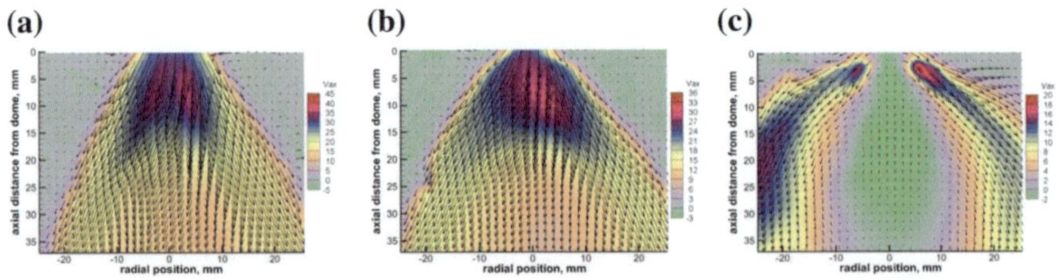

Fig. 8 Axial velocity distribution for water spray overlaid with vector field of the LDI-1 configuration with swirler vane angles of **a** 45°, **b** 52°, and **c** 60° [14]

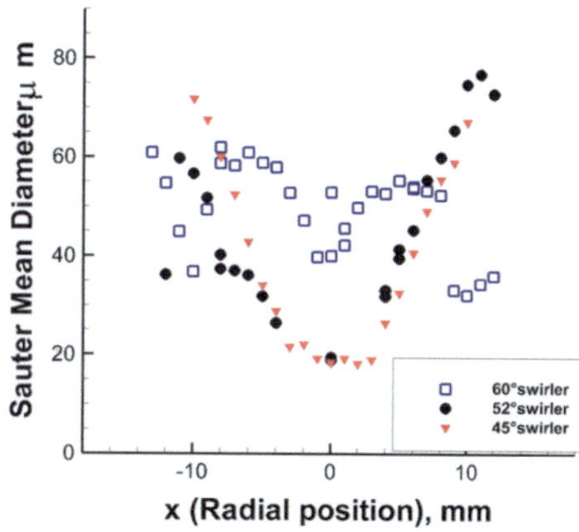

Fig. 9 SMD radial distributions for water spray of the LDI-1 configuration at 11 mm downstream of venturi for swirler vane angles of 45°, 52°, and 60° [14]

deemed the best swirl flow field characteristics considering the aspects of pressure loss and recirculation zone size, which aided fuel-air mixing for high combustion efficiency. Pourhoseini and Asadi [17] carried out experimental study of how the swirler vane angle affected thermal characteristics and pollutant emissions of a liquid fuel burner. The authors concluded that there could be an optimum swirler vane angle exhibiting both high combustion efficiency and low CO and NO emissions, which optimized the fuel-air contact time and area to maximize the mixing rate of inlet air and fuel jet. Wang et al. [18] reported that ground ignition performance for a 30° vane angle case was improved over that of a 20° vane angle case for the main stage of a LPP combustor due to an improved fuel-air distribution around the ignitor tip.

4 Impact of Flare Geometry on Swirling Flows

As discussed in Ren et al. [8], with or without flare is treated as the basic difference between conventionally recognized filming and simplex airblast fuel nozzles shown in Fig. 10 and the LDI concepts displayed in Fig. 11 for an air-atomizing device (AAD) with simple flare (F) and a pressure atomizing device (PAD) with complex flare (CF). In gas turbine combustion, the flare has significant impact on fuel-air mixing process, especially for liquid atomization.

Previous research has explored a variety of flare geometry parameters with respect to their impacts on swirling flows. Estefanos et al. [19] carried out experimental investigations on the effect of the flare expansion angle on the mean behavior of a non-reacting swirling flow. With larger flare angles, the width of the central recirculation zone increased, and the length of corner recirculation zones decreased. Wang et al. [20] studied the effect of venturi expansion angle on spray structure, finding that with a larger flare expansion angle, thinner films were promoted near the venturi exit

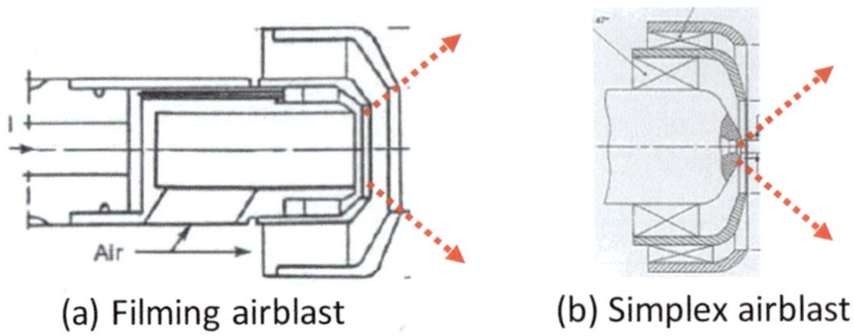

Fig. 10 Schematic of conventional **a** filming and **b** simplex airblast fuel nozzles [8]

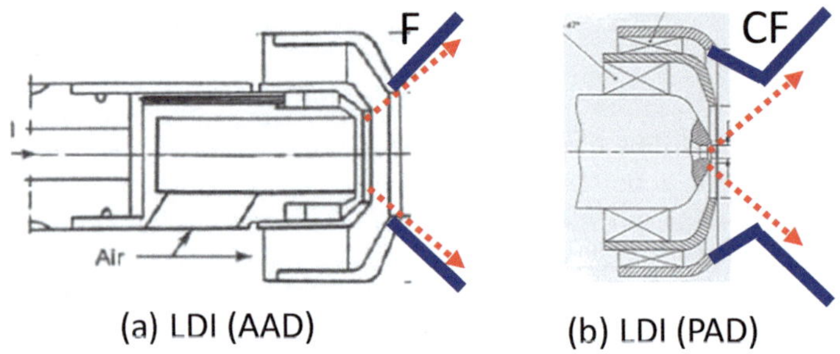

Fig. 11 Descriptive definitions of two LDI concepts: **a** air atomizing device with simple flare and **b** pressure atomizing device with complex flare [8]

and stronger interaction between the counter-rotating airflows was obtained. Li et al. [21] studied the influence of flare height on droplet size distribution, but found no significant effect on SMD distribution.

Besides these previous studies, for an LDI-2 configuration using an airblast injector, there has been little study of the swirler vane angle effects on the flow field and combustion characteristics. Ren et al. [22] reported preliminary results on the reacting flow field of an airblast SV-LDI injector with outer air swirler (OAS) vane angles of 45° and 60°, and with and without flare geometry. For all the 45° OAS cases, no central recirculation zone was observed. The flow field of the 60° OAS case without the flare (i.e., a diverging section) was also significantly affected and showed no central recirculation zone. Despite efforts such as the above, there remains much uncertainty regarding the practical impacts of specific design features of LDI swirl injectors. Given the prior research indicating that features such as the swirl vane angle significantly impact non-reacting flow field structure and spray dynamics, it is critical to future injector design efforts to understand the specific impact of each geometric choice within the LDI designs. As a result, the present study represents a continuing effort to explore these individual impacts at various reacting conditions using optical diagnostic resources to develop a robust understanding of geometric design effects on LDI swirl injector performance.

5 LDI-Related Fundamental Research on Combustion and Emissions Characteristics

The following sections detail the present investigations exploring the impact of OAS vane angle on lean blowout (LBO) limits and NO_x emissions for a counter-rotating axial-axial LDI swirler concept, including a 60° vane OAS (LDI-60°cw) configuration and a 45° OAS (LDI-45°cw) configuration. These two geometries serve as baselines for future studies which will investigate the impacts of various LDI-relevant swirler geometry features. The influence of the presence or absence of a venturi flare section on flame structure and operability is also explored and discussed. Key non-reacting and reacting flow field features are identified and characterized using time-resolved particle image velocimetry (TR-PIV), while the flame structure of each case is analyzed based on chemiluminescence results. Taken together, these results are expected to assist in identifying areas for design improvements of LDI configurations in terms of operability and emissions.

5.1 *Experimental Consideration*

The present experimental arrangement has been developed to investigate single element swirling flow/fuel injection systems in an optically accessible environment under atmospheric pressure conditions. A diagram of the experimental apparatus is shown in Fig. 12a. The swirler entrance manifold measures diameter × length of 80 mm × 90 mm. A set of three stacked fine-mesh inserts within the air inlet is designed to provide a uniform velocity profile to the entrance of the swirler manifold. The air inlet pressure (P_3) is measured in the entrance manifold 50 mm (L_{US}) upstream of the dump plate using an Omega PX303 pressure transducer with full-scale accuracy of 0.25% and 0.01 Torr resolution. Methane is used as fuel in this investigation. The air and methane mass flows are controlled using a set of calibrated flow orifices. The whole burner assembly is mounted to the dump plate such that the exit plane of the flare is flush with the dump plate surface. The test chamber with interior dimensions of 50.8 mm × 50.8 mm × 300 mm constructed from quartz is likewise secured to the dump plate, concentric with the swirler assembly. The assembled flow cell is mounted on a three-axis machine table to facilitate its movement relative to the diagnostic setup.

A schematic of the single element LDI injector used in this study is shown in Fig. 12b. It consists of a fuel tip with two concentric axial air swirlers, an inner air swirler (IAS) contained within the airblast-type fuel tip, and an OAS that is located within the converging-diverging venturi flare body. For the present study, helical 45°/60° OAS vanes are installed with clockwise rotation direction, while the IAS vanes are fixed at a 60° counterclockwise rotation, forming counter-rotating flows between OAS and IAS. The venturi throat diameter is $D_t = 0.52$ inch (13.2 mm), and its axial location is denoted as $x = 0$. The LDI flare exit has a full angle of 110°,

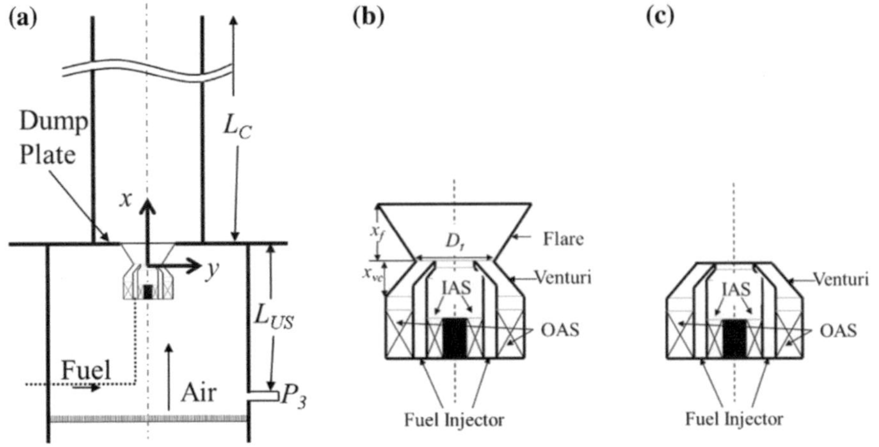

Fig. 12 Schematic of **a** test rig for fundamental experiments, **b** LDI nozzle, and **c** airblast nozzle

Table 2 Specifications of swirler/venturi sets used in the present fundamental research

Configuration	Swirler (IAS/OAS)	Venturi
LDI-60°cw	60°ccw/60°cw	With flare
Airblast-60°cw	60°ccw/60°cw	Without flare
LDI-45°cw	60°ccw/45°cw	With flare
Airblast-45°cw	60°ccw/45°cw	Without flare

a diameter of $D_f = 1$ inch (25.4 mm), and a length of $x_f = 0.37$ inch (9.4 mm). As the combustor length is $L_C = 12$ inch (304.8 mm), $L_C/D_f = 12$. A representative airblast injector as shown in Fig. 12c is created by simply removing the flare (the diverging section of the present LDI's venturi) from the LDI injector and is used to illustrate the differences between the fundamental characteristics of LDI and airblast injectors. Table 2 lists the specifications of the four configurations used in the current LDI-related fundamental studies. The air flow rate is kept at a constant 6.633 g/s. This air flow rate is for the LDI-60°cw, cold flow case with a pressure drop across the swirler of 3%. The Reynolds number based on the volumetric flow rate and D_t for each test condition is estimated to be 35,000.

Measurements of the velocity field were made using a Dantec Dynamics two-dimensional TR-PIV system. All axial-radial (U-V) data presented in this study were collected with the laser sheet of 1 mm thickness aligned with the center plane of the experiment. The laser repetition rate was set to 5 kHz and aluminum oxide seeding particles with 1 μm nominal diameter were used. Based upon the recommendation of Melling [23], seeding particle diameter should not exceed 1 μm to have sufficient frequency response at 10 kHz. Interrogation areas were set to 0.6 mm × 0.6 mm, with 50% overlap. Laser pulse delay times were set to follow the "1/4 rule" within the shear layer regions. The capture area for the mean velocity map was 50 mm × 60 mm, focused on the area from the dump plate to 60 mm downstream. The camera was located at a distance of 0.6 m from the laser plane, with a 55-mm focal length Nikkor lens. The F-stop of lens was set at 5.6 for all test cases. In addition, a 532 nm filter was mounted in front of the camera lens to eliminate unwanted light sources. For each test case, TR-PIV data were collected over a 1-s duration.

OH*, CH*, and NO_2* chemiluminescence signals were measured using a PI-MAX III intensified charge-coupled device camera, which was located 1.8 m from chamber center axis. A Nikon 105 mm UV lens was mounted in front of the camera with F-stop 4.5. The gain value was 15 with 100 ms exposure time. To compare differences caused by different OAS vane angles and various overall equivalence ratios, all camera settings were kept the same for all chemiluminescence tests. Species-specific bandpass filters were mounted in front of the lens for each radical, including 310 ± 2 nm with 10 nm full width at half maximum (FWHM) for OH*, 430 ± 2 nm with 10 nm FWHM for CH*, and 750 ± 10 nm with 70 nm FWHM for NO_2*. As NO_2 emissions are in the range of 608.8–851.5 nm [24], NO_2* was used here as a proxy indicator for the total NO_x emissions level, which would consist of both NO and NO_2. Each data set was averaged over 100 consecutive images at 10 Hz to obtain a mean radical distribution.

For non-dimensional analyses, the axial distance (x) and radial distance (y) were normalized by D_t, while all velocity components were normalized by the velocity magnitude at the venturi throat, U_t, defined by the following equation.

$$U_t = \frac{\dot{m}}{\rho A_t} \tag{2}$$

Here, $\dot{m}$ is the measured total mass flow rate, ρ is the air density, and A_t is the geometric area of the venturi throat.

5.2 *Impact of OAS Vane Angle on Flame Characteristics and LBO*

As is evident from the mean flame images presented in Fig. 13, several different flame modes emerge as overall equivalence ratio (ϕ) is reduced toward LBO for both LDI-60°cw and LDI-45°cw. For the LDI-60°cw configuration at $\phi = 0.85$, a vigorously burning flame is anchored at the venturi exit, with burning zones in both the central recirculation zone (CRZ) and corner recirculation zone (CNRZ). As overall equivalence ratio is reduced to $\phi = 0.65$, although the overall flame structure appears similar, the CNRZ flames disappear and the flame extends further downstream relative to the LDI-60°cw, $\phi = 0.85$ case. When further reducing overall equivalence ratio to $\phi = 0.62$, the flame weakens more and detaches from the venturi flare, instead anchoring within the tail of the CRZ near the axial centerline. Finally, as the overall equivalence ratio approaches LBO, the weakened flame spreads further downstream until flame lift-off is observed (cf., LDI-60°cw, $\phi = 0.57$). Moreover, reduction in ϕ beyond this point results in the flame exiting the chamber, namely

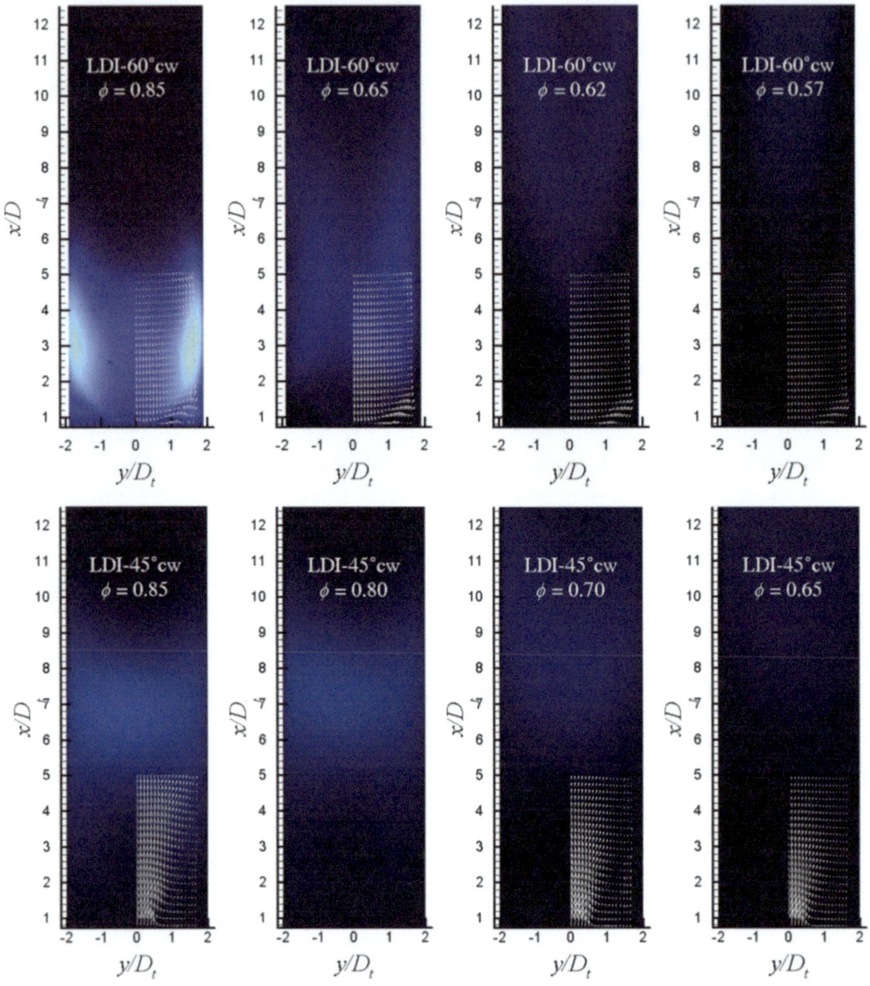

Fig. 13 Flame structure variations at 3% pressure drop as overall equivalence ratio approaches LBO for the LDI-60°cw (top) and LDI-45°cw (bottom) configurations, represented by averaged broadband flame images and mean *U*-*V* vector maps

LBO. On the other hand, the LDI-45°cw configuration exhibits very different flame structures from the LDI-60°cw configuration. In particular, under no conditions does the flame anchor at the venturi flare, but instead is lifted at some distance downstream of the dump plate, with no particular coherent flame structure. As ϕ is lowered for the LDI-45°cw configuration, while no dramatic structural changes are observed, the flame weakens, elongates, and stabilizes further downstream.

The LBO limits of each configuration as a function of overall pressure drop are plotted in Fig. 14. While following a roughly similar increasing trend of LBO with increasing overall pressure drop, the LDI-60°cw configuration consistently has

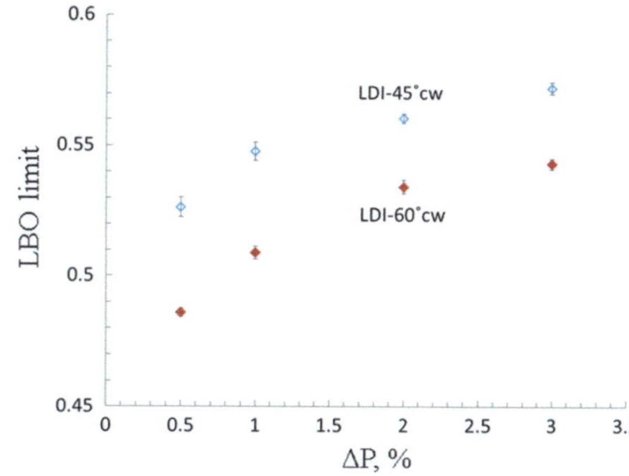

Fig. 14 Effect of OAS vane angle on LBO limits at varying overall pressure drops for the LDI-60°cw and LDI-45°cw configurations

lower LBO limits. Based on the broadband flame images, the stronger anchoring of the flame to the venturi flare for the LDI-60°cw configuration appears to result in improved LBO performance relative to the LDI-45°cw configuration.

5.3 *Impact of OAS Vane Angle on NO_2* Distribution*

Figure 15 compares NO_2* distributions from the two LDI configurations with $\phi =$ 0.65 and 0.62. It is seen that the LDI-60°cw, $\phi = 0.65$ case, exhibiting a venturi flare-anchored flame, produces the most NO_2* and shows significant chemiluminescence along the axial centerline starting within the CRZ at 0.5 D_t downstream of the dump plane and continuing downstream to 11–12 D_t. As ϕ drops to 0.62, the flame stabilization location moves downstream to the low-velocity region immediately following the CRZ—a "lifted" flame—and the overall NO_2* signal decreases by almost half. In fact, despite a slightly different flame structure, the LDI-60°cw, $\phi = 0.62$ and LDI-45°cw, $\phi = 0.65$ and 0.62 cases exhibit roughly similar NO_2* signal, signifying similar overall NO_x production. Taken together, the results of Fig. 15 suggest that, regardless of specific OAS vane angle, NO_x production within the LDI configuration may be largely driven by the flame primary zone location relative to long residence time streamlines. This further suggests that operability and NO_x emissions are in direct competition, thereby requiring design trade-offs for practical LDI applications.

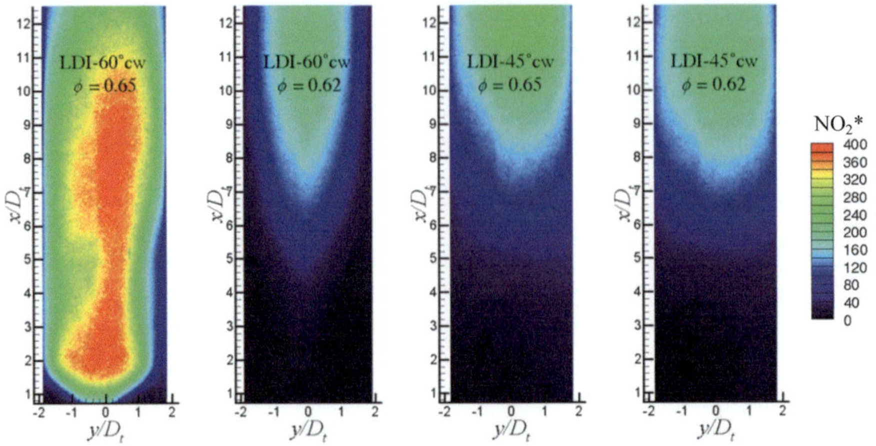

Fig. 15 Effect of OAS vane angle on NO_2* distribution at varying overall equivalence ratios for the LDI-60°cw and LDI-45°cw configurations

5.4 Impact of OAS Vane Angle on Reacting Flow Field and Flame Structure

To help elucidate the differences associated with the swirler vane angle variation, TR-PIV-measured mean axial velocity contours are obtained in addition to the chemiluminescence images of OH* and CH*. The mean axial velocity field at the 3% pressure drop condition is shown in Fig. 16. It is seen that the axial velocity contours in the LDI-60°cw configuration exhibit a strong CRZ near the venturi exit that convects burned products upstream toward the venturi flare. For the LDI-45°cw configuration, a swirling jet exists, despite no CRZ is found in the reacting flow field. Instead, this swirling jet flow achieves a maximum mean axial velocity at the flare exit and progressively weakens as the flow moves downstream and expands radially to fill the combustion chamber. This flow also results in CNRZs, which are both smaller and weaker compared to the CRZ, at the dump plane corners of the combustion chamber. As ϕ drops, the sizes of the CNRZs tend to shrink, thus providing even less opportunity for complete combustion.

In Fig. 17, mean radical distributions are overlaid with the mean axial-radial velocity vectors. To better delineate reaction zones and qualitatively compare the heat release of the reaction zone, the OH* and CH* chemiluminescence images are taken and compared at various test conditions. In the LDI-60°cw configuration, the CRZ results in a reaction zone anchored to the venturi flare. Specifically, at $\phi = 0.85$, high radical concentrations are observed in the high-shear/low-velocity region between the CRZ and CNRZs, in which unburned reactants exiting the flare meet hot recirculated product gases. As ϕ drops to 0.62, the heat release in the high-shear region is insufficient to sustain reactions in this zone. As a result, the flame stabilizes downstream in the low-velocity, post-CRZ region, while remaining anchored to the

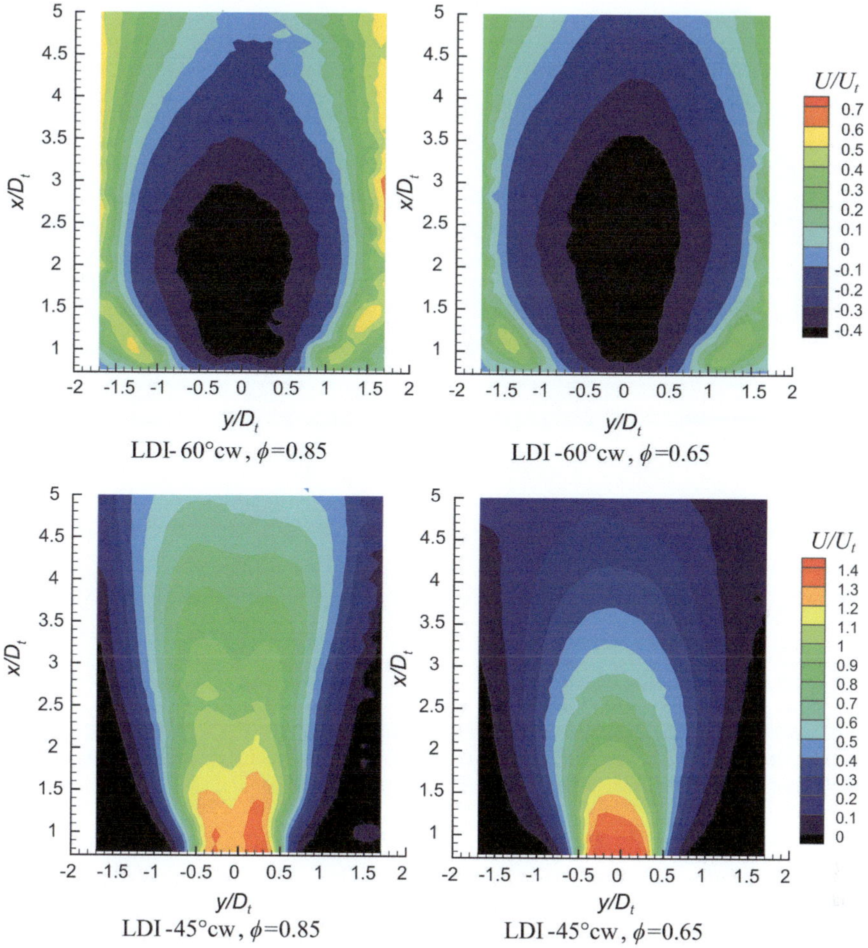

Fig. 16 Mean axial velocity fields of reacting flows at varying overall equivalence ratios in the LDI-60°cw (top) and LDI-45°cw (bottom) configurations

low-velocity downstream tail of the CRZ. In contrast, the LDI-45°cw flames are consistently stabilized far downstream of the swirling jet, where the expansion of the jet and the resultant reduced velocity allow a flame to be sustained. Lowering ϕ serves to weaken the flame and pushes it further downstream.

From these axial velocity fields and radical contours, it is also apparent why the LDI-60°cw and LDI-45°cw configurations exhibit different LBO limits. Whereas the flame may remain at least partially anchored by minor recirculation region within the low-velocity tail of the CRZ in the LDI-60°cw configuration, in the LDI-45°cw configuration the lack of a steady stabilization feature implies higher LBO limits.

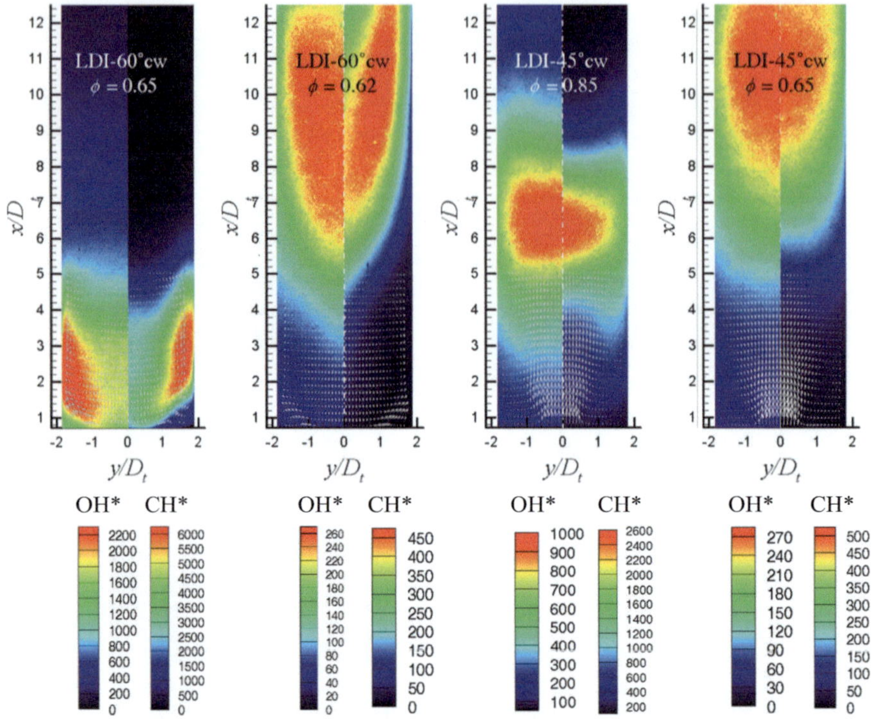

Fig. 17 Mean OH* (left) and CH* (right) overlaid with mean velocity vectors at varying overall equivalence ratios for the LDI-60°cw and LDI-45°cw configurations

This difference in stabilization mechanism is also apparent when comparing the mean centerline axial velocity profiles shown in Fig. 18 as a function of the overall equivalence ratio. While the CRZ weakens for the LDI-60°cw configuration as ϕ drops due to the decrease of thermal expansion—and the commensurate velocity gains—near the CRZ base, a negative velocity is nonetheless maintained along the axial centerline for a substantial distance from the dump plane, only reaching a stagnation condition ~5 D_t downstream. On the other hand, the LDI-45°cw configuration exhibits solely positive velocity along the centerline, decreasing steadily as the distance from the dump plane increases. The lack of recirculation of burned products in the LDI-45°cw configuration implies that the flame location is determined almost entirely by local flame propagation, and hence, the flame may only exist in the low-velocity regions far downstream of the dump plate, with minor anchoring at elevated overall equivalence ratios associated with the low velocity.

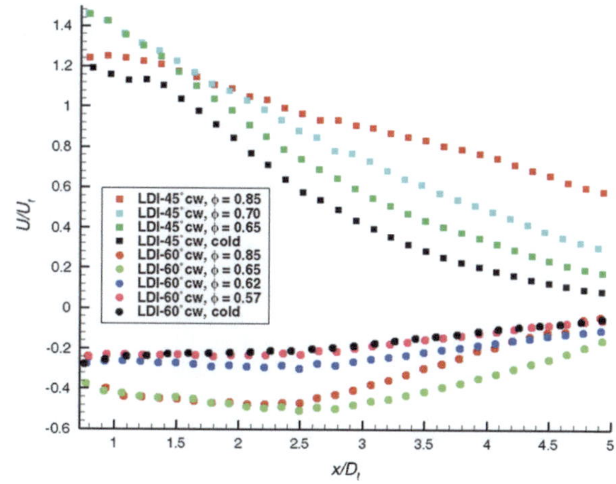

Fig. 18 Centerline mean axial velocity profile comparison of the LDI-60°cw and LDI-45°cw configurations at varying overall equivalence ratios

5.5 *Impact of Flare Section on Flame Characteristics*

Direct flame images overlaid with velocity vector maps for the baseline LDI configurations with and without the venturi flare section are shown in Fig. 19 for a pressure drop of 3% and $\phi = 0.65$. Without the venturi flare, both the Airblast-60°cw and Airblast-45°cw configurations exhibit lifted flames, although the Airblast-60°cw appears to have a stronger flame than Airblast-45°cw, with the primary flame zone of Airblast-60°cw located several D_t lower than that for the Airblast-45°cw case. Comparing the LDI-45°cw with its airblast counterpart, while the overall flame structure is similar, the flame sits closer to the dump plane for the LDI configuration. However, for the LDI-60°cw configuration with and without the venturi flare, the flame being either anchored to the dump plane or lifted is based solely on the presence of the flare.

The source of this change in flame behavior is apparent from the velocity field results, as shown in Fig. 20. The minor differences in flame structure between the two 45°-vane angle cases can be explained by the swirling jet structure, whereas for the LDI-45°cw case the jet is radially larger and axially shorter, removal of the venturi flare results in a thinner and dramatically longer jet. This behavior suggests that for a given flame propagation, the flame will not be able to stabilize until further downstream due to slower expansion of the swirling jet. More interesting, however, are the 60°cw-OAS cases. Removal of the venturi flare from the LDI-60°cw configuration results in a flow field much more similar to that found in the LDI-45°cw configuration, namely a swirling jet flow in lieu of a strong CRZ. Clearly, the removal of the flare section significantly weakens overall swirl strength to the point that vortex breakdown no longer occurs.

Considering the above in terms of swirl number, the results are somewhat intuitive. For a given vane configuration, tangential velocities are essentially fixed; the addition or removal of a flare section will alter this motion very little. However, due

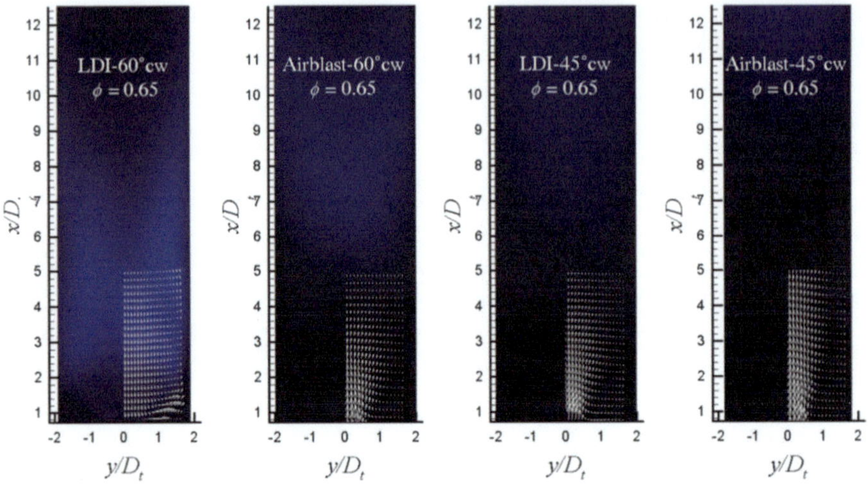

Fig. 19 Flame structure variations at 3% pressure drop and $\phi = 0.65$ for the LDI-60°cw, Airblast-60°cw, LDI-45°cw, and Airblast-45°cw configurations, represented by mean direct flame images overlaid with mean velocity vectors

to the Coanda effect—that fluid streamlines tend to follow solid boundaries—flow passing through the venturi will tend to expand, in the process reducing axial velocity. This in turn effectively increases the swirl number by decreasing the axial flux of axial momentum. One interesting implication is that the Coanda effect is typically influenced by the immersion depth of the mixer relative to the dome; a future study will explore the extent to which apparent swirl strength is influenced by swirl assembly insertion depth.

In terms of operability, the aforementioned alterations to the flow field greatly impact LBO limit. As shown in Fig. 21, across the pressure drop range tested, the LBO limits for the three configurations not exhibiting a CRZ are roughly similar, while the LDI-60°cw configuration exhibits LBO limits consistently 5–10% lower. Also shown in Fig. 22, NO_2* emissions are strongly tied to the presence of the CRZ. For each of the configurations resulting in a lifted flame, including Airblast-45°cw, Airblast-60°cw, and LDI-45°cw, NO_2* emissions are relatively low. Some correlation with swirl strength exists; the Airblast-45°cw configuration exhibits the lowest emissions, followed by the LDI-45°cw, and then the Airblast-60°cw case. However, when a CRZ is present, as for the LDI-60°cw case, high NO_2* signal is observed from just after the dump plane to 11–12 D_t downstream. This provides further evidence that for a single swirl injector, a design trade-off exists among operability, stability, and NO_x emissions.

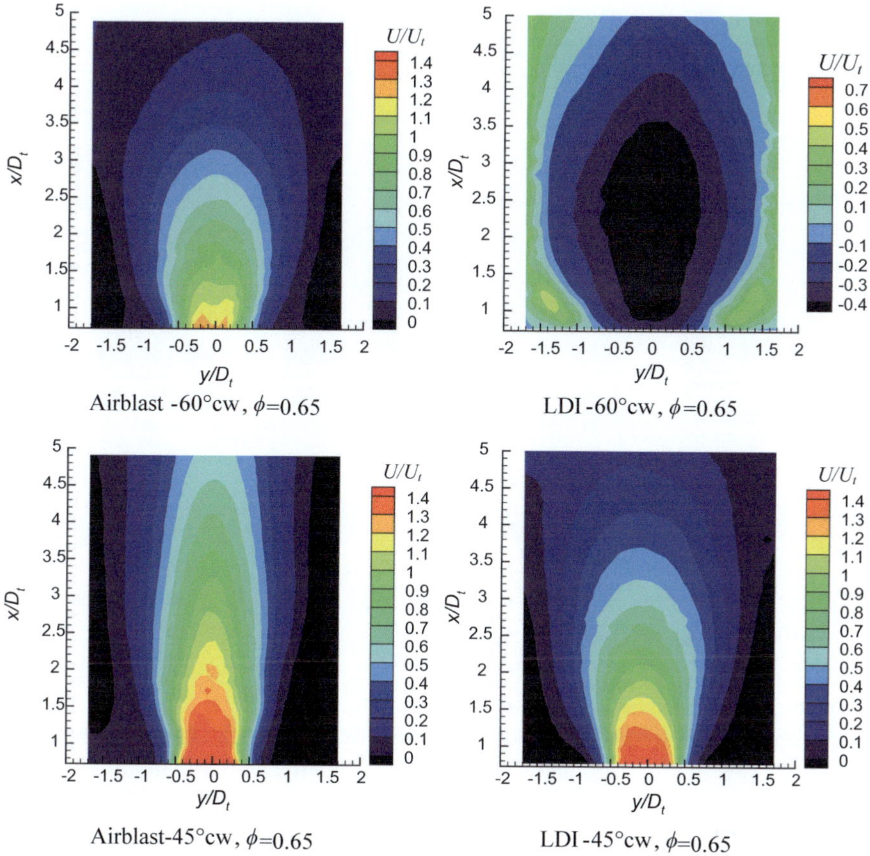

Fig. 20 Mean reacting axial velocity fields at $\phi = 0.65$ for the Airblast-60°cw, LDI-60°cw, Airblast-45°cw, and LDI-45°cw configurations

Fig. 21 Lean blowout limits of the four selected configurations at varying pressure drops

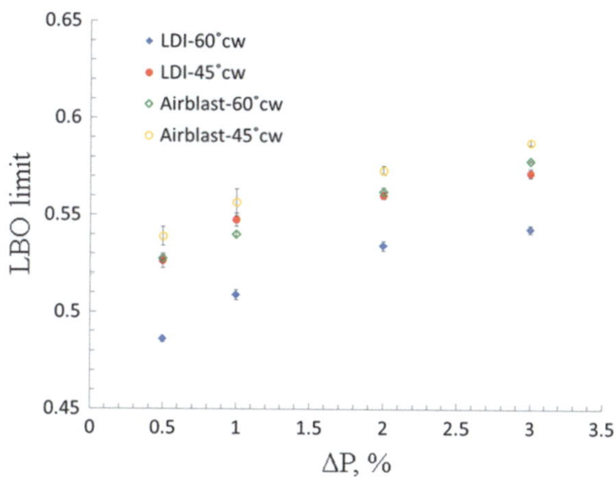

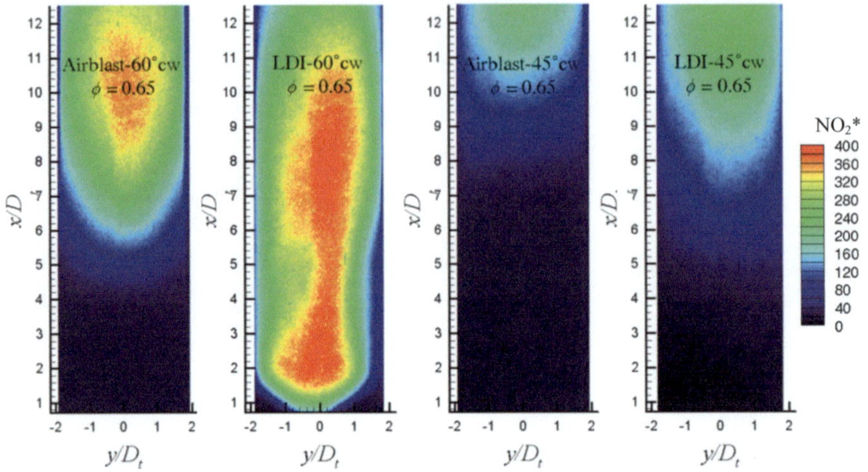

Fig. 22 NO_2* chemiluminescence contours at $\phi = 0.65$ for the Airblast-60°cw, LDI-60°cw, Airblast-45°cw, and LDI-45°cw configurations

6 Concluding Remarks

Lean direct injection is a promising, robust, and simple lean-front end combustion concept. In this chapter, recent technology development efforts to improve its operability are summarized and discussed. Additionally, a systematic fundamental research program has been carried out by characterizing velocity and radical fields using advanced diagnostic techniques to shed light on technology development paths for future LDI concepts, aimed at lowering emissions and expanding operational range. The present work has provided detailed reacting flow field, flame structure, and emissions information regarding the effects of swirler vane angle and overall equivalence ratio on the resulting flow field and flame response. The swirl strength was found to play an important role in determining the flame structure and the characteristic residence time of the reacting flow, with appreciable impact on operability and emissions. Furthermore, the impact of venturi geometry—with or without the flare section—on the flow structure and flame behavior was explored. Without the diverging venturi section, the flow pattern can change dramatically from a central recirculation zone to a swirling jet flow with corner recirculation zones, thereby impacting the average residence time within the combustion zone and the subsequent combustion characteristics. Therefore, optimization of the venturi's geometry may improve the operability and emissions performance of LDI.

Acknowledgements KBB would like to acknowledge the support of the National Research Council under the auspices of the Research Associateship Program at the Air Force Research Laboratory. The authors also thank Dr. Xin Hui of Beihang University for his help in the fundamental single element experiments.

References

1. Rizk NK, Mongia HC (1990) Ultra-low NO_x rich-lean combustion. In: ASME international gas turbine and aeroengine congress and exposition 90-GT-087
2. Rizk NK, Mongia HC (1991) Low NO_x rich-lean combustion concept application. AIAA1991-1962
3. Mongia HC (2003) TAPS—a fourth generation propulsion combustor technology for low emissions. AIAA2003-2657
4. Tacina R, Lee P, Wey C (2005) A lean-direct-injection combustor using a 9 point swirl-venturi fuel injector. ISABE-2005-1106
5. Liu YZ, Sun XX, Sethi V, Nalianda D, Li YG, Wang L (2017) Review of modern low emissions combustion technologies for aero gas turbine engines. Prog Aerosp Sci 94:12–45
6. Reddy DR, Lee CM (2016) An overview of low-emission combustion research at NASA Glenn. ASME Turbo Expo GT2016-56100
7. Tacina RR (1990) Low NO_x potential of gas turbine engines. AIAA 90-0550
8. Ren X, Sung CJ, Mongia HC (2018) In: Runchal AK, Gupta AK, Kushari A, De A, Aggarwal SK (eds) On lean direction injection research. Springer, Singapore, pp 3–26
9. Tacina KM, Podboy DP, He ZJ, Lee P, Dam B, Mongia HC (2016) A comparison of three second-generation swirl-venturi lean direct injection combustor concepts. AIAA 2016-4891
10. Tacina KM, Lee P, Chang C, He Z, Dam B, Podboy D (2015) An assessment of combustion dynamics in a low-NO_x second-generation swirl-venturi lean direct injection combustion concept. ISABE 2015-20249
11. ICAO Aircraft Engine Emissions Databank. https://www.easa.europa.eu/easa-and-you/environment/icao-aircraft-engine-emissions-databank
12. Fu Y (2008) Aerodynamics and combustion of axial swirlers. Doctoral dissertation, University of Cincinnati. Thesis advisor: Professor San-Mou Jeng
13. Gupta AK, Lilley DG, Syred N (1984) Swirl flows, abacus, 1st ed. Routledge, Tunbridge Wells, UK
14. Tedder SA, Tacina KM, Anderson RC, Hicks YR (2014) Fundamental study of a single point lean direct injector. Part I: Effect of air swirler angle and injector tip location on spray characteristics. AIAA2014-3435
15. Alkabie HS, Andrews GE (1991) Reduced NO_x emissions using low radial swirler vane angles. In: ASME international gas turbine and aeroengine congress and exposition 91-GT-363
16. Raj RTK, Ganesan V (2008) Study on the effect of various parameters on flow development behind vane swirlers. Int J Therm Sci 47(9):1204–1225
17. Pourhoseini SH, Asadi R (2016) An experimental study of optimum angle of air swirler vanes in liquid fuel burners. J Energy Resour ASME 139(3):032202-02-5
18. Wang B, Zhang C, Lin Y, Hui X, Li J (2016) Influence of main swirler vane angle on the ignition performance of TeLESS-II combustor. J Eng Gas Turbine Power 139(1):011501-01-08
19. Estefanos W, Tambe S, Jeng SM (2015) Effect of the flare expansion angle on the mean and dynamic behavior of the swirling flow generated by a counter rotating radial-radial swirler using a water test rig. ASME Turbo Expo GT2015-43470
20. Wang X, Lin Y, Hu H, Zhang C, Kang Y (2014) Effect of swirl cup's venturi shape on spray structure and ignition process. ASME Turbo Expo GT2014-25216
21. Li J, Hamza M, Kumaran A, Bhayaraju U, Jeng SM (2016) Study of development of a novel dual phase airblast injector for gas turbine combustor. ASME Turbo Expo GT2016-56340
22. Ren X, Xue X, Sung CJ, Brady KB, Mongia HC, Lee P (2016) The impact of venturi geometry on reacting flows in a swirl-venturi lean direct injection airblast injector. AIAA 2016-4560
23. Melling A (1997) Tracer particles and seeding for particle image velocimetry. Meas Sci Technol 8(12):1406–1416
24. Pearse RWB, Gaydon AG (1976) Identification of molecular spectra. Chapman and Hall

Dynamics of Non-reacting and Reacting Flows Past Bluff Bodies

Uddalok Sen, Sourav Sarkar, Sombuddha Bagchi, Achintya Mukhopadhyay and Swarnendu Sen

Keywords Bluff-body · Fluid injection · Vortex shedding · Flame stabilization · Proper orthogonal decomposition

List of Symbols

C_{L}	Lift coefficient
D	Diameter of the cylinder
f	Frequency of vortex shedding
F_{L}	Lift force
p	Pressure
Re	Reynolds number
St	Stokes number
T_{b}	Temperature of the burnt gas

U. Sen
Department of Mechanical and Industrial Engineering, University of Illinois at Chicago, Chicago, IL 60607, USA
e-mail: uddalok.sen.us@gmail.com

S. Sarkar · S. Bagchi · A. Mukhopadhyay (✉) · S. Sen
Mechanical Engineering Department, Jadavpur University, Kolkata 700032, India
e-mail: achintya.mukho@gmail.com

S. Sarkar
e-mail: souravsarkar.iitm@gmail.com

S. Bagchi
e-mail: sombuddha.bagchi@gmail.com

S. Sen
e-mail: sen.swarnendu@gmail.com

A. K. Gupta et al. (eds.), *Innovations in Sustainable Energy and Cleaner Environment*, Green Energy and Technology,
https://doi.org/10.1007/978-981-13-9012-8_18

T_u	Temperature of the unburnt gas
u	x velocity
U_∞	Free stream velocity
v	y velocity
ε	Ratio of the velocities of the injected stream to the free stream
η_i	Cumulative energy coefficient
ζ_i	Participant energy coefficient
λ^i	ith eigenvalue
μ	Dynamic viscosity
ν	Kinematic viscosity
ρ	Density
ω	Vorticity

1 Introduction

Fluid flow past bluff bodies is distinguishable from other types of flows due to its characteristic flow pattern. The most identifiable feature of such a flow is the anti-symmetric von Kármán vortex street, observed for flows within a certain range of Reynolds number. Although the flow past a circular cylinder [1–4] is the most studied scenario, there have been reports on flows past bluff bodies having square [5] and triangular [6] cross-sections as well. A brief overview of non-reacting and reacting flows past bluff bodies has been provided in the following sections. For incompressible, laminar, two-dimensional flow, the governing equations for the flow past a bluff body can be written as:

Continuity:

$$\frac{\partial u}{\partial x} + \frac{\partial v}{\partial y} = 0 \tag{1}$$

x-Momentum:

$$\frac{\partial u}{\partial t} + \frac{\partial (uu)}{\partial x} + \frac{\partial (uv)}{\partial y} = -\frac{1}{\rho}\frac{\partial p}{\partial x} + \nu\left(\frac{\partial^2 u}{\partial x^2} + \frac{\partial^2 u}{\partial y^2}\right) \tag{2}$$

y-Momentum:

$$\frac{\partial v}{\partial t} + \frac{\partial (vu)}{\partial x} + \frac{\partial (vv)}{\partial y} = -\frac{1}{\rho}\frac{\partial p}{\partial y} + \nu\left(\frac{\partial^2 v}{\partial x^2} + \frac{\partial^2 v}{\partial y^2}\right) \tag{3}$$

The lift force for incompressible, laminar, two-dimensional flow past a circular cylinder can be expressed as [7]:

$$F_{\mathrm{L}} = \frac{1}{2}\int_{0}^{2\pi} (-p\, \sin\theta + \nu\omega\, \cos\theta) D \mathrm{d}\theta \tag{4}$$

where

$$\omega = \frac{\partial v}{\partial x} - \frac{\partial u}{\partial y} \tag{5}$$

The corresponding lift coefficient is obtained as

$$C_{\mathrm{L}} = \frac{F_{\mathrm{L}}}{\frac{1}{2}\rho U_{\infty}^{2} D} \tag{6}$$

where U_{∞} is the free stream velocity and D is the diameter of the bluff body. For periodic vortex shedding in the case of two-dimensional, laminar, incompressible flow past a bluff body, the Strouhal number is expressed as:

$$St = \frac{fD}{U_{\infty}} \tag{7}$$

The free stream Reynolds number is given by

$$Re = \frac{\rho U_{\infty} D}{\mu} \tag{8}$$

A popular analysis technique used to study flow past bluff bodies is the proper orthogonal decomposition (POD) [8–10]. POD allows for the identification of the dominant mathematical modes in spatiotemporally evolving flows in terms of their energy and also facilitates the construction of a reduced order model [11–14] The technique, also referred to as principal component analysis (PCA) or Karhunen-Loéve decomposition, was first introduced by Kosambi [15] and can be used to rank the dominant modes of a spatiotemporally evolving system in terms of their energy contribution [16]. The snapshot POD procedure introduced by Sirovich [17] is by far the most popular one and can be briefly described as follows [18].

Any spatiotemporally evolving variable g can be expressed as

$$g(x, y, t) = \bar{g}(x, y) + g'(x, y, t) \tag{9}$$

where $\bar{g}$ and g' represent the temporal mean and the fluctuation part, respectively.

The quantity g is measured at M different spatial locations for N equally spaced intervals of time (or snapshots), and the snapshot matrix G (of order $M \times N$) is constructed from the g' values as

$$G = \begin{pmatrix} g_1'^1 & g_1'^2 & \cdots & g_1'^N \\ \vdots & \vdots & \vdots & \vdots \\ g_M'^1 & g_M'^2 & \cdots & g_M'^N \end{pmatrix} \tag{10}$$

The auto-covariance matrix Q is calculated as $Q = G^T G$, and the eigenvalues λ^i along with the corresponding eigenvectors A^i satisfying $QA^i = \lambda^i A^i$ are calculated. The number of eigenvalues equals the number of snapshots [19], and the eigenvalues are themselves arranged in descending order, i.e., $\lambda^1 > \lambda^2 > \cdots > \lambda^N > 0$. Alternatively, a singular value decomposition (SVD) of G can also be performed, which is expressed as $G = USV^T$, where U (of order $M \times M$) and V (of order $N \times N$) are orthogonal matrices with columns being the eigenvectors of GG^T and $G^T G$ (or Q), respectively. In other words, the ith column of V is A^i. S is an $M \times N$ diagonal matrix with elements θ_j being the singular values of G, arranged in descending order, $j = 1, 2, \ldots$, rank (G). Incidentally, the θ_j values are the square roots of the eigenvalues of Q, i.e., $\theta_j^2 = \lambda_j$. The normalized POD modes or eigenfunctions ϕ^i result from the projection of the spatial eigenvectors A^i and are given by

$$\phi^i = \frac{\sum_{i=1}^{N} A_n^i g'^n}{\left\| \sum_{i=1}^{N} A_n^i g'^n \right\|} \tag{11}$$

where $\|\cdot\|$ is defined as $\|z\| = \sqrt{z_1^2 + z_2^2 + \cdots + z_M^2}$. The temporal POD coefficients corresponding to each mode can be calculated as $\alpha^n = \psi^T g'^n$, where $\psi = [\phi^1, \ldots, \phi^N]$. The participant energy coefficient, ζ_i, denotes the proportion of total energy contained in the ith mode and is expressed as [20]

$$\zeta_i = \frac{\lambda_i}{\sum_{i=1}^{N} \lambda_i} \tag{12}$$

On the other hand, the cumulative energy coefficient η_i represents the proportion of the total energy contained in the first i number of modes and is given by [20]

$$\eta_i = \frac{\sum_{i=1}^{L} \lambda_i}{\sum_{i=1}^{N} \lambda_i} \quad L \le N \tag{13}$$

A reduced order model of the system can be constructed from the first L number of modes of the system that contain the bulk of the energy, and its spatial distribution at the kth time step is expressed as

$$g(x, y, t_k) = \bar{g}(x, y) + \sum_{i=1}^{L} \alpha_i^k(t)\phi^i(x, y) \tag{14}$$

In the present chapter, a brief review of the dynamics of non-reacting and reacting flows past bluff bodies is presented. The chapter primarily focuses on cylindrical bluff bodies having a circular cross section. The usage of the POD technique for analyzing such flows is highlighted as well. Some recent advancements on the dynamics of flow past bluff bodies when a second fluid is injected from the bluff body in both non-reacting and reacting scenarios have been presented as well. The chapter concludes by summarizing the key findings and laying out the future scope of research in this area.

2 Non-reacting Flow

Non-reacting flows past bluff bodies have immense engineering significance. The alternate shedding of vortices in the near wake of a circular cylinder is its most distinguishing feature, as shown in Fig. 1. Such a vortex shedding phenomenon is commonly referred to as the von Kármán vortex street. A vortex shedding in that fashion often leads to large transverse forces arising due to the pressure fluctuations, leading to structural vibrations, acoustic noise, or resonance [21], which may lead to structural failure. Zdravkovich [22] analyzed the flow past smooth and rough cylinders for a large range of Reynolds numbers and highlighted three transitions—in-the-near wake, along the free shear layers, and along the boundary layers. Each transition was also associated with several disturbance-sensitive flow regimes.

In practical situations, such as bluff body stabilized combustors, where there is a confinement present, the effect of lateral walls significantly affects the vortex shedding past the bluff body. The influence of the lateral boundaries for a flow Reynolds number of 100 was studied by Behr et al. [3], where it was concluded that the confinement increases the Strouhal number of vortex shedding as compared to the unconfined flow. Numerical simulations [24] also showed that for a fixed blockage ratio, the recirculation zone increases in length while the drag coefficient decreases with increasing Reynolds number. Increased confinement also results [25] in a shorter wake and a delayed transition from a steady attached wake to an unsteady detached wake flow. The presence of a two-dimensional instability in the presence of walls, leading to the beating phenomena of drag and lift coefficients, was also reported [26]. Brandon and Aggarwal [27] numerically investigated particle transport and deposition on a square cylinder placed in confined channel. Particle behavior in the presence of vortices downstream of the bluff body was analyzed, particle deposition was characterized as a function of the Stokes number.

3 Reacting Flow

Stabilization of flames in a reacting stream of high velocity has been a challenging problem for a long time. It has immense importance in a number of aerospace

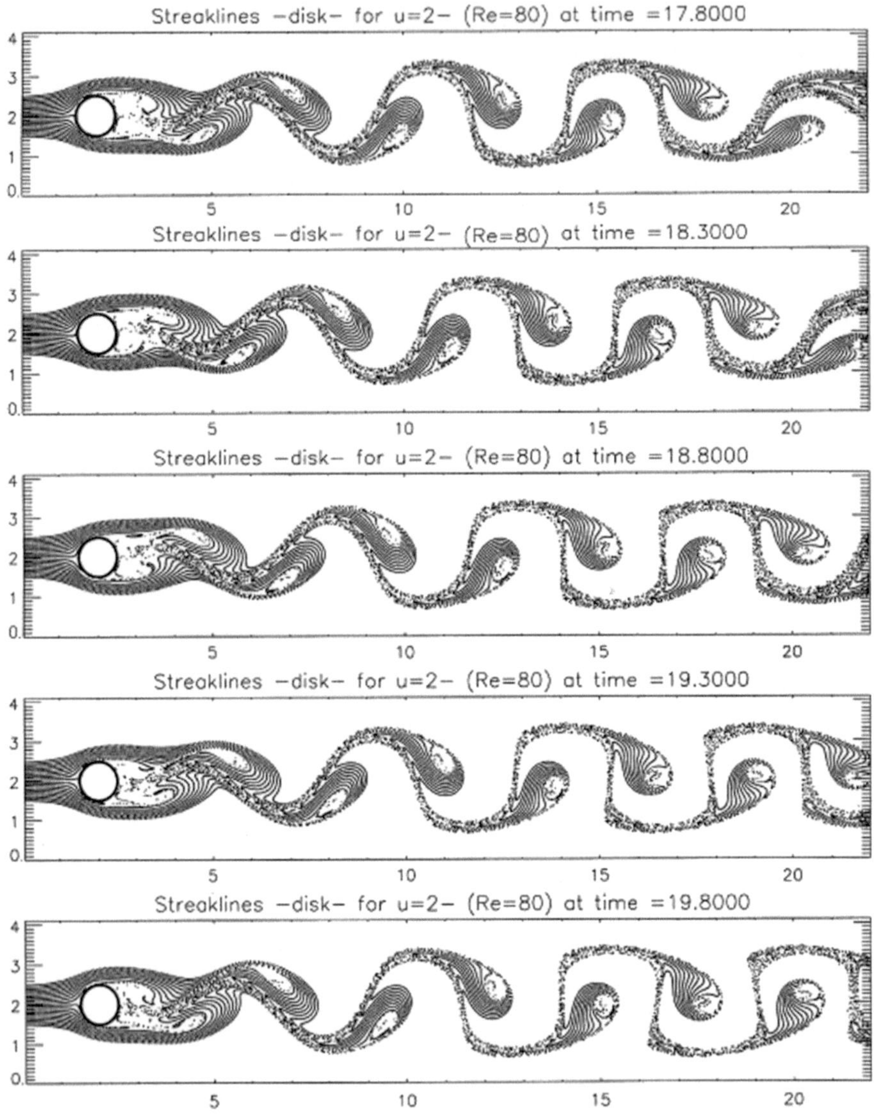

Fig. 1 Evolution of streaklines during one vortex shedding period about 2.0 time units starting at the time of 17.8 units. The Reynolds number is 80. The flow displays well developed von Kármán vortex street. The figure has been reproduced, with permission, from [23]

applications including ramjet and scramjet combustors, and turbojet afterburners. Improper flame stabilization often leads to flame blow-off, which is an undesirable event. Bluff body stabilization is one of the most significant and widely studied approaches for flame stabilization. Bluff body stabilized reacting flow is also a topic of academic interest as it provides a suitable platform to understand turbulent flame characteristics. It is often used as a computational test case to validate turbulence models.

Flame stabilization in a high velocity (such as ramjet and scramjet applications) combustible mixture is difficult to achieve due to two major reasons. Firstly, the incoming gas velocity is several orders of magnitude higher than the flame speed of the gas mixture. Secondly, the temperature of the incoming mixture is much less than the autoignition temperature of the mixture. Along with this, the residence time in the gas mixture is also too small to complete the combustion process in the combustor. The presence of a bluff body in a high-velocity stream addresses all these issues of flame stabilization. It creates a recirculation zone which simultaneously behaves as a momentum sink and heat source to the incoming gas mixture. The wake behind the bluff body also augments the mixing between the hot burnt gas and cold incoming mixture which leads to better combustion. The structure of a bluff body stabilized flame is presented in Fig. 2.

In the previous section, the dynamics of flow past a bluff body in the non-reacting situation has been highlighted. In the present section, characteristics of reacting flows past bluff bodies are emphasized. The heat release rate strongly influences the flow field in the reacting situation. It is found that heat release due to combustion stabilizes the flow past a bluff body. It brings down vorticity magnitudes and turbulence intensity values in comparison to the non-reacting situation. The effect of heat release rate on the stabilization effect can be interpreted from the vorticity transport equation

$$\frac{D\vec{\omega}}{Dt} = \underbrace{(\vec{\omega}\cdot\nabla)\vec{V}}_{\text{vortex stretching}} - \underbrace{\vec{\omega}(\nabla\cdot\vec{V})}_{\text{gas expansion}} - \underbrace{\frac{\nabla\rho\times\nabla p}{\rho^2}}_{\text{baroclinic torque}} + \underbrace{\nu\nabla^2\vec{\omega}}_{\text{viscous diffusion}}$$

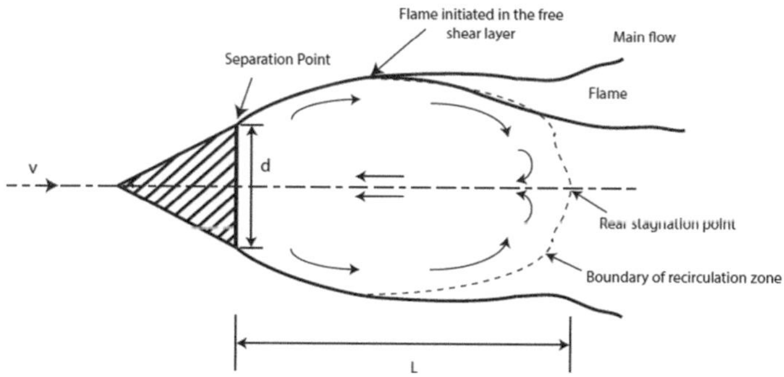

Fig. 2 Recirculation zone behind a bluff body

The last term on the right-hand side of the above equation represents the viscous diffusion of vorticity. This term significantly increases during reacting situations as viscosity increases with temperature. The second term on the right-hand side represents the vorticity destruction (as the negative sign is present) due to flow dilation. The gas expansion across the flame acts as a vorticity sink. This term is proportional to the ratio of unburnt gas density to burnt product density. The third term on the right-hand side of the equation represents the contribution of baroclinic torque. Baroclinicity, which is a measure of misalignment between the gradient of pressure from the gradient of density, also contributes to vorticity generation or destruction. It is found that the vorticity generation due to baroclinic effect for confined bluff body flow is opposite in direction to the vorticity in the shear layer, thus reducing the overall magnitude of the vorticity in the flow field. The aforementioned factors collectively stabilize the flow in the reacting situation.

The early research on bluff body stabilized flames is limited to experiments only. Researchers have investigated the effects of different flow parameters on flame stability for premixed flames. Nicholson and Fields [28] experimentally investigated the mechanism of flame stabilization in the wakes of the bluff body for both non-reacting and reacting premixed flames. They introduced particle tracking and high-speed imaging for flow characterization. It is found that the frequency of oscillation during ignition and blow-off is dependent on the duct dimension.

Williams and Shipman [29] explained the mechanism of flame stabilization of bluff body stabilized flames. They found that the composition of the mixture in the wake strongly affects the stability curve. The flame starts off in the shear layer initiated from the corner of the bluff body. Ignition delay of the mixture causes the flame to stabilize a small distance downstream of the bluff body. Zukoski and Marble [30] explained the flame stabilization mechanism by examining the shape of the recirculation zone for different incoming mixture velocities. It was reported that the size of the recirculation zone increases with the incoming mixture velocity. Longwell et al. [31] explained the stability of the bluff body stabilized flame by assuming the recirculation zone as a finite volume homogeneous chemical reactor. According to their hypothesis, flame extinction occurs when the heat transfer rate from the reaction zone to the unburnt mixture is sufficiently smaller than the chemical reaction rate. Winterfeld [32] experimentally measured the mass and heat exchange between the recirculation zone and the adjacent unburnt gas for an axisymmetric bluff body stabilized flame. These exchanges are strongly dependent on the geometry of the recirculation zone and flow residence time. The author investigated the effects of blockage ratio, equivalence ratio, and inflow velocity on the geometry of the recirculation zone and residence time. It was found that the flame stabilization correlated with the mass exchange through the shear layer. It was also reported that the flame holders with higher apex angle have a higher mass exchange. Winterfeld [33] further identifies that the Damkohler number, which is the ratio of the residence time to the chemical time, and the equivalence ratio are the two important controlling parameters of flame stability. It was observed that the mass and energy exchange reached the maxima close to the stoichiometric condition. Hence, it can be concluded that the highest velocity incoming mixture for a stable flame can be obtained for an

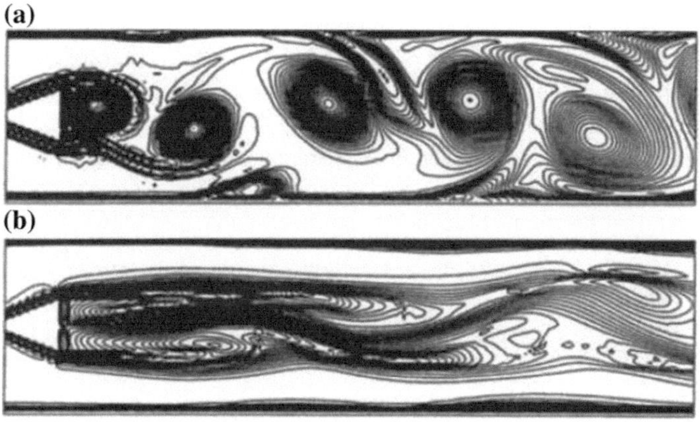

Fig. 3 Snapshot of vorticity in the **a** non-reacting and **b** reacting flow situations. The figure has been reproduced, with permission, from [41]

equivalence ratio equal or close to unity. A large number of experimental studies are available in literature where attempts have been made to correlate flame stability with the different parameters like pressure [34, 35], temperature [36], bluff body geometry [37, 38], etc.

In recent times, an extensive amount of both experimental and numerical studies have been performed to understand the dynamics of the bluff body stabilized flame. Fuji and Eguchi [39] reported experimental results of the comparison of cold and reacting flows around a bluff body flame stabilizer. They observed suppression of turbulence in the wake of the bluff body in the reacting situation. The flow dilatation due to the heat release rate is identified as the reason for this turbulence suppression. Ballal and Lefebvre [40] experimentally investigated the reasons of blow-off for a bluff body stabilized flame. They found that the incoming fresh mixture must be heated up to the ignition temperature for proper flame stabilization. Insufficient heat generation in the reaction zone fails to preheat the incoming mixture and that leads to blow-off. Fureby and Lofstrom [41] numerically investigated the unsteady flow pattern of non-reacting and reacting bluff body stabilized flows using LES. They found that the heat release rate made the flow less structured. The vorticity contours for the non-reacting and reacting flows have been presented in Fig. 3.

Erickson et al. [42] investigated the effect of the temperature ratio on vortex shedding behavior of a bluff body stabilized flame by varying the temperature of the incoming mixture. They found a suppression of the Bénard-von Kármán mode of vortex shedding for higher T_b/T_u. They also observed that the asymmetric vortex shedding becomes symmetric with increasing temperature ratio.

In recent years, blow-off mechanisms of bluff body stabilized flames have gained a lot of attention. Chaudhuri et al. [43] experimentally investigated the blow-off mechanism of two-dimensional bluff body stabilized turbulent premixed flames using simultaneous OH PLIF and PIV. They reported that for stable flames, the maximum OH intensity location coincides with the highest vorticity shear layer. Smith et al. [44] investigated the blowout of a premixed bluff body stabilized flame using large eddy simulation (LES) and compared the simulation with experimental results. Khosla et al. [45] investigated the role of the von Kármán vortex shedding in flame stabilization using LES. A number of experimental and numerical studies [46–49] are present in the literature which can be referred further. Shanbhogue et al. [50] reported a detailed review of bluff body stabilized flame and blow-off studies.

4 Non-reacting and Reacting Flow with Fluid Injection

In bluff body stabilized combustors, if the fuel injection system is so designed that the fuel is injected from the bluff body itself [27–30, 51–53], then the interaction between the vortices from the free stream and the injected stream play a significant role in the dynamics of the whole system. Moreover, suction and blowing on the cylinder surface have been used by other researchers [23] to control vortex shedding as well. An active control strategy of suppressing vortex shedding by suction and blowing was presented by Muralidharan et al. [54] and it was observed that, for a Reynolds number of 100, an injection velocity of 3 times the magnitude of the free stream velocity is able to completely suppress vortex shedding. Liu and Feng [55] used two synthetic jets having a top hat sinusoidal velocity profile placed at the two mean separation points to suppress lift fluctuations on a circular cylinder by inducing a symmetric vortex shedding mode. Recently, Sen et al. [56] investigated, with the help of numerical simulation methods, the flow past a circular cylinder in a rectangular confinement, with a second fluid being injected from two diametrically opposite slots on the cylinder. A rectangular flow domain (200 mm × 60 mm) was considered, and a cylinder (6 mm diameter) was placed centrally within the domain. An air free stream flow ($Re = 100$) was considered, and methane was injected through two diametrically opposite slots on the cylinder, as shown in Fig. 4.

In the present section, the comparison of a reacting flow and a non-reacting flow with injection has been presented. The vorticity contours depicted below (Fig. 5) show the von Kármán vortex street for the non-reacting flow. The vortices are separated downstream of the cylinder. For the reacting flow, the vortex shedding is highly suppressed for $\varepsilon = 1$, where ε is the ratio of the injection velocity of methane from the slotted cylinder to that of the free stream velocity, and no noticeable vortex shedding is observed.

In order to get a better understanding of the difference in flow structures, power spectral density analysis of the temporally fluctuating lift coefficient is performed and reported (Fig. 6). A peak frequency of 5.865 Hz is observed for $\varepsilon = 1$ non-reacting flow. No dominant frequency is observed for the corresponding reacting flow, which reiterates the highly suppressed vortex shedding observed from the CFD data (Fig. 7).

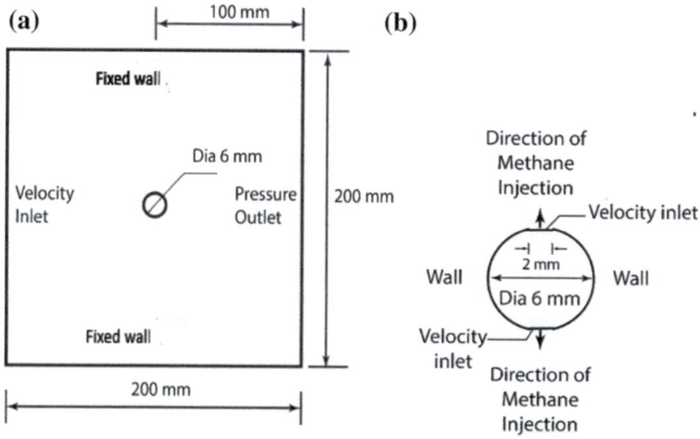

Fig. 4 Schematic of **a** computational domain and **b** bluff body

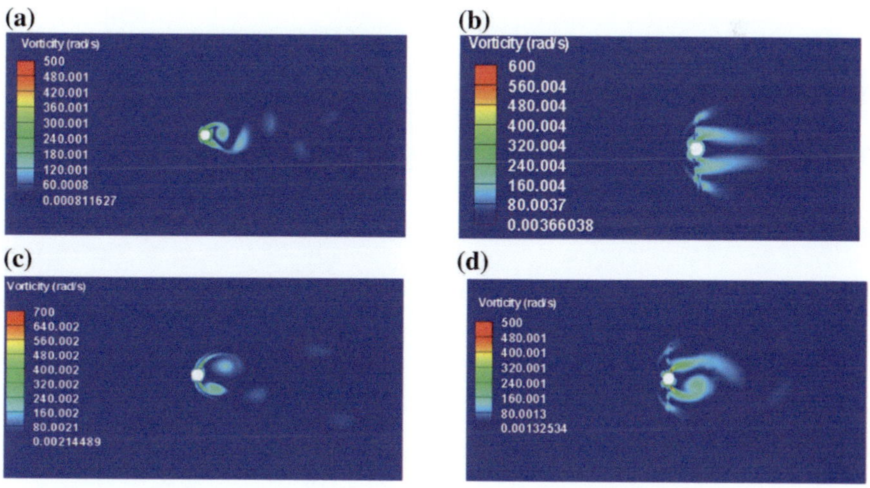

Fig. 5 Vorticity contours for **a** $\varepsilon = 1$ non-reacting flow, **b** $\varepsilon = 1$ reacting flow, **c** $\varepsilon = 2$ non-reacting flow, and **d** $\varepsilon = 2$ reacting flow. Taken from Bagchi et al. [57]. Reprinted with permission from Springer Nature

A similar kind of observation is found for $\varepsilon = 2$. From the PSD plots and the z-vorticity contours, it is apparent that there is spatial coherence in both the flows. In order to analyze it in detail, proper orthogonal decomposition is performed.

The proper orthogonal decomposition modes depict different flow features: some of them are associated with large-scale motion while the rest is due to the small-scale motion. The first mode of both the cases shows the boundary layer separation and is characteristic of the mean motion. The vortex shedding appears to be in the 2S mode, similar to that observed in the vorticity contours for both the flows (Figs. 8 and 9). It can be inferred that these frequencies are primarily due to mixing. For the

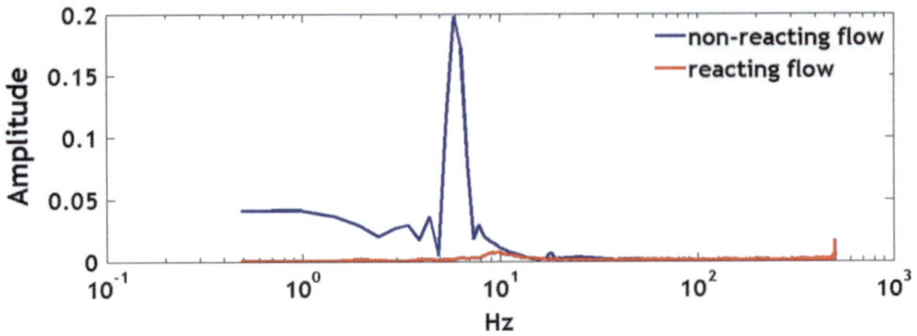

Fig. 6 Power spectral density plot of C_L for $\varepsilon = 1$

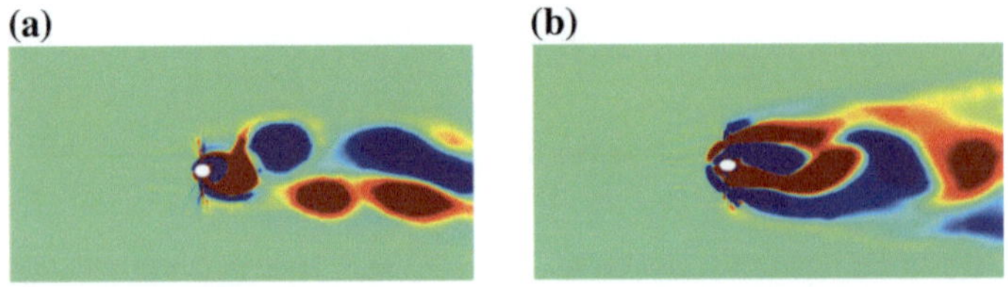

Fig. 7 z-vorticity (s^{-1}) contours for **a** non-reacting flow and **b** reacting flow for $\varepsilon = 2$

Fig. 8 POD modes of non-reacting flow for $\varepsilon = 2$

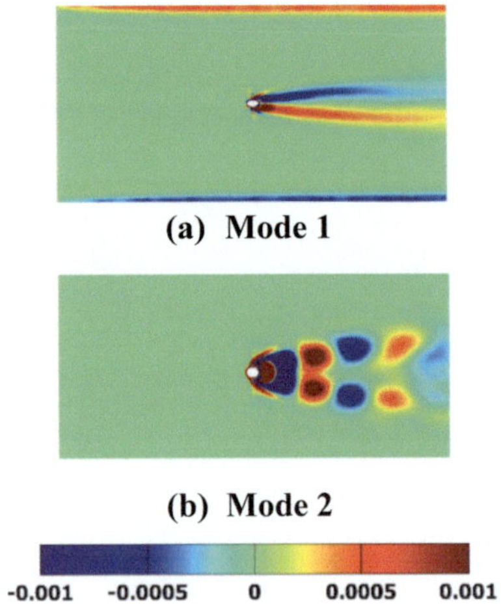

reacting flow, all the modes show distinct symmetry in the flow. The vortex shedding is suppressed by combustion. Mode 2 shows the von Kármán vortex shedding having a single peak frequency of 8.594 Hz (from the spectra of POD temporal coefficients). A 90° phase shift has been observed between all the modes for both the flows.

Furthermore, the effect of injection velocity in the reacting flow has been highlighted in this section. The vorticity contours illustrate the highly suppressed vortex shedding phenomena for the lower injection velocity. No vortices are observed at the wake of the bluff body. The von Kármán vortex street is observed for the higher

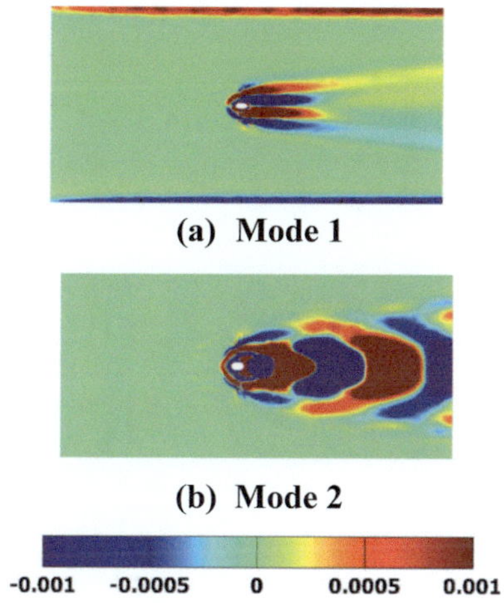

Fig. 9 POD modes of reacting flow for $\varepsilon = 2$

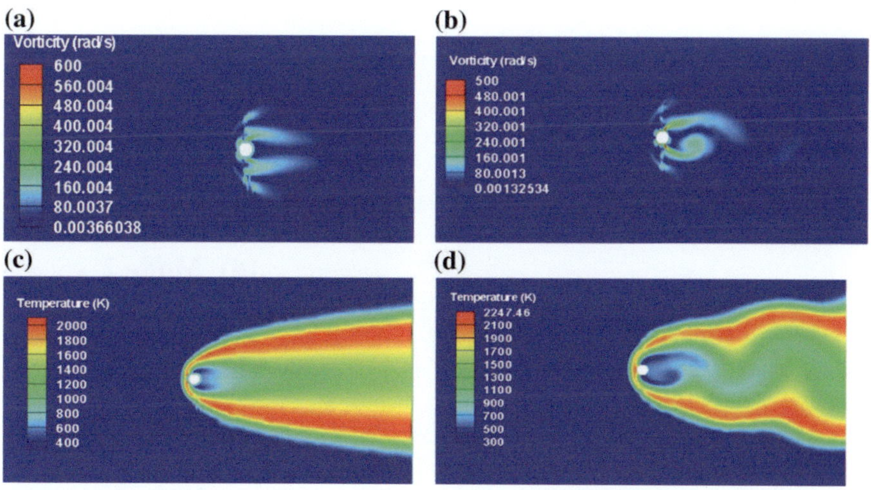

Fig. 10 Vorticity contours for **a** $\varepsilon = 1$ reacting flow and **b** $\varepsilon = 2$ reacting flow; Temperature contours for **c** $\varepsilon = 1$ reacting flow and **d** $\varepsilon = 2$ reacting flow. Taken from Bagchi et al. [57]. Reprinted with permission from Springer Nature

injection velocity. The temperature contour plots for $\varepsilon = 1$ and 2 in the reacting flow case depict the suppressed vortex shedding for the former. The plots show the presence of a comparatively cooler region where the vortices are observed at the wake of the bluff body (Fig. 10).

The suppression of vortex shedding in the lower injection velocity configuration can be observed from the power spectral density plots in Fig. 11, which show very

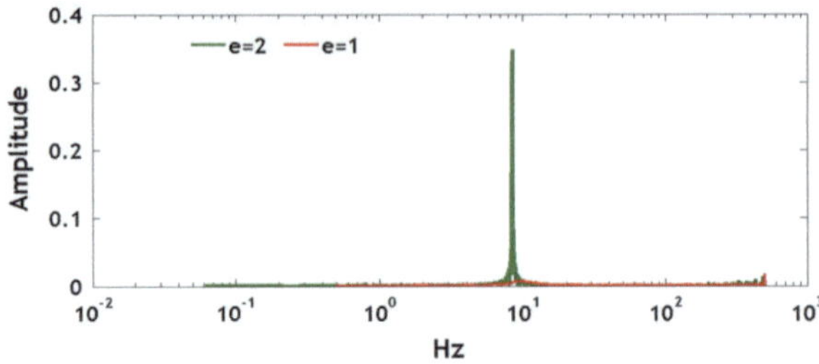

Fig. 11 Power spectral density plot of C_L for $\varepsilon = 1$ versus $\varepsilon = 2$

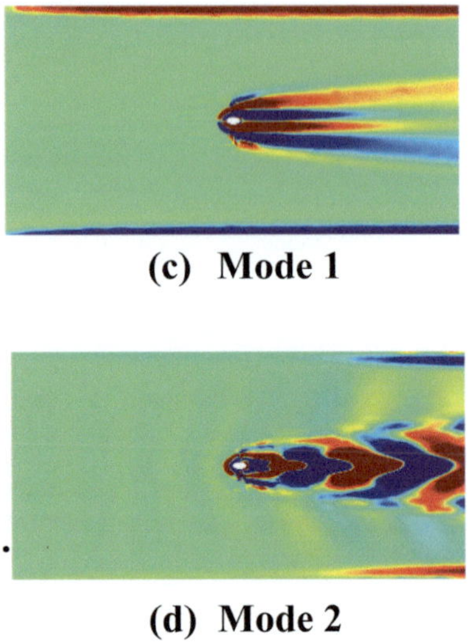

Fig. 12 POD modes for $\varepsilon = 1$ reacting flow

low amplitude of the vortex shedding frequency as compared to the higher injection velocity.

It is found that almost all the energy is contained in the first mode, i.e., the mean flow for $\varepsilon = 1$. The contribution of the small-scale structures is negligible compared to that of the large-scale structures. Hence, the mean flow can be a representation of the overall flow field. For $\varepsilon = 2$, there is a significant contribution of the higher modes on the flow field (Figs. 12 and 13).

The POD modes showing vortex shedding for $\varepsilon = 1$ have negligible energy, and hence have no impact on the overall flow field. Due to the lower injection velocity, the vortex shedding is suppressed. Higher injection velocity is required for better mixing.

Fig. 13 POD modes for $\varepsilon = 2$ reacting flow

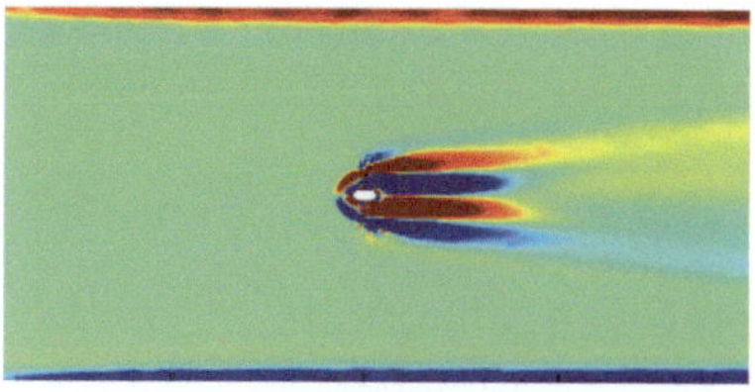

(b) Mode 1

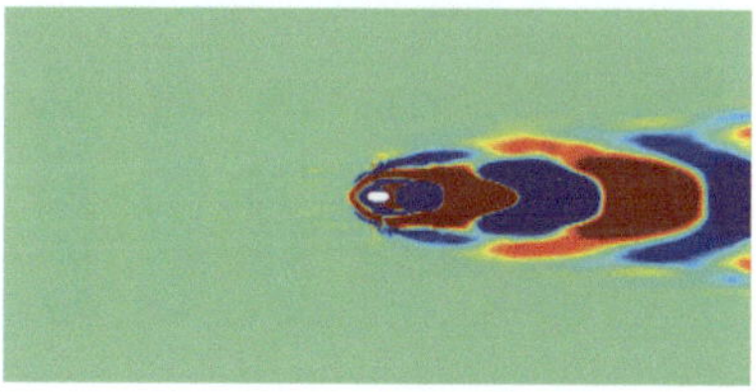

(b) Mode 2

5 Conclusions and Future Scope

The chapter presents an overview of the dynamics of non-reacting and reacting flows past bluff bodies. Researchers have observed different characteristics of the flow pattern in different conditions, the von Kármán vortex street being the most identifiable feature among them. The flow behavior changes significantly in the reacting situation as compared to the non-reacting one. A distinguishable vortex suppression effect is observed in the presence of heat release. Flow dilatation due to heat release is identified as the cause for this effect. The flow past bluff body with injection from the bluff body itself is reported in this chapter. The effect of injection velocity on the flow dynamics in the downstream wake of the cylinder was reported. It is found from the proper orthogonal decomposition that the wake was comprised of large-scale structures at lower injection velocities, while small structures were observed at higher injection velocities. From the power spectral density plots for the reacting flow at an injection velocity equal to free stream velocity, no sharp peak of the frequency of oscillating lift coefficient is extant, thereby indicating negligible and highly suppressed vortex shedding phenomena. However, with increase in the injection velocity, there are multiple peaks with no unique dominant frequency of vortex shedding in the reacting situation.

The effect of injection changes the flow dynamics significantly both in reacting and non-reacting situations. Further analysis using dynamic mode decomposition and spectral proper orthogonal decomposition can be performed to gain a more comprehensive and detailed understanding of the flows. The effect of fluid injection in reacting bluff body stabilized flow is not well reported in literature. Few studies are available in literature, and are only limited to laminar flows. The effect of fluid injection in turbulent reacting bluff body stabilized flow is an area for further investigation.

References

1. Fornberg B (1985) Steady viscous flow past a circular cylinder up to Reynolds number 600. J Comput Phys 61:297–320
2. Braza M, Chassaing P, Minh HH (2006) Numerical study and physical analysis of the pressure and velocity fields in the near wake of a circular cylinder. J Fluid Mech 165:79–130
3. Behr M, Hastreiter D, Mittal S, Tezduyar TE (1995) Incompressible flow past a circular cylinder: dependence of the computed flow field on the location of the lateral boundaries. Comput Methods Appl Mech Eng 123:309–316
4. Wu J, Sheridan J, Welsh MC, Hourigan K (2006) Three-dimensional vortex structures in a cylinder wake. J Fluid Mech 312:201–222
5. Jiang H, Cheng L, An H (2018) Three-dimensional wake transition of a square cylinder. J Fluid Mech 842:102–127
6. Ng ZY, Vo T, Sheard GJ (2018) Stability of the wakes of cylinders with triangular cross-sections. J Fluid Mech 844:721–745
7. Laroussi M, Djebbi M, Moussa M (2014) Triggering vortex shedding for flow past circular cylinder by acting on initial conditions: a numerical study. Comput Fluids 101:194–207
8. Gillies EA (1998) Low-dimensional control of the circular cylinder wake. J Fluid Mech 371:157–178
9. Feng L-H, Wang J-J, Pan C (2011) Proper orthogonal decomposition analysis of vortex dynamics of a circular cylinder under synthetic jet control. Phys Fluids 23:014106
10. Siegel SG, Seidel J, Fagley C, Luchtenburg DM, Cohen K, McLaughlin T (2008) Low-dimensional modelling of a transient cylinder wake using double proper orthogonal decomposition. J Fluid Mech 610:1–42
11. Rajaee M, Karlsson SKF, Sirovich L (2006) Low-dimensional description of free-shear-flow coherent structures and their dynamical behaviour. J Fluid Mech 258:1–29
12. Rowley CW, Marsden JE (2000) Reconstruction equations and the Karhunen-Lòeve expansion for systems with symmetry. Physica D 142:1–19
13. Willcox K, Peraire J (2002) Balanced model reduction via the proper orthogonal decomposition. AIAA J 40:2323–2330
14. Citriniti JH, George WK (2000) Reconstruction of the global velocity field in the axisymmetric mixing layer utilizing the proper orthogonal decomposition. J Fluid Mech 418:137–166
15. Kosambi DD (1943) Statistics in functions space. J Ind Math Soc
16. Ma XIA, Karniadakis GE (2002) A low-dimensional model for simulating three-dimensional cylinder flow. J Fluid Mech 458:181–190
17. Sirovich L (1987) Turbulence and the dynamics of coherent structures part I: coherent structures. Q Appl Math 45:561–571
18. Wang HF, Cao HL, Zhou Y (2014) POD analysis of a finite-length cylinder near wake. Exp Fluids 55:1790
19. Delville J, Ukeiley L, Cordier L, Bonnet JP, Glauser M (1999) Examination of large-scale structures in a turbulent plane mixing layer. Part 1. Proper orthogonal decomposition. J Fluid Mech 391:91–122
20. Ding P, Wu X-H, He Y-L, Tao W-Q (2008) A fast and efficient method for predicting fluid flow and heat transfer problems. J Heat Transfer 130:032502–032502-17
21. Williamson CHK (1996) Vortex dynamics in the cylinder wake. Annu Rev Fluid Mech 28:477–539
22. Zdravkovich MM (1990) Conceptual overview of laminar and turbulent flows past smooth and rough circular cylinders. J Wind Eng Ind Aerodyn 33:53–62

23. Li Z, Navon IM, Hussaini MY, Le Dimet FX (2003) Optimal control of cylinder wakes via suction and blowing. Comput Fluids 32:149–171
24. Chakraborty J, Verma N, Chhabra RP (2004) Wall effects in flow past a circular cylinder in a plane channel: a numerical study. Chem Eng Process 43:1529–1537
25. Singha S, Sinhamahapatra KP (2010) Flow past a circular cylinder between parallel walls at low Reynolds numbers. Ocean Eng 37:757–769
26. Griffith MD, Leontini J, Thompson MC, Hourigan K (2011) Vortex shedding and three-dimensional behaviour of flow past a cylinder confined in a channel. J Fluids Struct 27:855–860
27. Brandon DJ, Aggarwal SK (2001) A numerical investigation of particle deposition on a square cylinder placed in a channel flow. Aerosol Sci Technol 34:340–352
28. Nicholson HM, Field JP (1948) Some experimental techniques for the investigation of the mechanism of flame stabilization in the wakes of bluff bodies. In: Symposium on combustion and flame, and explosion phenomena, vol 3, pp 44–68
29. Williams GC, Shiman CW (1953) Some properties of rod-stabilized flames C homogeneous gas mixtures. Symp (Int) Combust 4:733–742
30. Zukoski EE, Marble FE (1955) Combustion research and reviews. Butterworth, London, p 167
31. Longwell JP, Frost EE, Weiss MA (1953) Flame stability in bluff body recirculation zones. Ind Eng Chem 45:1629–1633
32. Winterfeld G (1965) On processes of turbulent exchange behind flame holders. Symp (Int) Combust 10:1265–1275
33. Winterfeld G (1968) On the stabilization of hydrogen diffusion flames by flame-holders in supersonic flow at low stagnation temperatures. In: Smith IE (ed) Combustion in advanced gas turbine systems. Pergamon, pp 95–112
34. Scurlock AC (1948) Flame stabilization and propagation in high velocity gas streams. Mass Inst Technol Fuels Res Lab Meteor Report 19
35. De Zubay EA (1950) Characteristics of disk-controlled flames. Aeronaut Dig 61:54–56
36. Haddock GW (1951) Flame blowoff studies of cylindrical flameholders in channeled flows. Progress Report, pp 3–24
37. Maestre A (1955) Study of the stability limits with respect to obstacles flow resistance. Combust Res Rev
38. Wright FH (1959) Bluff-body flame stabilization: blockage effects. Combust Flame 3:319–337
39. Fujii S, Eguchi K (1981) A comparison of cold and reacting flows around a bluff-body flame stabilizer. J Fluids Eng 103:328–334
40. Ballal DR, Lefebvre AH (1975) The structure and propagation of turbulent flames. Proc R Soc Lond. A. Math Phys Sci 344:217
41. Fureby C, Löfström C (1994) Large-eddy simulations of bluff body stabilized flames. Symp (Int) Combust 25:1257–1264
42. Erickson R, Soteriou M, Mehta P (2006) The influence of temperature ratio on the dynamics of bluff body stabilized flames. In: 44th AIAA aerospace sciences meeting and exhibit, American Institute of Aeronautics and Astronautics
43. Chaudhuri S, Kostka S, Tuttle SG, Renfro MW, Cetegen BM (2011) Blowoff mechanism of two dimensional bluff-body stabilized turbulent premixed flames in a prototypical combustor. Combust Flame 158:1358–1371
44. Smith C, Nickolaus D, Leach T, Kiel B, Garwick K (2007) LES blowout analysis of premixed flow past V-gutter Flameholder. In: 45th AIAA aerospace sciences meeting and exhibit, American Institute of Aeronautics and Astronautics
45. Khosla S, Leach T, Smith C (2007) Flame stabilization and role of von Karman vortex shedding behind bluff body flameholders. In: 43rd AIAA/ASME/SAE/ASEE joint propulsion conference & exhibit, American Institute of Aeronautics and Astronautics
46. Kiel B, Garwick K, Lynch A, Gord J, Meyer T (2006) Non-reacting and combusting flow investigation of bluff bodies in cross flow. In: 42nd AIAA/ASME/SAE/ASEE joint propulsion conference & exhibit, American Institute of Aeronautics and Astronautics
47. Lieuwen T, Shanbhogue S, Khosla S, Smith C (2007) Dynamics of bluff body flames near blowoff. In: 45th AIAA aerospace sciences meeting and exhibit, American Institute of Aeronautics and Astronautics

48. Nair S, Lieuwen TC (2007) Near-blowoff dynamics of a bluff-body stabilized flame. J Propul Power 23:421–427
49. Dawson JR, Gordon RL, Kariuki J, Mastorakos E, Masri AR, Juddoo M (2011) Visualization of blow-off events in bluff-body stabilized turbulent premixed flames. Proc Combust Inst 33:1559–1566
50. Shanbhogue SJ, Husain S, Lieuwen T (2009) Lean blowoff of bluff body stabilized flames: scaling and dynamics. Prog Energy Combust Sci 35:98–120
51. Mondal S, Mukhopadhyay A, Sen S (2014) Dynamic characterization of a laboratory-scale pulse combustor. Combust Sci Technol 186:139–152
52. Shahi M, Kok JBW, Pozarlik AK, Roman Casado JC, Sponfeldner T (2013) Sensitivity of the numerical prediction of turbulent combustion dynamics in the limousine combustor. J Eng Gas Turbines Power 136: 021504–021504-12
53. Shahi M, Kok JBW, Roman Casado JC, Pozarlik AK (2018) Strongly coupled fluid–structure interaction in a three-dimensional model combustor during limit cycle oscillations. J Eng Gas Turbines Power 140:061505–061505-10
54. Muralidharan K, Muddada S, Patnaik BSV (2013) Numerical simulation of vortex induced vibrations and its control by suction and blowing. Appl Math Model 37:284–307
55. Liu Y-G, Feng L-H (2015) Suppression of lift fluctuations on a circular cylinder by inducing the symmetric vortex shedding mode. J Fluids Struct 54:743–759
56. Sen U, Mukhopadhyay A, Sen S (2017) Effects of fluid injection on dynamics of flow past a circular cylinder. Eur J Mech B Fluids 61:187–199
57. Bagchi S, Sarkar S, Sen U, Mukhopadhyay A, Sen S (2019) Numerical simulation of vortex shedding from a cylindrical bluff-body flame stabilizer. In: Sahoo P, Davim JP (eds) Advances in materials, mechanical and industrial engineering, lecture notes on multidisciplinary industrial engineering. Springer Nature, ISBN: 978-3-319-96967-1

Part V
Transport Processes in Energy Systems

Development of Advanced Cooling Technologies for Sustainable Future

Tao Cao and Yunho Hwang

Keywords High-efficiency cooling · Heat pump · Separate sensible and latent cooling

Abbreviations

AC	Air-conditioning
BDM	Bone-dry mass
CD	Clothes dryer
COP	Coefficient of performance
DBT	Dry bulb temperature
DOE	Department of Energy
DW	Desiccant wheel
EEV	Electronic expansion valve
EF	Energy factor
ERV	Energy recovery ventilation
HPCD	Heat pump clothes dryer
HPWH	Heat pump water heater
HVAC	Heating, ventilation, and air-conditioning
MRT	Mean radiation temperature
OU	Outdoor unit
PDM	Post-dry mass

T. Cao · Y. Hwang (✉)
University of Maryland, College Park, USA
e-mail: yhhwang@umd.edu

T. Cao
e-mail: taocao@umd.edu

A. K. Gupta et al. (eds.), *Innovations in Sustainable Energy and Cleaner Environment*, Green Energy and Technology,
https://doi.org/10.1007/978-981-13-9012-8_19

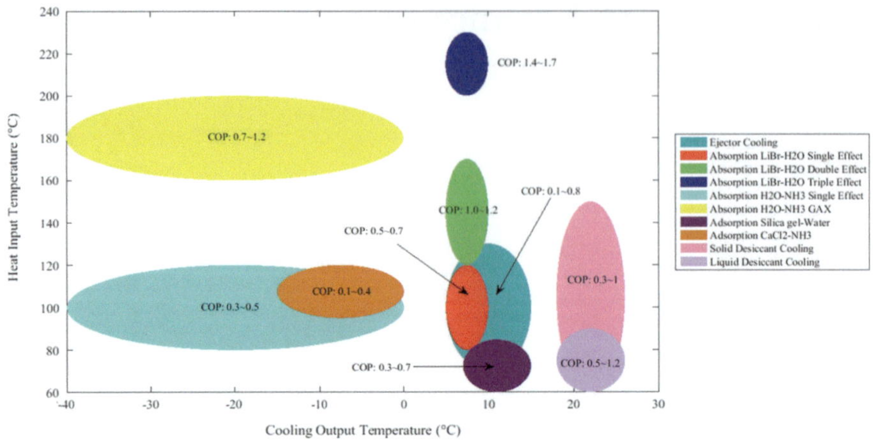

Fig. 1 Comparison of thermally activated cooling technologies

RHX	Radiant heat exchanger
SSLC	Separate sensible and latent cooling
TXV	Thermal expansion valve
VCC	Vapor compression cycle
VRF	Variable refrigerant flow
WBT	West bulb temperature

1 Overview

Cooling and heating demands have persistently existed in many sectors, such as building, transportation, and appliances, to name a few. In residential buildings, the majority of total energy use is for space heating and cooling, and water heating, which accounts for 50 and 18%, respectively (Department of Energy (DOE) [1]). It is expected heating and cooling demands will increase with higher living standards and indoor air quality demands.

Energy-efficient heating and cooling technologies have been surging as solutions to sustainable future. Heating and cooling technologies can be categorized based on their energy input formats. Two prevailing ones are thermal energy and electric energy. Vapor compression cycle-based cooling equipment is a representative of electric powered technology. Likewise, the absorption cycle represents thermal powered technology.

It is unrealistic to make a list and describe all technologies, considering many availabilities. Figure 1 compares some of thermally activated cooling cycles with required heat input temperatures, cooling output temperatures, and COP ranges [2]. It is observed each technology has its own applicable ranges.

This chapter will provide three case studies for energy-efficient cooling and heating technologies. The first one is utilizing an innovative heat pump cycle to fulfill heating demands in clothes dryers [3]. The second one is an innovative approach to combine electric and thermal driven cooling technologies to achieve a high-efficiency air conditioner [4]. The last one is an advanced integrated design of efficient cooling and heating system for a full-scale house [5]. The three cases are research outcomes of three projects the author led before.

2 Energy-Efficient Heat Pump Clothes Dryer

2.1 Background

This part is partially an outcome of a research project on designing ultra-efficient heat pump clothes dryer (HPCD). The team successfully developed a prototype with 59% energy saving over commercial electric dryers. The prototype won the first place in *MaxTech and Beyond* competition held by Lawrence Berkeley National Lab and Department of Energy in 2013.

In the USA, the annual electric consumption of residential electric clothes dryers (CDs) is approximately 71 TW h and four percent of total residential electricity consumption [6]. Electric or gas-powered CDs are prevailing in the US residential market. They are representing 79 and 21% of market share in 2010, respectively. HPCDs, which consume about one-third less electric energy, have been slowly stepping into the European and East Asia markets [7]. It is projected that HPCDs' market share in those regions should continue to gain in the next few years. To the author's best knowledge, there are no HPCD models available in the US market.

2.2 Technology

Figure 2 shows the schematic diagram of an electric CD. Room air flows through the electric heater and becomes hot air and dry air. It then takes off moisture from wet clothes in the drum. Electric CDs are usually vented, meaning hot air and humid air is dumped to the ambient. As the core component in the system, the electric heater consumes most of the energy. Due to the open-cycle operation, the energy in the discharge air is wasted.

Figure 3 shows the schematic diagram of a HPCD. A HPCD uses the condenser heat from the heat pump cycle as its heat source. The air leaving the drum passes through an evaporator where the moisture in the air is condensed and removed. The cooled air is later heated up again through the condenser before entering the drum. As shown in Fig. 3, HPCDs usually operate in a closed-cycle manner. Most energy in the HPCDs is consumed by compressors to drive the vapor compression cycle.

Figure 4 shows the schematic diagram of the two-stage HPCD, which is the concept for the award-winning prototype. The superheated vapor-phase refrigerant is discharged from the compressor and flows through the condenser, followed by the upper-stage expansion valve. After the expansion process to intermediate pressure level through the valve, the refrigerant becomes two-phase and then separates in the flash tank. From the tank, only the refrigerant in liquid phase enters the lower-stage expansion valve, is expanded, and enters the evaporator. After evaporation, it enters the compressor suction port. Meanwhile, the vapor-phase refrigerant coming out the flash tank is injected to the compressor through the injection port. The airflow path is the same as in the single-stage HPCD. The two-stage vapor injection cycle is expected to have higher heating capacities as compared to single-stage cycle, as more refrigerants flow through the condenser. In addition, compact heat exchangers were utilized for enhanced heat transfer performance. A high-efficiency brushless DC fan motor was selected to provide a designed airflow rate.

Fig. 2 Schematic diagram of electric clothes dryer

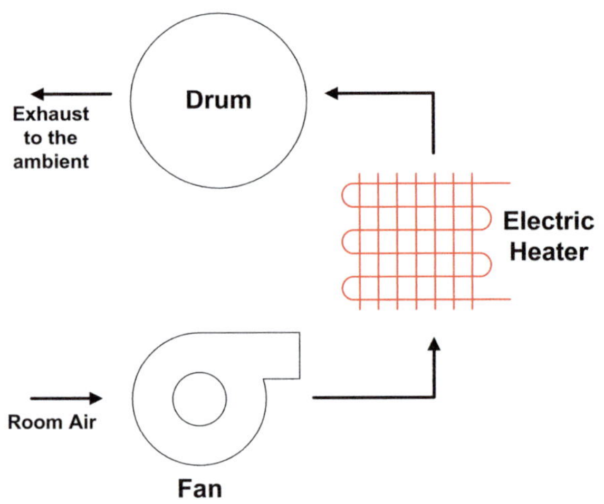

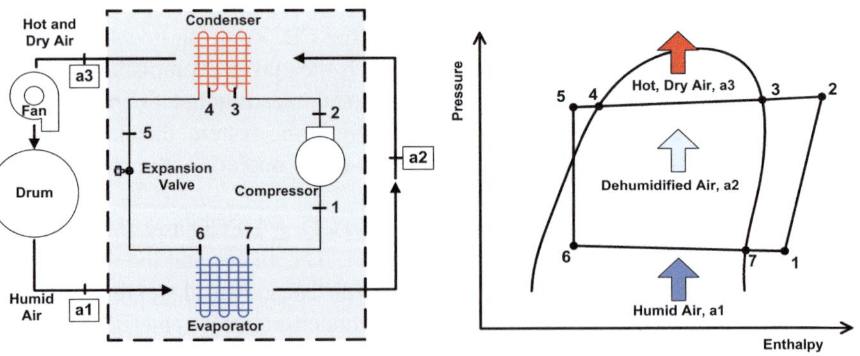

Fig. 3 Schematic diagram and P-h diagram of single-stage HPCD

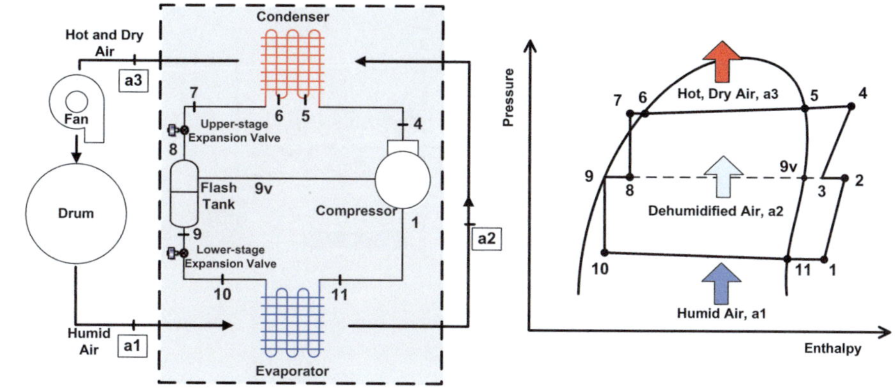

Fig. 4 Schematic diagram and P-h diagram of a two-stage HPCD [3]

2.3 *Experimental Analysis*

2.3.1 Test Facility

Figure 5 shows the schematic diagram of the two-stage HPCD test facility. The refrigerant loop works as described previously in Sect. 2.2. Note a heating tape is used after flash tank upper-stage outlet, shown as 8 in Fig. 5. The purpose is to ensure only vapor refrigerant is injected to the compressor. The heating tape has an energy consumption about 30 W. R134a was used as the working fluid. For air loop, hot air and humid air coming out the drum first passes through the evaporator, where moisture is removed. The resulting cold air and dry air then passes through the condenser, absorbs heat, and becomes hot air and dry air. It is then ready to pick up moisture from wet clothes again.

Several kinds of sensors were used in the test facility to enable optimum control strategy and in-depth performance monitoring. T-type thermocouples were used to measure the temperature of refrigerant and airflows. Pressure transducers were used to measure the absolute pressure of R134a. Mass flowmeter was installed at the condenser outlet to measure the mass flow rate of R134a. Two relative humidity sensors were installed at evaporator upstream and condenser downstream to calculate moisture removed by the vapor injection cycle. In addition, a watt meter was used to measure compressor power consumption and another one was used for fan motors.

Data acquisition software was programmed in LabVIEW to control the two-stage HPCD operations and monitor sensor readings. Motors were first turned on to drive the drum and blower. Compressor and solenoid valve were then turned on. To protect the compressor, only vapor-phase refrigerant should be injected into the compressor. Therefore, a control strategy based on the superheat degrees was developed to ensure no two-phase refrigerant flows into the compressor. Based on the absolute pressure readings from evaporator outlet and flash tank's vapor outlet, corresponding saturation temperatures can be calculated. Along with temperature readings from the

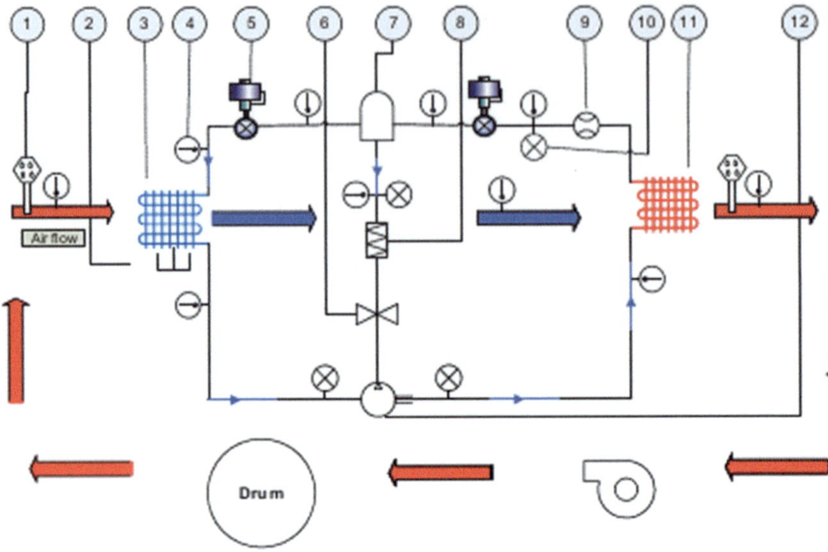

Fig. 5 Schematic diagram of the two-stage HPCD test setup [3]. (1) Relative humidity sensor; (2) Drainage collection system of evaporator; (3) Evaporator; (4) Thermocouple; (5) Electronic expansion valve; (6) Solenoid valve; (7) Flash tank; (8) Heating tape; (9) Mass flow meter; (10) Pressure transducer; (11) Condenser; (12) Vapor injection compressor

above two points, we can obtain the two superheat degrees: vapor injection superheat degree and evaporator outlet superheat degree. The vapor injection and evaporator outlet superheat degrees were set to 10 and 15 K, respectively, by controlling the two EEV openings (shown in Fig. 5). Manual control on EEV was also available as a backup option.

2.3.2 System Tested

Three clothes dryers listed in Table 1 were selected for experimental study. The baseline is an electric dryer from USA market. A hybrid HPCD was selected from European market. It has two working modes: eco mode and speed mode. In the eco mode, only the vapor compression cycle (VCC) works. In the speed mode, both the VCC and electric heater work to facilitate the drying process. The last one is the two-stage HPCD developed by University of Maryland team. Specifications of three products are summarized in Table 1. Costs of electric dryer and hybrid HPCD are market price. The price of the two-stage HPCD is an estimated total. More details on cost estimation are provided in cost analysis section.

Table 1 Specifications of three tested CDs

Products type	Capacity (ft^3)	Cost ($)
Electric CD (state-of the-art product in USA)	7.0	650
Hybrid HPCD (commercial product in Europe)	4.1	1,700
Two-stage HPCD (prototype)	7.8	900

2.3.3 Test Standard

Measuring the energy consumption of clothes dryer is described in test standards from Department of Energy (DOE) and applied in our tests [8]. For testing, blended fabric clothes of half cotton and half polyester were prepared. The total mass of dry clothes is referred to as bone-dry mass. Following standard instructions, we added 37.8 °C water to dry clothes to achieve 70% moisture of bone-dry mass in wet clothes. The mass of clothes after drying process or post-dry mass should be within three percent difference of bone-dry mass. The units were tested in chambers with controlled ambient conditions of 25 °C, 50% relative humidity. Though the standard specifies recommended testing loads or bone-dry mass (BDM) for dryers with different capacities, the studies indicate that increasing testing loads increases dryer performance [9]. Therefore, we test all three dryers at recommended loads and highest loads. The latter test enables comparison under the best performance for each unit.

2.3.4 Results Analysis

To evaluate clothes dryer performance, several metrics are commonly used. The DOE standard recommends energy factor (EF) based on bone-dry mass. It is defined as bone-dry mass over total energy consumption in a drying cycle. Some literature instead uses EF based on water removed from clothes. It is defined as total water removed over total energy consumption in a drying cycle. As we mentioned in the previous section, test standard specifies water added in bone-dry mass should be 70% of bone-dry mass, and post-dry mass should be within 3% deviation from bone-dry mass. Therefore, EF based on bone-dry mass varies proportionally to EF based on water removed. As a result, we only chose EF based on bone-dry mass in this study. In addition, we introduce a metric to better illustrate the energy saving potential of each dryer. It is defined as total energy consumption in a drying cycle over bone-dry mass.

Detailed test results of three clothes dryers are presented in Table 2. Note the commercial HPCD was tested under both eco mode and speed mode. For each mode, standard load and maximum load were both tested. The same two tests were also conducted for the two-stage HPCD. For the electric dryer, only the maximum load

Table 2 Summary of three CDs' test results [3]

Clothes dryer	Test No.	Bone dry-mass (g)	Wet mass pre-dry (g)	Dry mass post-dry (g)	Total water removed (g)	Power consumption (kWh)	Energy factor (BDW_{lbs}/kWh)
GE electric CD	1	13,620	23,160	13,633	9,527	6.97	4.30 ± 0.01
LG hybrid HPCD (speed mode)	1	3,360	5,711	3,336	2,375	1.89	3.91 ± 0.02
	2	3,347	5,680	3,337	2,343	1.80	4.09 ± 0.02
LG hybrid HPCD (eco mode)	1	3,336	5,678	3,347	2,331	0.94	7.83 ± 0.04
	2	3,337	5,708	3,341	2,367	0.95	7.67 ± 0.04
Two-stage HPCD	1	6,170	10,501	6,100	4,401	1.34	10.15 ± 0.03
	2	9,986	16,985	10,098	6,887	2.10	10.44 ± 0.03

test is shown here. Higher energy factor means better drying performance. It is observed that the two-stage HPCD performed best among all three dryers. Also, the EF increases with higher bone-dry mass (BDM). The best performance case for each dryer was selected for further comparison in Table 3. It is observed that the two-stage HPCD improved the energy factor by 143% and achieved 59% energy saving, as compared to electric clothes dryer. Compared to the commercial HPCD, the two-stage HPCD improved the energy factor by 33% and reduced energy consumption by 25%. Note residential electric clothes dryers consume about 71 TW h/year in the USA. Therefore, the two-stage HPCD could achieve up to 42 TW h/year energy saving, should it replace all electric clothes dryers in the market.

2.4 Cost Analysis

Comprehensive investigations on cost of major clothes dryers in the market were conducted. Table 4 presents average costs of CDs available in the market.

Electric dryers consume the most energy, averaging 6.41 kWh per cycle, but also present the most capacity (33 lbs on average) and the quickest drying time at 60 min. Heat pump dryers offer the opposite: minimum capacity (17 lbs), but the lowest energy consumption averaging 2.04 kWh. Heat pump clothes dryers have 167 min on average for a cycle, nearly 2.8 times compared to the electric dryers. Condenser

Table 3 Comparison of three CDs' energy performance [3]

Index	GE Electric CD	LG hybrid HPCD	Two-stage HPCD
Test load (lbs)	30	7.3	22
Drying time (h)	1.7	2.0	2.0
Power consumption (kWh)	6.97	0.94	2.10
Energy factor (BDW_{lbs}/kWh)	4.30	7.83	10.44
EF improvement	Baseline	82%	143% wrt electric CD 33% wrt hybrid HPCD
Energy consumption (kWh/BDW_{lbs})	0.233	0.128	0.096
Energy consumption reduction	Baseline	45%	59% wrt electric CD 25% wrt hybrid HPCD

Table 4 Average cost of CDs

Type	Cost (2010$)	Capacity (lbs)	Power (kW)	Energy consumed/cycle (kWh)
Electric	704.61	32.84	6.41	6.41
Condenser	545.90	17.38	2.50	3.92
Heat pump	1,026.72	16.95	0.84	2.04

dryers stand in between in all aspects except capacity in which lies closer toward that of heat pump dryers. Since condenser dryers have very small market shares, they were not considered in this study.

The two-stage HPCD is financially attractive compared to electric dryers only if the energy saved is able to offset capital investment difference. The payback period is introduced. It is defined as the amount of time the heat pump clothes dryer needs to operate in order for energy savings to pay off the capital investment difference, compared to electric dryers.

The two-stage HPCD consists of the following major components: drum, drum motor, vapor compression system, flash tank, and brushless direct current (DC) fan motors. For the cost estimation, we assumed that the cost of the vapor compression system is half the price of a similar 12,000 Btu capacity window-typed air-conditioning. We then performed market search to determine cost for all other components. We assumed total cost of the labor cost, marketing cost, and profit are the same as the material cost.

Table 5 summarizes payback period analysis for all three dryers. We used two baseline scenarios: electric dryer without bottom cabinet and electric dryer with bottom cabinet. Many customers are purchasing bottom cabinets, as many dryer doors are located inconveniently low. In the two-stage HPCD, the vapor compression sys-

Table 5 Payback period analysis of three CDs [3]

Type	Initial cost ($)	Initial cost difference ($)	Savings/year ($/year)	Payback period (year)
GE electric CD	650	Baseline 1		
GE electric CD with bottom cabinet	815	Baseline 2		
LG hybrid HPCD	1,700	1,050 (Baseline 1) 885 (Baseline 2)	29.1	36 (Baseline 1) 30.4 (Baseline 2)
Two-stage HPCD	900	250 (Baseline 1) 85 (Baseline 2)	38.0	6.5 (Baseline 1) 2.2 (Baseline 2)

tem is designed to locate at the bottom cabinet. The bottom cabinet cost is estimated to be $165 based on the available products for commercial dryers.

Several assumptions were used in payback period analysis. The estimated average annual energy use is 538 kWh for each electric dryer. Electricity price is $0.12/kWh. The two-stage HPCD is a highly competitive and promising new product, considering its 6.5 years payback period compared to electric dryers without bottom cabinet, and 2.2 years payback period compared to electric dryers with bottom cabinet.

2.5 *Summary*

We have successfully developed an energy-efficient two-stage HPCD utilizing advanced vapor injection cycle, compact heat exchangers, and brushless DC fan motor. The experimental evaluations of the two-stage HPCD and commercial clothes dryers demonstrated that the prototype could achieve 59% energy savings and improve the EF by 143% when compared to the electric dryer and achieve 25% energy savings and improve EF by 33% when compared to the heat pump dryer. The prototype has up to 42 TW h annual energy savings should it replace all electric dryers in the US market. It is also a financially attractive investment as it has only 2.2 years payback period compared to electric dryers with bottom cabinets or 6.5 years versus electric dryers without bottom cabinets.

3 A High-Efficiency Residential Air-Conditioning System

3.1 Background

This is partially the outcome of a research project on designing a high-efficiency air-conditioning system. The team delivered a prototype with 30% efficiency improvement over commercial products. The prototype is the winning design of the *MaxTech and Beyond* competition held by Lawrence Berkeley National Lab and Department of Energy in 2012.

To achieve both energy savings and thermal comfort, two unique technologies, separate sensible and latent cooling (SSLC) and radiant heat transfer, were applied in the design. The SSLC technology uses traditional vapor compression cycles (VCC) to provide only sensible cooling to the space, while the latent cooling is delivered by a desiccant wheel (DW). As the VCC provides only the sensible cooling, its evaporating temperature (and corresponding pressure) can be elevated above the dew point temperature of incoming air. For traditional air conditioners, the evaporating temperature is between 5 and 10 °C, but the VCC in the SSLC system can have an evaporating temperature over 15 °C. The increase of evaporating temperature reduces the pressure ratio of the compressor, which has the benefits of better compressor efficiency and a reduced compressor power input. However, to make a successful SSLC air conditioner, the only elevation of evaporating pressure in the VCC is not enough. This is because traditional DWs require electric heater or other forms of heat source to provide hot air over 80 °C for the regeneration of desiccant materials. This extra amount of energy consumption by the heat source can easily offset, if not exceed, that of compressor power savings. Gratefully, the latest research in the field of desiccant material leads to the reduction of regeneration temperature. The newly invented polymer material-based DW used can be regenerated at temperature of 50 °C or even lower. At such a low-temperature requirement, the waste heat from the condenser, which is otherwise released to the ambient air, can be utilized to regenerate the DW. Without the usage of any auxiliary heater, the energy saving from the compressor is maximized in the system level.

Any design of air-conditioning systems should also consider the thermal comfort of occupants. Most of the state-of-the-art air-conditioning (AC) systems can only control the indoor air temperature according to the settings. However, thermal comfort includes other factors such as air velocity and mean radiation temperature (MRT). To put it in a simple way, people can still experience thermally dissatisfactory in the places where although the temperature is within the comfort zone, the local air velocity and MRT are not. Such places can be those spots close to the air outlet of an AC (air velocity is high) or a hot wall (MRT is high). As a measure to enhance the air velocity and MRT controls by AC systems, the team applied a radiant heat exchanger (RHX) to the design. The RHX has an area of around 6 m^2 which can be installed against a wall. Compared to the traditional installation of radiant panels to the ceiling, a wall-attached RHX requires less effort for setup while still can provide good control of MRT inside the conditioned space. The natural convection process caused by the temperature difference between the RHX and other higher temperature surfaces guarantees the air movement at a comfortable speed compared with force convection process does.

3.2 Design Layout

The layout of the SSLC with DW design is demonstrated in Fig. 6. Inside the frame, it contains a set of vapor compression system using R410A as its refrigerant and fans. Figure 7 shows a computer-generated image of the AC unit installed through a wall and the wall mounted RHX. The compressor of the VCC was taken from a commercially available window-type AC system with a cooling capacity of 2.34 kW. An automotive AC component manufacturer supplied four identical micro-channel condensers used in the system. The four condensers were divided into two parts. The first part had one condenser, which was connected from the downstream of the compressor. Ambient air passing through the first part condenser was directed the DW for its regeneration. The refrigerant leaving the first part was then evenly distributed to the second part having three condensers. The second part condensers were connected in parallel with respect to the refrigerant flow direction before it entered the thermostatic expansion valve (TXV). Ambient air flowing through the second part was heated to a lower temperature than that of the first part because the high-temperature single-phase refrigerant was cooled by the first part, while a relatively low-temperature two-phase refrigerant was cooled by the second part. The reason for such an arrangement is to ensure that the desiccant material received an adequate amount of regeneration air at a sufficiently high temperature whereas the entire VCC would not be penalized by an excessive condensing pressure [10]. The refrigerant was then throttled by the TXV from the condensing pressure to the evaporating pressure. The low-pressure refrigerant was divided into two streams. One stream was fed to the conventional evaporator in the system, which was taken from another commercially available window-type AC system. The other stream was fed to the RHX. Then the two streams merged together after passing through the evaporator and RHX and were fed to the compressor. Due to the complication of airflows inside the system, three fans were installed. As shown in Fig. 6, they were used to direct flows toward the regeneration side of the DW, the process side of the DW and the low-temperature condensers. Because of the reduced air enthalpy difference across the evaporator, the SSLC system tends to have a higher airflow rate through the evaporator than that of the conventional system. Therefore, a special attention should be paid for fan selection in order to prevent the excessive fan power consumption, which may offset the compressor power savings. Hence, in the design, all three fans were driven by brushless direct current (brushless DC) motors, which are generally more efficient than the traditional AC motors.

The RHX was built using commercially available radiation panels and copper tubes. Each aluminum radiation panel is 0.1 m wide and 1.2 m long with a joist track on one side and a flat surface on the other side. Copper tubes were snapped

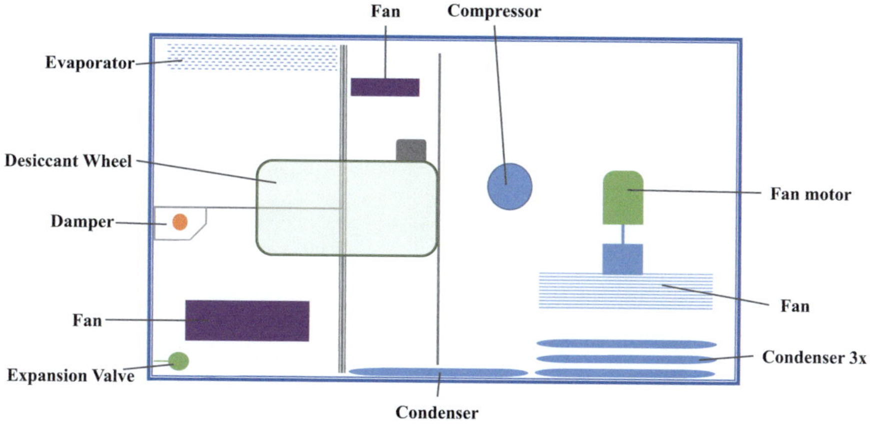

Fig. 6 System layout of SSLC system with DW

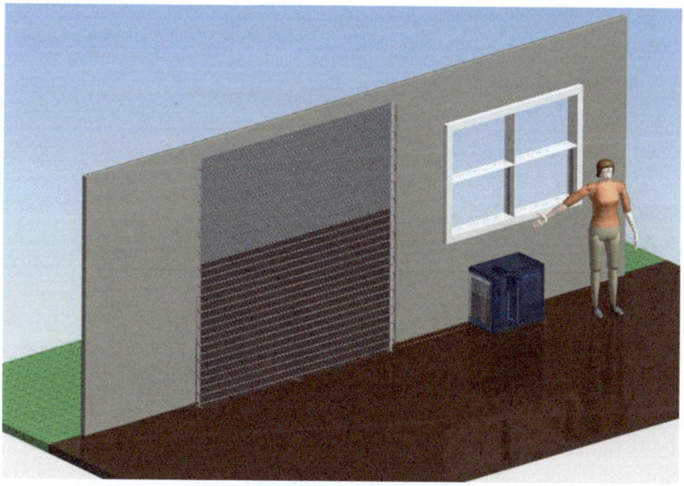

Fig. 7 Computer-generated image of the wall through installation of the AC and the wall mounted RHX

into the joist track and the track was pinched at the open end for a better thermal contact between the copper tube and the panel. All radiation panels were fixed on an aluminum panel and insulation was applied to the tube side of the panels so that the heat transfer only occurred toward the flat surface. Upon the completion of the RHX, six cross-flow fans were installed on the top of the panel to enhance the mixing of air near the RHX. Each fan consumes only nine watts of electricity and hence would not affect the total energy savings from the compressor.

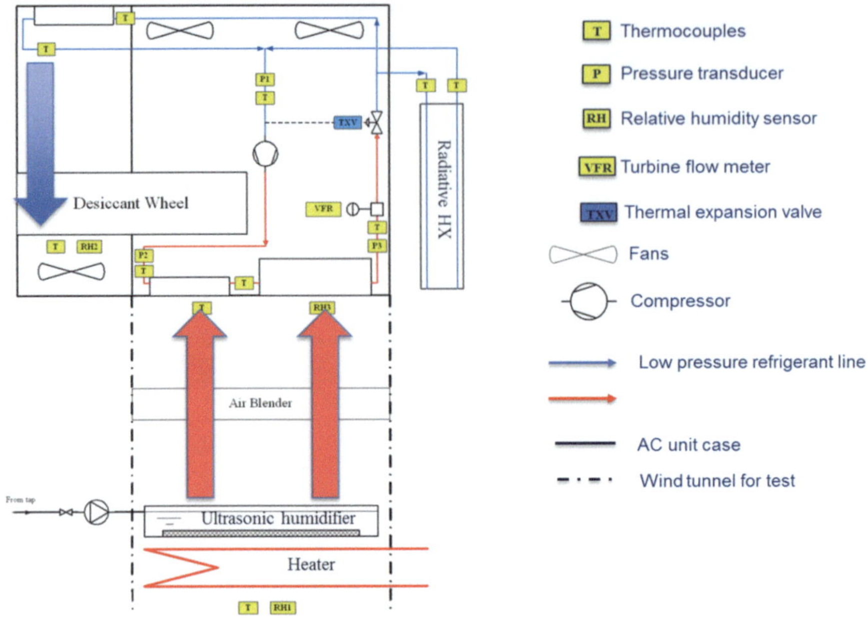

Fig. 8 SSLCAC system test setup and sensor locations

3.3 Test Facility and Results

An experimental facility was set up to test the performance of the *SSLC AC* + *RHX* system. The entire facility was placed in an environmental chamber which was set to the standard indoor unit condition defined in the AHRI standard 210/240 Standard [11] rating conditions (26.7 °C DBT, 19.4 °C WBT). A simple air duct made with polypropylene panels was built at the condenser side. An electric heater and an ultrasonic humidifier were installed inside the duct. The chamber air was heated and humidified to the outdoor unit test condition (35.0 °C DBT, 23.9 °C WBT) throughout the duct before entering the two parts of condensers. Thermocouples, humidity sensors, and pressure sensors were installed to the system as illustrated in Fig. 8. A turbine flowmeter was installed in the downstream of the condensers to measure the refrigerant mass flow rate. Due to an instrument limitation, only the total mass flow rate was measured and the individual mass rates through RHX and conventional evaporator remained unknown.

The system COP was calculated as the ratio of total airside cooling capacity over total power input. The total power input to the system includes compressor power consumption, three brushless DC fan power consumption, and six cross-flow fan power consumption. The airside cooling capacity cannot be directly measured because the SSLC AC system had neither space nor enough pressure lift for nozzles

Table 6 Test results of the baseline system and SSLC system

	Baseline system	SSLC and RHX	SSLC with DW and RHX
Evaporating pressure (kPa)	951	1,221	1,091
Condensing pressure (kPa)	2,481	2,456	2,310
Mass flow rate (g/s)	16.4	22.0	16.6
Cooling capacity (kW)	2.75	3.54	2.96
COP	2.91	4.15	3.80
COP improvement	1	1.43	1.31

to measure the airflow rate. Therefore, the airside cooling capacity was calculated by the refrigerant-side cooling capacity minus evaporator fan power consumption.

Table 6 compares the COP and other important parameters of the SSLC AC system at different operation modes as well as those of baseline system. The baseline system was defined as the SSLC AC system operating at conventional evaporating temperature with the DW and RHX shut off. The comparison shows the COP improvement due to an elevated evaporating pressure from 951 to 1,221 kPa in the case of SSLC and RHX mode and from 951 to 1,091 kPa in the case of SSLC with DW and RHX mode. The elevated evaporating pressure also made the fixed speed compressor deliver an increased refrigerant flow rates in both cases. The cooling capacities were affected by the mass flow rate and were increased for the SSLC AC system over the baseline system. The COP improvement of the SSLC AC system with DW and RHX reached 31%. This demonstrates that the SSLC AC met the energy saving expectation.

The thermal comfort aspect of the SSLC AC system was only evaluated by measuring the surface temperature of the RHX using an infrared camera due to the restricted time frame. The image was demonstrated in Fig. 9. The image shows the panel surface temperature was around 19 °C as the cold refrigerant flowed at the back side. The cross-flow fan motors' temperatures were higher than 30 °C. However, they should not affect the overall thermal comfort much because of their small power inputs.

3.4 Cost Analysis

The prototype was constructed using the parts from two commercially available window-type air conditioners and materials available in the laboratory. The two units cost $300 each. The RHX was constructed using aluminum cooling panels which cost roughly $70 per square meter and that lead to their total cost of $300. All the piping in the system were made of copper and cost around $100. The thermostatic expansion valve cost around $200. The low-temperature regenerated desiccant wheel valued at around $400. Four micro-channel heat exchangers cost around $200. Therefore,

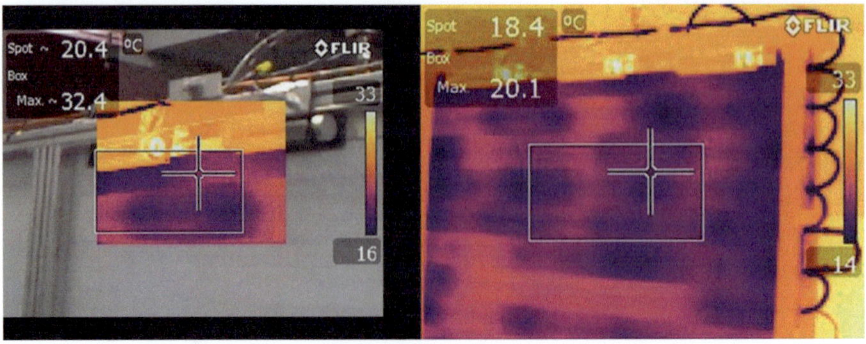

Fig. 9 Infrared images of the RHX and the cross-flow fans

Table 7 Cost analysis for prototype unit and the same unit if mass-produced

Component	Prototype cost ($)	Estimated cost if mass-produced ($)
Compressor	300	80
Heat exchangers (4 ea.)	50 ea.	20 ea.
Brushless DC fan motors (3 ea.)	100 ea.	20 ea.
TXV	200	15
Tubing	100	10
PCB and controller	100	30
Desiccant wheel	400	100
RHX paneling	300	50
Miscellaneous items	100	50
Total	2,000	500

the total cost of constructing the test unit was at around $2,000. Table 7 shows the breakdown of the prototype unit cost and the estimated cost for the same unit if mass-produced. A simple cost analysis predicted that the entire system could cost at around $500 if mass-produced.

3.5 *Summary*

The system presents the winner design of Max Tech and beyond competition held in 2012. It utilized the technologies of SSLC and RHX to achieve both energy savings and thermal comfort. A prototype AC system, which consists of an R410A VCC, a low-temperature driven DW, and a RHX, was fabricated and tested under

AHRI standard test condition. It shows a 31% COP improvement as compared to the baseline system. The surface temperature of RHX was well controlled to be around 19 °C (66 °F). A simple cost analysis was carried out, and it predicted that the entire system, if mass-produced, could cost at around $500.

4 Energy-Efficient Cooling and Heating System for a Residential House

4.1 Background

The reACT solar-powered home is the University of Maryland's entry into the Department of Energy Solar Decathlon event held in 2017. The reACT features 87.3 m^2 of built interior, surrounding 14.9 m^2 of glazed courtyard, with 37.2 m^2 of outdoor deck area, on about 743.2 m^2 of developed site. The courtyard aids in regulating light, temperature, and humidity in the home, operating in sync with a mechanical core at the heart of the interior, in addition to serving as additional spaces for dining and lounging. The core contains high-performance, environmentally sensitive systems that interact with one another, managing the flow of water, air, and energy throughout the home.

The Core of reACT is an integration of all technical systems in compact modular form. The objective and benefit of such design are to enable efficient waste heat and moisture harvesting, by facilitating interchanges of thermal energy and water. One of the representative examples is the interaction between the Core and Greencourt. In the summer, the Greencourt can be either closed to harvest excess heat in the hot water tank or opened and shaded to pre-cool the air flowing through variable refrigerant flow (VRF) outdoor unit. During the winter, the Greencourt can either preheat the air flowing through VRF outdoor unit or provide excess heat that can be harvested by a heat pump water heater (HPWH). Both VRF and HPWH locate at the attic of the Core. The design enables reACT to operate under wide range of climate zones and optimize its performance under various weather conditions.

4.2 Design Overview

4.2.1 Air-Conditioning and Heating System

VRF heat pump system is utilized to provide cooling and heating for living spaces. The system's capacity is 30,000 Btu. The seasonal energy efficiency ratio is 22 for up to four indoor units. Zoning control can be realized through each indoor unit. Therefore, it is possible to save energy while delivering thermal comfort for all zones. The VRF system is designed for operation in the heat mode when outdoor temperatures are as low as −20 °C.

4.2.2 Water Heating System

The HPWH utilizes an air source heat pump cycle. It consists of three major components: the outdoor units absorbing heat from the ambient air, the indoor unit heating up a 189 L capacity water tank, and control panels. The system's capacity is 3 kW. The nominal COP is 4.62, while the required power input is 0.65 kW. It can work in ambient temperature ranging from −20 to 30 °C.

4.2.3 Innovative Damper System for Energy Harvesting

The concept of pre-cooling and pre-heating was applied for VRF and HPWH by developing an innovative air damper system. The system directs air from the courtyard either to both VRF and HPWH systems or to HPWH first. In the cooling season, the air entering the VRF outdoor unit is pre-cooled by the HPWH. In the heating season, the preheated courtyard air, which is warmer than the ambient air, is used.

The simulation studies indicate that the pre-cooling in the summer reduces the VRF outdoor unit (OU) inlet air temperature by 7 °F, while the pre-heating increases the OU inlet air temperature by 7 °F. As the pre-cooling lowers the inlet air temperature, thus the OU operating pressure (condenser side), it reduces the VRF system energy consumption in the summer. Similarly, the pre-heating increases the OU operating pressure (evaporator side). It reduces VRF system energy consumption in the winter. The simulation studies show that preheating and pre-cooling can reduce energy consumption by 472 and 339 W, respectively.

4.2.4 Ventilation System

The energy recovery ventilation (ERV) is utilized as the main ventilation methods. The ERV exchanges the energy between two air streams. The ERV unit moves 118 cfm of air at an external pressure of 0.8 inH_2O. CO_2 sensors installed within the unit controls ventilation flow rates. The core component, cross-counter flow heat exchanger, has an efficiency up to 95%. The ERV is usually used between ambient air and indoor air streams. It can provide home with fresh air while pre-heat ambient air with indoor air in the heating season. The expelled indoor air also can be used for HPWH in the heating season for further energy harvesting.

Table 8 Comparison of VRF system energy consumption

Operation	8.8 kW w/o heat recovery	12.3 kW w/o heat recovery	Percent increase (%)
Cooling	471 kWh/year	680 kWh/year	44.7
Heating	1,817 kWh/year	2,057 kWh/year	13.2

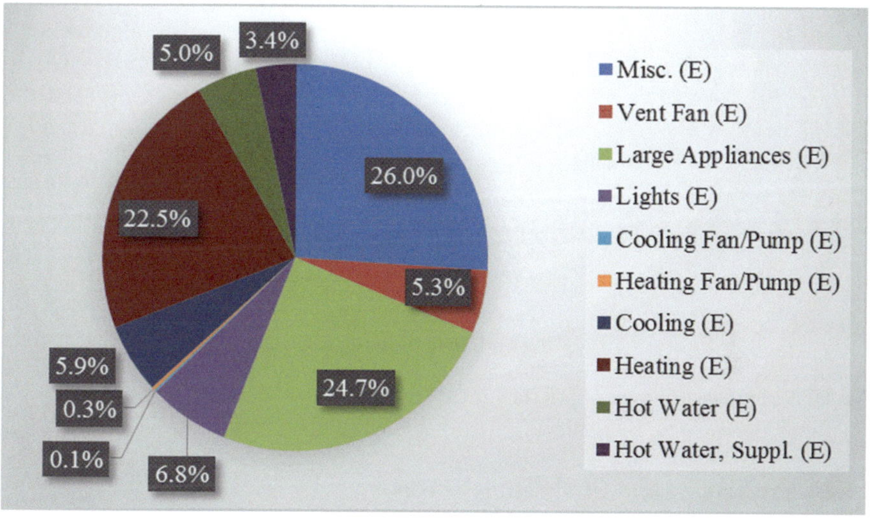

Fig. 10 Annual energy breakdown of reACT house

4.3 System Analysis

4.3.1 VRF System Analysis

One of the fundamental challenges for the reACT house design is VRF system selection and sizing. The smallest cooling capacity for commercially available VRF system is 12.3 kW with heat recovery function, and 8.8 kW without heat recovery function. While heat recovery function improves system efficiency, it is necessary to investigate if the benefits from heat recovery outweigh the inefficiencies from oversizing. Therefore, we conducted building modeling with BEOpt. All parameters were kept the same except for the VRF systems, one is 12.3 kW with heat recovery option and the other one is 8.8 kW without heat recovery option. The results are shown in Table 8, while Fig. 10 shows the energy consumption breakdown of the house.

It is observed that the 12.3 kW VRF system results in a higher cooling and heating energy consumption, by 44.7 and 13.2%, respectively. Apparently, the heat recovery function doesn't outweigh the inefficiencies from oversizing. As a result, we selected 8.8 kW VRF system for reACT house.

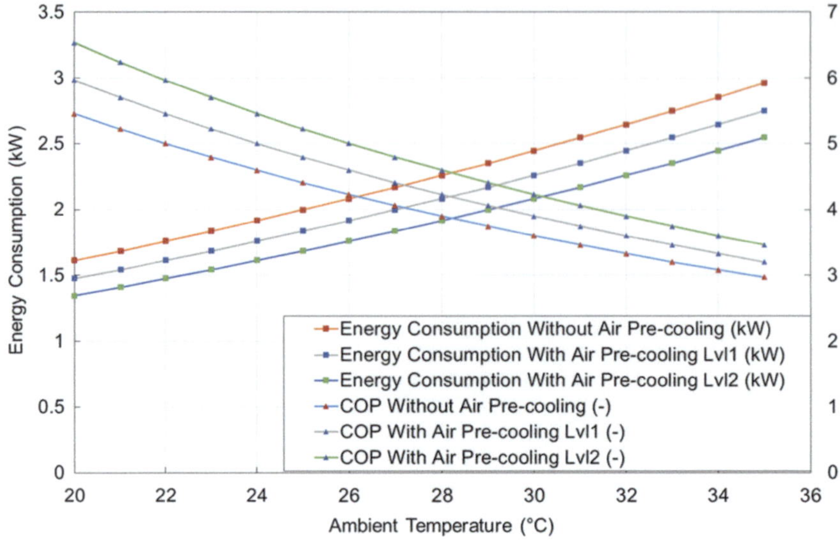

Fig. 11 Performance impacts of pre-cooling on VRF OU inlet air in cooling mode

4.3.2 Pre-cooling and Pre-heating Effects

As mentioned before, we reduced energy consumptions by pre-cooling and pre-heating of air flowing to the VRF outdoor unit. We performed parametric study to further evaluate the benefits. Figures 11 and 12 summarize the results for pre-cooling and pre-heating, respectively.

Figure 11 shows the effects of two levels of pre-cooling along with increasing ambient air temperatures. The first and second level reduces inlet outdoor air temperature by 2 and 4 °C, respectively. The VRF system is kept the same with a rated cooling capacity of 8.8 kW. It is observed that the energy consumption increases with ambient air temperatures. The pre-cooling reduces energy consumptions and increases COP. Level 1 and level 2 pre-cooling reduce energy consumption by 175 and 339 W, respectively.

Similarly, Fig. 12 demonstrates effects of two levels of pre-heating. The first and second level increases outdoor inlet air temperature by 2 and 4 °C, respectively. The VRF system is kept the same with a rated heating capacity of 9.4 kW. Note there is abrupt performance change around 5 °C inlet air temperature for the baseline case. This happens because the system needs to defrost when ambient air temperature is below 5 °C. Therefore, the performance degrades significantly. As such, when pre-heating is introduced, the system can operate without defrost at lower ambient temperatures. Level 1 and level 2 pre-heating reduce energy consumption by 239 and 472 W, respectively.

The HPWH can benefit from pre-heating in a similar way as the VRF system in heating mode. Figure 13 shows the effects of two levels of pre-heating. Level 1 and

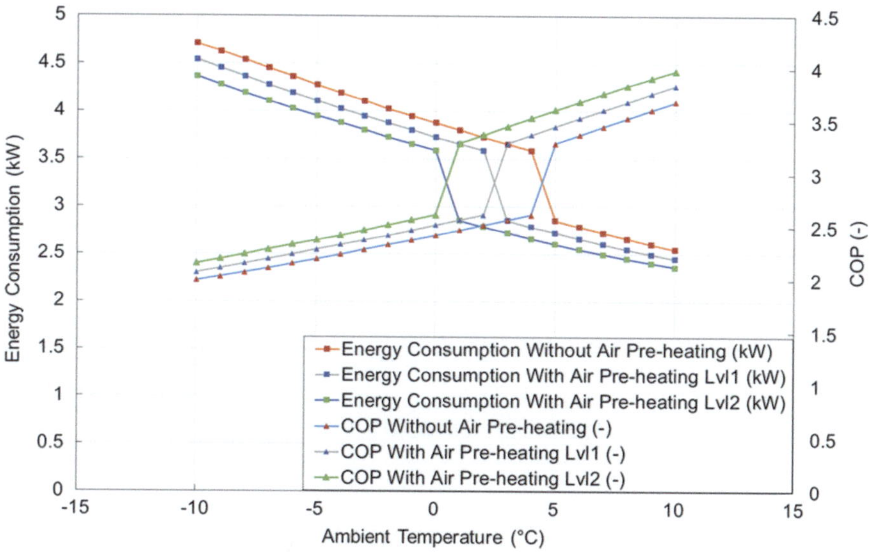

Fig. 12 Performance impacts of pre-heating on VRF OU inlet air in heating mode

level 2 increase ambient air temperature by 2 and 4 °C, respectively. The water heater is kept the same with a rated heating capacity of 1.2 kW and COP of 4.62. The hot water set point is 45 °C. Level 1 and level 2 could reduce energy consumption by 17 and 34 W, respectively.

4.3.3 Effect of Greencourt Control

The Greencourt has a movable rooftop, which makes solar energy harvesting feasible in the heating season. Therefore, air temperature in the Greencourt is expected to be higher than the ambient in winter. The air can be used for VRF system outdoor unit, achieving energy saving through pre-heating benefits. To quantify the pre-heating benefits from using Greencourt air, we performed simulation studies using EnergyPlus. The simulation period was set from January to March. Weather data from Denver, Colorado was used. Note the HPWH was assumed to be off to study pre-heating effects on the VRF system only. We also assume the VRF system draws 1,728 m^3/h of air from the Greencourt constantly, which is the upper limit for the system during heating mode. Figure 14 shows hourly temperature difference between the Greencourt air and ambient air. The highest difference is 5.6 K. Figure 15 presents daily VRF system performance with or without Greencourt control. As a summary, using the Greencourt air as a heat source reduces VRF system's monthly energy consumption from 583 to 572 kWh, compared to using ambient air.

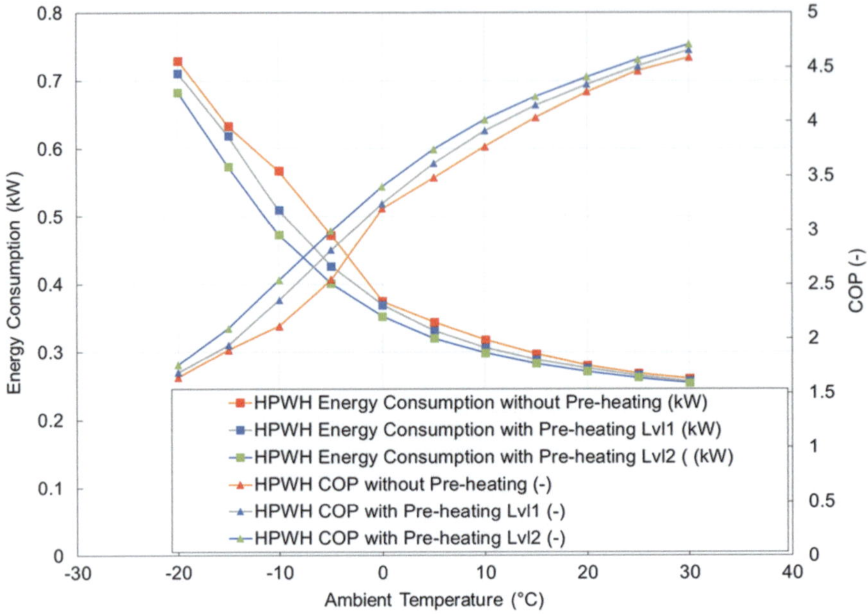

Fig. 13 Performance impacts of pre-heating on HPWH OU inlet air

4.3.4 Effect of Damper Control

As stated before, the VRF system and HPWH are co-located in the attic. While the HPWH absorbs heat from attic air, the VRF system rejects heat in the cooling mode. A lower attic air temperature means lower energy consumption for VRS system. Therefore, we introduced the damper control system to use the HPWH to pre-cool the attic air before flowing through the VRF outdoor unit.

Figure 16 shows the concept. Damper A and B can be closed to prevent unnecessary air leakage or opened for natural ventilation. In the cooling season, damper C and D are positioned to regulate ambient air first passing through HPWH outdoor unit, then VRF outdoor unit. Therefore, the inlet air temperature for VRF outdoor unit is lower than ambient. In the heating season, damper C and D are positioned to allow ambient air passing through HPWH and VRF outdoor units independently.

We then performed simulations to quantify benefits of the damper control in EnergyPlus. The simulation period was set from July to September. Weather data from Denver, Colorado was used. We assume the VRF system draws 1,431 m^3/h air from the ambient, which is the upper limit for the system during cooling mode. Figure 17 shows VRF system performance comparison with or without the damper control system. It is observed that the cooling performance is improved, and the overall energy consumption is reduced from 471.9 to 457 kWh. The annual energy savings of reACT is listed in Table 9.

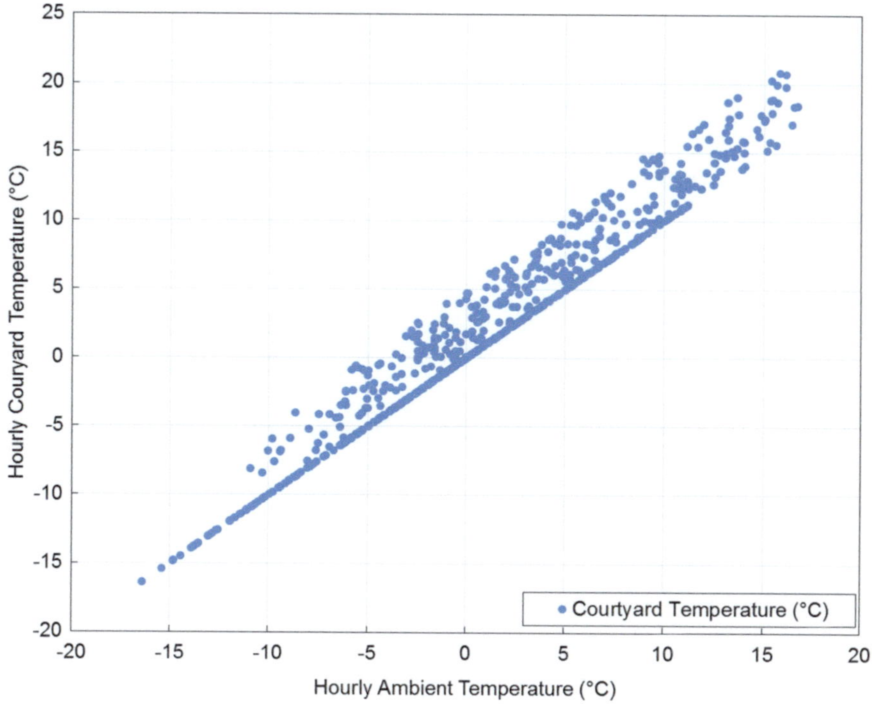

Fig. 14 Greencourt temperature with the rooftop radiation

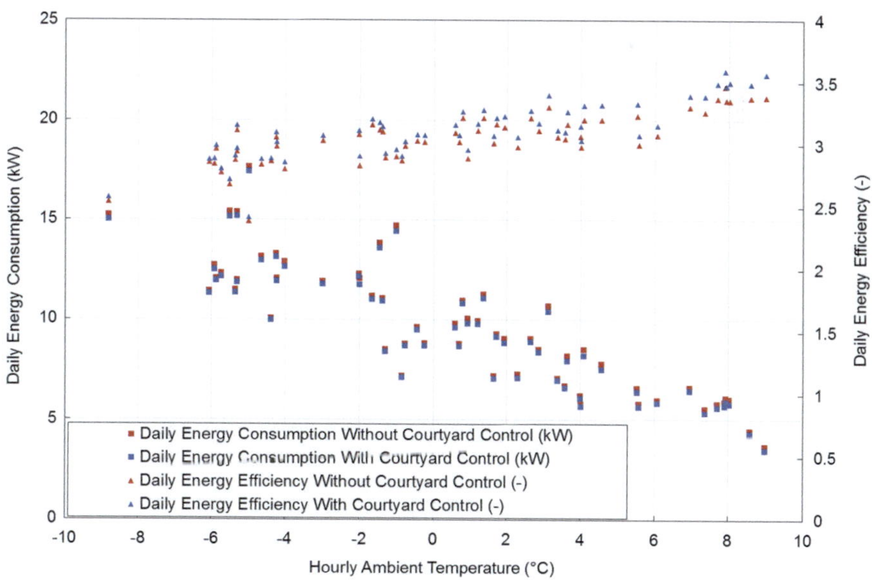

Fig. 15 Effects of Greencourt control on VRF system performance

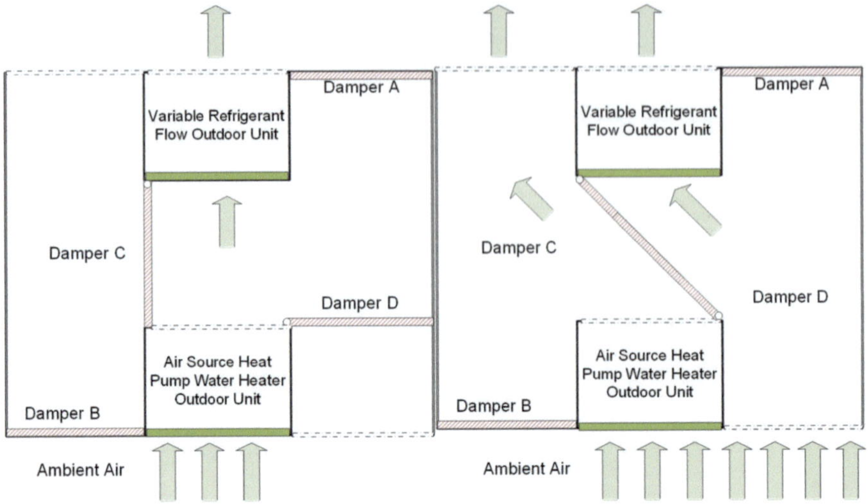

Fig. 16 Damper control: cooling season (left) and heating season (right)

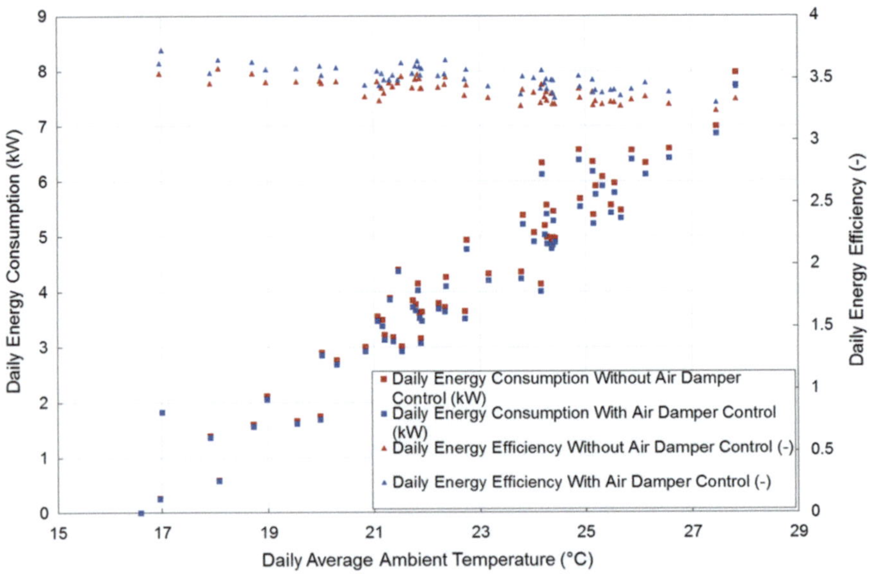

Fig. 17 Effects of air damper control on VRF system performance

Table 9 Energy savings from HVAC systems

Source	Energy use baseline (kWh/year)	Energy use reACT (kWh/year)	Change (kWh/year)
Misc.	2,092.6	2,092.6	N/A
Ventilation fan	427.9	427.9	N/A
Large appliances	1,987.1	1,987.1	N/A
Lights	548.1	548.1	N/A
Cooling fan/pump	11.7	11.7	N/A
Heating fan/pump	23.45	23.45	N/A
Cooling	471.86	457	−14.8
Heating	1,817.1	1,782.7	−34.4
Heating, Suppl.	275.5	0	−275.5
Hot water	404.5	384.2	−20.3
Total	8,060	7,714.7	−345.3
PV	12,547	12,547	N/A
Net (PV − Total)	4,487	4,832.3	345.3

4.4 Summary

The heating and cooling system for the reACT house is a well demonstration of advanced integrated design. The advanced heating and cooling system, combined with innovative energy harvesting design, delivers efficiency and sustainability. The simulation studies with BEOpt and EnergyPlus indicate significant energy savings. The pre-cooling and pre-heating from Greencourt, as well as coupled pre-cooling between water heater and VRF system, reduce cooling, heating, and hot water energy consumption by 14.8, 309.9, and 20.3 kWh, respectively. That yields 345.3 kWh annual energy savings.

5 Summary

While there are more technologies to explore, this chapter sheds light on efficient cooling and heating technologies and their applications in different fields. The three cases provide approaches for energy-efficient cooling and heating design: applying innovative technology (vapor injection cycle in heat pump clothes dryer), hybrid technology approach (separate sensible and latent cooling system for air-conditioning), integrated design approach (VRF, heat pump water heater and pre-heating/pre-cooling concept). The cooling and heating demands are around persistently. Energy-efficient technologies are core solutions for better design to cut off energy usage while maintaining or even enhancing thermal comfort.

References

1. DOE "Heat and cool". [Online]. Available: https://www.energy.gov/energysaver/heat-and-cool. Accessed 20 July 2018
2. Cao T (2016) Modeling and optimization of micro grid energy system for ship applications, College Park. University of Maryland, MD
3. Cao T, Ling J, Hwang Y, Radermacher R (2014) Development of a novel two-stage heat pump clothes dryer. In: Proceedings of the ASME 2014 international mechanical engineering congress and exposition, Montreal, Quebec, Canada
4. Ling J, Qian S, Li S, Huang L, Hwang Y, Radermacher R (2012) The winner design of the max tech and beyond competition: a high-efficient residential air-conditioning system. In: ASHRAE conference
5. U.S.D. Team (2017) reACT Team Maryland. University of Maryland. [Online]. Available: http://2017.solarteam.org/. Accessed 21 July 2017
6. D. EIA (2010) Annual energy outlook 2010 with projections to 2035. Washington DC
7. Meyer S, Victor F (2010 Aug) Do heat pump clothes dryers make sense for the US Market? In: ACEEE summer study on energy efficiency in buildings
8. Department of Energy (2011) 10 CFR Part 430 energy conservation program for consumer products: test procedures for clothes dryers and room air conditioners, Federal Register
9. Kao JY (1998 May) Energy test results of a conventional clothes dryer and a condenser clothes dryer. In: International appliance technical conference 49th proceedings
10. Ling J, Kuwabara O, Hwang Y, Radermacher R (2013) Enhanced options for separate sensible and latent cooling air-conditioning system. Int J Refrig 36(1):45–57
11. AHRI (2008) ANSI/ARI Standard 210/240 for performance rating of unitary air-conditioning and air-source heat pump equipment. Air-Conditioning and Refrigeration Institute, Arlington

Hybrid Approach in Microscale Transport Phenomena: Application to Biodiesel Synthesis in Micro-reactors

J. M. Costa Jr., P. C. Pontes, C. P. Naveira-Cotta, M. K. Tiwari, S. Balabani and R. M. Cotta

Keywords Hybrid methods · Microscale transport phenomena · Integral transforms · Inverse problems · Lumped models · Biodiesel synthesis · Micro-reactors

1 Introduction

Global demands for water, energy, and food are estimated to increase by 40, 50 and 35%, respectively, by 2030 [1], mainly due to global population increase and growth of middle class and its associated consumption pattern, with the inherent consequences on climate change and environmental impact as a whole. Thus, the

J. M. Costa Jr.
Federal Institute of Sertão Pernambucano, IFSERTÃO-PE, Serra Talhada, PE, Brazil

P. C. Pontes
Federal University of South and Southeast of Pará, UNIFESSPA, Santana do Araguaia, PA, Brazil

P. C. Pontes · C. P. Naveira-Cotta · M. K. Tiwari
Nanoengineered Systems Laboratory, UCL Mechanical Engineering, University College London, London WC1E 7JE, UK

C. P. Naveira-Cotta · R. M. Cotta
Laboratory of Nano and Microfluidics and Microsystems, Mechanical Engineering and Nanoengineering Departments, POLI & COPPE, Federal University of Rio de Janeiro, Rio de Janeiro, Brazil

S. Balabani · R. M. Cotta
UCL Mechanical Engineering, University College London, London WC1E 7JE, UK

R. M. Cotta (✉)
Interdisciplinary Nucleus for Social Development, NIDES/CT, UFRJ, Rio de Janeiro, Brazil
e-mail: cotta@mecanica.coppe.ufrj.br

A. K. Gupta et al. (eds.), *Innovations in Sustainable Energy and Cleaner Environment*, Green Energy and Technology,
https://doi.org/10.1007/978-981-13-9012-8_20

development of increasingly more sustainable processes is a driving force on most of the recent research efforts in engineering [2], including transport phenomenon-based processes where the design of improved heat and mass exchange equipment is an essential technological goal [3]. For instance, over the last few decades, heat exchangers employing micro- and mini-channels have been progressively preferred in applications that require high heat removal demands and/or limited dimensions [3–7].

The research reported here originated through a UK-Brazil Newton Fund-FAPERJ collaborative development project, offering a holistic approach to address the complex nexus challenges from the societal demand for food, water, energy and need for environmental care. The motivation involved a combined system, harnessing electrical energy from high-efficiency water-cooled HCPVs and using the resulting waste heat for water desalination and continuous biodiesel production, making use of nano- and microscale transport phenomena analysis tools recently employed in different thermal engineering applications [8–11]. The original aspects of the present research include problem formulation and simplification strategies, modern analytical–numerical solution methods for differential equations, parametric and functional identification through inverse analysis, symbolic computation tools, nano- and micro-fabrication, non-intrusive optical techniques in microscale heat and fluid flow, employed individually or jointly, as needed to answer the basic questions raised by the applications to be advanced. This chapter starts by describing the so-called *hybrid* approach that combines the computational–analytical–experimental tools to analyze a system and its behavior [12, 13]. The impact of nano- and microtechnologies on transport enhancement has already been illustrated through some fundamental analysis around the three major paths [8], namely miniaturization of heat exchange devices, thus bringing transport characteristic dimensions to the microscale, functionalization of the materials (solids and fluids) at the nanoscale, thus augmenting thermophysical properties, and geometrical structuring of micro-systems and passages, thus altering the transport or mixing mechanisms. These fundamental ideas and the hybrid analysis framework are here exemplified in a selected application involving the continuous synthesis of biodiesel in micro-reactors.

2 The Hybrid Approach

Figure 1 depicts the various steps in simulating a typical advanced engineering application [12, 13], starting with the construction of a physical model of the (usually intricate) real problem. Quite frequently, working on nano- and microtechnology implementation requires the problems to be modeled within a multiphysics and multiscale framework. The various stages depicted in Fig. 1 are interconnected; hence, in most simulation processes parallel improvements in each stage are necessary, in order to achieve a desirable combination of accuracy improvement and simulation cost reduction in the overall development.

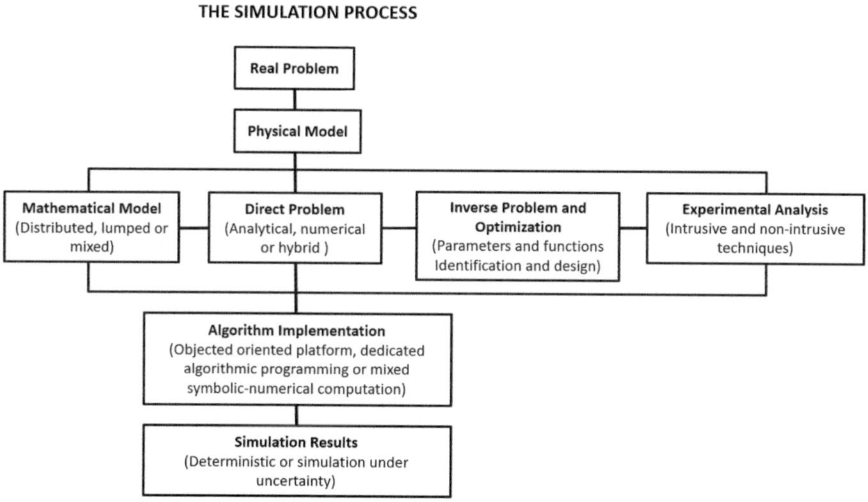

Fig. 1 Summary of the simulation process in engineering development

Starting at the mathematical model stage, the hybrid approach is already handy in providing mixed lumped-differential formulations, specifically employed to solve diffusion and convection–diffusion problems. The idea behind the partial or total lumping of partial differential formulations lies in the elimination of space variables from the formulation through an averaging process, once the local information in that particular coordinate or coordinates is not essential to the problem solution. However, the process involves an approximation, whereby the local boundary information in the coordinate to be eliminated, which reappears within the partial differential equation during the averaging process, needs to be expressed analytically in terms of the averaged potential. The classical lumped system analysis assumes negligible gradients along that coordinate and equates the average and boundary potential values. This is in general not sufficiently accurate and is thus of limited use. Higher-order approximations that provide a more informative relation between boundary and average potentials are thus quite useful in model reduction efforts. The Coupled Integral Equations Approach (CIEA) [14–20] is a systematic methodology for constructing improved lumped-differential formulations, based on approximating the integral expressions for the average potential and its associated flux through Hermite formulae for integration, which employ only the function and its spatial derivatives at the boundary points [21]. Then, more informative expressions relating average and boundary potentials are achieved, which are then incorporated in the averaged differential formulation. Accuracy is significantly improved over the classical lumping approach, while retaining the benefits such as the reductions in mathematical complexity and in simulation time/effort compared to the solution of the original fully differential problem formulation. Also of great relevance in the mathematical modeling stage is the reformulation of problems in heterogeneous media and/or complex geometrical arrangements as a single domain mathematical problem, which can be essential for a proper hybrid numerical–analytical treatment. The concept of a single domain formulation was formalized in the context of natural convection in porous media and conjugated heat transfer problems [22–25], and progressively employed to handle other classes of problems. The idea is simply that of rewriting a system of coupled equations for a composite/multi-component system, with various common interfaces, as a single region in which the functions associated with physical properties and sources are rewritten in the form of coefficients with space variability, accounting for the transitions through the different phases or components.

The direct problem solution stage in any simulation process is by far the most extensively studied in the literature. The advent of the electronic computer brought together a revolution on the numerical analysis and simulation tools for partial differential equations, through the best-known numerical techniques such as finite difference, finite element, and finite volume methods [26], which markedly changed the advanced engineering practice. Nevertheless, after decades of substantial scientific and technological progress on both software and hardware, such methodologies are still not universal or fully reliable, despite the considerable efforts of extension, generalization, and automatic error control. Particularly, from a computational cost perspective, they yet remain quite demanding and even prohibitive in computationally intensive tasks. As examples of situations when the direct problem solution needs to be obtained several thousands of times, one can recall the identification of parameters and functions in inverse problem analysis, the geometrical and parametric optimization of process performance, and the stochastic analysis of uncertainty propagation. While the classical analytical approaches for partial differential equations can be very restrictive in terms of classes of problems that can be dealt with, over the last three decades, hybrid numerical–analytical approaches have been developed in different physical contexts, at a much more modest pace in comparison to purely numerical methods, based on different analytical tools. The generalized integral transform technique (GITT), based on the classical integral transform method [27–29], has introduced a hybrid nature to this eigenfunction expansion approach [30–32], yielding error-controlled solutions to a large number of linear and nonlinear convection–diffusion problems, as reviewed in various sources [33–39]. The interest in analytical solution methodologies for PDEs might not be so evident to the less experienced researcher in the field, but it has been experiencing a noticeable revival, mainly for the following reasons:

1. **Exact solutions**: analytical methods lead to computationally costless solutions, with user-prescribed accuracy, though such solutions might not be obtainable when the formulation complexity increases.
2. **Trends and asymptotic behaviors**: analytically obtained solutions offer minimal computational effort, besides yielding parametric analysis and limiting behavior identification even with very little computational work.
3. **Reference benchmark results**: independent analytical benchmark results are essential in the verification of numerical methods and codes.

4. **More straightforward and feasible**: symbolic computation platforms have enabled a number of ideas on analytically based methodologies that were either abandoned or never even considered.
5. **Intensive computational tasks**: the advantage of analytical solutions is even more evident in costly computations, such as parameters and functions' inverse analysis identification, process optimization, and stochastic simulation for uncertainty propagation.
6. **Bridge to build hybrid numerical-analytical approaches**: They are instrumental in the search for higher precision and robustness in classes of problems that are not solvable through classical analytical methods.

The basic steps in the GITT approach can be summarized as follows [33–39]:

1. **Select an analytical filtering solution**: to improve convergence behavior of the eigenfunction expansions, if required. Filtering to achieve homogeneous boundary conditions and/or reduce the importance of equation source terms is always helpful. Filtering can also be applied recursively if necessary, or implicitly, such as in the case of a nonlinear filter.
2. **Choose the associated eigenvalue problem**: which should desirably incorporate characteristic linear behaviors of the original problem formulation represented by the coefficients of the differential operators. Both diffusive or convective eigenvalue problems may be adopted. More recently, nonlinear eigenvalue problems have also been employed with marked improvement on convergence.
3. **Develop the integral transform pair**: obtain the transform and inversion, that define the integral transform operation and explicitly recover the potential.
4. **Solve the eigenvalue problem**: either in analytical form and symbolic computation [40], or through the GITT approach itself, transforming the chosen differential eigenvalue problem into an algebraic one.
5. **Integral transform the original PDE**: obtain the transformed differential system, which shall be an ODE system for a total transformation, when all the independent variables are eliminated except one. It can result in an initial value problem, for an original parabolic or hyperbolic-type problem, or in a boundary value problem, for an original elliptic formulation. A partial transformation may also be applied, when two independent variables are kept in the transformed system, yielding a partial differential transformed system. The partial transformation option may also be adopted in solving a boundary value problem via a pseudo-transient formulation.
6. **Compute transformed system coefficients**: the coefficients are integrals, single or multiple, which involve the eigenfunctions. When analytical expressions are not obtainable through symbolic computation [40], there is a marked advantage in promoting semi-analytical integrations, when the oscillatory nature of the eigenfunctions is analytically handled while the non-oscillatory portion of the integrand is approximated by piecewise polynomial interpolation. Traditional numerical integration may also be employed, and again, the knowledge about the oscillatory behavior of the eigenfunctions can be advantageous in defining appropriate domain partitions to reduce computational costs in adaptive integration,

especially for higher-order eigenvalues, which represent the higher oscillation frequencies.

7. **Solve transformed system**: numerically (or analytically when applicable), solve the resulting coupled ODE or PDE system for the transformed potentials, which needs to be truncated to a sufficiently large order for numerical evaluation. Reliable automatic solvers are readily available for the numerical solution of stiff ODE systems, and even for one-dimensional systems of PDEs with adaptive remeshing, such as in the routine ND Solve of the *Mathematica* system [40].
8. **Recall inversion formula**: to analytically reconstruct the hybrid solution of the desired potential. At this point, the inversion formula can be employed for accuracy testing, once the intermediate tasks were accomplished under error control, and thus allowing for the best selection of the transformed system truncation orders. Error estimates are then automatically provided at any desired position and time.

This hybrid approach has been extensively used as a benchmarking tool, aiding in the verification of more general and flexible numerical codes, but also as a production tool itself, as previously discussed, especially in highly intensive computational jobs and/or precision-demanding applications. The various classes of problems so far treated by the GITT include linear and nonlinear diffusion, convection–diffusion, and reaction–convection–diffusion problems, moving boundary problems, eigenvalue problems of unknown analytical solution, irregular domains, heterogeneous media, boundary layer equations, and Navier–Stokes equations [30–39]. Fairly recently, an open-source general-purpose symbolic–numerical code was implemented, coined as the UNIT code (Unified Integral Transforms) [41, 42] and implemented in the *Mathematica* platform [40]. It offers all of the basic analytical and numerical steps in the GITT implementation, allowing for a more straightforward utilization of this methodology.

The next stage in the simulation process involves using both the direct problem solution and the experimental data from the adjacent stage in Fig. 1, to perform the inverse problem analysis, as required, for parameter and function identification or even state estimation [43, 44]. However, this analysis may also be employed ahead of the experimentation, aimed at achieving an optimal design for the experimental procedure and setup [43]. In the identification of properties and source terms, the direct problem solution is employed first for the sensitivity analysis of the quantities to be estimated and to produce simulated data, so as to evaluate the inverse problem solution ahead of employing actual experimental data. Bayesian inference has been preferred for the inverse analysis in the present context [45–52], first of all due to its inherent robustness, but also due to advantages in being able to account for a priori information on the parameters. The drawback of this statistical approach, associated with the usually large number of evaluations of the direct problem required for an appropriate statistical inference through the Monte Carlo–Markov Chain method, is alleviated by the use of the present hybrid solution pathways with the reduced computational cost. Also, in two other aspects the hybrid direct problem solution is particularly beneficial to the cost reduction in the inverse problem analysis. Firstly, the experimental data acquired by the non-intrusive experimental techniques, to be discussed later, is often available in large quantities, in both space and time, resolving a certain region in the overall domain in great detail. Therefore, it is feasible and interesting to work within the integral transformed domain, by performing the same integral transformation process of the direct problem solution, onto the spatially distributed experimental data. Such is the case, for example, with temperature measurement fields obtained by infrared thermography (IRT) or laser-induced fluorescence (LIF) methods, velocity and concentration distributions obtained with particle image velocimetry (PIV) and LIF methods. A marked data reduction effect then occurs, with an additional regularization effect on the inverse problem solution. Secondly, for a function estimation process, the unknown function can itself be expressed as an eigenfunction expansion, as a choice of parametrization, which permits obtaining analytical expressions for the transformed system coefficients, thus further reducing computational costs by avoiding numerical integration at the direct problem solution stage. Figure 2 summarizes the interconnections that characterize the hybrid approach advocated here, involving the four stages of mathematical model, direct problem, inverse problem, and experiments.

In parallel to the advancement of inverse problem analysis tools, experimental techniques have experienced significant progresses along recent years, such as in the case of infrared thermography for non-intrusive temperature measurements [53–59]. Microscopic lenses can also be employed to refine further the spatial information, and pixels of the order of 20 μm are fairly common in the present commercial IRT systems. However, these temperature measurements are in general restricted to external surfaces of the micro-systems, unless IR transparent windows are employed to capture the internal fluid interface temperatures. Nevertheless, the inverse problem analysis may provide estimates of internal temperatures, working as a virtual sensor. Other non-intrusive experimental techniques have been employed in the context of this joint research effort, such as microparticle image velocimetry (μPIV) for velocity measurements and micro-laser-induced fluorescence (μLIF) for tempera-

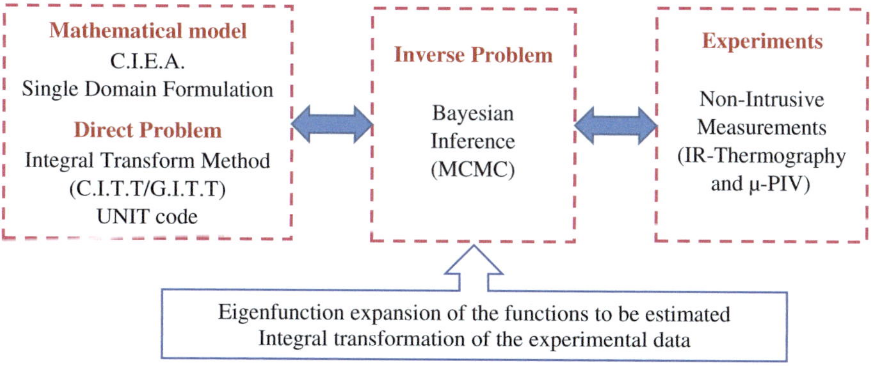

Fig. 2 Hybrid approach in mathematical modeling and direct–inverse problem analysis

ture and concentration measurements [59–65]. μPIV is a non-intrusive quantitative imaging technique that uses statistical cross-correlation algorithms to detect the particle motion of tracer particles seeded in the microscale flows. By determining the particle displacement in the flow, μPIV enables measurements of the actual velocity field of the imaged microfluidic systems. Fluid temperature measurements in microscale devices can be obtained using a μLIF system by means of a suitably selected fluorescent dye injected in the flow to sense local temperature.

The stage of algorithm implementation follows the simulation process shown schematically in Fig. 1. At least three possibilities are shown in the graph, namely the employment of an object-oriented platform, such as the well-known commercial multiphysics simulation software systems, the preparation of dedicated algorithmic codes through one of the well-established high-level programming languages, and the use of modern symbolic computation environments. All of these tools add up to a successful engineering development effort. In the present research, particular emphasis is given on the mixed symbolic–numerical *Mathematica* platform [40].

3 Application to Microscale Transport Phenomena

Microsystems based on flowing fluids have been extensively developed along the last few decades, including among others, micro-mixers, micro-pumps, micro-reactors, and micro-heat exchangers, benefiting from the larger surface area/volume ration aimed at enhancing and accelerating transport phenomena. Such technological development efforts have required advancing not only experimental techniques at the microscale, but also formulation and methodological extensions for the simulation of these micro-devices. While the continuum mechanics hypotheses are usually valid within the range of dimensions of such systems, there are sensible differences in the relative importance of the various possible effects in comparison to the more usual macro-systems. For instance, conjugated heat transfer with the device substrates, axial diffusion along the fluid currents, viscous dissipation effects, fluid slip in liquid flows over superhydrophobic surfaces, and flow rarefaction at the channel walls for gaseous flows are examples of phenomena that may play a major role when modeling fluid flow and heat and mass transfer at the microscale. Since the computational analysis of transport phenomena in microsystems usually involves multiple scales in geometry and physical parameters, the costs associated with traditional numerical methods tend to markedly increase, thus making the present hybrid method more attractive in such classes of problems, as illustrated in [10].

In this chapter, the hybrid methodology will be illustrated by considering a microreactor analysis for continuous synthesis of biodiesel. The findings presented are part of an ongoing collaborative research aiming to improve the design of the microreactors recently demonstrated for biodiesel synthesis [66], targeting residence times of around one minute or less. For this purpose, the hybrid approach has been employed throughout the development and design of this microfluidic system [67–71] and then validated through experimental tests employing high speed and thermal imaging infrared cameras [72]. Advanced additive micro-manufacturing platforms were also used to propose a more economical fabrication of multiple micro-reactors to assemble a modular prototype [72]. There are key process parameters, such as oil and alcohol flow rates, geometry, concentration and type of catalyst, and substrate temperature, that can be optimized to reach increased efficiency of the basic unit within the prototype, which will then lead to the parallel arrangement of a sufficiently large number of micro-reactors for demonstration. It has also been employed to develop an innovative device for recovery of rejected heat from other processes, such as combustion engines or high-concentration photovoltaic panels [73], through an integrated micro-heat exchanger mounted on the same micro-reactor substrate. For the sake of brevity, these aspects are not fully addressed here.

3.1 Problem Formulation

Biodiesel is a biodegradable product that can be used as an alternative fuel since its physicochemical properties are very similar to those of conventional diesel, but with the advantages of being a clean and renewable source of energy and contributing to the reduction of pollution and greenhouse effect. The most common biodiesel production route is via the transesterification reaction that occurs in three successive steps [74]. Fatty acid esters (biodiesel) are obtained in each step of the reaction, ethyl ester in case ethanol is used as the alcohol, and methyl ester if methanol is used [74]. The three steps of the transesterification reaction are then written as:

$$\text{Triglyceride (TG)} + \text{Alcohol(A)} \underset{k_2}{\overset{k_1}{\rightleftarrows}} \text{Diglyceride (DG)} + \text{Ester(B)}$$

$$\text{Diglyceride (DG)} + \text{Alcohol(A)} \underset{k_4}{\overset{\mathrm{k}_3}{\rightleftarrows}} \text{Monoglyceride (G)} + \text{Ester(B)}$$

$$\text{Monoglyceride (MG)} + \text{Alcohol(A)} \underset{k_6}{\overset{k_5}{\rightleftarrows}} \text{Glycerol (GL)} + \text{Ester(B)}$$

where $k_1, k_2, k_3, k_4, k_{5,}$ and k_6 are the kinetic coefficients that need to be determined for the proper simulation of the biodiesel synthesis process.

The design of micro-devices for biodiesel production via transesterification reaction is a complicated task due to the complex liquid–liquid reactant flow that takes place within the micro-channels [75–78]. Thus, the computational simulation is an important tool that helps to understand the physical–chemical phenomenon and, consequently, to determine the suitable conditions that maximize the conversion of triglycerides during the biodiesel synthesis. The mathematical model is obtained through a steady-state mass balance that considers diffusive–convective–reactive effects to describe the transesterification route for biodiesel synthesis in a micro-reactor of rectangular section [67–72], as represented in Fig. 3.

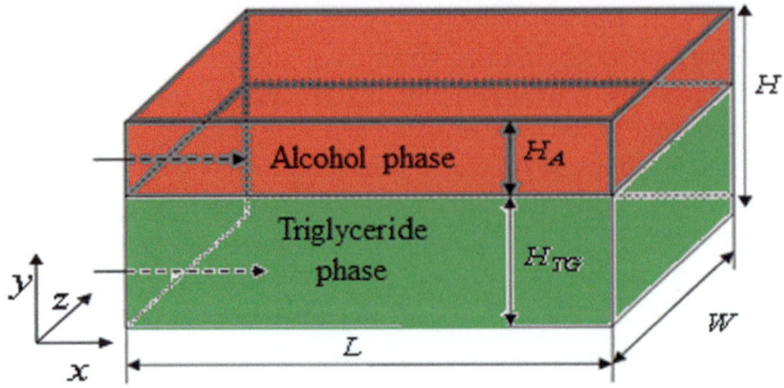

Fig. 3 Geometry and coordinates system of micro-reactor for biodiesel synthesis (triglyceride—green region, alcohol + catalyst—red region)

The dimensionless mass transfer equations for the species concentrations, considering second-order consecutive elementary and reversible reactions preferably in the triglyceride phase, under isothermal conditions and with constant physical properties, can be written as:

$$U_{TG}(Y,Z)\frac{\partial F_s(X,Y,Z)}{\partial X} = \frac{1}{Pe_s}\frac{\partial^2 F_s(X,Y,Z)}{\partial X^2} + \xi_s\left(\frac{\partial^2 F_s(X,Y,Z)}{\partial Y^2} + \delta\frac{\partial^2 F_s(X,Z)}{\partial Z^2}\right) + \varsigma G_s \quad (1a)$$

$$F_{TG}(0,Y,Z) = 1; \quad F_s(0,Y,Z) = 0, \quad s = \mathrm{A, DG, MG, GL, B} \quad (1b, c)$$

$$\left.\frac{\partial F_s}{\partial X}\right|_{X=1} = \left.\frac{\partial F_s}{\partial Y}\right|_{Y=0} = \left.\frac{\partial F_s}{\partial Z}\right|_{Z=0} = \left.\frac{\partial F_s}{\partial Z}\right|_{Z=1} = 0, \quad \text{for all species s} \quad (1d–g)$$

$$F_A(X,1,Z) = F_{A_0}; \quad \left.\frac{\partial F_s}{\partial Y}\right|_{Y=1} = 0, \quad s = \mathrm{TG, DG, MG, GL, B} \quad (1h, i)$$

where F_s are the dimensionless concentrations, and the kinetic expressions for each species, G_s, are given by [69]:

$$G_{TG} = -k_1 F_{TG} F_A + k_2 F_{DG} F_B \quad (1j)$$

$$G_A = (-k_1 F_{TG} - k_3 F_{DG} - k_5 F_{MG})F_A + (k_2 F_{DG} + k_4 F_{MG} + k_6 F_{GL})F_B \quad (1k)$$

$$G_{DG} = (k_1 F_{TG} - k_3 F_{DG})F_A + (-k_2 F_{DG} + k_4 F_{MG})F_B \quad (1l)$$

$$G_{MG} = (k_3 F_{DG} - k_5 F_{MG})F_A + (-k_4 F_{MG} + k_6 F_{GL})F_B \quad (1m)$$

$$G_{GL} = k_5 F_{MG} F_A - k_6 F_{GL} F_B \quad (1n)$$

$$G_B = (k_1 F_{TG} + k_3 F_{DG} + k_5 F_{MG}) F_A + (-k_2 F_{DG} - k_4 F_{MG} - k_6 F_{GL}) F_B \quad (1o)$$

while the velocity profiles for the vegetable oil and the alcohol, assumed fully developed and in laminar stratified regime, are given by Pontes et al. [69]:

$$\begin{aligned} u_{TG}(y,z) = \sum_{i=1}^{\infty} \left(\frac{2}{w}\right)^{1/2} \sin\left(\frac{i\pi}{w} z\right) S_i \Bigg\{ \left(1 + \frac{\mu_A}{\mu_{TG}}\right) \left[-\sinh\left(\frac{Hi\pi}{w}\right) + \sinh\left(\frac{i\pi(H-y)}{w}\right)\right] \\ + \left(\frac{\mu_A}{\mu_{TG}} - 1\right) \left[\sinh\left(\frac{(H-2H_{TG})i\pi}{w}\right) + \sinh\left(\frac{i\pi(H - H_{TG} + y)}{w}\right) \right. \\ \left. - \sinh\left(\frac{i\pi(H - H_{TG} - y)}{w}\right) - \sinh\left(\frac{i\pi(H - 2H_{TG} + y)}{w}\right)\right] + 2\sinh\left(\frac{i\pi y}{w}\right)\Bigg\} \end{aligned} \quad (2a)$$

$$\begin{aligned} u_A(y,z) = \sum_{i=1}^{\infty} \left(\frac{2}{w}\right)^{1/2} \sin\left(\frac{i\pi}{w} z\right) S_i \Bigg\{ \left(1 + \frac{\mu_{TG}}{\mu_A}\right) \left[-\sinh\left(\frac{Hi\pi}{w}\right) + \sinh\left(\frac{i\pi y}{w}\right)\right] \\ + \left(1 - \frac{\mu_{TG}}{\mu_A}\right) \left[\sinh\left(\frac{(H-2H_{TG})i\pi}{w}\right) + \sinh\left(\frac{i\pi(2H_{TG} - y)}{w}\right) \right. \\ \left. - \sinh\left(\frac{i\pi(H - H_{TG} - y)}{w}\right) - \sinh\left(\frac{i\pi(H + H_{TG} - y)}{w}\right)\right] + 2\sinh\left(\frac{i\pi(H-y)}{w}\right)\Bigg\} \end{aligned} \quad (2b)$$

with $i = 1, 3, 5, \ldots$ and

$$S_i = \frac{-2\sqrt{2}w^{5/2}\Delta P}{i^3 L\pi^3 \left[(\mu_{TG} - \mu_A)\sinh\left(\frac{(H-2H_{TG})i\pi}{w}\right) + (\mu_{TG} + \mu_A)\sinh\left(\frac{Hi\pi}{w}\right)\right]} \quad (2c)$$

The associated dimensionless groups are given by:

$$F_{TG} = \frac{C_{TG}}{C_{TGo}};\ F_A = \frac{C_A}{C_{TGo}};\ F_{DG} = \frac{C_{DG}}{C_{TGo}};\ F_{MG} = \frac{C_{MG}}{C_{TGo}};\ F_B = \frac{C_B}{C_{TGo}};\ F_{GL} = \frac{C_{GL}}{C_{TGo}};$$
$$F_{Ao} = \frac{C_A^*}{C_{TGo}};\ Y = \frac{y}{H_{TG}};\ X = \frac{x}{L};\ Z = \frac{z}{w};\ U_{TG}(Y,Z) = \frac{u_{TG}(Y,Z)}{u_{TG,AV}};\ \varsigma = \frac{LC_{TGo}}{u_{TG,AV}};$$
$$\xi_s = \frac{LD_s}{u_{TG,AV} H_{TG}^2};\ Pe_s = \frac{u_{TG,AV} L}{D_s};\ \delta = \frac{H_{TG}^2}{w^2} \quad (3a\text{–}o)$$

where C_s and D_s are the dimensional concentrations and the mass diffusion coefficients of each species, $u_{TG,AV}$ is the average flow velocity for the layer containing the triglyceride species, μ represents the dynamic viscosities of the fluids, H is height of the microchannel, H_{TG} is the height of the fluids interface, L represents the length and W the width of the micro-reactor, and ΔP is the total pressure drop along the micro-reactor.

H_{TG}, the height of triglyceride layer, can be numerically computed for specific experimental values from the ratio between the volumetric flow rates of the two phases present in the microsystem (triglyceride and alcohol), using the following relation:

$$\frac{Q_{\mathrm{TG}}}{Q_{\mathrm{A}}} = \frac{\int_0^w \int_0^{H_{\mathrm{TG}}} u_{\mathrm{TG}}(y, z)\mathrm{d}y\mathrm{d}z}{\int_0^w \int_{H_{\mathrm{TG}}}^H u_{\mathrm{A}}(y, z)\mathrm{d}y\mathrm{d}z} \quad (4)$$

where Q_{TG} and Q_{A} are the triglyceride and the alcohol volumetric flow rates, respectively, while the pressure drop ΔP is obtained from the equation for the volumetric flow rate of the triglyceride phase.

In order to evaluate the influence of temperature on the reaction conversion rate under isothermal conditions, the Arrhenius equation can be used to estimate the kinetic constants at different temperature levels, as follows [70]:

$$k_{i,1} = \alpha_i \mathrm{Exp}\left[-\frac{E_i}{RT_1}\right], \quad i = 1, 2, \ldots, 6 \quad (5a)$$

where $k_{i,1}$ is the kinetic constant at the temperature T_1. Manipulating Eq. (5a), the ratio of the kinetic constants at different temperature levels can be readily obtained from:

$$\ln\left(\frac{k_{i,2}}{k_{i,1}}\right) = \frac{E_i}{R}\left[\frac{1}{T_1} - \frac{1}{T_2}\right], \quad i = 1, 2, \ldots, 6 \quad (5b)$$

where $k_{i,2}$ is the kinetic constant at the temperature T_2.

The consumption of triglyceride along the biodiesel synthesis is influenced by the time the species remain inside the micro-channel, also called the residence time. Since it is assumed that the reaction occurs only in the triglyceride phase, then the residence time may be determined from the ratio of the triglyceride layer volume and its volumetric flow rate, as:

$$\tau = \frac{LWH_{\mathrm{TG}}}{Q_{\mathrm{TG}}} \quad (6a)$$

The triglyceride consumption is evaluated in terms of the fractional conversion, according to:

$$\text{Conversion TG } (\%) = \left(\frac{F_{\mathrm{TG,AV}}(0) - F_{\mathrm{TG,AV}}(1)}{F_{\mathrm{TG,AV}}(0)}\right) \times 100 \quad (6b)$$

3.2 GITT Solution of the 3D Micro-reactor Problem

Equation (1) constitutes a nonlinear system of three-dimensional partial differential equations, coupled through the kinetic expressions for each species, which has been solved by the generalized integral transform technique (GITT) [69, 70]. First of all, in order to homogenize the boundary condition of the alcohol species in the *Y* direction and ensure improved convergence behavior, a simple filter is proposed as:

Table 1 Auxiliary eigenvalue problems for eigenfunction $\Gamma(Y, Z)$: (a) eigenvalue problems; (b) normalized eigenfunction; (c) norm; (d) eigenvalues

	Coordinate Y	Coordinate Z
(a)	$\frac{d^2\phi_{1,m}}{dY^2} + \beta_{1,m}^2\phi_{1,m}(Y) = 0$ $\left.\frac{d\phi_{1,m}}{dY}\right\|_{Y=0} = 0,\ \phi_{1,m}(1) = 0$	$\delta\frac{d^2\phi_{2,n}}{dZ^2} + \beta_{2,n}^2\phi_{2,n}(Z) = 0$ $\left.\frac{d\phi_{2,n}}{dZ}\right\|_{Z=0} = 0,\ \left.\frac{d\phi_{2,n}}{dZ}\right\|_{Z=1} = 0$
(b)	$\tilde{\phi}_{1,m}(Y) = \frac{\cos(\beta_{1,m}Y)}{\sqrt{N_{\phi 1}}}$	$\tilde{\phi}_{2,n}(Z) =$ $\frac{1}{\sqrt{N_{\varphi 2}}}\begin{cases} 1 & \text{if } \beta_{2,n} = 0 \\ \cos\left(\frac{\beta_{2,n}}{\sqrt{\delta}}Z\right) & \text{if } \beta_{2,n} \neq 0 \end{cases}$
(c)	$N_{\phi 1} = \int_0^1 \phi_{1,m}^2(Y)dY = \frac{1}{2}$	$N_{\phi 2} = \int_0^1 \phi_{2,m}^2(Y)dZ =$ $\begin{cases} 1 & \text{if } \beta_{2,n} = 0 \\ 1/2 & \text{if } \beta_{2,n} \neq 0 \end{cases}$
(d)	$\beta_{1,m} = \frac{(2m-1)\pi}{2},\ m = 1, 2, \ldots$	$\beta_{2,n} = \sqrt{\delta}n\pi,\ n = 0, 1, 2, \ldots$

$$F_A(X, Y, Z) = F_{Ah}(X, Y, Z) + F_{Ao} \tag{7}$$

To satisfy the boundary conditions and establish the basis to represent the concentrations as eigenfunction expansions, two different eigenvalue problems were adopted, one for the alcohol species and the other for the remaining species. In [70], two different possibilities were investigated, either accounting for the velocity profile in the eigenvalue problem or just considering the simplest constant coefficients' formulation. Here, only the second alternative is detailed to illustrate the methodology, as in [69], but results are compared later on for both solution paths. Thus, the eigenvalue problem for the alcohol concentration equation can be defined as:

$$\frac{\partial^2\Gamma_i}{\partial Y^2} + \delta\frac{\partial^2\Gamma_i}{\partial Z^2} + \beta_i^2\Gamma_i(Y, Z) = 0 \tag{8a}$$

$$\left.\frac{\partial\Gamma_i}{\partial Y}\right|_{Y=0} = 0, \quad \Gamma_i(1, Z) = 0, \quad \left.\frac{\partial\Gamma_i}{\partial Z}\right|_{Z=0} = 0, \quad \left.\frac{\partial\Gamma_i}{\partial Z}\right|_{Z=1} = 0 \tag{8b–e}$$

Problem (8) is then solved by separation of variables, leading to a couple of one-dimensional problems to represent the original eigenfunction $\Gamma(Y, Z)$ in accordance with Table 1. One can then express the normalized eigenfunction $\widetilde{\Gamma}(Y, Z)$ and the corresponding eigenvalues as:

$$\widetilde{\Gamma}_{m,n}(Y, Z) = \tilde{\phi}_{1,m}(Y)\tilde{\phi}_{2,n}(Z) \tag{9a}$$

$$\beta_{m,n}^2 = \beta_{1,m}^2 + \beta_{2,n}^2 \tag{9b}$$

Table 2 Auxiliary eigenvalue problems for eigenfunction $\Psi(Y, Z)$: (a) eigenvalue problems; (b) normalized eigenfunction; (c) norm; (d) eigenvalues

	Coordinate Y	Coordinate Z
(a)	$\frac{d^2\varphi_{1,m}}{dY^2} + \mu_{1,m}^2\varphi_{1,m}(Y) = 0$ $\left.\frac{d\varphi_{1,m}}{dY}\right\|_{Y=0} = 0, \left.\frac{d\varphi_{1,m}}{dY}\right\|_{Y=1} = 0$	$\delta\frac{d^2\varphi_{2,n}}{dY^2} + \mu_{2,n}^2\varphi_{2,n}(Y) = 0$ $\left.\frac{d\varphi_{2,n}}{dY}\right\|_{Y=0} = 0, \left.\frac{d\varphi_{2,n}}{dY}\right\|_{Y=1} = 0$
(b)	$\tilde{\varphi}_{1,m}(Y) = \frac{1}{\sqrt{N_{\varphi 1}}}\begin{cases} 1 & \text{if } \mu_{1,m} = 0 \\ \cos(\mu_{1,m}Y) & \text{if } \mu_{1,m} \neq 0 \end{cases}$	$\tilde{\varphi}_{2,n}(Z) = \frac{1}{\sqrt{N_{\varphi 2}}}\begin{cases} 1 & \text{if } \mu_{2,n} = 0 \\ \cos\left(\frac{\mu_{2,n}}{\sqrt{\delta}}Z\right) & \text{if } \mu_{2,n} \neq 0 \end{cases}$
(c)	$N_{\varphi 1} = \int_0^1 \varphi_{1,m}^2(Y)dY = \begin{cases} 1 & \text{if } \mu_{1,m} = 0 \\ 1/2 & \text{if } \mu_{1,m} \neq 0 \end{cases}$	$N_{\varphi 2} = \int_0^1 \varphi_{2,m}^2(Z)dZ = \begin{cases} 1 & \text{if } \mu_{2,n} = 0 \\ 1/2 & \text{if } \mu_{2,n} \neq 0 \end{cases}$
(d)	$\mu_{1,m} = m\pi, m = 0, 1, 2, \ldots$	$\mu_{2,n} = \sqrt{\delta}n\pi, n = 0, 1, 2, \ldots$

The eigenvalue problem for the other remaining species can be defined as:

$$\frac{\partial^2\Psi_i}{\partial Y^2} + \delta\frac{\partial^2\Psi_i}{\partial Z^2} + \mu_i^2\Psi_i(Y, Z) = 0 \tag{10a}$$

$$\left.\frac{\partial\Psi_i}{\partial Y}\right|_{Y=0} = 0, \quad \left.\frac{\partial\Psi_i}{\partial Y}\right|_{Y=1} = 0, \quad \left.\frac{\partial\Psi_i}{\partial Z}\right|_{Z=0} = 0, \quad \left.\frac{\partial\Psi_i}{\partial Z}\right|_{Z=1} = 0 \tag{10b–e}$$

Again, problem (10) is readily solved analytically and the solution is expressed through the one-dimensional eigenvalue problems as presented in Table 2. Then, the normalized eigenfunction $\widetilde{\Psi}$ and its eigenvalues can be written as:

$$\widetilde{\Psi}_{m,n}(Y, Z) = \tilde{\varphi}_{1,m}(Y)\tilde{\varphi}_{2,n}(Z) \tag{11a}$$

$$\mu_{m,n}^2 = \mu_{1,m}^2 + \mu_{2,n}^2 \tag{11b}$$

After solving the related eigenvalue problems, the following integral transform pairs for each of the species involved in the transesterification reaction can be defined:

– Transforms

$$\overline{F}_{\mathrm{TG}_i}(X) = \int_0^1 \int_0^1 \widetilde{\Psi}_i(Y, Z) F_{\mathrm{TG}}(X, Y, Z)\mathrm{d}Y\mathrm{d}Z$$

$$\overline{F}_{\mathrm{Ah}_i}(X) = \int_0^1 \int_0^1 \widetilde{\Gamma}_i(Y, Z) F_{\mathrm{Ah}}(X, Y, Z)\mathrm{d}Y\mathrm{d}Z$$

$$\overline{F}_{\mathrm{DG}_i}(X) = \int_0^1 \int_0^1 \widetilde{\Psi}_i(Y, Z) F_{\mathrm{DG}}(X, Y, Z)\mathrm{d}Y\mathrm{d}Z$$

$$\overline{F}_{\mathrm{MG}_i}(X) = \int_0^1 \int_0^1 \widetilde{\Psi}_i(Y, Z) F_{\mathrm{MG}}(X, Y, Z)\mathrm{d}Y\mathrm{d}Z$$

$$\overline{F}_{\mathrm{B}_i}(X) = \int_0^1 \int_0^1 \widetilde{\Psi}_i(Y, Z) F_{\mathrm{B}}(X, Y, Z)\mathrm{d}Y\mathrm{d}Z$$

$$\overline{F}_{\mathrm{GL}_i}(X) = \int_0^1 \int_0^1 \widetilde{\Psi}_i(Y, Z) F_{\mathrm{GL}}(X, Y, Z)\mathrm{d}Y\mathrm{d}Z \qquad \text{(12a–f)}$$

– Inverses

$$F_{\mathrm{TG}}(X, Y, Z) = \sum_{i=1}^{\infty} \widetilde{\Psi}_i(Y, Z)\overline{F}_{\mathrm{TG}_i}(X), \; F_{\mathrm{Ah}}(X, Y, Z) = \sum_{i=1}^{\infty} \widetilde{\Gamma}_i(Y, Z)\overline{F}_{\mathrm{Ah}_i}(X)$$

$$F_{\mathrm{DG}}(X, Y, Z) = \sum_{i=1}^{\infty} \widetilde{\Psi}_i(Y, Z)\overline{F}_{\mathrm{DG}_i}(X), \; F_{\mathrm{MG}}(X, Y, Z) = \sum_{i=1}^{\infty} \widetilde{\Psi}_i(Y, Z)\overline{F}_{\mathrm{MG}_i}(X)$$

$$F_{\mathrm{B}}(X, Y, Z) = \sum_{i=1}^{\infty} \widetilde{\Psi}_i(Y, Z)\overline{F}_{\mathrm{B}_i}(X), \; F_{\mathrm{GL}}(X, Y, Z) = \sum_{i=1}^{\infty} \widetilde{\Psi}_i(Y, Z)\overline{F}_{\mathrm{GL}_i}(X)$$

(13a–f)

It can be noticed that the eigenfunction expansions in Eq. (13) have been represented with a single index i, thus combining the eigenvalues and eigenfunctions of the one-dimensional auxiliary problems by reordering the summation terms in decreasing order of importance, adopting the sum of the squared eigenvalues as the sorting criterium [16, 69]. Finally, the integral transformation of Eqs. (1a–1o), after filtering with Eq. (7), is accomplished employing the integral transform operator with the respective eigenfunction for each species along both the Y and Z domains. The transformed ODE system then becomes:

$$\sum_{j=1}^{\infty} O_{ij}^{1} \frac{d\overline{F}_{s_j}}{dX} = \frac{1}{Pe_s} \frac{d^2\overline{F}_{s_j}}{dX^2} - \xi_s \mu_i^2 \overline{F}_{s_i} + \varsigma \overline{G}_{s,ijk}, \text{ where s} = \text{TG, DG, MG, B, GL} \tag{14a}$$

$$\sum_{j=1}^{\infty} O_{ij}^{2} \frac{d\overline{F}_{Ah_j}}{dX} = \frac{1}{Pe_A} \frac{d^2\overline{F}_{Ah_j}}{dX^2} - \xi_A \beta_i^2 \overline{F}_{Ah_i} + \varsigma \overline{G}_{Ah,ijk} \tag{14b}$$

$$\overline{F}_{TG_i}(0) = \bar{f}_{1,i}, \quad \overline{F}_{Ah_i}(0) = -F_{Ao}\bar{f}_{2,i}, \quad \overline{F}_{s_i}(0) = 0, \text{ where s} = \text{DG, MG, B, GL} \tag{14c–e}$$

$$\left.\frac{d\overline{F}_{s_i}}{dX}\right|_{X=1} = 0, \quad \text{for all species s} \tag{14f}$$

The term $\overline{G}_{s,ijk}$ represents the transformed reaction kinetic terms, according to:

$$\overline{G}_{TG,ijk} = \sum_{k=1}^{\infty}\sum_{j=1}^{\infty}\left(-k_1 O_{ijk}^{3} \overline{F}_{TG_j}\overline{F}_{Ah_k} + k_2 O_{ijk}^{4} \overline{F}_{DG_j}\overline{F}_{B_k}\right) - k_1 F_{Ao}\overline{F}_{TG_i} \tag{14g}$$

$$\begin{aligned}\overline{G}_{A,ijk} = &\sum_{k=1}^{\infty}\sum_{j=1}^{\infty}\left[O_{ijk}^{5}\left(-k_1\overline{F}_{TG_j} - k_3\overline{F}_{DG_j} - k_5\overline{F}_{MG_j}\right)\overline{F}_{Ah_k}\right.\\ &\left. + O_{ijk}^{6}\left(k_2\overline{F}_{DG_j} + k_4\overline{F}_{MG_j} + k_6\overline{F}_{GL_j}\right)\overline{F}_{B_k}\right]\\ &+ F_{Ao}\sum_{j=1}^{\infty} O_{ij}^{7}\left(-k_1\overline{F}_{TG_j} - k_3\overline{F}_{DG_j} - k_5\overline{F}_{MG_j}\right)\end{aligned} \tag{14h}$$

$$\begin{aligned}\overline{G}_{DG,ijk} = &\sum_{k=1}^{\infty}\sum_{j=1}^{\infty}\left[O_{ijk}^{3}\left(k_1\overline{F}_{TG_j} - k_3\overline{F}_{DG_j}\right)\overline{F}_{Ah_k} + O_{ijk}^{4}\left(-k_2\overline{F}_{DG_j} + k_4\overline{F}_{MG_j}\right)\overline{F}_{B_k}\right]\\ &+ F_{Ao}\left(k_1\overline{F}_{TG_i} - k_3\overline{F}_{DG_i}\right)\end{aligned} \tag{14i}$$

$$\begin{aligned}\overline{G}_{MG,ijk} = &\sum_{k=1}^{\infty}\sum_{j=1}^{\infty}\left[O_{ijk}^{3}\left(k_3\overline{F}_{DG_j} - k_5\overline{F}_{MG_j}\right)\overline{F}_{Ah_k} + O_{ijk}^{4}\left(-k_4\overline{F}_{MG_j} + k_6\overline{F}_{GL_j}\right)\overline{F}_{B_k}\right]\\ &+ F_{Ao}\left(k_3\overline{F}_{DG_i} - k_5\overline{F}_{MG_i}\right)\end{aligned} \tag{14j}$$

$$\overline{G}_{GL,ijk} = \sum_{k=1}^{\infty}\sum_{j=1}^{\infty}\left(k_5 O_{ijk}^{3}\overline{F}_{MG_j}\overline{F}_{Ah_k} - k_6 O_{ijk}^{4}\overline{F}_{GL_j}\overline{F}_{B_k}\right) + k_5 F_{Ao}\overline{F}_{MG_i} \tag{14k}$$

$$\begin{aligned}\overline{G}_{B,ijk} = &\sum_{k=1}^{\infty}\sum_{j=1}^{\infty}\left[O_{ijk}^{3}\left(k_1\overline{F}_{TG_j} + k_3\overline{F}_{DG_j} + k_5\overline{F}_{MG_j}\right)\overline{F}_{Ah_k}\right.\\ &\left. + O_{ijk}^{4}\left(-k_2\overline{F}_{DG_j} - k_4\overline{F}_{MG_j} - k_6\overline{F}_{GL_j}\right)\overline{F}_{B_k}\right]\\ &+ F_{Ao}\left(k_1\overline{F}_{TG_i} + k_3\overline{F}_{DG_i} + k_5\overline{F}_{MG_i}\right)\end{aligned} \tag{14l}$$

where the coefficients produced by the integral transformation procedure are given by:

$$O_{ij}^1 = \int_0^1\int_0^1 U_{TG}\widetilde{\Psi}_i\widetilde{\Psi}_j dYdZ,\quad O_{ij}^2 = \int_0^1\int_0^1 U_{TG}\widetilde{\Gamma}_i\widetilde{\Gamma}_j dYdZ,\quad O_{ijk}^3 = \int_0^1\int_0^1 \widetilde{\Psi}_i\widetilde{\Psi}_j\widetilde{\Gamma}_k dYdZ,$$

$$O_{ijk}^4 = \int_0^1\int_0^1 \widetilde{\Psi}_i\widetilde{\Psi}_j\widetilde{\Psi}_k dYdZ;\quad O_{ijk}^5 = \int_0^1\int_0^1 \widetilde{\Gamma}_i\widetilde{\Psi}_j\widetilde{\Gamma}_k dYdZ;\quad O_{ijk}^6 = \int_0^1\int_0^1 \widetilde{\Gamma}_i\widetilde{\Psi}_j\widetilde{\Psi}_k dYdZ$$

$$O_{ij}^7 = \int_0^1\int_0^1 \widetilde{\Gamma}_i\widetilde{\Psi}_j dYdZ;\quad \bar{f}_{1,i} = \int_0^1\int_0^1 \widetilde{\Psi}_i dYdZ;\quad \bar{f}_{2,i} = \int_0^1\int_0^1 \widetilde{\Gamma}_i dYdZ \qquad (15a–i)$$

Once the transformed concentrations have been numerically determined from solving the ODE system (14), the original concentrations can be recovered through the inverse formulas defined in Eq. (13a–f) and in the filter, Eq. (7). The concentrations can also be presented in terms of flow averaged potentials, defined as follows:

$$F_{s,AV}(X) = \frac{\int_0^1\int_0^1 U_{TG}(Y,Z)F_s(X,Y,Z)dYdZ}{\int_0^1\int_0^1 U_{TG}(Y,Z)dYdZ},\quad s = \text{TG, A, DG, MG, GL, B} \qquad (16)$$

By introducing the eigenfunction expansions for the dimensionless concentrations of the species, the following average concentrations can be obtained:

$$F_{s,AV}(X) = \frac{\sum_{i=1}^{\infty}\int_0^1\int_0^1 U_{TG}(Y,Z)\widetilde{\Psi}_i(Y,Z)dYdZ\overline{F}_{s_i}(X)}{\int_0^1\int_0^1 U_{TG}(Y,Z)dYdZ},\quad s = \text{TG, DG, MG, GL, B} \qquad (17a)$$

$$F_{A,AV}(X) = \frac{\sum_{i=1}^{\infty}\int_0^1\int_0^1 U_{TG}(Y,Z)\widetilde{\Gamma}_i(Y,Z)dYdZ\overline{F}_{Ah_i}(X)}{\int_0^1\int_0^1 U_{TG}(Y,Z)dYdZ} + F_{Ao} \qquad (17b)$$

The built-in function ND Solve of the *Mathematica* system [40] is a suitable computational tool for numerically handling the coupled nonlinear transformed system, Eq. (14), since it provides automatic error-controlled solutions and allows for expressing the dependence of the transformed fields along the variable X as continuous interpolation functions.

3.3 CIEA for Improved Lumped-Differential Formulation

The Coupled Integral Equations Approach (CIEA) [14–20] aims at reducing the number of independent variables in partial differential equations, via averaging (lumping) operations in the coordinates selected to be eliminated, but partially preserving the local information in these space variables through the corresponding boundary conditions. The use of Hermite formulae to approximate the averaging integrals [21] provides improved formulations when compared to the classical lumped system analysis [20], which is the simplest possible lumped approximation. The partial differential three-dimensional model shown in Eq. (1) is here reformulated using the

CIEA technique, yielding a nonlinear coupled mathematical model composed only of first-order ODEs that enables simulations with reduced computational costs [14–20].

Following the CIEA procedure, the lumping was applied in the transverse and vertical directions, coordinates Z and Y, respectively [71]. The approximation $H_{0,0}$, which corresponds to the trapezoid rule, was used to approximate the average mass flow of the species, while the approximation $H_{1,1}$, which corresponds to the corrected trapezoid rule, was used to approximate the average species concentrations. In general form, the approximations $H_{0,0}$ and $H_{1,1}$ are given by:

$$H_{0,0} \Rightarrow \int_0^h f(x)\mathrm{d}x \approx \frac{h}{2}(f(0) + f(h)) \tag{18a}$$

$$H_{1,1} \Rightarrow \int_0^h f(x)\mathrm{d}x \approx \frac{h}{2}(f(0) + f(h)) + \frac{h^2}{12}\big(f'(0) + f'(h)\big) \tag{18b}$$

The dimensionless average concentrations in the transverse direction (Z-coordinate), $\overline{F}_{\mathrm{s}}$, and the Z-averaged velocity field, $\overline{U}_{\mathrm{TG}}(Y)$, are defined as:

$$\overline{F}_{\mathrm{s}}(X, Y) \equiv \frac{\int_0^1 U_{\mathrm{TG}}(Y, Z) F_{\mathrm{s}}(X, Y, Z)\mathrm{d}Z}{\int_0^1 U_{\mathrm{TG}}(Y, Z)\mathrm{d}Z} \tag{19a}$$

$$\overline{U}_{\mathrm{TG}}(Y) = \int_0^1 U_{\mathrm{TG}}(Y, Z)\mathrm{d}Z \tag{19b}$$

The species balance equations, Eq. (1), are first averaged in the transverse Z-coordinate, neglecting the axial diffusion term in light of the high values of the Peclet number encountered in this application, as reported in [70]. Next, the resulting equations are averaged in the vertical direction (Y-coordinate) and a system of first-order ordinary differential equations is obtained for the transversally averaged concentrations, given by Costa Junior and Naveira-Cotta [71]:

$$\overline{\overline{U}}_{\mathrm{TG}} \frac{\mathrm{d}\overline{\overline{F}}_{\mathrm{s}}(X)}{\mathrm{d}X} = \varsigma \overline{\overline{G}}_{\mathrm{s}} \quad \text{where s} = \text{TG, DG, MG, GL, B} \tag{20a}$$

$$\overline{\overline{U}}_{\mathrm{TG}} \frac{\mathrm{d}\overline{\overline{F}}_{\mathrm{A}}(X)}{\mathrm{d}X} = \xi_{\mathrm{A}}\Big(3P^* \overline{\overline{F}}_{\mathrm{A}}(X) + Q^*\Big) + \varsigma \overline{\overline{G}}_{\mathrm{A}} \tag{20b}$$

$$\overline{\overline{F}}_{\mathrm{TG}}(0) = 1, \quad \overline{\overline{F}}_{\mathrm{s}}(0) = 0, \quad \text{where s} = \text{A, DG, MG, GL, B} \tag{20c, d}$$

where

$$\overline{\overline{U}}_{\mathrm{TG}} = \int_0^1 \overline{U}_{\mathrm{TG}}(Y)\mathrm{d}Y, \quad P^* = \frac{-6A^*(1)\overline{\overline{U}}_{\mathrm{TG}}}{\overline{U}_{\mathrm{TG}}(1)}, \quad Q^* = A^*(1) F_{\mathrm{A}_0}\left(4 - \frac{1}{\overline{U}_{\mathrm{TG}}(1)} \left.\frac{\partial \overline{U}_{\mathrm{TG}}}{\partial Y}\right|_{Y=1}\right),$$

$$A^*(Y) = 3\overline{U}_{\mathrm{TG}}\left(\frac{1}{\left.\frac{\partial U_{\mathrm{TG}}}{\partial Z}\right|_{Z=0}} - \frac{1}{\left.\frac{\partial U_{\mathrm{TG}}}{\partial Z}\right|_{Z=1}}\right) \tag{21a–d}$$

The averaged expressions for the nonlinear source terms $\overline{\overline{G}}_{\mathrm{s}}$, for the different species in the transesterification reaction, are given by Costa Junior and Naveira-Cotta [71]:

$$\overline{\overline{G}}_{\mathrm{TG}} = -k_1\left(M\overline{\overline{F}}_{\mathrm{TG}}\overline{\overline{F}}_{\mathrm{A}} + N\overline{\overline{F}}_{\mathrm{TG}}\right) + k_2 O\overline{\overline{F}}_{\mathrm{DG}}\overline{\overline{F}}_{\mathrm{B}} \tag{22a}$$

$$\overline{\overline{G}}_{\mathrm{A}} = \left(-k_1\overline{\overline{F}}_{\mathrm{TG}} - k_3\overline{\overline{F}}_{\mathrm{DG}} - k_5\overline{\overline{F}}_{\mathrm{MG}}\right)M\overline{\overline{F}}_{\mathrm{A}} + \left(-k_1\overline{\overline{F}}_{\mathrm{TG}} - k_3\overline{\overline{F}}_{\mathrm{DG}} - k_5\overline{\overline{F}}_{\mathrm{MG}}\right)N + \left(k_2\overline{\overline{F}}_{\mathrm{DG}} + k_4\overline{\overline{F}}_{\mathrm{MG}} + k_6\overline{\overline{F}}_{\mathrm{GL}}\right)O\overline{\overline{F}}_{\mathrm{B}} \tag{22b}$$

$$\overline{\overline{G}}_{\mathrm{DG}} = \left(k_1\overline{\overline{F}}_{\mathrm{TG}} - k_3\overline{\overline{F}}_{\mathrm{DG}}\right)M\overline{\overline{F}}_{\mathrm{A}} + \left(k_1\overline{\overline{F}}_{\mathrm{TG}} - k_3\overline{\overline{F}}_{\mathrm{DG}}\right)N + \left(-k_2\overline{\overline{F}}_{\mathrm{DG}} + k_4\overline{\overline{F}}_{\mathrm{MG}}\right)O\overline{\overline{F}}_{\mathrm{B}} \tag{22c}$$

$$\overline{\overline{G}}_{\mathrm{MG}} = \left(k_3\overline{\overline{F}}_{\mathrm{DG}} - k_5\overline{\overline{F}}_{\mathrm{MG}}\right)M\overline{\overline{F}}_{\mathrm{A}} + \left(k_3\overline{\overline{F}}_{\mathrm{DG}} - k_5\overline{\overline{F}}_{\mathrm{MG}}\right)N + \left(-k_4\overline{\overline{F}}_{\mathrm{MG}} + k_6\overline{\overline{F}}_{\mathrm{GL}}\right)O\overline{\overline{F}}_{\mathrm{B}} \tag{22d}$$

$$\overline{\overline{G}}_{\mathrm{GL}} = k_5 M\overline{\overline{F}}_{\mathrm{MG}}\overline{\overline{F}}_{\mathrm{A}} + k_5 N\overline{\overline{F}}_{\mathrm{MG}} - k_6 O\overline{\overline{F}}_{\mathrm{GL}}\overline{\overline{F}}_{\mathrm{B}} \tag{22e}$$

$$\overline{\overline{G}}_{\mathrm{B}} = \left(k_1\overline{\overline{F}}_{\mathrm{TG}} + k_3\overline{\overline{F}}_{\mathrm{DG}} + k_5\overline{\overline{F}}_{\mathrm{MG}}\right)M\overline{\overline{F}}_{\mathrm{A}} + \left(k_1\overline{\overline{F}}_{\mathrm{TG}} + k_3\overline{\overline{F}}_{\mathrm{DG}} + k_5\overline{\overline{F}}_{\mathrm{MG}}\right)N + \left(-k_2\overline{\overline{F}}_{\mathrm{DG}} - k_4\overline{\overline{F}}_{\mathrm{MG}} - k_6\overline{\overline{F}}_{\mathrm{GL}}\right)O\overline{\overline{F}}_{\mathrm{B}} \tag{22f}$$

where

$$M = \frac{-3\overline{\overline{U}}_{\mathrm{TG}}^2\left(B^*(1)\left(\left.\frac{\partial\overline{U}_{\mathrm{TG}}}{\partial Y}\right|_{Y=0}\right)^2 + 6B^*(0)\overline{U}_{\mathrm{TG}}(1)\left(\left.\frac{\partial\overline{U}_{\mathrm{TG}}}{\partial Y}\right|_{Y=1} + 2\overline{U}_{\mathrm{TG}}(1)\right)\right)}{\left(\left.\frac{\partial\overline{U}_{\mathrm{TG}}}{\partial Y}\right|_{Y=0}\right)^2\overline{U}_{\mathrm{TG}}(1)\left(\left.\frac{\partial\overline{U}_{\mathrm{TG}}}{\partial Y}\right|_{Y=1} - 4\overline{U}_{\mathrm{TG}}(1)\right)} \tag{22g}$$

$$N \Im \frac{\otimes F_{A_0}\overline{\overline{U}}_{\mathrm{TG}}\left(12B^*(0)\left(\overline{U}_{\mathrm{TG}}(1)\right)^2\left(\left.\frac{\partial\overline{U}_{\mathrm{TG}}}{\partial Y}\right|_{Y\Im 1} \otimes 2\overline{U}_{\mathrm{TG}}(1)\right) \Downarrow B^*(1)\left(\left.\frac{\partial\overline{U}_{\mathrm{TG}}}{\partial Y}\right|_{Y\Im 0}\right)^2\left(\left.\frac{\partial\overline{U}_{\mathrm{TG}}}{\partial Y}\right|_{Y\Im 1} \Downarrow 2\overline{U}_{\mathrm{TG}}(1)\right)\right)}{2\left(\left.\frac{\partial\overline{U}_{\mathrm{TG}}}{\partial Y}\right|_{Y\Im 0}\right)^2\overline{U}_{\mathrm{TG}}(1)\left(\left.\frac{\partial\overline{U}_{\mathrm{TG}}}{\partial Y}\right|_{Y\Im 1} \otimes 4\overline{U}_{\mathrm{TG}}(1)\right)} \tag{22h}$$

$$O = \frac{18\overline{\overline{U}}_{\mathrm{TG}}^2\left(B^*(1)\left(\left.\frac{\partial\overline{U}_{\mathrm{TG}}}{\partial Y}\right|_{Y=0}\right)^2 + B^*(0)\left(\left.\frac{\partial\overline{U}_{\mathrm{TG}}}{\partial Y}\right|_{Y=1} - 2\overline{U}_{\mathrm{TG}}(1)\right)^2\right)}{\left(\left.\frac{\partial\overline{U}_{\mathrm{TG}}}{\partial Y}\right|_{Y=0}\right)^2\left(\left.\frac{\partial\overline{U}_{\mathrm{TG}}}{\partial Y}\right|_{Y=1} - 4\overline{U}_{\mathrm{TG}}(1)\right)^2} \tag{22i}$$

$$B^*(Y) = 18\overline{U}_{\mathrm{TG}}^2\left(\frac{1}{\left(\frac{\partial U_{\mathrm{TG}}}{\partial Z}\Big|_{Z=0}\right)^2} - \frac{1}{\left(\frac{\partial U_{\mathrm{TG}}}{\partial Z}\Big|_{Z=1}\right)^2}\right) \tag{22j}$$

3.4 Inverse Problem Analysis

Inverse problem analysis is an essential tool in engineering design for the identification of parameters, such as physical properties, and functions, such as equation and boundary condition source terms. Besides, such methodologies have evolved toward experimental planning and optimum design procedures [43–52].

The statistical approach for the solution of inverse problems is briefly reviewed here, following the Markov chain–Monte Carlo (MCMC) method [79–81]. The MCMC method is used in conjunction with the Coupled Integral Equations Approach for the solution of the inverse problem of estimating the kinetic coefficients in the transesterification reaction, in order to obtain low-cost inverse problem solutions, as detailed in [71].

The MCMC method represents a Bayesian approach to parameter estimation. Bayes theorem is given by:

$$\pi_{\mathrm{posterior}}(\mathbf{P}) = \pi(\mathbf{P}|\mathbf{Y}) = \frac{\pi_{\mathrm{prior}}(\mathbf{P})\pi(\mathbf{Y}|\mathbf{P})}{\pi(\mathbf{Y})} \tag{23}$$

where $\mathbf{P}$ is the parameter vector, $\mathbf{Y}$ is the vector of measurements, $\pi_{\mathrm{posteriori}}(\mathbf{P})$ is the posterior probability density, $\pi_{\mathrm{prior}}(\mathbf{P})$ is the prior density, $\pi(\mathbf{Y}|\mathbf{P}))$ is the likelihood function, and $\pi(\mathbf{Y})$ is the probability density of the measurements, which represents a normalization constant.

Assuming that the measurement errors are additive, uncorrelated, with Gaussian distribution of zero mean, and known covariance matrix, the likelihood function is given by:

$$\pi(\mathbf{Y}|\mathbf{P}) = (2\pi)^{-1/2}|\mathbf{W}|^{-1/2}\exp\left[-\frac{1}{2}[\mathbf{Y} - \mathbf{Y}_e(\mathbf{P})]^{\mathrm{T}}\mathbf{W}^{-1}[\mathbf{Y} - \mathbf{Y}_e(\mathbf{P})]\right] \tag{24}$$

where $\mathbf{Y}_e$ $(\mathbf{P})$ is the solution of the direct problem, obtained with a sample of the parameter vector $\mathbf{P}$ and $\mathbf{W}$ is the covariance matrix of the errors in $\mathbf{Y}$.

It is in general not possible to analytically obtain the posterior probability density; thus, the MCMC method is used to draw samples of all possible parameters. One of the most widely used algorithms for implementing the MCMC method is known as the Metropolis–Hasting algorithm, as described in [79].

3.5 Results and Discussion

The reagents chosen in the present illustration are soybean oil (triglycerides) and methanol, with sodium hydroxide as catalyst. The hybrid approach here reviewed is then employed to report the direct–inverse problem analysis through the evolution of triglyceride conversion in terms of residence time and reaction temperature, with available data for the chosen reagents [69–71, 75], as summarized in Table 3. The generalized integral transform technique (GITT) was employed to handle both the two- and three-dimensional formulations of the complete mass transfer problem, Eq. (1) [67–70]. Both situations of a constant coefficients eigenvalue problem and the alternative of including the velocity profile in the eigenvalue problem formulation have been dealt with [69, 70], with relative merits that shall be briefly discussed in what follows. Also, the results obtained through the reduced model obtained with the Coupled Integral Equations Approach were verified comparing the reformulated first-order ordinary differential equation system with the three-dimensional model results for micro-reactors with square cross section (200 × 200 μm and 400 × 400 μm) and length of 2.33 cm.

The eigenfunction expansions for the average concentrations of the triglyceride and biodiesel species were first analyzed in terms of convergence rate, taking the temperature $T = 298$ K and residence time $\tau = 30$ s, for each of the eigenvalue problem choices, as shown in Table 4. The GITT solution obtained with the eigenvalue problem accounting for the velocity profile presents an improved behavior in terms of convergence rate, since it needs around 50 terms to reach up to four fully converged significant digits, while the simpler eigenvalue problem with constant coefficient requires, for the same case, around 100 terms. These results highlight

Table 3 Data for the direct–inverse problem analysis of biodiesel synthesis in micro-reactors [69–71, 75]

Parameters	Values	Parameters	Values
Alcohol	Methanol	k_1 [m³/(mol.s)]	4.368×10^{-6}
Oil	Soybean	k_2 [m³/(mol.s)]	9.623×10^{-6}
Q_{TG}/Q_A	3.4	k_3 [m³/(mol.s)]	1.88×10^{-5}
μ_{TG} (Pa s)	5.825×10^{-2}	k_4 [m³/(mol.s)]	1.074×10^{-4}
μ_A(Pa s)	5.47×10^{-4}	k_5 [m³/(mol.s)]	2.117×10^{-5}
D_{TG} (m²/s)	1.58×10^{-9}	k_6 [m³/(mol.s)]	9.0×10^{-7}
ρ_{TG} (kg/m³)	885	Reaction	Activation energy (J mol^{-1})
D_A (m²/s)	1.182×10^{-10}	TG → DG	55.0355
D_{DG}, D_{MG}, D_{GL}, D_B (m²/s)	1.38×10^{-9}	DG → TG	41.5833
C_{TGo} (mol/m³)	1014	DG → MG	83.1498
F_{Ao}	4.4	MG → DG	61.2906
R (J mol^{-1} K^{-1})	8.314	MG → GL	26.8834
		GL → MG	40.1430

Table 4 Convergence of eigenfunction expansions for the average concentrations of triglyceride and biodiesel in a square micro-reactor ($L = 233 \times 10^{-4}$ m, $H = W = 100 \times 10^{-6}$ m)

NT	$X = 0.3$		$X = 0.5$		$X = 0.7$	
	$C_{TG,AV}$	$C_{B,Av}$	$C_{TG,AV}$	$C_{B,Av}$	$C_{TG,AV}$	$C_{B,Av}$
Eigenvalue problem: constant coefficients						
20	0.955411	0.049206	0.904022	0.117023	0.844553	0.210476
40	0.955648	0.048923	0.904303	0.116622	0.844850	0.209973
60	0.955815	0.048725	0.904513	0.116322	0.845079	0.209585
80	0.955802	0.048740	0.904492	0.116351	0.845054	0.209627
100	0.955752	0.048799	0.904426	0.116445	0.844981	0.209751
Eigenvalue problem: variable coefficients						
10	0.955449	0.049150	0.904122	0.116853	0.844666	0.210240
20	0.955705	0.048851	0.904378	0.116504	0.844923	0.209831
30	0.955751	0.048798	0.904427	0.116438	0.844977	0.209748
40	0.955761	0.048785	0.904438	0.116423	0.844988	0.209729
50	0.955766	0.048781	0.904444	0.116415	0.844997	0.209717

the gain, in terms of convergence rates that can be achieved considering more informative eigenvalue problems, which in the present case is the one that considers the velocity profile in its formulation.

Figure 4 illustrates the evolution with residence time of the dimensionless average concentration of each species at the 400×400 μm micro-reactor outlet, as obtained by GITT with increasing truncation orders of the eigenfunction expansions, namely $NT = 1, 5$, and 40. It can be noticed that to the graph scale, the results for fairly low truncation orders, such as $NT = 5$, are already in excellent agreement with the fully converged results with the higher truncation order of $NT = 40$ terms. This observation opened up the possibility of using another reduced model, based on low truncation order approximations of the integral transform solution, as successfully accomplished in [70]. The triglyceride concentration, as expected, reduces with the increase of the residence time, while the other species are formed. Thus, the intermediate species reach a maximum value and then start to decrease as well, leading to the formation of biodiesel.

Figure 5a, b presents the triglyceride conversion as a function of residence time for the 200×200 μm and the 400×400 μm cross sections, respectively, with $T = 45$ °C. The dashed lines are for the reduced model, Eq. (22), solved through the ND Solve algorithm [40]. The dotted lines are for the GITT solution of the three-dimensional model, Eq. (1) [69, 70]. One can observe the reasonable agreement of the two solutions, corresponding to the reduced model and to the GITT simulations of the three-dimensional model, especially for smaller and larger residence times. After the reduced model has been verified, it may be used in the estimation of the kinetic constants, with considerable reduction on the computational effort in the analysis of the inverse problem.

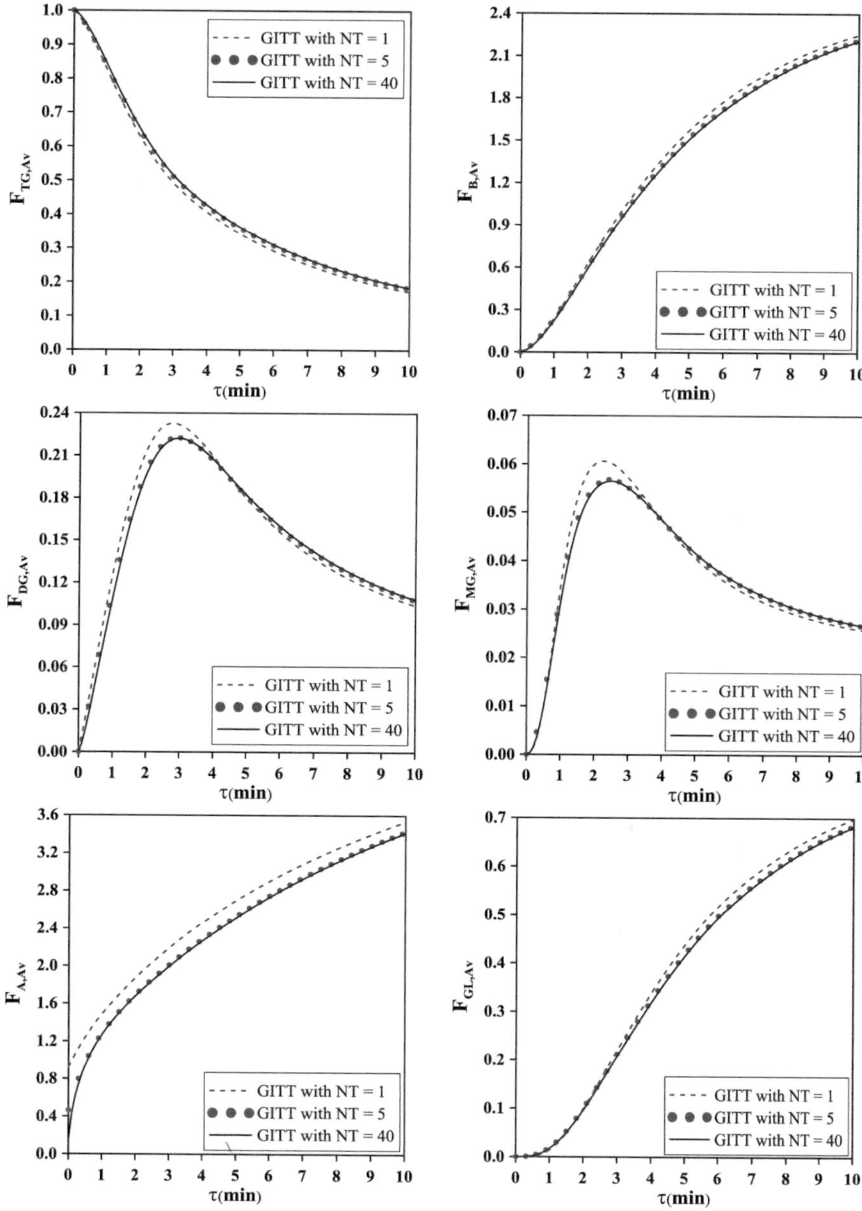

Fig. 4 Average dimensionless concentrations of the different species at the micro-reactor outlet as computed from the three-dimensional model via GITT for the truncation orders $NT = 1, 5,$ and 40

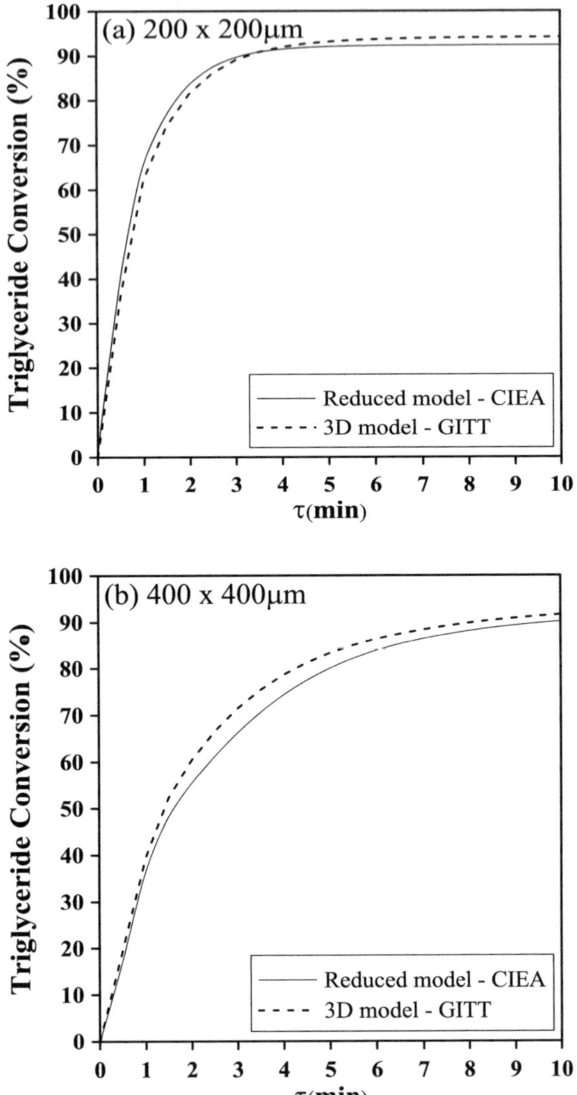

Fig. 5 Comparison of triglyceride conversion in terms of residence time as obtained by the improved lumped-differential formulation (CIEA) and the complete three-dimensional model solved by GITT

In the estimation of the kinetic constants, emphasis is given to the use of the non-linear coupled first-order ordinary differential system, Eq. (22), using the Coupled Integral Equations Approach (CIEA) for model reduction [71]. The Markov chain—Monte Carlo method (MCMC) is the tool in the inverse problem analysis, using simulated experimental data to estimate the kinetic constants, including low triglyceride conversion rates at shorter residence times and lower temperatures. Simulated data is taken at these less efficient conditions so as to achieve higher concentrations of the intermediate species and provide sensitivity to the parameters to be estimated in the inverse problem solution [71].

A few results are provided below to demonstrate the MCMC method. The GITT solution for the fully converged three-dimensional model, Eq. (1), is employed to construct this synthetic experimental data, obtained with the input data from the literature provided in Table 5, while maintaining the CIEA reformulation for the direct problem solution in the inverse analysis. For the construction of the vector of simulated experimental measurements, results were obtained at twenty different residence times (0.5, 1.0, 1.5, 3.0, 3.5, 4.0, 5.5, 6.5, 8.0, 8.5, 9.5, 10.0, 12.0, 13.5, 14.0, 14.5, 16.0, 17.5, 18.5, and 19.0 min), for the corresponding volumetric flow rates of the triglyceride species of 0.399, 0.199, 0.133, 0.0660, 0.0570, 0.0498, 0.0362, 0.0307, 0.0249, 0.0234, 0.0210, 0.0199, 0.0166, 0.0148, 0.0142, 0.0137, 0.0125, 0.0114, 0.0108, and 0.0105 ml/h, respectively. The simulated experimental results were obtained for four species, triglyceride, diglyceride, monoglyceride, and biodiesel. To capture the effect of uncertainties, experimental standard deviation for the synthetic measurements was taken as 5% of each concentration maximum value, as computed from the three-dimensional model [71]. Table 5 presents the estimated kinetic constants with their relative deviations from the exact values, now employing the simulated experimental data from the complete three-dimensional model and introducing the experimental error. The parameters k_1 and k_2 lead to the largest deviations (14 and 13%, respectively), but overall a good agreement with the exact values is achieved. It should be recalled that the deviations in the present results also incorporate the errors in the simplified reduced model, which provides a marked reduction in the computational cost [71]. For instance, the inverse analysis for 100,000 states in the MCMC algorithm, using the CIEA for the direct problem solution, leads to a CPU time reduction of around 9 times with respect to the GITT solution of the three-dimensional model with $N = 5$ terms in the eigenfunction expansions. Nevertheless, the GITT solution is still more accurate than the CIEA solution, even for this reduced truncation order. The concentration profiles at the micro-reactor outlet (at $X = 1$) were then computed with the exact and estimated kinetic constants, for all four species, TG, DG, MG, and B, as shown in Fig. 6. An overall good agreement of the estimated and exact concentrations was achieved, even though a reduced model has been adopted for the direct problem solution. For the 99% confidence intervals, as expected, larger bounds are achieved upon including the experimental error.

As mentioned above, the kinetic constants are essential parameters to be identified in this reaction–convection–diffusion model. The present hybrid approach is thus able to simultaneously estimate all the unknown parameters in the transport phenomena modeling and design the micro-device to meet specific requirements. In the present illustration, methanol has been selected as the alcohol phase, in accordance with the majority of previous works in biodiesel synthesis. However, the final goal of the research was to develop a microsystem for biodiesel synthesis employing the ethanol route [72]. In fact, after obtaining dedicated experimental results and undertaking the proper estimation of the associated kinetic constants, new promising synthesis results have demonstrated a high convergence rate of 99.9% of TG in 99.6% of biodiesel in

Table 5 Estimated kinetic constants and relative errors for the case of simulated data with experimental error

Kinetic constants	Exact [m³/(mol.s)]	Estimated [m³/(mol.s)]	Relative error (%)
k_1	4.368×10^{-6}	3.729×10^{-6}	14.62
k_2	9.623×10^{-6}	8.313×10^{-6}	13.61
k_3	1.880×10^{-5}	2.038×10^{-5}	8.39
k_4	1.074×10^{-4}	1.174×10^{-4}	9.28
k_5	2.117×10^{-5}	1.942×10^{-5}	8.25
k_6	9.000×10^{-7}	8.507×10^{-7}	5.48

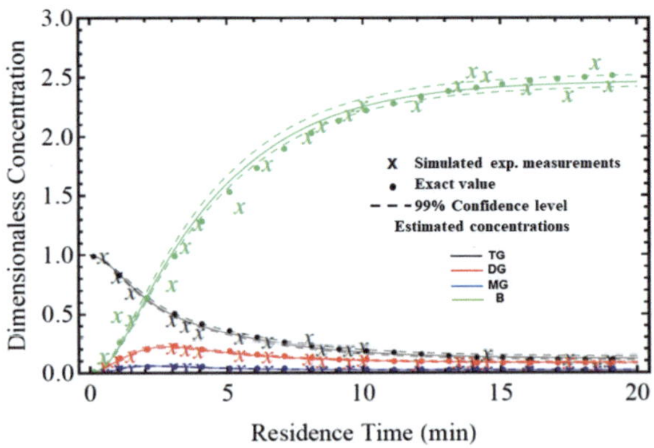

Fig. 6 Estimated and exact concentration distributions at the micro-reactor exit ($X = 1$) in terms of residence time, including the 99% confidence level bounds (with experimental error)

less than 35 s, using a molar ratio of 20:1 (ethanol/soybean oil) and NaOH as catalyst [72].

4 Conclusions

A hybrid computational–analytical–experimental approach has been reviewed for the analysis of heat and mass transfer phenomena, with emphasis on microscale applications. The approach combines various tools such as a lumped-differential problem reformulation strategy for model reduction, known as the Coupled Integral Equations Approach (CIEA), a hybrid numerical–analytical solution method for partial differential equations based on the generalized integral transform technique (GITT), and an inverse problem solution implemented in the Bayesian inference framework. Following an overall description of the approach and the pertinent literature review, the potential of the proposed hybrid approach is illustrated through an actual application dealing with continuous biodiesel synthesis in micro-reactors.

Acknowledgements The authors acknowledge the financial support by CAPES/INMETRO, CNPq, and FAPERJ, from Brazil, and the UK Newton Fund. RMC is also grateful to the Leverhulme

Trust for the Visiting Professorship (VP1-2017-028) and CNPC acknowledges the postdoctoral fellowship provided by CAPES/Brazil. RMC, CPNC, and PCP acknowledge the hospitality of the Department of Mechanical Engineering, University College London (UCL), UK.

References

1. US NIC (United States National Intelligence Council) (2012) Global trends 2030: alternative worlds. US NIC, Washington DC, USA, pp 137
2. United Nations, The Sustainable Development Goals Report 2016, New York, https://unstats.un.org/sdgs/report/2016/The%20Sustainable%20Development%20Goals%20Report%202016.pdf (2016)
3. Ohadi MM, Choo K, Dessiatoun S, Cetegen E (2013) Next generation microchannel heat exchangers. Springer Briefs in Thermal Engineering and Applied Science, Springer
4. Steinke ME, Kandlikar SG, Single-phase heat transfer enhancement techniques in microchannel and mini-channel flow. In: Proceedings of the international conference on microchannels and mini-channels. ASME, Rochester, NY, June 2004
5. Yener Y, Kakaç S, Avelino M, Okutucu T (2005) Single-phase forced convection in microchannels—a state-of-the-art review. In: Kakaç S, Vasiliev LL, Bayazitoglu Y, Yener Y (eds) Microscale heat transfer—fundamentals and applications. NATO ASI Series, Kluwer Academic Publishers, Netherlands, pp 1–24
6. Sobhan CB, Peterson GP (2008) Microscale and nanoscale heat transfer: fundamentals and engineering applications. CRC Press, Boca Raton
7. Kockmann N (2008) Transport phenomena in micro process engineering. Springer, Berlin
8. Cotta RM, Mota CAA, Naveira-Cotta CP, Nunes JS, Avelino MR, Castellões FV, Quaresma JNN (2009) Heat transfer enhancement in laminar forced convection: nanofluids, microchannels, structured surfaces, invited keynote lecture, ICHMT digital library. In: Proceedings of international symposium on convective heat and mass transfer in sustainable energy—CONV-09, 38p. Yasmine Hammamet, Tunisia
9. Rebay M, Kakaç S, Cotta RM (eds) (2016) Microscale and nanoscale convective heat transfer: concepts, analysis, and applications. CRC Press, Boca Raton. ISBN 978-1-498-73630-5
10. Cotta RM, Knupp DC, Naveira-Cotta CP (2016) Analytical heat and fluid flow in microchannels and microsystems. Springer, Mechanical Eng. Series. ISBN 978-3-319-23311-6
11. Maitra T, Zhang S, Tiwari MK (2017) Thermal transport in micro and nanoscale systems. In: Kulacki FA et al (eds) Handbook of thermal science and engineering. Springer
12. Cotta RM, Orlande HRB (2003) Hybrid approaches in heat and mass transfer: A Brazilian experience with applications in national strategic projects. Heat Transf Eng (Invited Editorial) 24(4):1–5
13. Cotta RM, Su J, Pontedeiro AC, Lisboa KM (2018) Computational-analytical integral transforms and lumped-differential formulations: benchmarks and applications in nuclear technology. Invited Plenary Lecture, international symposium on turbulence, heat and mass transfer, THMT-ICHMT. Begell House, Rio de Janeiro, 10th–13th July 2018
14. Aparecido JB, Cotta RM (1989) Improved one-dimensional fin solutions. Heat Transf Eng 11(1):49–59
15. Cotta RM, Ozisik MN, Mennig J (1990) Coupled integral equation approach for phase-change problem in two-regions finite slab. J Franklin Inst 327(2):225–234
16. Cotta RM, Mikhailov MD (1997) Heat conduction: lumped analysis, integral transforms, symbolic computation. Wiley Interscience, New York
17. Correa EJ, Cotta RM (1998) Enhanced lumped-differential formulations of diffusion problems. Appl Math Model. 22:137–152
18. Cotta RM (1998) Chap. 10: Improved lumped-differential formulations in heat transfer (Invited chapter, volume 2). In: Sunden B, Faghri M (eds) Modelling of engineering heat transfer phenomena. Heat transfer series. Computational Mechanics Publication, UK, pp 293–325

19. Alves LSB, Sphaier LA, Cotta RM (2000) Error analysis of mixed lumped-differential formulations in diffusion problems. Hybrid Meth Eng 2(4):409–435
20. Sphaier LA, Su J, Cotta RM (2017) Macroscopic heat conduction formulation. In: Kulacki FA et al (eds) Handbook of thermal science and engineering. Springer
21. Hermite MC (1878) Sur la Formule d'Interpolation de Lagrange. J Crelle 84
22. Hirata SC, Goyeau B, Gobin D, Carr M, Cotta RM (2007) Linear stability of natural convection in superposed fluid and porous layers: influence of the interfacial modelling. Int J Heat Mass Transf 50:1356–1367
23. Hirata SC, Goyeau B, Gobin D, Chandesris M, Jamet D (2009) Stability of natural convection in superposed fluid and porous layers: equivalence of the one- and two-domain approaches. Int J Heat Mass Transf 52:533–536
24. Knupp DC, Naveira-Cotta CP, Cotta RM (2012) Theoretical analysis of conjugated heat transfer with a single domain formulation and integral transforms. Int Comm Heat Mass Transf 39(3):355–362
25. Knupp DC, Cotta RM, Naveira-Cotta CP (2015) Fluid flow and conjugated heat transfer in arbitrarily shaped channels via single domain formulation and integral transforms. Int J Heat Mass Transf 82:479–489
26. Minkowycz WJ, Sparrow EM, Murthy JY (eds) (2006) Handbook of numerical heat transfer, 2nd edn. Chapter 16, Wiley, New York
27. Luikov AV (1968) Analytical heat diffusion theory. Academic Press, New York
28. Ozisik MN (1980) Heat conduction. Wiley, New York
29. Mikhailov MD, Ozisik MN (1993) Unified analysis and solution of heat and mass diffusion. Wiley, 1984; also, Dover Publications
30. Cotta RM (1986) Diffusion in media with prescribed moving boundaries: application to metals oxidation at high temperatures. In: Proceedings of the II Latin American congress of heat & mass transfer, V. 1. São Paulo, Brasil, pp 502–513
31. Cotta RM, Ozisik MN (1986) Laminar forced convection in ducts with periodic variation of inlet temperature. Int J Heat Mass Transf 29(10):1495–1501
32. Cotta RM (1990) Hybrid numerical-analytical approach to nonlinear diffusion problems. Num Heat Transf Part B 127:217–226
33. Cotta RM (1993) Integral transforms in computational heat and fluid flow. CRC Press, Boca Raton
34. Cotta RM (1994) Benchmark results in computational heat and fluid flow: -the integral transform method. Int J Heat Mass Transf (Invited Paper) 37:381–394
35. Cotta RM (1998) The integral transform method in thermal & fluids sciences & engineering. Begell House, New York
36. Cotta RM, Mikhailov MD (2006) Hybrid methods and symbolic computations. In: Minkowycz WJ, Sparrow EM, Murthy JY (eds) Handbook of numerical heat transfer, 2nd ed. Chap. 16. Wiley, New York
37. Cotta RM, Knupp DC, Naveira-Cotta CP (2016) Analytical heat and fluid flow in microchannels and microsystems. Mechanical Eng. Series, Springer
38. Cotta RM, Knupp DC, Quaresma JNN (2017) Analytical methods in heat transfer. In: Kulacki FA et al (eds) Handbook of Thermal Science and Engineering. Springer
39. Cotta RM, Naveira-Cotta CP, Knupp DC, Zotin JLZ, Pontes PC, Almeida AP (2018) Recent advances in computational-analytical integral transforms for convection-diffusion problems. Heat Mass Transf (Invited Paper) 54:2475–2496
40. Wolfram S (2017) Wolfram mathematica, version 11. Wolfram Research Inc., Champaign
41. Cotta RM, Knupp DC, Naveira-Cotta CP, Sphaier LA, Quaresma JNN (2013) Unified integral transforms algorithm for solving multidimensional nonlinear convection-diffusion problems. Num Heat Transf Part A-Appl 63:1–27
42. Cotta RM, Knupp DC, Naveira-Cotta CP, Sphaier LA, Quaresma JNN (2014) The unified integral transforms (UNIT) algorithm with total and partial transformation. Comput Thermal Sci 6(6):507–524

43. Ozisik MN, Orlande HRB (2000) Inverse heat transfer: fundamentals and applications. Taylor and Francis, New York
44. Orlande HRB, Fudym O, Maillet D, Cotta RM (2011) Thermal measurements and inverse techniques. CRC Press, Boca Raton
45. Naveira-Cotta CP, Orlande HRB, Cotta RM (2010) Inverse analysis of forced convection in microchannels with slip flow via integral transforms and Bayesian inference. Int J Thermal Sci 49:879–888
46. Naveira-Cotta CP, Orlande HRB, Cotta RM (2010) Integral transforms and Bayesian inference in the identification of variable thermal conductivity in two-phase dispersed systems. Numer Heat Transf Part B- Fundam 57:173–202
47. Naveira-Cotta CP, Cotta RM, Orlande HRB (2011) Inverse analysis with integral transformed temperature fields for identification of thermophysical properties functions in heterogeneous media. Int J Heat Mass Transf 54(7-8):1506–1519
48. Naveira-Cotta CP, Cotta RM, Orlande HRB (2011) Estimation of space variable thermophysical properties. In: Orlande HRB, Fudym O, Maillet D, Cotta RM (eds) Thermal measurements and inverse techniques, Chap. 20. CRC Press, Boca Raton, pp 675–707
49. Knupp DC, Naveira-Cotta CP, Ayres JVC, Orlande HRB, Cotta RM (2012) Space-variable thermophysical properties identification in nanocomposites via integral transforms, Bayesian inference and infrared thermography. Inverse Prob Sci Eng 20(5):609–637
50. Knupp DC, Naveira-Cotta CP, Ayres JVC, Cotta RM, Orlande HRB (2012) Theoretical-experimental analysis of heat transfer in nonhomogeneous solids via improved lumped formulation. Integral transforms and infrared thermography. Int J Thermal Sci 62:71–84
51. Knupp DC, Naveira-Cotta CP, Fonseca HM, Orlande HRB, Cotta RM, Fudym O (2014). Thermal characterization of non-homogeneous media. In: Basu SK, Kumar N (eds) Modelling and simulation of diffusive processes. Springer Series: simulation foundations, methods and applications, Chapter 6, pp 119–139
52. Abreu LAS, Orlande HRB, Colaço MJ, Kaipio J, Kolehmainen V, Cotta RM (2018) Detection of contact failures with the Markov Chain Monte Carlo method by using integral transformed measurements. Int J Thermal Sci 132:486–497
53. Hetsroni G, Mosyak A, Pogrebnyak E, Yarin LP (2005) Heat transfer in microchannels: comparison of experiments with theory and numerical results. Int J Heat Mass Transf 48:5580–5601
54. Naveira-Cotta CP, Orlande HRB, Cotta RM, JS Nunes (2010) Integral transforms, Bayesian inference, and infrared thermography in the simultaneous identification of variable thermal conductivity and diffusivity in heterogeneous media. In: 14th international heat transfer conference. Paper no. IHTC14-22511, V. 3. Washington, pp 341–350
55. Knupp DC, Naveira-Cotta CP, Ayres JVC, Orlande HRB, Cotta RM (2010) Experimental-theoretical analysis in a transient heat conduction setup via infrared thermography and unified integral transforms. Int Rev Chem Eng IRECHE 2(6):736–747
56. Knupp DC, Naveira-Cotta CP, Orlande HRB, Cotta RM (2013) Experimental identification of thermophysical properties in heterogeneous materials with integral transformation of temperature measurements from infrared thermography. Exp Heat Transf 26:1–25
57. Abreu LA, Orlande HRB, Kaipio J, Kolehmainen V, Cotta RM, Quaresma JNN (2014) Identification of contact failures in multi-layered composites with the Markov Chain Monte Carlo Method. ASME J Heat Transf 136(10):101302–101311
58. Costa Junior JM, Naveira-Cotta CP, Fudym O (2014) Image processing of temperature fields from infrared thermography of micro-mixers with polymeric substrates. Thermal Eng (RETERM) 13(1):71–79
59. Knupp DC, Naveira-Cotta CP, Renfer A, Tiwari MK, Cotta RM, Poulikakos D (2015) Analysis of conjugated heat transfer in micro-heat exchangers via integral transforms and non-intrusive optical techniques. Int J Num Meth Heat Fluid Flow 25(6):1444–1462
60. Renfer A, Tiwari MK, Brunschwiler T, Michel B, Poulikakos D (2011) Experimental investigation into vortex structure and pressure drop across microcavities in 3D integrated electronics. Exp Fluids 51:731–741

61. Renfer A, Tiwari MK, Tiwari R, Alfieri F, Brunschwiler T, Michel B, Poulikakos D (2013) Microvortex-enhanced heat transfer in 3D-integrated liquid cooling of electronic chip stacks. Int J Heat Mass Transf 65:33–43
62. Cagney N, Balabani S (2014) Streamwise vortex-induced vibrations of cylinders with one and two degrees-of-freedom. J Fluid Mech 758:702–727
63. Ma S, Sherwood JM, Huck WTS, Balabani S (2014) On the flow topology inside droplets moving in rectangular microchannels. Lab Chip 14:3611–3620
64. Sharma CS, Tiwari MK, Zimmermann S, Brunschwiler T, Schlottig G, Michel B, Poulikakos D (2015) Energy efficient hotspot-targeted embedded liquid cooling of electronics. Appl Energy 138:414–422
65. Zhang S, Tiwari MK, Balabani S, Naveira-Cotta CP, Cotta RM (2017) Residence time control in micromixers based on vortex shedding. In: 9th world conference on experimental heat transfer, fluid mechanics and thermodynamics, Iguaçu Falls, Brazil, 12–15 June 2017
66. Costa Junior JM, Naveira-Cotta CP, Nunes JS, Tostado CP (2015) Design, fabrication and characterization of micro-reactor for biodiesel synthesis. Heat Pipe Sci Technol 6(3–4):135–153
67. Pontes PC, Chen K, Naveira-Cotta CP, Costa Junior JM, Tostado CP, Quaresma JNN (2016) Mass transfer simulation of biodiesel synthesis in microreactors. Comput Chem Eng 93:36–51
68. Pontes PC, Naveira-Cotta CP (2016) Inverse problem analysis for identification of reaction kinetics constants in microreactors for biodiesel synthesis. J Phys: Conf Ser 745
69. Pontes PC, Naveira-Cotta CP, Macêdo EN, Quaresma JNN (2015) Integral transforms analysis of three-dimensional mass transfer in the transesterification process in micro-reactors. In: Proceedings of the international symposium on advances in computational heat transfer—ICHMT, Piscataway, USA
70. Pontes PC, Naveira-Cotta CP, Quaresma JNN (2017) Three-dimensional reaction-convection-diffusion analysis with temperature influence for biodiesel synthesis in micro-reactors. Int J Thermal Sci 118:104–122
71. Costa Junior JM, Naveira-Cotta CP (2019) Estimation of kinetic constants in micro-reactors for biodiesel synthesis: Bayesian inference with reduced mass transfer model. Chem Eng Res Des 141:550–565
72. Costa Junior JM (2017) Theoretical-experimental analysis of microreactors for biodiesel synthesis with rejected heat recovery. D.Sc. thesis, Mechanical Eng. Dept., PEM/COPPE, Federal University of Rio de Janeiro, UFRJ
73. de Souza IF, Guerrieri D, Naveira-Cotta CP, Tiwari MK (2018) Experimental-theoretical analysis of a micro heat exchanger for waste heat recovery from HCPV Systems. ASME J Thermal Sci Eng Appl. https://doi.org/10.1115/1.4041439. Published on line 3rd Sept 2018
74. Schwarz S, Borovinskaya ES, Reschetilowski W (2013) Base catalyzed ethanolysis of soybean oil in micro-reactors: experiments and kinetic modeling. Chem Eng Sci 104:610–618
75. Al-Dhubabian AA (2005) Production of biodiesel from soybean oil in a micro scale reactor. M.Sc. thesis, Oregon State University, Corvallis, OR, USA
76. Billo RE, Oliver CR, Charoenwat R, Dennis BH, Wilson PA, Priest JW, Beardsley H (2015) A cellular manufacturing process for a full-scale biodiesel microreactor. J Manufact Syst 1:409–416
77. Santana HS, Tortola DS, Reis EM, Silva JL Jr, Taranto OP (2016) Transesterification reaction of sunflower oil and ethanol for biodiesel synthesis in microchannel reactor: experimental and simulation studies. Chem Eng J 302:752–762
78. Tiwari A, Rajesh VM, Yadav S (2018) Biodiesel production in micro-reactors: a review. Energy Sustain Dev 43:143–161
79. Kaipio J, Somersalo E (2004) Statistical and computational inverse problems. Appl Math Sci 160
80. Gamerman D, Lopes HF (2006) Markov Chain Monte Carlo: stochastic simulation for Bayesian inference, 2nd edn. Chapman & Hall/CRC, Boca Raton
81. Orlande HRB (2012) Inverse problems in heat transfer: new trends on solution methodologies and applications. ASME J Heat Transf 134:031011

Part VI
Sustainable Energy Technologies

Biofuels: Past, Present, Future

G. Abdulkareem-Alsultan, N. Asikin-Mijan, H. V. Lee and Y. H. Taufiq-Yap

Keywords Fossil fuel · Biodiesel · Bioethanol · Biobutanol · Lignocellulosic · Algae biofuels

1 Introduction

Biofuels are liquid fills obtained from plants. First-generation biofuels are currently broadly delivered and utilised and are produced using starch, sugar, vegetable oils and fats by using traditional innovation with high cost [1]. The method of ethanol production from cellulosic and lignocellulosic materials is being opposed because of the insufficient development of efficient and cost-effective techniques to separate the multipolymeric crude material. Accordingly, useful frameworks for the generation of cellulases and other cellulose-corrupting proteins should be developed. Lignocellulosic biofuels decrease high vitality costs and include the efficient use of other elective powers [2, 3].

G. Abdulkareem-Alsultan · N. Asikin-Mijan · Y. H. Taufiq-Yap (✉)
Catalysis Science and Technology Research Centre (PutraCat), Faculty of Science, Universiti Putra Malaysia, 43400 UPM Serdang, Selangor, Malaysia
e-mail: taufiq@upm.edu.my

G. Abdulkareem-Alsultan
e-mail: kreem.alsultan@yahoo.com

H. V. Lee
Nanotechnology & Catalysis Research Centre (NanoCat), Institute of Postgraduate Studies, University Malaya, 50603 Kuala Lumpur, Malaysia

A. K. Gupta et al. (eds.), *Innovations in Sustainable Energy and Cleaner Environment*, Green Energy and Technology,
https://doi.org/10.1007/978-981-13-9012-8_21

2 Present: First, Second and Third Biofuel Generations

Biofuels can be classified based on their production technologies, as follows: first-, second-, third- and fourth-generation biofuels.

2.1 *First- and Second-Generation Biofuels: Bioethanol and Biodiesel*

First-generation biofuels incorporate ethanol and biodiesel. Biodiesel is used to replace petroleum diesel, and bioethanol is used to replace gasoline [4]. Ethanol derived from biomatter via the fermentation process is typically referred to as bioethanol. Bioethanol is generally synthesised from a biological origin that has considerable amounts of sugars and materials that can be converted into starch or cellulose [5]. Sugar cane, sugar beetroot and sugar sorghum are raw materials that contain sugar and hence can be used for ethanol production. Wheat, barley and corn are also raw materials that contain starch which can easily be converted into sugar by using available technologies. However, the increasing demand for these food-based materials, especially sugar cane and corn, induces competition within the food industry (food vs. fuel debate).

Biodiesel is a diesel-equivalent biofuel which consists of a methyl ester with long-chain fatty acids (FAME) derived from transesterification or esterification reaction of lipid feedstocks, such as vegetable oils or animal fats. Given that biodiesel contains 10% O and 0% S, its consumption induces proficient vehicle motor. Biodiesel can be used directly as fuel with some engine modifications or can be blended with petroleum diesel and used in diesel engines with few or no modifications. Biodiesel is an environmentally friendly material that can reduce 40% of the brown haze issue and substantially diminish CO level [6]. Blending biodiesel in mineral diesel during combustion reduces the emission of particulate matters, hydrocarbons and CO to an extent with only minor power loss [7]. Although the use of biodiesel is highly effective and attractive, the main obstacle for its commercialisation is the high cost of feedstock. Introducing waste cooking oil as a feedstock for the viability of continuous transesterification and recovering high-quality glycerol as a biodiesel by-product are primary options to decrease the biodiesel cost [4].

2.2 *Second Biofuel Generation: Lignocellulosic Processing*

Lignocellulosic materials for biofuel generation were obtained from switch grass, prairie grasses, small revolution woody harvests, agrarian build-ups and ranger service materials and deposits [8]. Lignocellulosic biomass converts into gas, solid char and liquid products when subjected to high temperature in the absence of O (i.e. pyrolysis). The cost for this biomass is considerably not as high as the costs of vegetable oil, corn and sugar cane. This property of lignocellulosic biomass is an advantage. After the biomass undergoes lignocellulosic processing process, the liquid product called bio-oil or pyrolysis oil is typically brown, dark red or black in colour with a density of approximately 1.2 kg/L. Bio-oil has a typical water content of 14–33 wt% which cannot be easily removed by conventional methods (e.g. distillation). In fact, bio-oil may phase separate above certain water content. The high heating value of bio-oil is typically in the range of 15–22 MJ/kg which is lower than that of conventional fuel oil (43–46 MJ/kg) mainly due to the presence of oxygenated compounds in bio-oil, such as acids, alcohols, ketones, aldehydes, phenols, ethers, esters, sugars, furans, N compounds and multifunctional compounds [4]. The pH of bio-oil is commonly within the range of 2–4 which is primarily related to the acetic acid and formic acid contents. The acidic nature of bio-oil may result in further problems related to the corrosion of the equipment used for storage, transport and processing [9]. The literature suggested that bio-oil poses significant issues because this substance is different from crude oil which consists of pure hydrocarbon fuel-like structure [10–12]. Although the bio-oil is less CO_2/GHG neutral, clean and causes less pollution because bio-oil usage emits no NO_x and less SO_x [13], the potential of substituting bio-oil to petroleum fuel, and chemical feedstock is limited due to its high viscosity, high water and ash contents, low heating value, instability and high corrosiveness [14].

This biomass is difficult to change, and its creation is subject to new developments. Biofuel creation is connected to ware showcase. Thus, the cost of replacing the original feedstock to the final item should be low for its transportability. Various biomass (e.g. corn with the ethanol/domesticated animals) also create a variety of items out of a similar feedstock due to biorefinery. The transformation procedure for the second-generation biofuel is typically done by using two diverse methodologies, namely thermal and biopathways [2].

2.3 *Third-Generation Biofuel: Algae Biofuels*

Microalgae are attractive biomass feedstocks used to produce biofuels due to their high cellular lipid contents and photosynthetic rates. Some microalgae have >60% oil content by dry biomass weight, whilst the average oil content is approximately 20–50 wt%. Microalgae present a variety of advantages because many algae species exhibit high growth rates and can be grown on nonarable land by using wastewater, thereby avoiding undesirable competition with food crops for land and water resources [15]. Microalgae grow 12-fold faster and yield 30-fold more triglycerides per ha than the conventional oil-producing land plants, such as rape and sunflower. Given that microalgae can grow in wastewater, it is a N and industrial CO_2 (flue gases) source. Hence, microalgae cultivation may be synergistic with biological wastewater recycling, as well as an active CO_2 sink. These features make microalgae a promising green energy bioresource [5].

Green growth delivers biomass faster and on less land surface than lignocellulosic biomass. In any case, the creation of algal biomass presents specialised difficulties, such as lipid extraction and dewatering, in addition to the geological difficulties in areas such as Canada, where temperature is frosty majority of the year. Biofuels may not depend entirely on one age but rather may be a mix of the three ages to adapt to the expanded overall request because of the consumption on the planet's oil assets.

3 Future: Fourth-Generation Biofuel

3.1 Policy Overview

Fourth-generation biofuel exploits the engineered science of cyanobacterial growth which is a new but continuously developing field of study. Engineered science includes the outline and development of new organic parts, gadgets and frameworks and replanning of existing conventional natural structures for essential purposes. A photosynthetic/nonphotosynthetic frame, either characteristic or engineered, is needed to deliver biofuels with high PFCE. For the first-, second- and third-generation biofuels, the crude material is either biomass or waste which are both influenced by the after-effects of yesterday's photosynthesis (but not from fossil assets).

Although these biofuels regularly are beneficial in a specific locale or nation, they are constantly constrained by the accessibility of the natural crude materials, that is, the biomass which confines their application on a worldwide scale. Fourth-generation biofuels will be founded on crude materials that are essentially unlimited, shabby and broadly accessible. Photosynthetic water part (water oxidation) into its constituents by sun-powered vitality can turn into a huge supporter of fuel creation on a worldwide scale both by simulated photosynthesis and coordinate sunlight-based biofuel generation advances. H generation, as well as the creation of decreased C-based biofuels, is possible by the improved associative obsession of climatic CO_2 and the original outline of the manufactured metabolic pathways for fuel generation. The age of 'planner microscopic organisms' with new valuable properties requires progressive logical advances in few areas of research.

The European Union economy technique features the significance of disclosure about the foundation of functional bioeconomy, and engineered science is anticipated as a critical empowering innovation for the acknowledgment of the bioeconomy in substitution to petroleum derivatives. Engineered science can make the bioeconomy significantly extensive by creating various natural mixes, including a variety of different biofuels. A stress over the decline in EU's biocapacity is also observed. When no moves are made, bioresource utilisation will surpass their recharging limit by 2030. Subsequently, creating biofuels utilising insignificant crude material assets during their generation is critical [16]. The biofuels from the fourth era are produced (i) by creating photosynthetic microorganisms to deliver photobiological sun-oriented powers, (ii) by consolidating photovoltaic and microbial fuel generation (electro-biofuels) or (iii) by engineering cell manufacturing plants or engineered organelles to create the desired high-esteem chemicals (generation of petroleum products at present) and biofuels.

3.2 *Advanced Biochemical Conversion*

3.2.1 Advanced Biochemical Ethanol Production

Bioethanol creation is a pathway for unrefined petroleum attenuation and natural consistency. Bioethanol can be utilised as fuel with considerable attributes, such as high octane number, low cetane number and high vaporisation temperature. Its primary disadvantages are destructiveness, low fire iridescence, decrease in vapour weight, miscibility with water and lethality to biological systems. One significant issue with bioethanol fuel is crude material accessibility. The feedstock supply for bioethanol creation can fluctuate from season to season and relies upon geographic areas. For example, lignocellulosic biomass, timberland-based woody materials, rural deposits and municipal waste are important feedstocks for bioethanol generation due to its high accessibility and ease, although business has not been started [2]. The supply and the conscious utilisation of organisms render the bioethanol creation process profoundly impossible to miss.

Numerous innovations and strategies for biomass-based ethanol generation are in progress and anticipated to be illustrated. In the present work, an innovative investigation of the biochemical technique that can be utilised to deliver bioethanol is completed, and a review of current patterns and issues is led. With the increasing worldwide interest for vitality, vitality deficiency will be a worldwide issue. Bioethanol is a critical sustainable fuel used mostly to supply fossil-inferred fills. Bioethanol generation worldwide increased from 50 million m^3 in 2007 to more than 100 million m^3 in 2012. Brazil and the USA utilise approximately 80% of the world supply of corn or sugar cane. In creating economies, nutrition-related feedstock is ideally supplemented by nonfood crude materials, such as sweet sorghum and cassava.

The utilisation of regular biomass can fundamentally expand the bioethanol generation, and the present chapter aims to review lignocellulose-based bioethanol. The EBB and flow mechanical improvement and bottlenecks characterise the short- and medium-term inquiries about needs. Modern ethanol is primarily delivered petrochemically through the acid-catalysed hydration of ethylene. Ethanol used in mixed refreshments and by far the most used ethanol as biofuel is delivered by ageing, where certain yeast species (e.g. *Saccharomyces cerevisiae*) or microscopic organisms (e.g. *Zymomonas mobilis*) utilise sugars in O-led conditions to create ethanol and CO_2. The primary explanations behind the improvement in bioethanol are its utilisation as a good carbon-impartial sustainable fuel, diminishing CO_2 outflows and related environmental change; its utilisation as octane enhancer in unleaded gas and its utilisation as an oxygenated fuel blend for a clean gas burning, thereby subsequently decreasing tailpipe poison emissions and enhancing the encompassing air quality [16]. Ethanol has suitable properties for start IC motors. Its engine octane (MON) and research octane numbers (RON) are 90 and 109 compared with 99 and 91 for general fuel, respectively.

Considering its low cetane number, ethanol is not consumed efficiently by the pressure start and is also not effortlessly miscible with diesel fuel. To enhance ethanol utilisation in pressure start (IC) motor vehicles, measures, such as the expansion of an emulsifier with a specific end goal to build the ethanol–diesel miscibility, ethylhexyl nitrate or dibutyl peroxide, can be expanded to improve the cetane number, a double fuel operation can be utilised, where ethanol and diesel are brought independently into the barrel, and diesel motors can be changed to adjust their autoignition attributes. The ageing of sugar-based crude materials is the original bioethanol, whilst the utilisation of lignocellulose crude materials is generally called second-generation bioethanol. The third era of algal bioethanol is the one at a beginning of the examination. The original procedures are examined in detail. Given that the present paper studied the second era of bioethanol in 2010, writing is compressed [17]. The second era of bioethanols utilises cellulose-discharged sugars despite the cost of the expected chemicals to hydrolyse cellulose. The advancement of this innovation can manage various cellulose-containing farming side effects, such as straw, wood trimmings, sawdust and bamboo.

The specific procedures of utilising lignocellulosic crude materials will be managed. Regardless of whether the first, second or third era of feedstock is utilised, ageing produces a lean liquor stock. However, this liquor stock is unusable in modern fuel applications. Ethanol should be subsequently cleansed. Partial refining can thin ethanol to 95.6 vol% (89.5 mol%) compared with the azeotropic arrangement with a breaking point of 78.2 °C. Ethanol enhancement by conventional refining is impossible, but different choices have been examined, as detailed in a previous study [1]. Accessible biomass can be arranged into essential sources and created as either yield or key item (sugar stick, short revolution vitality manors), auxiliary sources as deposits from the generation forms (bagasse, rice husks and straw) and tertiary sources as build-ups delivered amid and after the end of application (the natural division of city strong waste (MSW), sewage treatment slop and wood trimmings). The accessibility of natural squanders and deposits may vary and is influenced by showing the development despite the fact that atmosphere and different elements have effects particularly with regard to the essential sources. The vitality of biomass build-ups and natural squanders depends upon the yield, the accessible aggregate land zone and the type of generation.

The natural portion of city strong waste (MSW) is a modest source of biomass and spreads local and mechanical wastes gathered in a particular zone. The general capability of the natural portion of civil waste and its waste wood division is dependent upon the monetary development and the utilisation of biomaterials. It is evaluated at the vicinity with 5 and 50 EJ (i.e. estimated time of arrival in joule)/year. The unit EJ is 1018 J. Agroindustrial biomass deposits are the results of horticulture or related industries, including cotton stalks, wheat and rice straw, coconut shells, maize cobs, jute sticks and rice husks. Horticultural build-ups are delivered decentralised and are thin. Given the high transportation cost, the application of agrarian build-ups as the fundamental fuel in control stations is costly. The capability of farming deposits changes from 15 EJ/year up to 70 EJ/year as an element of provincial creation, collecting handling elements and recoverability factors. Inside the agroindustrial build-ups, dried compost is considered as a tertiary source. An aggregate overall gauge is difficult to make with 5–55 EJ/year with low appraisal because of its present use as a compost, whilst the high gauge considers the specialised potential of the aggregate. Ranger service build-ups incorporate biomass that are not reaped or expelled from the arranging areas in hardwood and softwood businesses through woodland administration operations, such as precommercial diminishing, death expulsion and biting the dust trees. Ranger service squander incorporates wood chips, sawdust and bark, thereby resulting in 65% biomass vitality potential.

The extraction costs and the expected transportation to the concentrated handling plants make the application of wood powers costly. Few examinations concentrated on the utilisation of wood powers for vitality creation at an area level in applying suitable outlines of decentralised littler plants have been conducted. The vitality capability of the backwoods worldwide is again difficult to determine. The conceivable commitment by 2050 is evaluated at 98 EJ/year of abundance of regular wood development and 32–52 EJ/year handling deposits. Although these lignocellulosic biomass assets show critical vitality esteem of 150 EJ/year, only a part of this asset can be utilised as feedstock for bioethanol because of the reasons mentioned above. Lignocellulose which is the central part of the plant cell dividers is mostly made of cellulose (40–60% of the aggregate's dry weight), hemicellulose (20–40%) and lignin (10–25%). Cellulose comprises long chains of β-glucose monomers assembled into microfibril groups. The hemicelluloses, such as xyloglucans or xylans, are connected to the microfibrils by H bonds. Lignins are phenolic mixes framed by the polymerisation of three monomers, such as p-coumaryl, coniferyl and sinapyl alcohols.

Lignin adds to the compressive quality and firmness of the cell divider. Once the lignocellulosic biomass is pretreated and hydrolysed, the discharged sugars can mature, and the downstream procedure is similar to that of the original feedstock. High lignin or potentially high fiery debris indicates that fixations are unfavourable for bioethanol creation. Particularly, softwood can subsequently be prohibited. The broad H linkages among cellulose particles prompt a crystalline and solid network structure. Starch requires a temperatures ranging from 60 to 70 °C to be changed from crystalline to shapeless surface, whilst cellulose requires a temperature of 320 °C and high weights of up to 25 MPa to change the inflexible crystalline structure into an undefined structure in water. Cotton, flax and concoction pulps speak to the purest sources of cellulose, whilst delicate hardwoods contain lower than half of cellulose [18]. Hemicellulose is any of various heteropolymers, including hexoses (D-glucose, D-galactose and D-mannose) and pentose (D-xylose and L-arabinose), with nebulous structure. Hemicellulose may contain sugar acids (uronic acids), and its spine chain is fundamentally made out of xylan linkages, including α-xylose (90%) and L-arabinose (10%). The levels of fanning and xylan synthesis differ from the nature and the origin of the crude materials. To be completely hydrolysed into free monomers, hemicellulose requires an extensive variety of chemicals in a variety of sugars. Lignin is an inflexible biopolymer with sweet smell that covalently clings into hemicellulosic xylans and in charge of the inflexibility and abnormal state of minimisation of the plant cell divider.

Lignin consists of monomers of phenyl propionic liquor, that is, coumaryl, coniferyl and sinapyl liquor. The lignin portion in biomass sources changes significantly. Lignin segments are significant as a result of their weakening effect on the procedures of hydrolysis and maturation. The phenolic gatherings framed from the debasement of lignin considerably deactivate cellulolytic proteins and consequently hamper enzymatic hydrolysis. However, lignin alteration through hereditary building can extensively decrease lignin arrangement and enhance ethanol yield. However, this process can be tricky because lignin segments fill in as the real plant safeguard framework to pathogens and bugs, and its change can disturb the plants' common insurance. Holding the lignin in addition to the advantage of the vitality economy of the procedure, once recuperated, it can be connected in a joined warmth and power unit (CHP), hence being a potential vitality self-managing source of the procedure. Biomass feedstock with high lignin content is unsuitable as a crude material for bioethanol ageing.

This absolutely disposes of the large part of the delicate woods. The diverse organisation of biomass feedstock (dry) is additionally reflected in its basic piece. The correct structure relies to a remarkably on the biomass sources. C, H and O are the key segments of biomass and decide their calorific esteem. The slag substance of biomass sources is very low, and its synthesis should not withstand and be mulled over because a large part of the powder will gather in the lignin build-up, thereby hampering further vitality age by fouling or sintering. Biomass contains considerable amounts of K, Cl and Si and low Ca, Mg, Al, Fe and Na concentrations. The fiery debris substance and its synthetic structure have a solid capacity as biomass species. The components in the fiery remains are O, Ca, K, Si, Mg, Al, S, Fe, P, Cl, Na, Mn and Ti.

3.2.2 Biobutanol

Biobutanol or bio-based butanol fuel is a second era of alcoholic fuel with a higher vitality thickness and lower instability than ethanol. Over 12 organisations are centred around creating biobutanol on a business scale. Biobutanol is a true biofuel for masses with practically no effect on sustenance supply and capacity to contend positively with $80 BBL oil. Exertion is centred on both the ageing of sugars, starch and different biomass and through the pyrolysis and reformulation of biomass [4].

Biobutanol in business utilises as a part of a current market worth over $5 billion. Ethanol has increased colossal ubiquity in the USA as an inexhaustible option to conventional gas. Rather than relying upon reducing stores of oil, ethanol enables the development our fuel source to transform plant material into fluid fuel. However, ethanol considerably reduces the fuel proficiency of the auto. Consumer reports showed that E-85 fuel which utilises 85% ethanol and 15% gas causes a 20–30%

decrease in gas mileage. Ethanol is additionally more destructive to few metals and more elastic than customary fuel. Thus, it can break down the auto's motor. Hence, unless the auto has been extraordinarily adjusted, the utilised gas in the tanks can only contain approximately 10% ethanol. Fortunately, for the sustainable power source, another biofuel is present on the block butanol. Butanol–ethanol is greater and more effective than other ethanols and high vitality as a fuel. Therefore, butanol–ethanol will not diminish the gas mileage as much as ethanol, and it does not harm the auto. Hence, it can be directed straight into the tank without making any adjustment to the motor [1].

However, the question on why butanol is not sold in a pump although it is an impressive fuel remains. To answer this question, we verify how biofuels are made. Biofuel production starts with a sugar origin which is usually corn (albeit different plants, e.g. sugarcane, switchgrass and sugar beets can be also utilised). The parts are ground into a fine powder and blended with water to shape of a dull crush. Catalysts are added to separate this starch considerably further and transform them into sugar. In cooking squash, microorganisms, such as yeast, are added to the blend, and ageing starts. During maturation, the yeast transforms sugars into liquor, thereby making a corn-based lager. Considering that we cannot simply brew our gas tanks, extra refining is required [19]. Yeast is sifted; after the brew is sent to a refining segment where the liquor is isolated from the water, the liquor is additionally refined by sending it through an atomic strainer (a channel), thereby resulting in a blend that is approximately 200 proof. Ethanol is blended with a small amount of gas, rendering it undrinkable and delivering it out to service stations across the nation. The fuel we at last put into our autos is blended with gas to bring it down to 10% ethanol by volume. The refining procedure mentioned above should be considered. This procedure requires enormous measures of vitality, thereby making it very costly.

High cost can be avoided when ethanol mystically isolated out from the water blend individually without being refined. Hence, butanol becomes an integral factor. Although ethanol is extremely dissolvable (i.e. it blends well with water), butanol is not [20]. At the point when the brew has an 11% liquor by volume (ABV) or higher, butanol isolates out from water and buoys to the surface similar to oil [21]. Hence, if vodka is made of butanol rather than ethanol, then the majority of the liquor will sit in the best 40% of the container, and the remaining 60% of the base will be unadulterated water. Although this phenomenon will be a problem to vodka merchants, it will benefit the biofuel business. Butanol can make the process of refining easy and decrease its cost by essentially isolating the lager-like blend by its own particular gadgets [1]. Biofuel makers effortlessly have the capacity to skim unadulterated butanol off the highest point of the arrangement which can be dropped specifically into our fuel tanks without being weakened to a low rate. Man-made butanol may completely supply conventional fuel. However, liquor is known for its capacity to sanitise (e.g. rubbing liquor and liquor swabs) because of its ability to kill microorganisms.

Similarly, butanol is poisonous to every creature that makes it over an ABV, such as bud light which is approximately 4%. Given that butanol only buoys to the highest point of the arrangement at a centralisation of 11% or higher, the microorganisms undergoing ageing kick the bucket before the brew achieves this edge. This test aims to study how to keep small animals live sufficiently long for them to produce the desired volume of butanol. Liu et al. portray the advances that have been made in the butanol generation process. These researchers reported that a wide range of microorganisms that can utilise mature sugar into butanol are present. *S. cerevisiae* is used to blend brew and produce bread because of its qualities for more than 40 years. Now, we have a total guide of the yeast's DNA for the simple improvement of its qualities to upgrade its execution. For instance, configuration yeast strains yield considerably more butanol than they would normally do. In 2016, researchers developed a strain of yeast that produces approximately 130 mg/L of butanol, and we prepared the same amount of titre. However, yeast cannot endure sufficiently high levels of butanol to proficiently deliver biofuels [21]. *Synechococcus* stretches which is a cyanobacteria (also called blue–green growth), and a 400 mg/L titre is used. However, similar to yeast, cyanobacteria cannot deal with the lethality of butanol.

3.2.3 Biochemical Drop-in Hydrocarbon

Few approaches can be used to deliver drop-in biofuels that are practically equal to petroleum-derived transportation fuel blendstocks. Three areas that should be studied include the following: oleochemical forms, that is the hydroprocessing of lipid feedstocks obtained from oil seed products, green growth or fat; thermochemical structures, that is the thermochemical transformation of biomass to liquid intermediates (gas or oil) trailed by synergist updating and hydroprocessing to hydrocarbon powers and biochemical procedures, that is the organic change in biomass (sugars, starches or lignocellulose-inferred feedstocks) to extended chain alcohols and hydrocarbons [21]. The fourth class consists of half and half of thermochemical and biochemical innovations, such as ageing of blend gas and reactant changing of sugars/starches, respectively [22]. One test showed that the advancement of drop-in biofuel shares with the worldwide oil industry is necessary to expand the measures of H_2 [20].

In oil, high amount of H is needed to update raw petroleum feedstocks with declining quality (i.e. progressively heavy and harsh in terms of the oils sourced from Venezuela and Canada) [19]. Prominent measures of H are expected to create thick vitality and exceedingly diminish drop-in biofuels. In the future, hydrogen may utilise H obtained from petroleum gas [23]. Oil refineries utilise H to overhaul poor quality unrefined petroleum both to evacuate risky S and other heteroatom polluting influences (hydrotreating) and to break long carbon affix particles to short-chain atoms whilst improving them with H (hydrocracking). One consequence of these H-consuming forms (known as hydroprocessing) is increasing the H/C proportion of the second-rate unrefined oils. The H/C proportion is a decent pointer of oil feedstock quality due to high S content, or the nearness of dense carbon chains (e.g. in coal) decreases the H/C proportion [20]. As the nitty-gritty in principle body of the report, the H/C proportion can be imagined as a staircase. Using the staircase comparison, the more steps that must scale H/C staircase to accomplish an H/C proportion reasonable for liquid gas, diesel and stream transportation fuel applications, the more hydrogen data sources and handling endeavours are required [23].

For non-H-devouring procedures, synergist or warm breaking can be utilised to build the H/C proportion of oil feedstocks by evacuating C as tars and roast (coke) [17]. In any case, this approach consumes feedstock and decreases yields because of evasion, especially when costs for raw petroleum feedstocks are high [19]. Developing drop-in biofuel requires H_2 inputs or other compound lessening procedures to overhaul O-rich starch, lignin or lipid feedstocks to H-rich hydrocarbons that are essentially proportional to oil determined by fluid fills [19]. In the initial report, some hydroprocessing steps may be required for most drop-in biofuel innovation stages, with imported H used to expel oxygen (as H_2O) from oxygenated lignocellulose intermediates or lipid feedstocks. Meanwhile, non-H-expending forms (regardless of whether concocted or natural) should oxidise critical measures of C feedstock to deliver the required H or effectively decrease power transporters (e.g. NADPH utilised as a part of organic procedures). Similar to oil feedstocks, these elective courses to deoxygenation decreases yields by expending a part of the feedstock. Hence, it is less attractive than hydroprocessing utilising imported H [4].

3.3 Advanced Thermochemical Conversion Method

3.3.1 Fischer–Tropsch (FT) Method

Fluid transportation hydrocarbon fills and other compound items can be delivered from syngas via a set-up of synergist synthetic process called FT combination which was named after the original German innovators Franz Fischer and Hans Tropsch in the 1920s [1]. During World War II, FT union provided the required fluid hydrocarbon fills for the German war exertion. Afterwards, confronting disengagement amid the politically sanctioned racial segregation time, South Africa diverted to FT union from coal gasification to supply considerable amounts of hydrocarbon fuel and compounds needed. Afterwards, numerous refinements and acclimations to the innovation have been made; these improvements include impetus advancement and reactor plan. Contingent upon the origin of syngas, the change is frequently alluded to as coal-to-fluids (CTLs) or potentially gas-to-fluids (GTLs) [24]. Currently working CTL plants, such as Shell's plant in Bintulu, Malaysia (GTL FT process) [17], incorporate Sasol's Sasolburg I and II plants. Few world-class GTL and CTL plants at different phases of building, development and generation in Nigeria, Qatar and China, as well as USA with the declaration of front-end building and configuration on Sasol's Lake Charles GTL and ethane cracker complex in Louisiana [23] are available at present.

3.3.2 Gasification

Biofuels from biomass gasification are investigated here and shown to be an appealing alternative. Late advancement in gasification procedures and critical age pathways for biofuel creation, process outline and joining and socionatural effects of biofuel age are studied to determine the gasification-to-biofuel qualifications as a reasonable and eco-accommodating innovation [25].

The amalgamation of vital biofuels, such as biomethanol, bioethanol and higher alcohols, bio-dimethyl ether, FT fills, biomethane, biohydrogen and green growth-based energies, is investigated, together with late innovations, impetuses and reactors [26]. Thermodynamic investigations for each biofuel are also analysed. Syngas cleaning is an essential issue for biofuel creation, and inventive pathways (e.g. those utilised by Choren Industrietechnik in Germany and BioMCN from the Netherlands) allow productive methanol age. The transformation of syngas to FT transportation energies, such as fuel and diesel, over Co or Fe impetuses is investigated and shown to be a promising choice for biofuel development. Biomethane has become a lucrative option for regular transportation of fuel with the top sides of gaseous petrol, including thick circulation, exchange and supply arrangements.

Approaches to deliver H_2 are also discussed. However, important issues, such as stockpiling, are costly with low efficiency. Green growth-based fills are innovative but may be economically critical in terms of their capacity to settle many CO_2 atoms to quickly develop in numerous situations for flexible use [17]. In any case, rational processing set-ups resulting in ideal plant outlines are vital. Hence, point-to-point reconciliation process is a useful device for the efficient use of current and new procedures. LCA and moral issues are also discussed. The utilisation of nourishment crops, rather than sustenance squanders, is still difficult and must be settled on a case-to-case basis.

3.3.3 Pyrolysis

The quick pyrolysis of bio-oils created in the exhibit and semi-business plants have potential as a fuel for stationary power generation utilising boilers or turbines. However, they require critical adjustment to end up noticeably an adequate transportation fuel. The synergist overhauling of pyrolysis vapours by using zeolites is a technique that can be used to expel O from natural mixes and changing them to hydrocarbons. This work assessed the arrangement of business and research facility blended impetuses for their hydrocarbon creation execution via pyrolysis/synergist breaking course [20]. Three biomass feedstocks, namely cellulose, lignin and wood, were pyrolysed (bunch tests) in quartz watercrafts in physical contact with the impetuses at a temperature extending from 400 to 600 °C and impetus/biomass proportions with the weight in the range of 5–10. Molecular beam mass spectrometry (MBMS) is utilised to investigate the item vapour and gas structure. The most highest yield of hydrocarbons (approximately 16 wt%, including 3.5 wt% of toluene) was accomplished by utilising Ni-, Co-, Fe- and Ga-substituted ZSM-5. Tests performed using a semi-nonstop stream reactor showed the adjustment in the structure of the volatiles delivered by the pyrolysis/synergist vapour breaking responses as an element of the impetus time-on-stream [4]. The deoxygenation movement diminishes with time in terms of the coke stores shaped on the impetus.

3.3.4 Hydroprocessing of Catalysts and Upgrade

The interest for hydroprocessing impetuses is increasing due to their applications in the refining oil and biofuels which agrees with strict natural directions controlling discharges from transportation vehicles. Transportation fuel is overwhelmed by nonrenewable energy sources with carbon emanation dangerous creation strategies. To move far from these sources, fluid powers should be delivered from agrarian stocks crops, edit squander, ranger service waste or green growth. Changing these feedstocks into top-notch skills is a significant test [22]. Portraying the present status in preparing agricultural feedstock into brilliant fluid transport powers showed that creators use catalysis for transformation. This paper provides an insight into the systems and conventions associated with new hydroprocessing impetuses and procedures and spread the techniques for redesigning these fluids to present-day transport vehicles appropriate for operation in current fuel and diesel motors. This paper also provides a prologue to the system of hydroprocessing responses, the use of various metals in hydroprocessing, the effect of impetus bolsters, applications in refining new feedstock and inexhaustible fill models, the administration of used hydroprocessing impetuses and H creation [17]. Hydroprocessing catalysts and processes will aid analysts from the academe and industry about future energy advancement and treatment to create current and future fluid transport powers.

3.4 Market Forecast and Unit Economics

Biofuels are supporters of the sustainable power source for a very long time under the Renewable Fuel Standard. These powers have numerous attractive properties, including the advancement of vitality freedom, the rustic improvement and the decrease in national carbon emissions. Nonetheless, few uncertain ecological and financial concerns remain [25]. Ecologically, a significant part of the biomass is required to originate from agricultural extension, as well as improvement, which may remarkably influence the net natural effect. Monetarily, the absence of a created foundation and bottlenecks along the store network may influence the business' monetary imperativeness. Approximately 30 million sections of land (12 million ha) under the Conservation Reserve Program (CRP) speak to one land base for conceivable development [22].

Here, we analyse the possible part of the CRP in biofuel industry advancement by evaluating the scope of natural impacts on six end purposes of concern and re-enacting contrasts in possible industry development by broadly utilising the frameworks [17].

The model studies seven land-utilised situations (different rates of CRP development for biofuel) and five financial situations (sponsorship plans) to investigate the advantages of utilising CRP. The subjective evaluation uncovered a wide variety of potential effects. Lignocellulosic feedstocks had the best potential to enhance the ecological condition with respect to push crops. However, the most possible results are thought to be nonpartisan or relatively negative. Reproductions shown indicated that industry development is significantly more touchy to financial situations than the utilised conditions equal to the volumes of biofuels can be delivered with no CRP with 100% use [25, 27]. The blend of the subjective evaluation and the inventory network display altogether proposes that the substantial-scale transformation of the CRP to push yields would result in considerable ecological cost, without an accompanying advantage as far as biofuel creation [4, 28].

4 Conclusion

The utilisation of petroleum product as a source of vitality has aid the improvement for a long time. This innovation has taken into consideration many advantages, such as power, transportation and fabrication hardware. The world is looked at the complex financial and natural issues related to utilise vitality to keep up and enhance our way of life. Our economy relies upon the ease of vitality [1]. The possibility of the sustainable power source has been around for a long time but did not receive considerable attention mainly because of low cost oil 50 years ago. However, this topic cannot be disregarded again as we enter another level of cognisance about our fuel utilisation and the absence of nonrenewable energy sources due to the consciousness about their effect on the environment. The strain with low cost-elective vitality is essential, particularly on a worldwide scale. Given that these monetary issues are progressing, prominent natural outcomes may be observed if we cannot change the design of vitality utilisation [4].

CO_2 and other supposed ozone-harming substances in the environment which trap the warmth transmit from the earth and cause environmental change worldwide may be developed [22]. Accordingly, creating contrasting options to fuel is one of the unique needs of numerous countries. With the recorded costs of oils, biofuels produced using biomass are receiving considerable attention. Biofuel generation worldwide has expanded from 4.8 billion ga in 2000 to approximately 16.0 billion ga in 2007. However, this record is still <3% of the fuel supply for transportation worldwide [26, 29, 30]. This paper reviews the advancement of biofuels, their generation, utilisation and effect on a global scale.

Acknowledgements The authors would like to sincerely appreciate the Universiti Putra Malaysia (UPM), Malaysia, for the funding through the research group project GP-IPD/2016/9490400.

References

1. Schenk PM et al (2008) Second generation biofuels: high-efficiency microalgae for biodiesel production. Bioenergy Res 1:20–43
2. Agarwal AK (2007) Biofuels (alcohols and biodiesel) applications as fuels for internal combustion engines. Prog Energy Combust Sci 33:233–271
3. Ajanovic A (2011) Biofuels versus food production: does biofuels production increase food prices? Energy 36(4):2070–2076
4. Demirbas A (2011) Competitive liquid biofuels from biomass. Appl Energy 88(1):17–28
5. Escobar JC, Lora ES, Venturini OJ, Yáñez EE, Castillo EF, Almazan O (2009) Biofuels: environment, technology and food security. Renew Sustain Energy Rev 13(6–7):1275–1287
6. Tang Q, Zheng YY, Wang T, Wang J, Alcala A, Bridgwater AV et al (2017) Biorefineries for biofuel upgrading: a critical review. Renew Sustain Energy Rev 58(6):520–594
7. Nair P, Singh B, Upadhyay SN, Sharma YC (2012) Synthesis of biodiesel from low FFA waste frying oil using calcium oxide derived from Mereterix mereterix as a heterogeneous catalyst. J. Cleaner Prod 29–30:82–90
8. Scharlemann JP, Laurance WF (2008) How green are biofuels? Science 319(5859):43–44
9. Oasmaa A, Kuoppala E, Elliott DC (2012) Development of the basis for an analytical protocol for feeds and products of bio-oil hydrotreatment. Energy Fuels 26(4):2454–2460
10. Mortensen PM, Grunwaldt J, Jensen PA, Knudsen KG, Jensen AD (2011) Applied catalysis A: general A review of catalytic upgrading of bio-oil to engine fuels. Appl Catal A 407(1–2):1–19
11. Venderbosch R, Prins W (2010) Fast pyrolysis technology development. Biofuels Bioprod Biorefin 4:178–208
12. Lu Q, Li WZ, Zhu XF (2009) Overview of fuel properties of biomass fast pyrolysis oils. Energy Convers Manag 50(5):1376–1383
13. Xiu S, Shahbazi A (2012) Bio-oil production and upgrading research: a review. Renew Sustain Energy Rev 16(7):4406–4414
14. Jacobson K, Maheria KC, Dalai AK (2013) Bio-oil valorization: a review. Renew Sustain Energy Rev 23:91–106
15. Santillan-Jimenez E, Pace R, Marques S, Morgan T, McKelphin C, Mobley J, Crocker M (2016) Extraction, characterization, purification and catalytic upgrading of algae lipids to fuel-like hydrocarbons. Fuel 180:668–678
16. Hannon M, Gimpel J, Tran M, Rasala B, Mayfield S (2010) Biofuels from algae: challenges and potential. Biofuels 1(5):763–784
17. Ragauskas AJ, Williams CK, Davison BH, Britovsek G, Cairney J, Eckert CA et al (2006) The path forward for biofuels and biomaterials. Science 311(5760):484–489
18. Hildebrand M, Davis AK, Smith SR, Traller JC, Abbriano R (2012) The place of diatoms in the biofuels industry. Biofuels 3(2):221–240
19. Gouveia L, Oliveira AC (2009) Microalgae as a raw material for biofuels production. J Ind Microbiol Biotechnol 36:269–274
20. Brennan L, Owende P (2010) Biofuels from microalgae—a review of technologies for production, processing, and extractions of biofuels and co-products. Renew Sustain Energy Rev 14:557–577
21. Himmel ME, Ding SY, Johnson DK, Adney WS, Nimlos MR, Brady JW, Foust TD (2007) Biomass recalcitrance: engineering plants and enzymes for biofuels production. Science 315(5813):804–807
22. Hill J, Nelson E, Tilman D, Polasky S, Tiffany D (2006) Environmental, economic, and energetic costs and benefits of biodiesel and ethanol biofuels. Proc Natl Acad Sci USA 103(30):11206–11210
23. Rawat I, Kumar RR, Mutanda T, Bux F (2011) Dual role of microalgae: phycoremediation of domestic wastewater and biomass production for sustainable biofuels production. Appl Energy 88(10):3411–3424
24. Stephanopoulos G (2007) Challenges in engineering microbes for biofuels production. Science 315:801–804

25. Lee SK, Chou H, Ham TS, Lee TS, Keasling JD (2008) Metabolic engineering of microorganisms for biofuels production: from bugs to synthetic biology to fuels. Curr Opin Biotechnol 19(6):556–563
26. Singh J, Gu S (2010) Commercialization potential of microalgae for biofuels production. Renewable Sustainable Energy Rev 14:2596–2610
27. Asikin-Mijan N, Lee HV, Taufiq-Yap YH, Abdulkrem-Alsultan G, Mastuli MS, Ong HC (2017) Optimization study of SiO_2-Al_2O_3 supported bifunctional acid–base NiO-CaO for renewable fuel production using response surface methodology. Energy Convers Manag 141:325–338
28. Zhao C, Brück T, Lercher JA (2013) Catalytic deoxygenation of microalgae oil to green hydrocarbons. Green Chem 15(7):1720
29. Alsultan GA, Asikin-Mijan N, Lee HV, Albazzaz AS, Taufiq-Yap YH (2017) Deoxygenation of waste cooking to renewable diesel over walnut shell-derived nanorode activated carbon supported CaO-La_2O_3 catalyst. Energy Convers Manag 151:311–323
30. Abdulkareem-Alsultan G, Asikin-Mijan N, Lee HV, Taufiq-Yap YH (2016) A new route for the synthesis of La-Ca oxide supported on nano activated carbon via vacuum impregnation method for one pot esterification-transesterification reaction. Chem Eng J 304:61–71

Advanced Solar Power Technology-Multiple Junction Photovoltaics

Ray Y. Lin

Keywords Concentration photovoltaics · HCPV · Multiple junction solar cells · Modules · Trackers · HCPV systems

1 Introduction and Background

Smog in many parts of the world including India is so severe that around many industrialized cities no clear sky can be seen during the day most of the time. The smog level was so heavy that even at high noon, it is hard to see bright sunlight. Air pollution from fossil fuel burning of power plants, industrial fossil fuel usages as well as automobile gas or diesel engine exhausts obviously contribute to heavy smog. In the meantime, the earth receives 174 petawatts (PW, or 1015 W) of incoming solar radiation at the upper atmosphere. Roughly, 30% is reflected back to space (by scattering) 19% is absorbed by the clouds and atmosphere and the rest strikes the earth surface which amount to 89.3 PW [1]. Should only a small portion of this amount of energy be used as our power source with photoelectric devices in place of fossil fuel, most smog would no longer cover the earth surface. Over the last couple of decades, on top of the long existing hydropower, a number of renewable energy techniques have been investigated and developed in the attempt to reduce the greenhouse gas emission: solar photovoltaics, solar thermal power, wind power, biofuels, geothermal, ocean power, to name a few.

R. Y. Lin (✉)
University of Cincinnati, Cincinnati, OH 45221, USA
e-mail: ray.lin@uc.edu

TaiCrystal International Technologies Co., Pingzhen, Taoyuan, Taiwan

A. K. Gupta et al. (eds.), *Innovations in Sustainable Energy and Cleaner Environment*, Green Energy and Technology,
https://doi.org/10.1007/978-981-13-9012-8_22

Table 1 World potential renewable energy resources to satisfy 15 TW needs [2]

Energy resource	Theoretical potential (TW)	Extractable potential (TW)	Technical potential (TW)
Hydropower	12	3.5	1.2
Ocean wave	34	8.5	0.62
Ocean surface currents	8.1	2.0	0.012
Ocean thermal gradient	3.9	0.033	0.0033
Ocean tidal	2.4	0.6	0.037
Geothermal	44	2.8	1.9
Wind	1,000	250	14
Solar PV	89,000	58,000	7500
Solar biofuel	89,000	61,000	2500
Solar thermal	89,000	19,000	5600

1.1 Potential Renewable Energy Technologies

According to the DOE projection back in 2006 [2], the world short-term energy need could grow by 15 TW until 2050 and 30 TW until 2100. If the CO_2 concentration in the atmosphere was to keep at the 550 ppmv level to avoid the disaster of irreversible global warming, this amount of extra energy need must be satisfied with C-neutral energy sources. The projected energy demand comes from the anticipated economical growth resulting from the world population increase as well as the GDP level that accompanies with the population. This is resulted from the analysis by the experts brought together by the Intergovernmental Panel on Climate Change (IPCC) which is an organization jointly established by the World Meteorological Organization (WMO) and the United Nations Environment Programme (UNEP). In 2001, the total world energy consumption was 13.5 TW. It was projected that by 2050 the total energy consumption would reach 27.6 TW or an increase of roughly 14–15 TW from the 2001 level assuming a moderate energy intensity decrease of 0.8%/year, which is similar to the rate over the last 100 years. The DOE analysis showed that, other than solar power, there is no C-neutral resource on earth that could support such energy demands (Table 1).

As shown in Table 1, the only technically potential C-neutral energy resource on earth is solar to meet the 15 TW or even 30 TW power needs, especially from solar PV. PV industry has grown rapidly over the last 10 years.

Many technologies have evolved, but silicon photovoltaics is the only technology dominating the world installation since the world started seriously considering PV as an important power source for our utility needs both in our daily life and in industry.

1.2 World PV Industries

The world total cumulative PV installation has reached 320 GW in 2016 and 415 GW by the end of 2017. The total PV installation in 2017 alone was around 94.6 GW with the most being installation in Asia. Although the yearly installation steadily grew over the years (Fig. 1) [3, 4], it is still far too low to be an important power resource if we are to keep the earth atmospheric greenhouse effect from running out of the control. Ideally, to completely replace our power source to renewable, we would need 10,000–20,000 GW power installation.

The 2007 number in the figure accounts for everything installed up to 2007. The 2018 figure is a projection by IEA, International Energy Agency. Of the 95 GW PV installation in 2017, 93% used silicon PV technology. It is expected that, in the near future, the primary technology for PV installation is still most likely silicon PV unless the price and reliability of the CPV(concentration photovoltaics) system are drastically improved soon.

1.3 Solar Energy Technologies

The solar energy technologies include those of crystalline silicon, thin film, multiple junction III–V compounds, and DSSC (dye-sensitive solar cell). Most installations in recent years use panels of crystalline silicon of mono and poly forms. Figure 2 shows the historical solar cell efficiency evolution of most technologies compiled by NREL, National Renewable Energy Laboratories [5]. Although silicon PV technologies dominate the PV installation over the years, their efficiencies remain rather unchanged in the last decade, suggesting that these technologies had reached their limits. On the other hand, the multiple junction PV technologies had improved with its conversion efficiency increased from around 38% in 2006 to 46% in 2015. At the present time, the record cell efficiencies of various types of the solar cell remain roughly the same as those in recent years. Green et al. [6] have compiled the efficiency data yearly and their latest compilation reported in June 2018 is shown

Fig. 1 World annual new PV installation. Author's drawing with data from [3, 4]

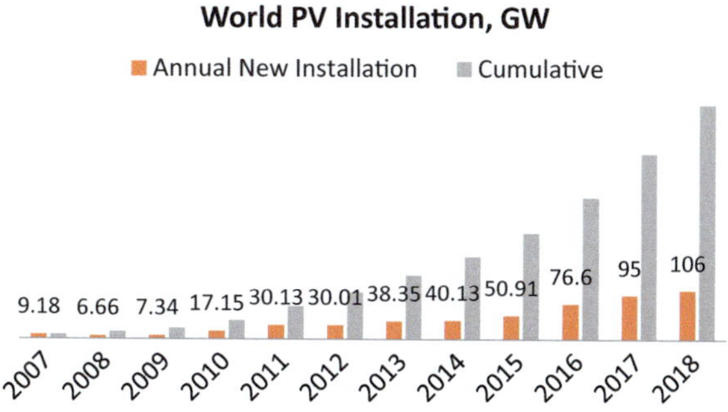

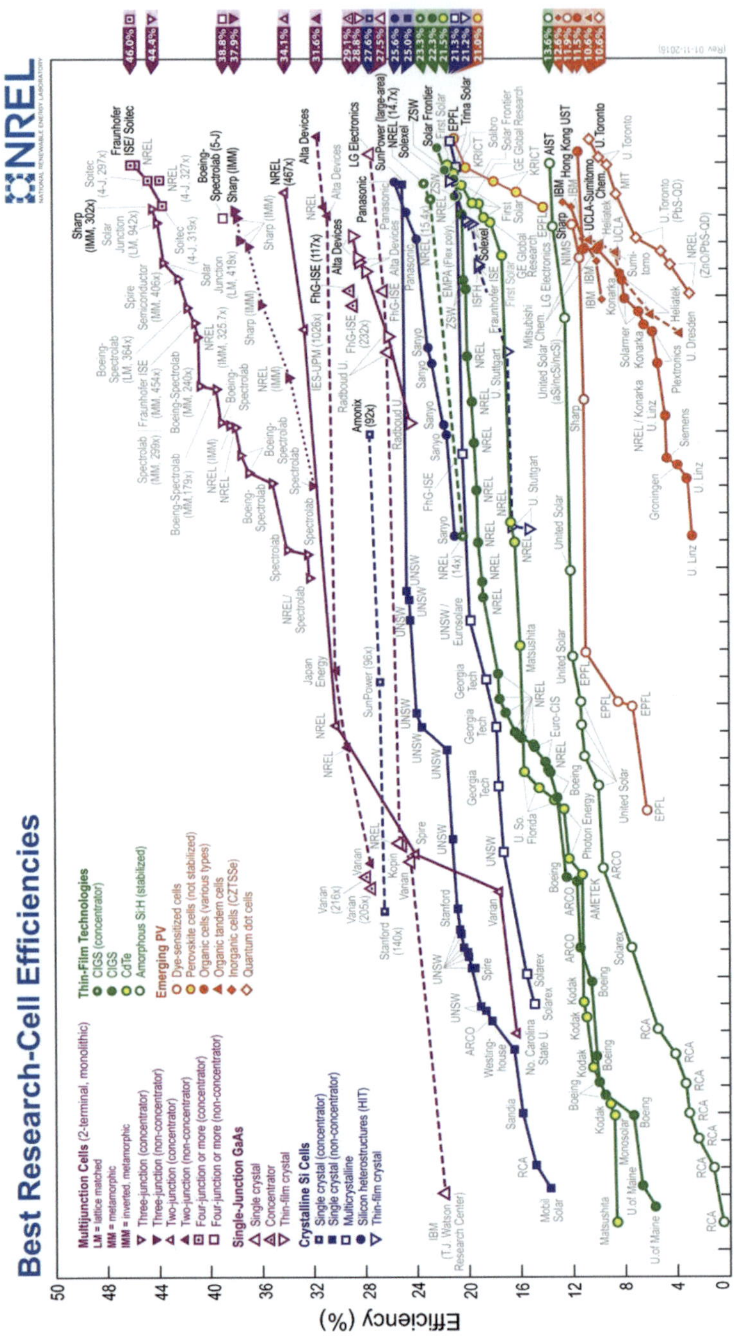

Fig. 2 Historical solar cell records [5]. https://www.energy.gov/eere/solar/downloads/research-cell-efficiency-records

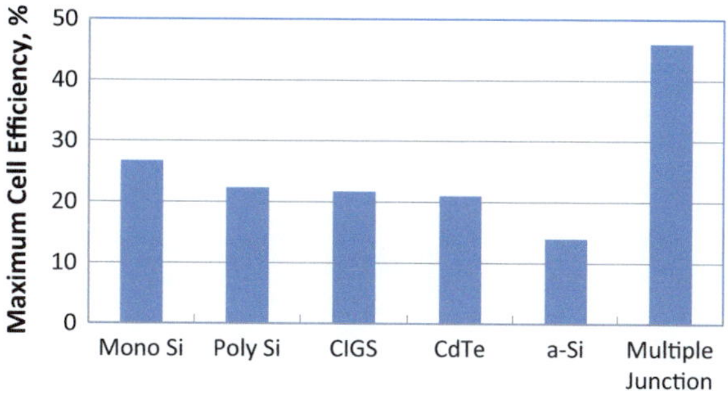

Fig. 3 Top solar cell efficiencies of various PV technologies as of 2018. Data extracted from [6]

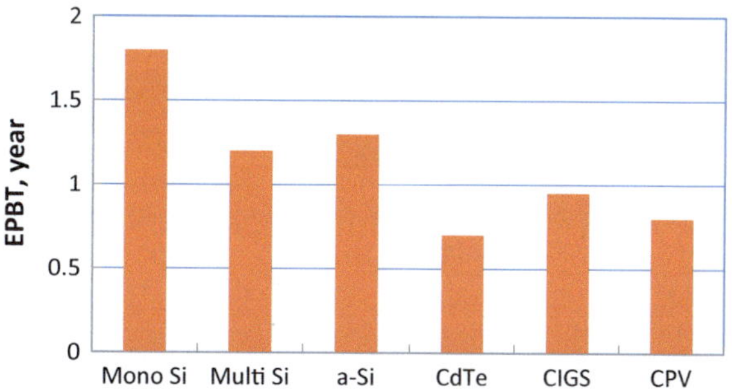

Fig. 4 Energy payback time for PV technologies as compiled by Fraunhofer in 2018. Data extracted from [3]

in Fig. 3. Multiple junction solar cells stay far more efficient than all other types of photovoltaic cells by about twofold.

It is important to note that on the top of the conversion efficiency, the energy payback time (EPBT) is another major factor that future energy developments must consider. EPBT is defined as the time that the amount of energy collected by a certain technology over the energy required to create the complete system of that specific technology including every component in the system. Figure 4 shows the summary of EPBT of various PV technologies installed at southern Italy. Based on data collected at Catania, Sicily, Italy and compiled by Fraunhofer [3], the EPBT of CPV technology is superior to most PV technologies except CdTe thin-film photovoltaics. It takes less than one year for CPV to generate more energy than the energy needed for making all components.

2 Cell Development of High-Concentration Photovoltaics, HCPV

Because of its high solar energy conversion rate, concentration photovoltaics (CPV) has been closely studied over the last 10 years using multiple junction solar cells. The cell efficiency has increased from around 35 to 46% during this 10-year period of development. The first 1 MW high-concentration PV (HCPV) installation was com-

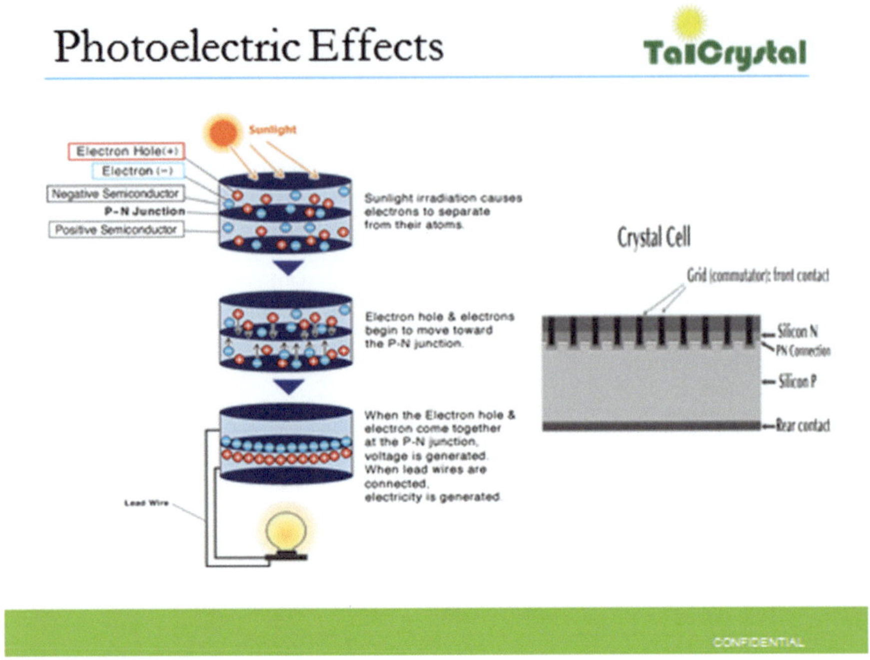

Fig. 5 Photovoltaic effect illustration

pleted in Spain in 2006. Starting from 2010, all HCPV installations were megawatt levels in China, USA, South Africa, and Australia. Due to mass production, HCPV installation cost has drastically reduced from $12/W in 2010 down to about $2.50/W in 2016. With new technologies being continuously developed, the future CPV installation cost is expected to further reduce.

2.1 *Photoelectric Effects*

All photovoltaic technologies are based on photoelectric phenomena of semiconductors. As shown in Fig. 5, when photons arrive at the surface of photovoltaic materials, electrons are excited and electron-hole pairs form. Due to the P–N junction, electron holes migrate to the P side and electrons migrate to the N side creating the voltage drop across the P–N junction reflecting the energy gap of the semiconductor material. When the external lead wires are connected, electrons will move through the wires to combine with the electron holes and creating electric power.

Multiple junction (MJ) solar cells have the highest energy conversion efficiency among all photovoltaic technologies. The world efficiency record of MJ cells was reported to have reached 46% in December 2014 by Fraunhofer. Under outdoor

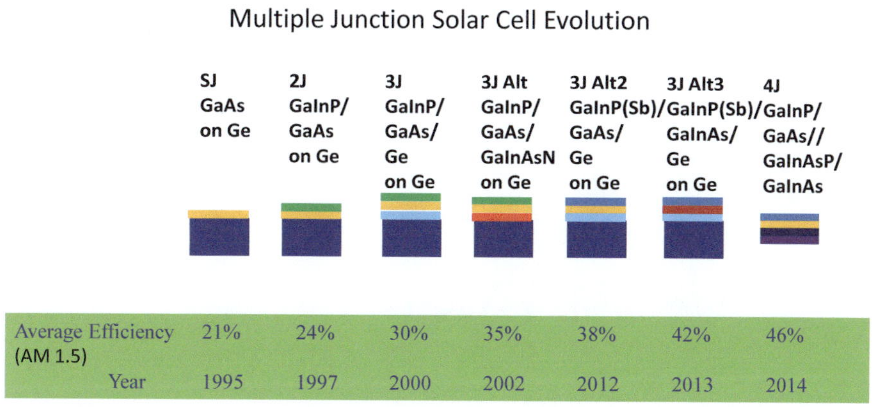

Fig. 6 Development of multiple junction solar cells. Author's drawing with data from [7, 8]

operating conditions, CPV module efficiencies have exceeded 33% ("one-third of a sun"). MJ cells are designed to have a number of cells stacking together with tunnel junctions connecting adjusting cells. To minimize light scattering between cells, all compounds used should have similar lattice constants.

2.2 *HCPV Cell History*

Multiple junction photovoltaics started in 1995 with a single-junction cell of GaAs over germanium and 21% efficiency. The two-junction tandem cell was quickly developed in 1997 with 24% efficiency. Figure 6 shows the roadmap of multiple junction solar cells up to four junctions [7, 8]. The reason that the conversion efficiency of multiple junction cells exceeds that of the silicon cells is due to the fact that the amount of solar irradiation absorbed and converted to electricity by the MJ cell structure is far more than that of a silicon cell.

2.3 *Solar Irradiation*

The sun sends its energy to earth continuously for the last billions of years. The solar spectrum arriving at the earth atmosphere surface can be represented by the curve AM0 in Fig. 7. Going through the earth atmosphere, the solar energy is reflected and absorbed with the net solar irradiation reaching the earth ground as shown in the AM1.5 curves for both global and direct irradiation. The AM0 spectrum stands for the solar irradiation that solar cells in space will see such as those on the satellites. AM1.5G would be the spectrum that silicon and thin-film solar cells can gather

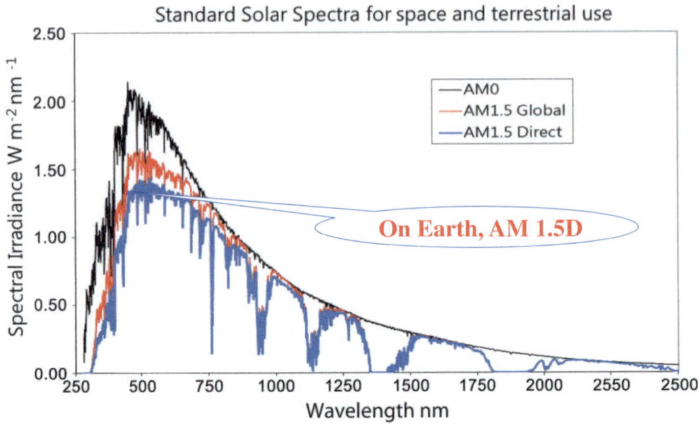

Fig. 7 Solar irradiation in space and on the earth surface [9]. AM0 stands for zero air mass or in space and AM1.5 stands for air mass 1.5 as an average of the solar irradiance covering the daytime from morning to evening

whereas AM1.5D would be the solar spectrum that multiple junction cells under high solar concentration would be facing.

For a silicon PV cell, the amount of solar energy absorbed is limited by the silicon energy gap of 1.1 eV and photons exceeding 1.1 eV would result in excess heat without providing extra voltages (Fig. 8a). For a triple-junction solar cell, the solar energy absorbed by each cell can be represented by the schematic diagram (Fig. 8b), in which the upper cell, GaInP cell, absorbs the short wavelength portion (blue), the GaInAs cell absorbs the middle wavelength portion (green) and the Ge cell absorbs the long wavelength portion (red) of the solar spectrum. Comparing the red area (Si PV power) in Fig. 8a and all red, green, and blue areas MJ PV power) in Fig. 8b, it is clear that with the same sunlight exposure, MJ PV cell would generate about double the power Si PV generates.

2.4 Multiple Junction Cell Structure Design

In fact, for a tandem solar cell, many different kinds of junctions can be designed to maximize the solar irradiation absorption and to optimize the power generation [10–12]. The idea is to select proper semiconductor materials and stacking them in a proper sequence. To minimize light scattering between cells, materials selected must have a similar structure with almost identical lattice constants. Lattice matching is one of the critical criteria in the multiple junction solar cell design.

Figure 9 shows a semiconductor material chart with a number of materials of various energy gaps. The triple-junction CPV cells mostly used in industry today contain Ge, GaInAs, and GaInP three layers.

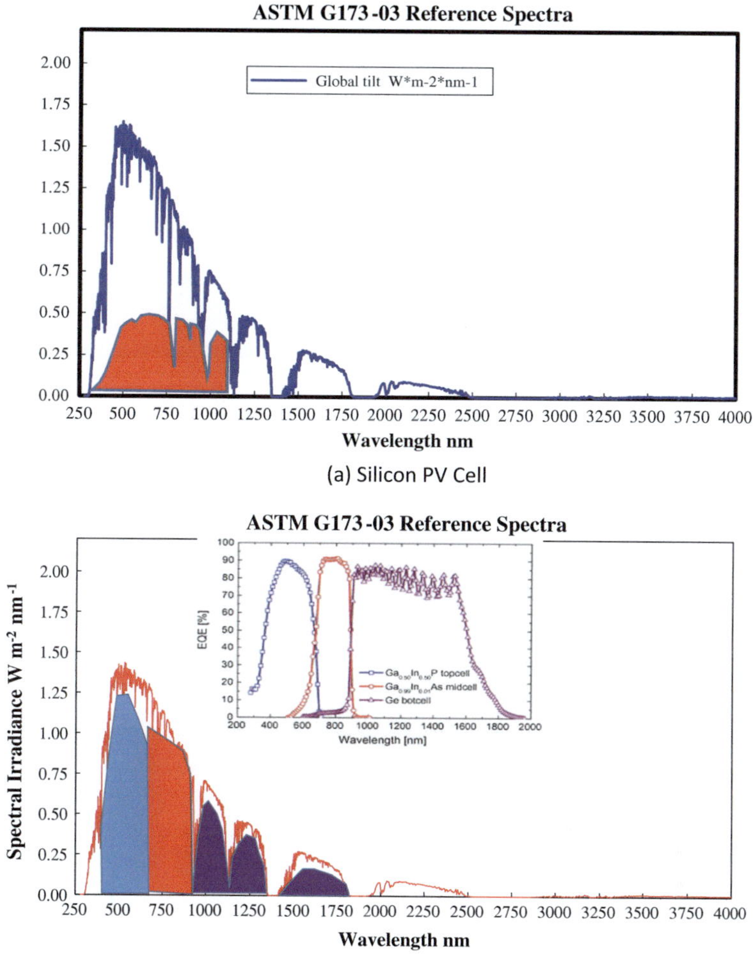

Fig. 8 Schematic drawings of solar spectrum absorbed and power generated by **a** a silicon PV cell; and **b** a triple-junction solar cell. The small insert figure shows measured external quantum efficiency (EQE) of a triple-junction solar cell. The colored areas represent the portions of the irradiance converting to electricity as a result of EQE

From Fig. 9, it can be seen that these three layers have a similar lattice constant of around 5.65. Both GaInAs and GaInP layers are adjusted to have lattice constants matching that of germanium. The corresponding energy gaps of these two layers are, therefore, fixed to be 1.48 and 1.84 eV, respectively. Four and higher junction cell structures can be designed with the same concepts. Of course, the actual materials available are limited due to the lattice constant constraint. Although researchers had tried to avoid the lattice restriction by concepts such as the inverted metamorphic

Table 2 Theoretical multiple junction solar cell efficiencies

# junctions in solar cell	One sun-efficiency (%)	Max. concentration-efficiency (%)
1 junction	30.8	40.8
2 junctions	42.9	55.7
3 junctions	49.3	63.8
4 junctions	52.6	68.7
6 junctions	56.2	72.7
…	…	…
1000 junctions and more	68.2	86.8

The last column represents efficiencies of different number of junctions under 500 sun solar concentration

(IMM) structural design, in actual practice, it showed more limitations in the way cell stacking can be done. Figure 10 shows schematic drawings of three-, four-, and six-layer tandem cells. Both three- and four-junction cells are available commercially. The six-junction cell, however, is still only a concept although, theoretically, a six-junction solar cell can have 72.7% conversion efficiency. Table 2 shows the theoretical calculation of efficiency for MJ PV cells with different amounts of cell stacking. The calculation is based on the available photon counts at each wavelength of AM 1.5D spectrum and the ideal external quantum efficiency of various materials. Just for academic interests, should one develop a 1000 layer tandem cell, one would have a cell of 86.8% conversion rate.

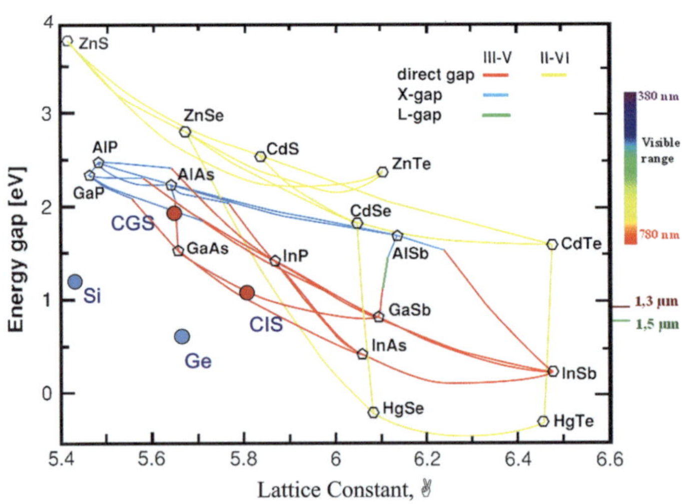

Fig. 9 Energy gap and lattice constant of semiconductor materials. Figure modified by author

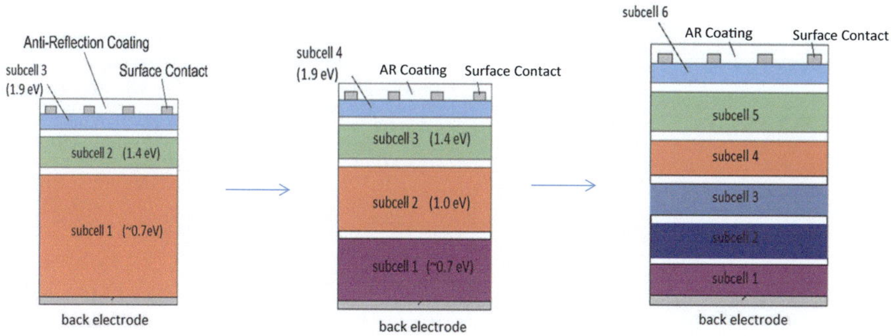

Fig. 10 Schematic drawings of multiple junction solar cell structures: three junctions, four junctions, and six junctions

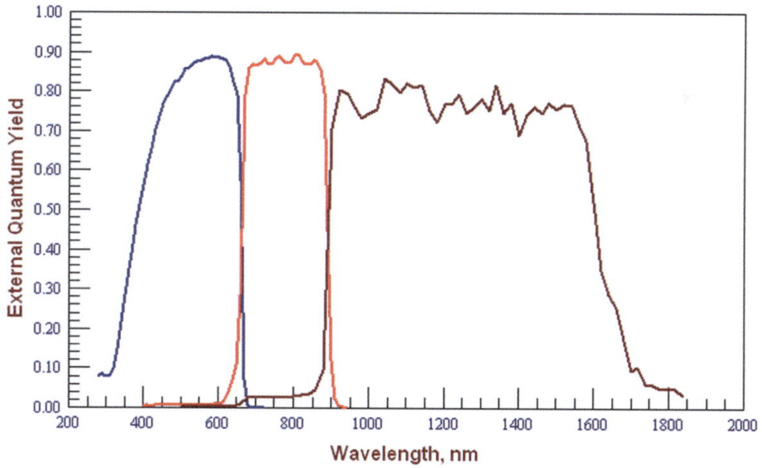

Fig. 11 External quantum efficiency of a triple-junction solar cells. Author's drawing

After a cell design is realized by actually making the cell with MOCVD (metal oxide chemical vapor deposition, a commercial production process), the cell must be tested for its external quantum efficiency (EQE). Figure 11 shows the result of an EQE measurement for a triple-junction solar cell prepared at TaiCrystal International Technologies using a flash tester at IOFFE similar to that reported by Shvarts and his coworkers [13]. The cell total surface area was 5.6 mm × 6.5 mm, and the testing temperature was at 25 °C. Table 3 lists currents determined for each layer. It is quite clear that the bottom cell has the most current under the simulated solar irradiation.

The external quantum efficiency results obtained suggest that, under all types of solar irradiation, the present triple-junction design provides far more current in the bottom layer than the currents generated at the middle and the top layers. Since the MJ cell current flows through all three layers, the excess current generated in the bottom layer would not be reflected during the entire cell current measurement. The

Table 3 Individual cell current of a triple-junction solar cells under different solar irradiations

Jsc (mA/cm^2)	Top cell (280–720 nm)	Middle cell (400–940 nm)	Bottom cell (500–1840 nm)
AM0 (1366.1 W/m^2)	17.90	15.30	26.61
AM1.5G (1000 W/m^2)	14.09	13.80	19.05
AM1.5D (1000 W/m^2)	13.56	14.01	20.45

Photocurrent density (mA/cm^2)
Cell ID: Cell-01
Cell type: Triple-junction SC
Cell structure: GaInP/GaInAs/Ge
Cell size: 5.62 mm × 6.51 mm
Temperature: 25 °C
Producer: TaiCrystal International Technologies Co., Ltd.

determined cell current would be the lowest current of all three layers. For example, under AM 1.5D irradiation, the current of the cell would only be 13.56 mA/cm^2, which is the current of the top layer. New cell designs can focus on increasing the bottom layer voltage to achieve a high overall triple-junction voltage while maintaining the overall current.

2.5 *Temperature Effect*

High-temperature environments often lower solar photovoltaic performance. One of the advantages of multiple junction cells over tradition silicon solar cells is the less efficiency reduction in the high-temperature environments than that of silicon cells. It is known that silicon PV modules in the desert stop generate electricity as the ambient temperature reaches 45–50 °C. At such a high-environmental temperature, PV modules can easily reach 60 °C or higher in the sun. To determine the cell performance at elevated temperatures, we have measured the cell I–V responses for commercially monocrystalline silicon cells and in house triple-junction cells. Figure 12 shows the effect of temperature on the power output of solar cells.

The regression analysis of power output determined can be represented as (for the temperature 25–100 °C):

for MJ solar cells

$$P_{\mathrm{MJ}}(\%\text{ of }40\,^{\circ}\mathrm{C}) = 100 - 0.071(T - 40);\quad \text{and}$$

for Si monocrystalline cells,

$$P_{\mathrm{Si}}(\%\text{ of }40\,^{\circ}\mathrm{C}) = 100 - 0.41(T - 40)$$

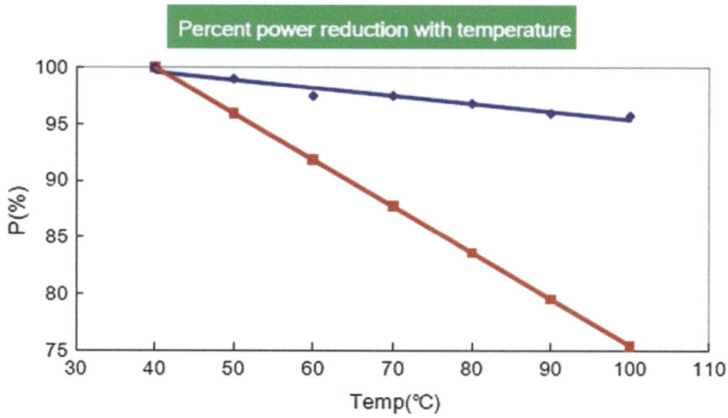

Fig. 12 Percent power generation of CPV cells versus silicon monocrystalline cells under various temperatures [8]

In the above equations, P_{MJ} and P_{Si} are normalized power of MJ cells and Si cells, respectively, using the measured 40 °C cell power as the reference. T stands for cell temperature in °C during measurement. Test data appear linear against the test temperature for both types of cells. The temperature coefficients (rates of deterioration) for Si cells are roughly 5.8 times as high as that of MJ cells. This result might be interpreted as that the Si cell efficiency deteriorates roughly six times faster than that of MJ cells as the cell temperature increases. In actual field applications, as the environmental temperature increases to 50 °C, we can expect that the Si cells would stop function under the sun as heat is collected by the modules.

2.6 *High Sun Concentration Applications*

To reduce the cost of cells per unit power output, one is easily considering increasing the solar concentration for the module design. Indeed, by increasing the solar concentration, one would obtain a high-power output with the same size of solar cells. A typical 4 mm × 4 mm cell under 500× sun concentration can provide about 2.5 W power under a direct normal incidence of 1000 W/m^2 sun power. The same cell installed with a 1200× concentration lens would supply about 6 W power. The proper cooling system must be designed so that no overheating occurs during practical applications. The cell cost in the design with 1200× lenses would be about only 5/12 of the cell cost in the module with 1200× lenses. Of course, modules with high-concentration lenses would require more accurate lining so that the sunspot can precisely drop on the cell surface. The cost of the 1200× lenses will also be more than that of the 500× lenses piece by piece. But if the lens cost per watt is being calculated, the per watt lens cost will be lower with 1200× design. Although high solar concentrations require sophisticated skill in module design and tracking accu-

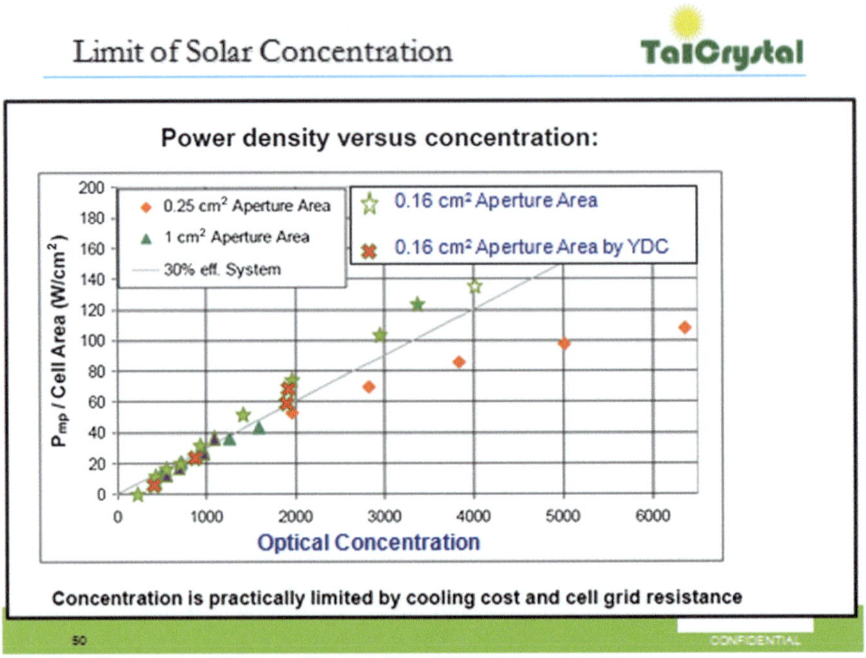

Fig. 13 Power generation of multiple junction solar cells under different solar concentrations [8]

racy, the overall financial and technological advantages are significant. Laboratory scale studies have revealed that by increasing the solar concentration, the cell power output increases linearly. Figure 13 shows the strong dependence of the power output from increased solar concentration. The linear power production with the solar concentration suggests that up to a 3000× solar concentration, the current cell technology can still provide proper power generation. For the power concentration to push even higher in the module design, the stability of tunnel junction between layers of the multiple junctions must be carefully examined. The difficulty is the lens to cell alignment will also be rather critical to assure proper module performance.

3 HCPV Modules

For power generation, solar cells must be assembled to HCPV modules. There are two main HCPV module designs: light transmission concentrators and light-reflecting concentrators. Both have pros and cons in terms of power generation, controllability, assembling easiness, cost and maintenance complications. With transmission-type concentrators, certain energy losses exist for solar irradiation going through transparent lenses depending on the lens materials and the focusing curvatures. The focusing

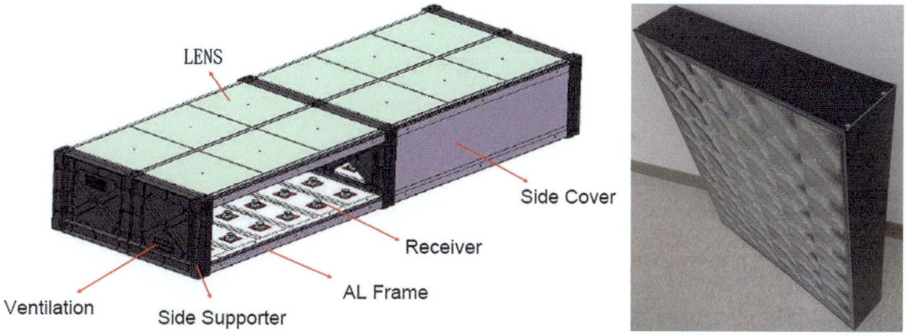

Fig. 14 HCPV modules of transmission type

mechanism also plays an important role in the power generation along with the cooling design, especially for systems with solar concentration being greater than 500×. Figure 14 shows a schematics of the transmission-type HCPV module optical system and a picture of an actual module available commercially. The lenses are SOG (silicon on glass) with specific transmission design. The geometric concentration ratio is 1200× between the lens and the cell. There are 42 receivers with 4 mm × 4 mm triple-junction cells. Specially designed secondary optics are placed on the top of the cells to increase the light collection and homogenize the concentrated light distribution.

Thermal management is another important feature emphasized in the module design. To achieve the best heat dissipation, cell bonding void control, material selection, and thermal conduction simulation must all be optimized. Weather sustainability including moisture management, module sealing, and corrosion prevention should be carefully prepared. For the best power performance, the tracking tolerance of modules is designed up to ±0.6°.

The reflective-type modules use parabolic dishes to collect parallel light beams and focus them to a spot where the solar cell is located. The light reflection mechanism typically results in less energy loss than that through the transmission. However, it is rather difficult to have a low-cost commercial parabolic reflector for massive installation. Weather tarnishing on the reflective surface is another issue for long duration application of the system. On the other hand, since the solar cell is not enclosed in a box, it is possible to have a cooling system installed along with the cell support. With proper arrangement, the cooling system can be used to withdraw thermal energy from the coolant. This is the concept of CPV-Thermal (CPVT) by extracting both photovoltaic energy (electricity) and thermal energy (heat) from a single CPV system. By combining both PV and thermal energy, the conversion rate can easily reach 50% or even higher. Figure 15 shows an example of a reflective-type CPV system installed near Yingchuan, China in 2017.

CPV modules have achieved a conversion rate of 33–35%. Researchers are continuously searching for new designs to make CPV modules of efficiency up to 50% and higher. DARPA-E had sponsored researches including those from MIT, Cal Tech,

Fig. 15 HCPV modules of reflective type. Author's photograph

University of Pennsylvania and a few industrial partners on a MOSAIC program (Micro-scale Optimized Solar-cell Arrays with Integrated Concentration program) for advanced CPV systems starting 2016 [14].

4 Tracking Systems

HCPV power systems require direct sunlight due to the high-concentration focusing mechanism. Typically, a dual-axial tracker is used to accurately facing the modules to the sun with $\pm 1^\circ$ tracking accuracy.

One commercially available tracker is shown in Fig. 16. This type of tracker has a dual-axis tracking system which can track sunlight in multiple angles. It includes one slew driver, in charge of horizontal movement, one linear actuator, in charge of vertical movement, and a controller unit with the light sensor, in charge of tracking direction. The tracker is also equipped with safety mode. When encountering a strong blast, the tracker system could automatically switch (with "Auto mode") to a safety position with the module array lied flatly during hurricane or typhoon. In the case, a severe hurricane or typhoon is predicated, and safety prearrangement can be done by fastening the steel rings attached on the truss and on the foundation surface. Due to the rotation requirement, the tracker to tracker distance must be maintained so that no tracker will be touching each other during operation. Another factor to be considered for tracker placement is the shadowing effect both in the morning and in the late afternoon when the sun is at the low angles. The tracker location design must be so that the solar panels on each tracker are not in the shades of panels from other trackers nearby.

Fig. 16 Dual-axial tracker. Author's photograph collection

5 HCPV System Integration

To install a complete HCPV system, one would need the modules, the trackers, the power cables, the DC power boxes, the inverters, the AC power boxes, and monitoring systems. Figure 17 shows the physical connections of these units. The HCPV systems can be connected to the local grid system. They can also be a stand-alone installation. In the case of a stand-alone installation, storage batteries must be installed at the same time. It is a typical practice to equip the HCPV system with 5-day equivalent storage batteries so that there would be sufficient power for the load in case the out of sunlight condition lasts for 5 days continuously.

The HCPV system can be installed both on the rooftop and in the ground. The rooftop installation typically is for small power systems and can be either grid connected or stand-alone. The in-ground installations usually are for large utility power in an open field. Due to the sun tracking need, the spacing between trackers must be carefully designed depending on the longitude and latitude of the installation site so that no shadowing effects hinder the power generation. Figure 18 shows a single-tracker system installed on a rooftop. The building structure engineer had done proper design so that the roof floor strength is sufficient for the weight of the HCPV system as well as against a strong wind.

5.1 Alamosa HCPV Project

Figure 19 shows a picture of a 30 MW HCPV installation located at Alamosa, Colorado, USA. It is located in the San Luis Valley. It consists of over 500 dual-axis CPV Amonix 7700 tracker assemblies. The system was built on a 225-acre plot. Each

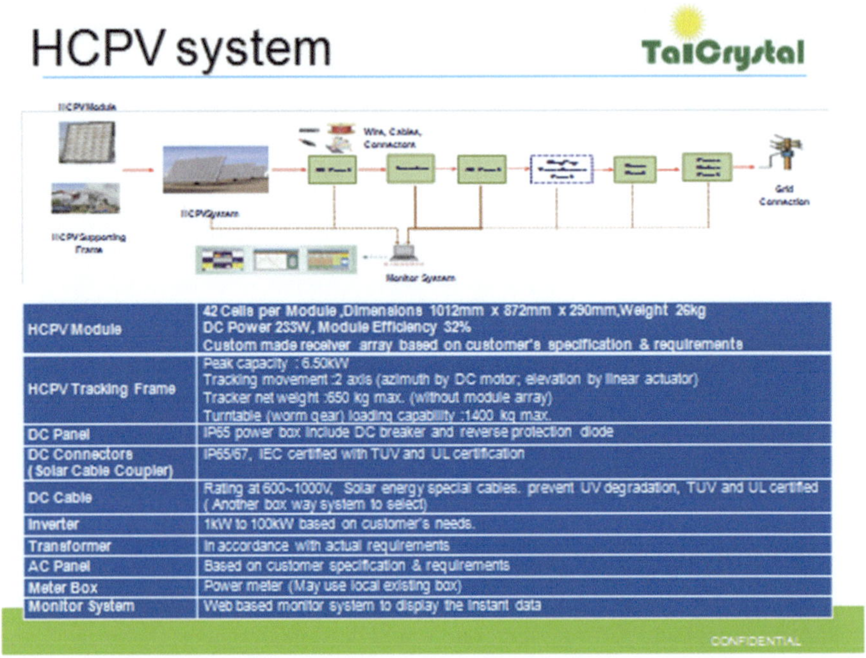

HCPV Module	42 Cells per Module ,Dimensions 1012mm x 872mm x 290mm,Weight 26kg DC Power 233W, Module Efficiency 32% Custom made receiver array based on customer's specification & requirements
HCPV Tracking Frame	Peak capacity : 6.50KW Tracking movement :2 axis (azimuth by DC motor; elevation by linear actuator) Tracker net weight :650 kg max. (without module array) Turntable (worm gear) loading capability :1400 kg max.
DC Panel	IP65 power box include DC breaker and reverse protection diode
DC Connectors (Solar Cable Coupler)	IP65/67, IEC certified with TUV and UL certification
DC Cable	Rating at 600~1000V, Solar energy special cables. prevent UV degradation, TUV and UL certified (Another box way system to select)
Inverter	1KW to 100KW based on customer's needs.
Transformer	In accordance with actual requirements
AC Panel	Based on customer specification & requirements
Meter Box	Power meter (May use local existing box)
Monitor System	Web based monitor system to display the instant data

Fig. 17 HCPV system integration

Fig. 18 A rooftop HCPV system of transmission type. Author's photograph

Fig. 19 HCPV 30 MW installation at Alamosa, Colorado. Author's photograph

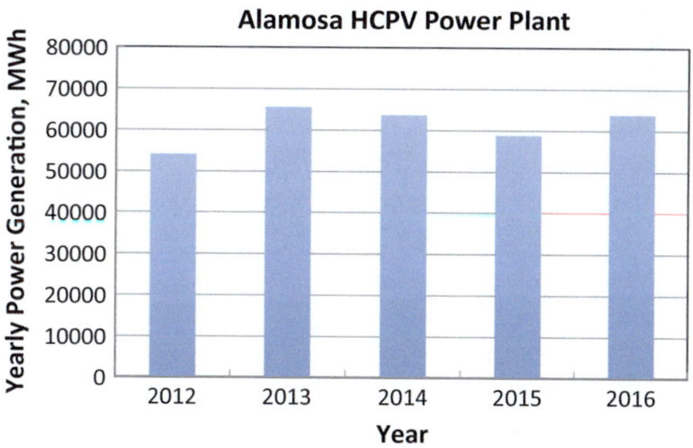

Fig. 20 Yearly power generation at 30 MW HCPV installation at Alamosa, Colorado. Author's drawing from data [15]

tracker is 70 feet wide and 50 feet tall. Each has 7560 Fresnel lenses that concentrate sunlight by a multiple of 500 onto multi-junction gallium arsenide photovoltaic cells. Dual-axis tracking, concentrating photovoltaics, and multi-junction cells were the three techniques that increased the system power output and not typically seen in regular silicon PV systems. The HCPV systems are certified for operation for 50 years. The output is being sold to Public Service of Colorado, a subsidiary of Xcel Energy, under a long-term power purchase agreement.

Yearly electricity generation at the Alamosa HCPV power plant can be seen in Fig. 20. Although this was only an experimental project installed in 2012, the stability of the HCPV power plant can be observed. The Alamosa HCPV project has proven that HCPV is a viable technique for PV power supply. However, the installation cost remains high even today. New technique to further reduce the production cost must be developed.

5.2 USDOE MOSAIC Program

The MOSAIC (Micro-scale Optimized Solar-cell Arrays with Integrated Concentration) Program funded potentially low cost and high efficiency technologies and related system concepts to achieve new performance and cost benchmarks for solar-electric generation from photovoltaics (PV). Starting in 2016, there were 11 institutions sponsored by the USDOE ARPA-e program to develop low-cost high-efficiency PV technology based on CPV achievements so far [14]. These 11 institutions included both industrial and academic partners: (1) Panasonic Boston Laboratory—Newton, MA, with the research entitled "Low Profile CPV Panel with Sun Tracking for Rooftop Installation"; (2) The Pennsylvania State Univ—University Park, PA on the subject of "Wide Angle Planar Microtracking Microcell CPV"; (3) University of Rochester—Rochester, NY, on "Planar Light Guide Concentrated Photovoltaics"; (4) California Institute of Technology—Pasadena, CA, on "Micro-Optical Tandem Luminescent Solar Concentrator"; (5) MIT—Cambridge, MA, on "Integrated Micro-Optical Concentrator Photovaltics with Lateral Multijunction Cells"; (6) MIT—Cambridge, MA, on "Wafer-Level Integrated Concentrating Photovoltaics"; (7) George Washington University (GWU) on "Micro-Scale Ultra-High Efficiency CPV/Diffuse Hybrid Arrays Using Transfer Printing"; (8) University of Arizona, Tucson, AZ, on "A High-Efficiency Flat Plate PV with Integrated Micro-PV atop a 1-Sun Panel"; (9) Texas A&M University Engineering Experiment Station—College Station, TX, on "Waveguiding Solar Concentrator"; (10) Palo Alto Research Center—Palo Alto, CA on "Micro-Chiplet Printer for MOSDAIC"; and (11) Glint Photonics, Inc—Burlingame, CA, on "Stationary Wide-Angle Concentrator PV System."

In the present concepts of CPV technologies, multiple junction III–V compound PV cells are being used to minimize the material cost by utilizing optical concentration to reduce the amount of PV material required. Such CPV systems, however, are limited in their application since, as a result of solar concentrating mechanism, only the direct component of the solar irradiation is effective for power generation, as well as the need for expensive mechanical tracking systems. ARPA-e's MOSAIC program was structured to develop technologies and concepts that could lower the cost of solar photovoltaic (PV) power systems and improve their performance. The MOSAIC teams were to develop microscale concentrated photovoltaic systems (CPV) that were similar in cost and size to conventional solar PV systems, but with greatly increased performance levels. As shown above, participants of the project covered expertise in conventional flat-plate PV, CPV, manufacturing, optical engineering, and material science for the purpose to produce a new class of PV panels. It was the goal of the joint effort to facilitate cost-effective solar power systems across a wide range of geographical locations, lowering US greenhouse gas emissions and reducing dependence on imported energy.

On top of the USDOE MOSAIC programs, others also studied micro-CPV technologies. Panasonic, Sharp, Sandia National Laboratories, Semprius, Insolight, Morgan Solar, and Institute of Solar Energy (Polytechnical University of Madrid) are a few examples [16]. Multiple junction cells are the base they used for micro-CPV developments.

6 Concluding Remarks

It is highly possible that the need of 10,000–20,000 GW worldwide energy consumption be met by PV power system installations over a 20–30 year span. However, PV systems cannot produce power at night and are vulnerable to the weather conditions. For PV to become the main power supply, reliability issues must be overcome. A low-cost and reliable power storage system is a must for us to seriously consider PV as the main power need. The latest development in electric vehicles on the power storage system may also help the advancement of PV power storage. Lithium-ion batteries appear to have made significant improvement during the last few years. Other types of high-density batteries are also being carefully investigated including programs such as graphene-nanosilicon material development. With the effort in electric vehicles and PV system needs, it is hopeful that a reasonably acceptable energy storage system be developed within the foreseeable future.

7 Summary

1. If we are to keep the greenhouse effect of CO_2 emission from running out of control, we must cut down the fossil fuel usage in all aspects of our activities.
2. Among many types of renewable energy resources, photovoltaics is the only one resource sufficient to meet the growing need of power due to the projected global population growth or 15 TW by 2025.
3. Despite its low-energy conversion rate, Silicon PV is the primary photovoltaic technology installed worldwide to date. The total PV installation is expected to reach 500 GW by the end of 2018.
4. Future renewable energy technology is most likely the solar photovoltaic technology with much higher efficiency than that of silicon technology. Multiple-junction photovoltaics is probably the one to be used for our future energy growth.
5. Multiple-junction photovoltaics has the potential of reaching an energy conversion efficiency of more than 50% with increasing layers of P–N junction as well as dual-energy conversion from both photovoltaics and thermal energy absorption.
6. Future multiple junction photovoltaic technology development may include higher conversion efficiency, thinner and more effective module design to reduce the need of tracking accuracy and thus, reducing the overall installation cost.
7. Reaching the grid parity is the target for all future photovoltaic development.
8. For the PV technology to become the primary power resource for human civilization, energy storage technology must be simultaneously grown so that renewable energy can be available to the need without interruption due to weathering effects.

9. Solar PV had another landmark year in 2017. The world added more capacity from solar PV than from any other type of power generating technology, and more solar PV was installed than the net capacity additions of fossil fuels and nuclear power combined. Solar PV was the top source of new power capacity in several countries, including China, India, Japan, and the USA [17].

Acknowledgements Parts of this work were supported from Taiwan's Industrial Development Bureau, Ministry of Economic Affairs, under the project entitled "High Efficiency III–V Compound Solar Cells for Ultra High Solar Concentration Applications." The following individuals had contributed to the content of this work and are greatly appreciated: Elton F. Lien, Sam H. Wang, Frank F. Huang, Ryan C. Chi of TaiCrystal International Technologies Co., Taiwan; Mr. Kubo, Mr. Nakamura, Mr. Yamashita of Yamashita Denso Co., Japan; Kuni Takahashi of Holonix International Co. Ltd, Japan; and Dr. Maxim Shvarts, Ioffe Physical Technical Institute, Russia.

References

1. Wallace JW, Hobbs PV (1977) Atmospheric science: an introductory survey. Academic Press, San Diego, pp 320–321. N 978-3
2. Tsao J (n.d.) Solar FAQs. n. pag. Web. 2 Nov 2014. http://www.sandia.gov/~jytsao/Solar%20FAQs.pdf
3. Photovoltaics report, Fraunhofer ISE, updated: 19 June 2018
4. Snapshot of Global PV Markets 2016 (PDF). IEA-PVPS. p 11. Retrieved 27 Oct 2017; and "Global Market Outlook 2017–2021" (PDF). SolarPower Europe. 13 June 2017. p 7. Retrieved 13 Nov 2017
5. Wayne Hicks, Claims for solar cell efficiency put to test at NREL, Feb 8, 2016, National Renewable Energy Laboratory, https://phys.org/news/2016-02-solar-cell-efficiency-nrel.html; https://www.energy.gov/eere/solar/downloads/research-cell-efficiency-records
6. Green MA, Hishikawa Y, Dunlop ED, Levi DH, Hohl-Ebinger J, Ho-Baillie AWY (2018 June) Solar cell efficiency tables (version 52). Prog Photovolt Res Appl **26**(7)
7. Lin RY (2013) HCPV cells for high optical concentration (>1000 suns) applications. CPV Summit USA 2013, San Jose, CA, Sept 23–24, 2013; Lin RY, Lien E (2014) Effect of light distribution on the HCPV solar cell performance for ultra high solar concentration applications. In: CPV-10 proceedings, 10th international conference on concentrator photovoltaic systems, Albuquerque, New Mexico, Apr 7–10, 2014
8. Lin RY (2015) The future of solar PV-ultimate energy for human civilization. In: Proceedings on world industrial technology summit 2015, Apr 1–4, 2015, Kaoshiung, Taiwan
9. NREL Solar Spectra., https://rredc.nrel.gov/solar//spectra/am1.5/ASTMG173/ASTMG173.html
10. Kinsey GS, Bagienski W, Nayak A, Liu M, Gordon R, Garboushian V (2013-04-01) Advancing efficiency and scale in CPV arrays. IEEE J Photovolt 3(2): 873–878
11. Wiesenfarth M, Philipps SP, Bett AW, Horowitz K, Kurtz S (2017) Current status of concentrator photovoltaic (CPV) technology. Fraunhofer ISE I NREL CPV Report 1.3 Apr 2017 TP-6A20-63916
12. Honsberg CB (2005) Paths to ultra-high efficiency (>50% efficient) photovoltaic devices. In: 20th European photovoltaic solar energy conference
13. Andreev VM, Ionova EA, Larionov VR, Rumyantsev VD, Shvarts MZ, Glenn G (2006) Tunnel diode revealing peculiarities at I–V measurements in multijunction III-V solar cells. In: Proceedings of the IEEE 4th world conference on photovoltaic energy conversion, Hawaii, May 7–12, 2006, pp 799–802

14. Micro-scale optimized solar-cell arrays with integrated concentration program, 11 projects funded for high performance solar power, Aug 24, 2015. https://arpa-e.energy.gov/?q=news-item/11-projects-funded-high-performance-solar-power
15. Alamosa solar generating project, 20190323. https://en.wikipedia.org/wiki/Alamosa_Solar_Generating_Project
16. Domínguez C, Jost N, Askins S, Victoria M, Antón I (2017) A review of the promises and challenges of micro-concentrator photovoltaics. In: AIP conference proceedings 1881, 080003
17. REN21, Advancing the global renewable energy transition: highlights of the REN21 renewables 2018 global status report in perspective, renewables global status report (GSR). www.ren21.net/GSR, www.ren21.net/GFR

Study of Biofuel Animal Manure

Ryoichi S. Amano and Mohamed Saeed Ibrahim Hussein

Keywords Biomass · Biofuel · Pyrolysis · Gasification · Renewable energy

1 Introduction

Pyrolysis is a method that decomposes organic material into gases using heat in the absence of oxygen. When an organic material is subjected to high amounts of temperature and no oxygen, the molecules start to vibrate, stretching, and vibrating the molecules until it breaks down into smaller molecules [1].

Gasification is a process that converts hydrocarbons into flammable gas. Biomass is heated in a controlled chamber, and a specific amount of oxygen is fed into the room. The amount of oxygen does not be enough to set the biomass on fire, so instead, the mass gives off flammable fumes or syngas [1] (Fig. 1).

The chapter contributes to the understanding of the feasibility of animal waster can be used for a useful source of energy. Some detailed analysis is presented in this chapter for determining which temperature and gas convert the most biomass into syngas in the shortest amount of time.

R. S. Amano (✉) · M. S. I. Hussein
Mechanical Engineering Department, University of Wisconsin-Milwaukee, Milwaukee, WI 53211, USA
e-mail: amano@uwm.edu

A. K. Gupta et al. (eds.), *Innovations in Sustainable Energy and Cleaner Environment*, Green Energy and Technology,
https://doi.org/10.1007/978-981-13-9012-8_23

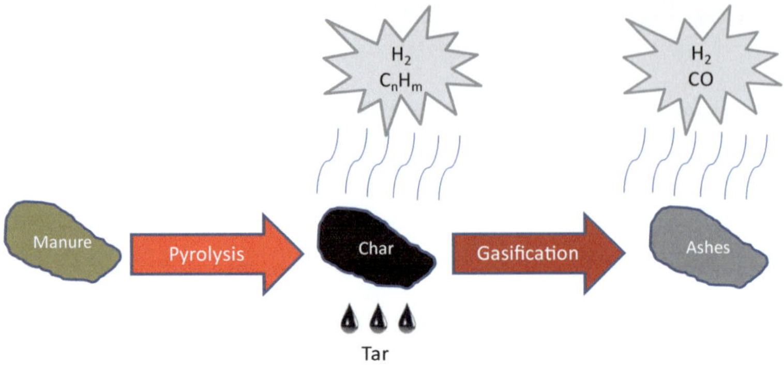

Fig. 1 Biochemical processes used in the study

2 Analysis of Study

2.1 Design Analysis

The topic of biomass has contracted much attention in recent years, mainly due in part to the environmental aspects of using organic waste to produce energy and power. The Biomass Power Association (BPA) names themselves as America's leading organization in gaining power from biomass and for a good reason: They operate 80 biomass power plants across 20 states. [2] Their power plants use a variety of different waste materials, including scrap lumber, forest debris, farm plant waste, and landfill waste. They even have a power plant in Louisiana that runs on rice hulls. These plants take these waste materials and decompose them to produce methane gas [2] and carbon dioxide. These gases are turned into fuel and immediately burned to generate electricity for local communities.

Another leading organization in the biomass industry is the Renewable Fuels Association (RFA), a collaboration of ethanol fuel production companies with the purpose of expanding the ethanol fuel industry in the USA [3]. Their prominence in recent years has been influenced by the Renewable Fuel Standard, which is a federal program introduced in 2005 that is discussed more later in this paper. Essentially, the standard says that a certain percentage of all US gas produced for the automotive industry has to come from renewable sources. In 2015, the total number was 20.5 billion gallons, and the Renewable Fuels Association's efforts in ethanol biofuel production are a large part of why that number can be reached [4].

As can be seen, there is already a prevalent biomass industry already in place in the USA. The Biomass Power Association also uses pyrolysis techniques to decompose their fuel. So then, What is the point of this senior design project? Well, there are still ways this project can potentially influence the renewable fuel industry. This study uses chicken litter and the idea that farm animal manure can be harnessed for fuel. The RFA is focused on the production of ethanol biofuel. Ethanol, short for ethyl alcohol, does not use pyrolysis or gasification and instead produces their fuel by making alcohol from corn waste. The Renewable Fuel Standard program puts an increase in ethanol production. The BPA can also gain from this project. While they do use pyrolysis and gasification to decompose their waste, they do not use animal waste in their power plants yet. Animal waste is a massive source of biomass that has yet to be tapped by any major organization. There are many farms in the USA that take care of animals, and all of them could provide biomass for the biomass industry.

3 Experimental Study

This reduction in biomass project is a continuation of the same project built upon over the last three semesters. The reforming chamber used for this study is shown in Fig. 2. The first component is the apparatus base; made from wood; it is the foundation. The load cell is used to measure the mass of the biomass during the experiment and is attached to the top of the adjustable wooden frame. Next is the MTAirtorch electric heater. It is used to heat the inert gasses. The Waldor electrical current temperature controller has attached to the heater. It is unviewable in Fig. 2 but sits behind the ventilation. This device is used to control the temperature of the input inert gas to the chamber. The most critical component is the atmosphere chamber, made from stainless steel and ceramic insulation. The box is sealed, except the outlet valve, to control the environment. The thermocouples are mounted inside the chamber and inserted through the outlet valve. Both the thermocouples and the load cell are attached to the data acquisition (DAQ) device, which itself is connected to a computer running the LabVIEW program. The last item is the gas vacuum, which is used to remove the gas so that it does not contaminate the building. The DAQ itself is an NI 9219 with four universal input channels and sits inside an NI cDAQ-9171 USB chassis that takes the data and imports it into the LabVIEW program.

3.1 Other Instrumentation

Shimadzu TGA-50 Thermogravimetric Analyzer.

A thermogravimetric analysis (TGA) machine was employed for this analysis. TGA measures physical and chemical properties in materials that decompose with an increase in heat. This machine was used to test the chicken manure and many different types of animal manure to determine which should produce the most gasses (Fig. 3).

Fig. 2 Reforming chamber

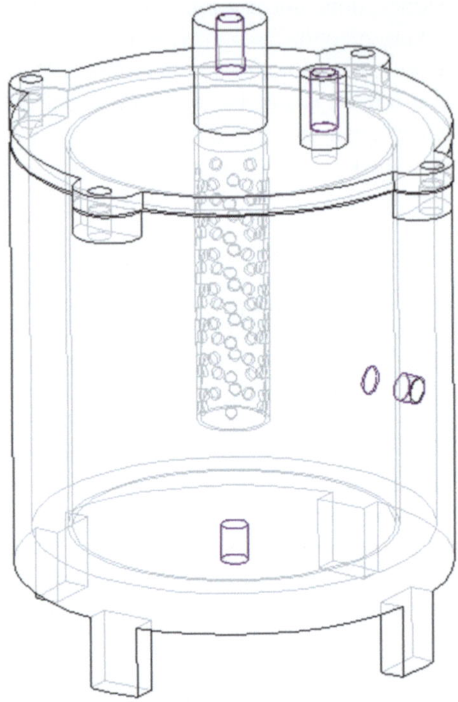

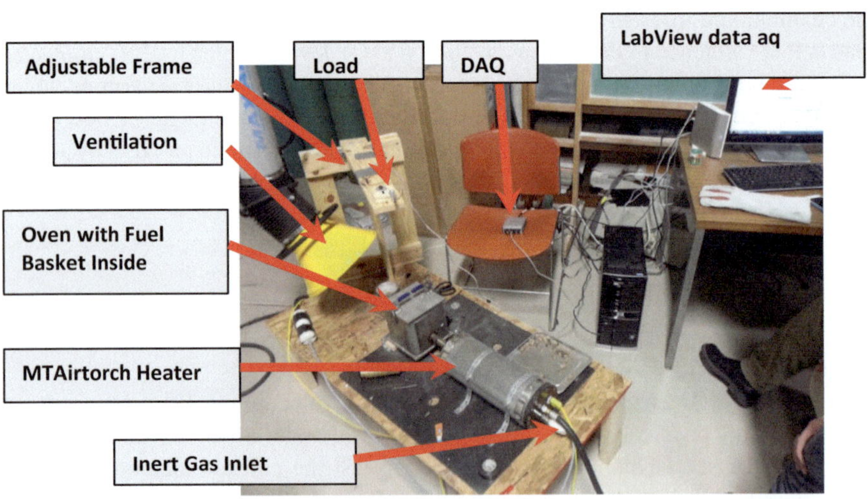

Fig. 3 Experimental setup

3.2 Thermogravimetric Analysis (TGA)

Thermogravimetric Analysis (TGA) is the technique to find the effect of different thermal processes on the mass of the substance. The test sample is raised temperatures and the mass of the sample and time are monitored while the different processes are taking place. If the first derivative of the mass-time curve is calculated ($\mathrm{d}m/\mathrm{d}t$), a derivative thermogravimetry (DTG) is derived which can provide the maximum reacting rates and the corresponding temperatures.

White et al. [5] performed an extensive review of the different thermal analyses; the kinetics of pyrolysis and different kinetics models, in their study, they referred to the famous Arrhenius rate expression that was used as the first step of almost any kinetics model,

$$k(T) = A\ \exp\left(\frac{-E_\mathrm{a}}{RT}\right) \tag{1}$$

Even though A is slightly dependent on the temperature, it is usually considered as a constant. Two main techniques are then utilized to find the reaction kinetics: the isothermal and the non-isothermal methods.

In the isothermal method, the temperature is fixed, and the following canonical equation is used:

$$\frac{\mathrm{d}\alpha}{\mathrm{d}t} = k(T)f(\alpha) = A\exp\left(\frac{-E_\mathrm{a}}{RT}\right)f(\alpha) \tag{2}$$

where $f(\alpha)$ is a function depending on the reaction mechanism and $\mathrm{d}\alpha/\mathrm{d}t$ is the rate of the isothermal process, and the extent of reaction α is given by:

$$\alpha = \frac{w_\mathrm{o} - w_t}{w_\mathrm{o} - w_f} = \frac{v_t}{v_f} \tag{3}$$

In the non-isothermal method, Eq. (2) is written as:

$$\frac{\mathrm{d}\alpha}{\mathrm{d}T} = \frac{\mathrm{d}\alpha}{\mathrm{d}t}.\frac{\mathrm{d}t}{\mathrm{d}T} \tag{4}$$

$$\frac{\mathrm{d}\alpha}{\mathrm{d}T} = \frac{k(T)}{\beta}f(\alpha) = \frac{A}{\beta}\exp\left(\frac{-E_\mathrm{a}}{RT}\right)f(\alpha) \tag{5}$$

The reaction order models are often used where:

$$\frac{\mathrm{d}\alpha}{\mathrm{d}T} = k(T)(1-\alpha)^\mathrm{n} \tag{6}$$

In their review, they also discuss more kinetics models, and they show the results of these models on agricultural biomass pyrolysis. They also describe the different thermal degradation steps, starting with the evaporation of the free moisture followed by the decomposition of less stable polymers at lower rates than the more refractory components at higher temperatures. At a temperature around 400 °C, only char residue is present after what they called the primary decomposition phase. Then at higher temperatures, the second slow stage of aromatization takes place.

Mansaray and Ghaly [6] used Eq. (2) with the reaction order model to calculate the kinetic reaction parameters for the P/G of rice husk using a controlled environment of oxygen. The used equation was written as:

$$\frac{\mathrm{d}X}{\mathrm{d}t} = -A\exp\left(\frac{-E_{\mathrm{a}}}{RT}\right)X^{\mathrm{n}} \tag{7}$$

where X here is equal to $(1-\alpha)$ and was denoted as the weight of the sample undergoing reaction. Four varieties of rice husk were tested using pure oxygen from ambient temperature to 700 °C at a heating rate of 20 °C/min. The percentage of thermal decomposition was higher in the first phase than the second phase. The different rates made it necessary to divide the kinetics into two discrete global reactions. The response of the four rice husk varieties was very similar. The TGA and DTG analysis showed the two most critical phases of the gasification overlapping between 206 and 467 °C. An intermediate temperature was chosen on the TGA curves to separate these two regions. The highest degradation rates were observed at a temperature close to 290 °C for the four different varieties. Approximately 20% of the mass was a residue after the rate of reaction was almost zero. The kinetic parameters, activation energy, Arrhenius constant, and order of reaction were calculated from Eq. (7). The authors then found a necessity for studying the effect of the heating rate on the kinetic parameters.

Yanik et al. [7] studied the pyrolysis of different algal biomass from the Black Sea, using the TGA method. They heated the biomass in a nitrogen environment to an 800 °C temperature and monitored the change in the mass and the rate of mass conversion. The total percentage of weight conversion ranged between 55 and 70%, and the percentage of conversion was highest between 250 and 450 °C, depending on the algae tested. The main contents of the algal biomass are carbohydrates and proteins whose degradation temperatures lie between 190 and 390 °C which justifies their findings.

Mermoud et al. [8] studied the steam gasification of a single particle of charcoal at different temperatures, steam concentration, flow velocity, and particle size both numerically and experimentally. The gasification time was proportional to the particle size and inversely proportional to the heat and the steam concentration. They then utilized a particle mechanism to simulate the gasification process, and their results were acceptable up to 60% conversion. The mass fraction of the gas yields was calculated using the numerical model. For larger particle sizes, the numerical results were not accurate beyond 60% that they concluded to be due to asymmetry and fractures in the sample after certain conversion percentage.

Van de Steen et al. [9] studied the effect of changing the reacting gas on wood gasification. In their study, steam, carbon dioxide, and oxygen gasification at temperatures ranging between 800 and 1050 °C were analyzed both experimentally and numerically. It was found that the most critical parameter was the particle thickness, and they were able to modify a particle mechanism using their experimental data. The conversion was faster as the temperature, and the gas concentration increased. Oxygen showed the highest conversion rate while carbon dioxide had the lowest conversion rates.

Wang et al. [10] used the three most essential components of biomass: hemicellulose, cellulose, and lignin as their samples and mixed the three components with different percentages. The samples were heated in a nitrogen atmosphere from 30 to 800 °C. The weight of the sample was measured over time. Hemicellulose decomposed in the lower temperature range 200–350 °C with maximum conversion rate at 260 °C, cellulose in a higher range 260–430 °C with maximum conversion rate at 360 °C, while lignin in the most upper range 200–500 °C with maximum conversion rate at 370 °C. For the mixture samples, the DTG curves showed more than one peak at the 260 and 370 °C. It was also found that the presence of the three components promotes the gasification process due to the interaction between them.

Mermoud et al. [11] used steam in the gasification of large wood char particles. They implemented different pyrolysis heating rates (2.6–900 °C/min.) and studied the effect on mass, density, and porosity. The higher heating rate decreased the apparent density and increased the porosity. The apparent density followed a linear evolution versus the log of the heating rate. The ratio of the initial to final volume was approximately the same for different heating rates which means a higher volatile matter yield at higher heating rates. The gasification rate of the char prepared at the 900 °C/min.; the heating rate was 2.6 times higher than the less porous char made at the 2.6 °C/min when both were gasified at the same 1200 K temperature with 20% by volume steam-nitrogen mixture gas.

4 Experimental Procedures

First, we use the mold to shape and standardize manure samples. Then, put the samples in the dehydration oven and put on low heat to remove water from the samples. After that, we put the desired amount of samples onto the wire tray. Then, we insert the thermocouple and plate into the chamber. Then, we add the thermocouple and plate into the room. After that, we place ceramic insulation and metal chamber cover on top, threading the wire from the plate through the holes on top. Then, we hang basket wire from the load cell. It is followed by connecting the load cell and thermocouple to the DAQ module. We program a LabVIEW workspace to accept and interpret data from the load cell and thermocouple. Finally, we open up the inert gas valve and set MTAirtorch to 440 °C (Fig. 4).

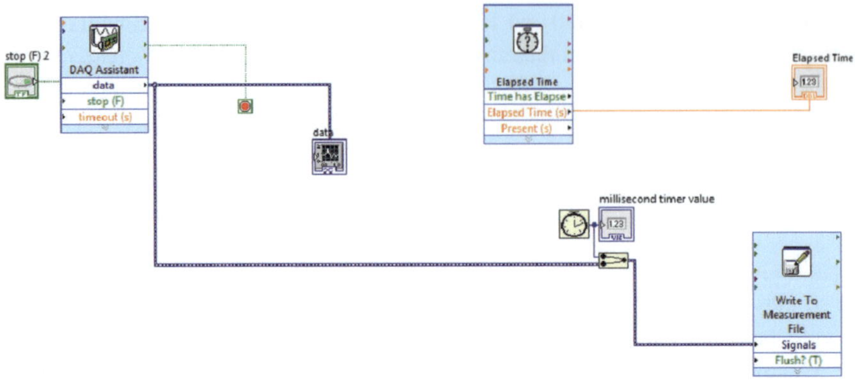

Fig. 4 LabView workspace diagram

5 Results

5.1 Initial Testing

Initial testing was done to see if the transducers were working correctly and data was being sent successfully to the computer. Each sensor was first calibrated using the LabVIEW program. Boiling water calibrated the thermocouple. Using a thermometer, the correct temperature was entered in the program alongside the voltage difference taken by the thermocouple and DAQ. This was done at five different temperatures to set up a scale that the LabVIEW program would use to interpret voltage difference from the thermocouple into heat. The load cell was similarly calibrated using five different weights.

Figure 5 shows one of the 12 data runs taken from the thermocouple. The MTAirtorch was set for 440 °C. It can be seen that the temperature of the gas flowing through the chamber is not 440 °C but slightly less. This is to be expected. There are some heat loss between the MTAirtorch and the room, and it is impossible to heat a flowing fluid correctly. The temperature itself does not change much as the lowest point in this test was 432.8 °C, and the highest variation was 434.4 °C. All the other tests produced similar results.

Fig. 5 Initial thermocouple testing

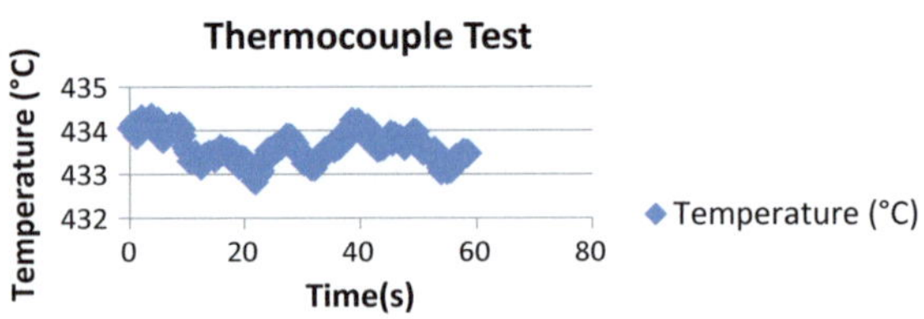

Figure 6 shows one of the 12 runs of testing done for the load cell. As can be seen, the weight of the basket and chicken manure remained stable, as was expected as pyrolysis and gasification weren't happening at this point. The variation of data around the 110 to 130-second mark is when the basket hanging wire was pulled on to see if the load cell could detect changes in mass.

5.2 *Testing*

After the transducers were properly calibrated and tested, runs were performed, and reduction of the chicken manure's mass was attempted. Figure 6 shows the data from one of those runs. It became apparent, however, that there was a problem: The mass was diminishing much to slow, and the mass itself was wrong. The number of chicken manure samples placed on the basket was more than 20 g, and it should have reduced more than just 3 g over the period of 7 min shown in Fig. 7.

After the problem was noticed, another run was done. This time a 50 g weight was added to the basket. Figure 8 shows the results of that test. It became more apparent that there was a problem. The mass itself should still have been more than 61 g, and it should not have increased at the 150- to 200-second mark. After that, the mass did reduce but still much too slowly.

Figure 9 shows the case of the nitrogen pyrolysis with the evolution of different gases at 1000 °C. The maximum peak of CO_2 is 23% of the total evolving gas volume after 2 min from the start of the reaction. CO peaks value increased significantly to 27% in the fourth minute. H_2 peak is 2% in magnitude. Other hydrocarbons contribute with the lower percentage of the volume: 10.5, 1.0, and 6.1% for CH_4, C_2H_6, and C_2H_4, respectively.

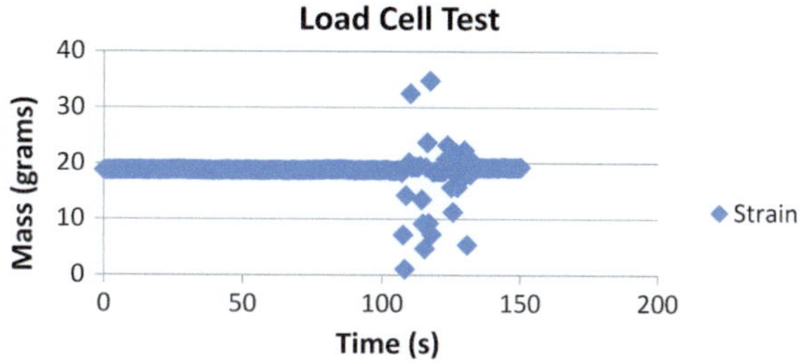

Fig. 6 Initial load cell test

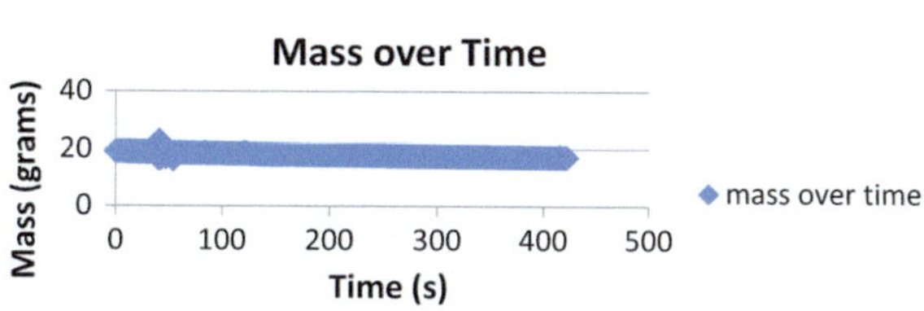

Fig. 7 Reduction of biomass data

Fig. 8 Reduction of biomass with weight added

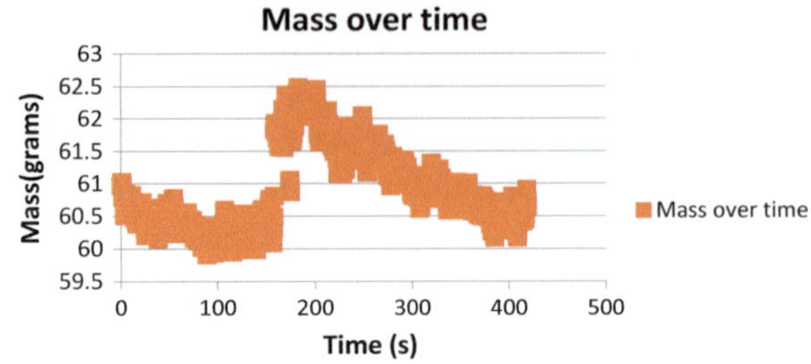

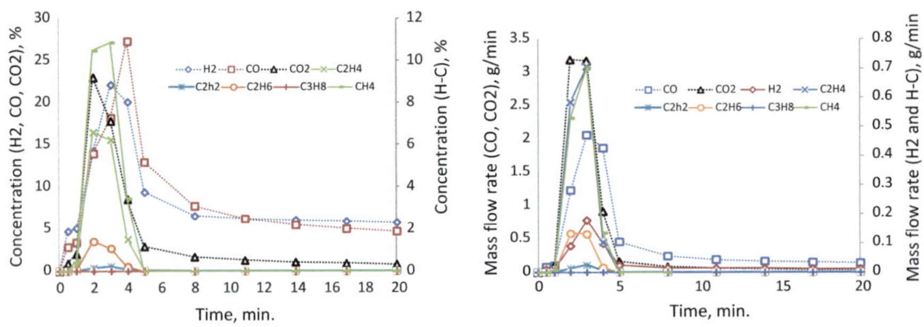

(a) Mole concentration **(b) Mass concentration**

Fig. 9 The evolution of different gas at 1000 °C

6 Conclusions

The thermogravimetric and differential thermal analyses were conducted for chicken manure using three different gasifying media (nitrogen, air, and carbon dioxide) with varying rates of heating from room temperature to high temperature. For N_2 and CO_2, the reactions were endothermic and thus energy must be supplied regarding the heating of the sample to maintain the reaction. Air gasification was exothermic, and ignition was observed at 600 °C. Thus, the reaction has the potentials to be self-sustainable with no external heating. The chicken manure thermal degradation implied the presence of the three components: hemicellulose, cellulose, and lignin. The highest reaction rates were observed at temperatures corresponding to known peak characteristics of the three components. The only air had an extra peak for ignition at 600 °C. The reaction kinetic parameters for the conversion reaction were calculated for the different media. The nitrogen pyrolysis was divided into two regions at 360 °C with the order of reaction of five for both regions. Kinetic parameters for air gasification were calculated using a third order single region reaction. CO_2 had the most complicated mechanism of the three cases and was divided into three regions at 360 and 630 °C. The kinetic parameters varied with the heating rate. It is recommended to utilize values generated by the lowest heating rate because the slow heating rate allows a quasi-equilibrium state and thus decreasing the effects of measurements errors due to delay in response or any transient condition error.

References

1. What Is Pyrolysis? Meaning of Pyrolysis. Pyrocrat Systems. Web. 26 Feb 2015. http://www.pyrolysisplant.com/what-is-pyrolysis/
2. Biomass Power Association. Biomass Power Association. N.p., 2015. Web. 07 May 2015
3. RFA: Renewable Fuels Association. RFA: Renewable Fuels Association. Renewable Fuels Association. 2015. Web. 07 May 2015
4. Greenhouse Gas Emissions Standards and Fuel Efficiency Standards for Medium- and Heavy-Duty Engins and Vehicles. Washington D.C.: US Government Printing Office, 15 Sept 2011
5. White JE, Catallo WJ, Legendre BJ (2011) Biomass pyrolysis kinetics: a comparative critical review with relevant. J Anal Appl Pyrol 91:1–33
6. Mansaray KG, Ghaly AE (1999) Determination of kinetic parameters of rice husks in oxygen using thermogravimetric analysis. Biomass Bioenerg 17:19–31
7. Yanik J, Stahl R, Troeger N, Sinag A (2012) Pyrolysis of algal biomass. J Anal Appl Pyrol
8. Mermoud F, Golfier F, Salvador S, Van De Steene L, Dirion JL (2006) Experimental and numerical study of steam gasification of a single charcoal particle. Combust Flame 145:59–79
9. Van de steen L, Tagutchou JP, Escudero Sanz FJ, Salvador S (2011) Gasification of woodchip particles: experimental and numerical study of char–H2O, char–CO2, and char–O2 reactions. Chem Eng Sci (66): 4499–4509
10. Wang S, Guo X, Wang K, Luo Z (2011) Influence of the interaction of components on the pyrolysis behavior of biomass. J Anal Appl Pyrol 91:183–189
11. Mermoud F, Salvador S, Van de steen L, Golfier F (2006) Influence of the pyrolysis heating rate on the steam gasification rate of large wood char particles. Fuel (85): 1473–1482

Sustainable Production of Bioenergy

Y. H. Taufiq-Yap, M. S. Ahmad Farabi, O. N. Syazwani, M. Lokman Ibrahim and T. S. Marliza

Keywords Sustainability · Biodiesel · Renewable energy · Economics and politics

1 Introduction

Recently, one of the major global concerns is none other than adverse environmental impact resulting from the burning of coals, fossil fuels and compressed natural gas that lead to global warming [1, 2]. Environmental climate started to

Y. H. Taufiq-Yap (✉) · M. S. Ahmad Farabi · O. N. Syazwani · T. S. Marliza
Catalysis Science and Technology Research Centre, Faculty of Science, Universiti Putra Malaysia, 43400 UPM Serdang, Selangor, Malaysia
e-mail: taufiq@upm.edu.my

Y. H. Taufiq-Yap · M. S. Ahmad Farabi · O. N. Syazwani
Department of Chemistry, Faculty of Science, Universiti Putra Malaysia, 43400 UPM Serdang, Selangor, Malaysia

M. S. Ahmad Farabi
Faculty of Science and Technology, Universiti Sains Islam Malaysia, 71800 Nilai, Negeri Sembilan, Malaysia

M. L. Ibrahim
School of Chemistry and Environment, Faculty of Applied Science, Universiti Teknologi MARA, 40450 Shah Alam, Selangor, Malaysia

Centre for Nanomaterials Research, Institute of Science, Universiti Teknologi MARA, 40450 Shah Alam, Selangor, Malaysia

T. S. Marliza
Department of Science and Technology, Faculty of Agriculture, Science and Technology, Universiti Putra Malaysia Bintulu Campus, P.O. Box 396, Nyabau Road, 97008 Bintulu, Sarawak, Malaysia

A. K. Gupta et al. (eds.), *Innovations in Sustainable Energy and Cleaner Environment*, Green Energy and Technology,
https://doi.org/10.1007/978-981-13-9012-8_24

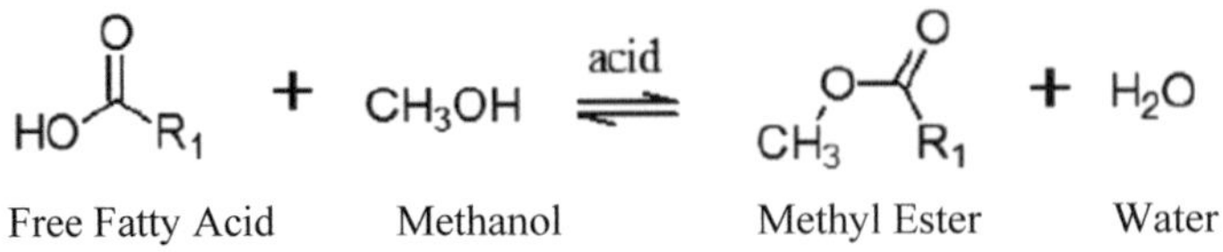

Fig. 1 Esterification reaction of free fatty acid for the formation of methyl ester or biodiesel

change and become more unpredictable due to rapid accumulation of free CO_2 gases resulted from human activities since the industrial revolution in the eighteenth century. Increase in the greenhouse gases can be seen from sudden droughts and extreme colds in a number of countries [3]. One of the alternative methods to slow down the emission of greenhouse gases is by transform into low carbon density of society. In achieving these scenarios, new alternative of energy needs to be considered in order to decrease the percentage of irreversible climate changes and also to reduce the petroleum consumption. As reported by Aguilera et al. [4], fossil fuel will be exhausted in several decades if excessively used. Nowadays, global energy demand is greatly exceeding the total energy sources that can be provided from the fossil fuels. Because of that, replacing non-renewable energy with green renewable energy is imperative in order to fulfill the needs of energy demand.

For the past few years, a number of alternative renewable energy such as wind turbines, solar panels, river dams, geothermal power and biofuels are being utilized to reduce the consumption of non-renewable energy [5]. Although the utilization of new energy still cannot support the whole energy consumption, however, since past decades the production of biofuel, ethanol and hydro-treated vegetable oil is increased significantly and became the most common and high-potential alternative fuels.

1.1 Biofuel as Green Renewable Energy

In past decade, biodiesel or also known as "first generation diesel" has become one of the most potential green renewable energy sources that exclusively derived from vegetable oil or animal fats. Generally, fatty acid methyl esters or FAMEs is a product of chemically reacting renewable lipid or fatty feedstock with short chain alcohol in the presence of acid or basic catalysts. Figure 1 shows the chemical equation of esterification reaction. At equilibrium state, 1 mol of methyl ester and 1 mol of water are produced when 1 mol of fatty acid is reacted with 1 mol of methanol. However, due to the reversible process of the reaction, an excess of alcohol or methanol is required to drive the reaction forward (desired product). FAMEs and water produced from the reaction can be separated either by using the reduced pressure distillation or by using rotary vacuum evaporation technique [6].

The chemical equation of transesterification reaction process was shown in Fig. 2. The 3 mol of methanol reacts with 1 mol of triglyceride (TG) to produce 3 mol of

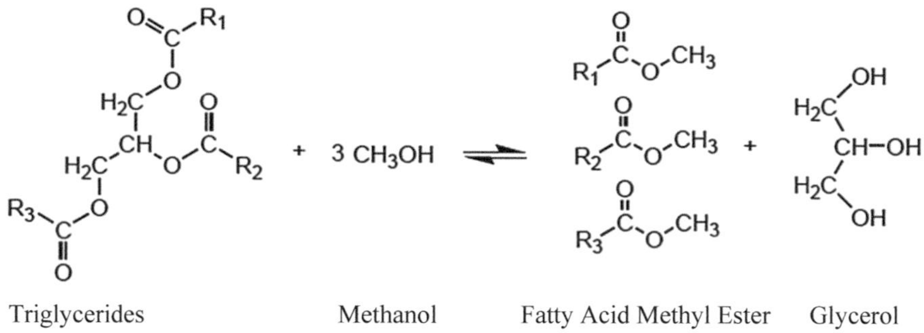

Fig. 2 Transesterification of triglycerides with methanol

FAMEs and 1 mol of glycerol. As been reported by Rashid et al. [7] and Fukuda et al. [8], an optimum amount of methanol is crucial to drive the reaction forward to increase the yield of FAME as the reaction is reversible process.

1.2 Biodiesel Production: Obstacle and Issues

Despite the fact that biodiesel gives an excellent combustion profile and is able to decrease CO_2 emission substantially, however, the full utilization of biodiesel is still not possible due to several issues and obstacles. In order to expand the utilization of biodiesel, the industry needs guarantee of continuous supply chain between raw material suppliers, industry and end consumers. Furthermore, international law and regulations on biofuels need to be liberalized and imposed free market trade in order to ensure that all exporter and importer countries get same benefit [9]. Research found that, delay in domestic utilization of biodiesel is due to poor infrastructure, economic impediment, government policy and also regulatory framework. In Southeast Asian Nation, issues and obstacles that become hindrance in large-scale production of biodiesel were in exporting the product. In 2008, EU started to implement Renewable Energy Directive (RED), which is disqualified most of the palm oil-based biodiesel company due to the requirement of minimum carbon life cycle [10].

Nevertheless, the main obstacles experienced by most of the country are the production cost and availability of feedstock. As mentioned above, the cost of biodiesel production mainly comes from the feedstock. At the beginning of biodiesel production, people have a dispute about the “food versus fuel” issues as the feedstock for the first-generation biodiesel directly came from edible raw materials. Figure 3 shows the share of feedstock used in the production of biodiesel. Almost 82% of world biodiesel production comes from the vegetable oil which is an edible oil. Aside from food vs fuel debate, biodiesel production also blamed for the excessive agricultural practice that leads to the sudden decrease in ecological resources [11].

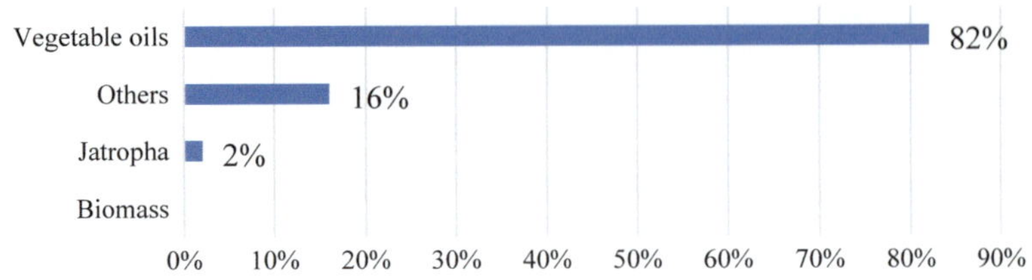

Fig. 3 Percentage of feedstock's share for biodiesel production

Specifically in Malaysia and Indonesia, rapid palm oil plantation enlargement was being linked with deforestation along with the uncontrolled CO_2 emission. A number of importer countries also being held back by the issues of obliteration of Orangutan habitat which lead to the extinction of that species [10]. In the last few years, increasing in global food price due to the fact that biodiesel sector has diverted huge percentage of food crops from the global food market also became one of the issues that hindering the larger-scale biodiesel production.

Henceforward, in order to lessen the feedstock cost and to gain the social acceptance biofuel feedstock, the biodiesel needs to be derived from non-edible feedstock. Researchers introduced the second-generation feedstock that mainly derived from non-edible material such as waste cooking oil (WCO) and palm fatty acid distillate (PFAD).

To ensure the high quality and quantity of biodiesel with cutting-edge technology advancement, biodiesel industry needs to become the technology-intensive industry. For example, in Malaysia, advancement in DNA decoding has led the biodiesel industry with new oil palm breed that can increase the palm oil extraction by 100%. Malaysian Palm Oil Board (MPOB) along with Sime Darby Group is the example of research institute that brings the oil palm industry to a new level and at the same time provides enough data and infrastructure to secure the sustainability of biodiesel industry in Malaysia. The data such as the optimal time to harvest oil palm fruit and the best genetic trait of oil palm tree that can be planted is one of the techniques that introduced by the outstanding research institute for helping the larger-scale biodiesel production to reduce the production cost [10].

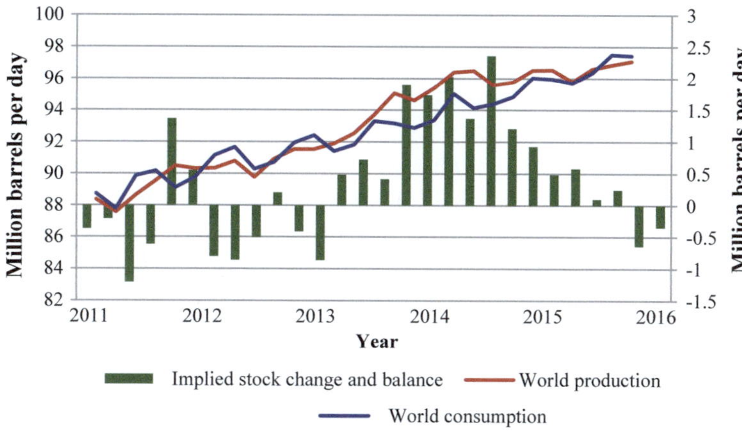

Fig. 4 Statistics of world oil production and consumption balance from 2011 to 2016

2 Sustainability of Bioenergy

Currently, transportation sectors demand on fuel resource at extremely high quantity and continue exponentially every year. Due to that scenario, the sustainability of the biofuel productivity is very important; therefore, several aspects must be considered, including (1) the biofuel's net energy balance as compared to other fuels like petroleum, (2) the relation between the bioenergy feedstock and soil-building exercises, (3) the economic and technological compatibility and (4) the complete preparation and acceptance by the government. The particular points are fundamentally related to each other and come to be the main factors to sustain the biofuel productions.

Figure 4 shows the latest statistics of world oil production and their consumption in 5 years starting from 2011. It was observed that within 5 years, the oil production and consumption experienced the drastic increment on supply and use demonstrated the high demand of the energy resource due to the vastly economic growth and the introduction of advanced technology in various industries. However, according to the BP's statistical review in 2009, the oil discoveries already started to reduce since 1960 and continuing declining around 4.5–6% per annum, meaning that our world will enter the new phase of depletion of oil resource.

Because of that reason, the production of biofuel is crucial to balance the demand of energy in various industries, especially transportation sector. As a result, from that case, the sustainability of biofuel production becomes a prominent and the most important aspect. It was reported that there are few ways to increase and to maintain the production of biodiesel such as increase the quantity of the feedstock supply, lower the cost of the biodiesel production and increase the acceptance by the user and government.

3 The Application of Biodiesel as Transportation Fuel

The transportation sector plays an important role in daily activities around the world. In fact, it is also the major components of globalization and gives an enormous contribution to the economy. The energy used in the transportation sector includes the energy consumed in transporting people and goods by road, rail, marine, air, water and pipeline. Road vehicles such as small trucks, light-duty vehicles, automobiles, minivans, sport utility vehicles and motorbikes as well as medium and heavy-duty vehicles, such as buses used for passenger travel and large trucks used for freight moving are one of the fastest growing energy sources and monopolize global oil consumption [12]. The number of cars on the road has increased from five million after the Second World War to nearly one billion today and is expected to reach two billion in the following 20 years [13].

Figure 5 shows the world's total transportation energy consumption for OECD nations and the non-OECD nations accounted for 55 and 45% in 2012, respectively. It was expected that by the year 2020, both nations will have an equal projection of world transportation energy. Due to the high rise of living cost and use of freight transport to meet consumer requests for goods in non-OECD nations, the non-OECD demand for fuels continues to overcome OECD demand and is expected to account for 61% of global transportation energy consumption in 2040. Meanwhile, OECD nations only show an average increase in total transportation energy use with only 0.2% per year from 2012 to 2040. This is may be due to the well-established consuming patterns, comparatively poor economic and population growth rates and vehicle efficiency improvements [14].

Basically, the transportation sector is accountable for more than 50% of total world oil demand and will be the greatest growing energy consumption sector in the future. Between 2006 and 2030, around three quarters of the projected increase in oil demand is mainly dominated by transportation sector. Total world liquid consumption in the

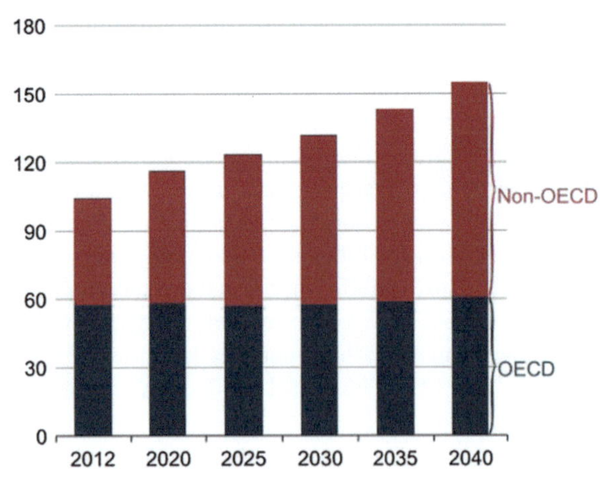

Fig. 5 Delivered transportation energy consumption by country grouping, 2012–2040 (quadrillion Btu) [14]

transportation sector increased to 53% in 2007 and projected to reach about 61% in 2035 accounting for 87% of the total world liquid consumption between 2007 and 2035 [12].

To be general, this activity is highly dependent on fossil, especially petroleum-based fuels, for instance, gasoline, diesel fuel, liquefied petroleum gas (LPG) and NG. Petroleum and other liquid fuels maintain as the biggest energy's sources, although their energy consumption declines from 33% in 2012 to 30% in 2040. International Energy Outlook (IEA) 2016 stated that liquid fuel consumption grows by 1.1% per year in the transportation sector, and by 1.0% per year in the industrial sector, from 2012 to 2040 worldwide [14].

Unfortunately, the use of petroleum-based fuel was not sustainable and green to our environment since it is not renewable, not biodegradable and produces emissions. There are two sectors that produced nearly two-thirds of world CO_2 emissions in 2013: electricity and heat generation, by far the largest, which accounted for 42%, while transport accounted for 23%. In fact, the road transport which majorly dependent on petroleum-based fossil fuel shows the increasing CO_2 emissions from 1990 to 2013 [15].

Therefore, in order to bring sustainability into the environment as well as reduce the dependency of petroleum-based fuel, there are vital to find alternative fuel such as biofuel. There are two potential liquid biofuels that can be used for transportation and considered to replace petroleum-based fuel: bioethanol and biodiesel [16]. The report from IEA's Energy Technology Perspectives suggested that, demand of bioethanol and biodiesel could contribute about 13% fuel in global transportation and 6% of global emission reductions by 2050 [15].

Biodiesel has gained a lot of attention since they are renewable, biodegradable, non-toxic and produce less sulfur dioxide with lower (GHG) emissions compared to fossil fuels [17, 18]. The most advantage of biodiesel is that it comes from renewable sources, unlike petroleum products whose underground sources are being rapidly depleted. Coming from vegetable and animal fat, biodiesel can be produced according to demand and helps to recycle waste oils, hence reducing wastage. Besides, biodiesel also produces less emission of greenhouse gases such as CO_2, HC, SO_x and No_x [19] and leads to the complete combustion and reduced emission due to having a lot of free oxygen [20].

Over hundred years, Rudolf Diesel, a French thermal engineer designed a prototype of the diesel engine and tested it with peanut (Arachis hypogaea) oil for the first time. Direct use of vegetable oils in the engine will lead to poor atomization due to low volatility and high viscosity. There are many ways to reduce the vegetable oil viscosity in order to be used in diesel engine such as microemulsion, pyrolysis, dilution and transesterification. The most facile and productive method is transesterification or esterification of vegetable oils and animal fats with simple alcohol producing biodiesel and a by-product glycerol [21]. This simple reaction can take place either with or without the presence of catalysts or enzymes.

Biodiesel can be used as a pure fuel (B100) or blended with petroleum in any percentage. It offers full blending potential with conventional diesel where a high cetane number giving improved combustion in compression ignition engines, and low

Table 1 Parameters of ASTM D6751-09 quality standards

Property of the fuel	Biodiesel
Standard method	ASTM D6751
Fuel composition	FAME ($C_{12}-C_{22}$)
Density (g/cm^3)	0.878
Pour point (°C)	−15–16
Cloud point (°C)	−3–12
Flash point (°C)	100–170
Cetane number	48–60
Water (vol.%)	0.05
Carbon (wt%)	77
Hydrogen (wt%)	12
Oxygen (wt%)	11
Sulfur (wt%)	0.05

emissions of sulfur and particulates. B20 (a blend of 20% by volume biodiesel with 80% by volume petroleum diesel), B2 (a blend of 2% by volume biodiesel with 98% by volume petroleum diesel) and B5 (a blend of 5% by volume biodiesel with 95% by volume petroleum diesel) are common fuel blends used today. This biofuel works in any diesel engine with few or no modifications to the engine or the fuel system. Most major engine companies have stated formally that the use of blends up to B20 will not void their parts and workmanship warranties. This includes blends below 20% biodiesel, such as the 2% biodiesel blends that are becoming more common. Shahid and Jamal have reviewed a range of vegetable resources to investigate their usefulness for biodiesel production. They concluded that using a mixture of petroleum diesel and biodiesel, at an 80:20 ratio (B20) was the most preferable [22]. Table 1 represents the parameters that must be met to achieve ASTM D6751-09 quality standards.

4 Current Production of Biodiesel

To date, commercial biodiesel production has been well established and is available to be purchased as turn-key plants in many countries [19]. The world's largest biodiesel producer, European Union, has produced biodiesel for an industrial scale since 1992. To date, there are more than 256 biodiesel plants in the EU producing about more than 11 million tonnes of biodiesel yearly with an average capacity consumption of 53%. These plants are dominantly located in Germany and France [23]. In 2015, EU has increased their biodiesel production with 11.8 billion liters of biodiesel of top 16 countries in the world for biodiesel production as presented in Table 2. The highest producer of biodiesel in the world was led by the USA, which produced 4.8 billion liters of biodiesel followed by Brazil with 4.1 billion liters in

Table 2 Biodiesel global production, top 16 countries and EU-28 in 2015 [24]

Country	Biodiesel production (billion liters)
United States	4.8
Brazil	4.1
Germany	2.8
Netherlands	1.5
France	2.4
China	0.4
Argentina	2.1
Thailand	1.2
Singapore	1.0
Canada	0.3
Indonesia	1.7
Spain	0.6
Colombia	0.6
Belgium	0.4
India	0.1
Malaysia	0.7
EU-28	11.5
World	30.1

2015. Other countries with significant biodiesel markets in 2015 included Germany, France, Indonesia and Brazil. Meanwhile, in Asia countries, Indonesia is the largest biodiesel producer followed by Thailand, and Malaysia as shown in Table 2. The total biodiesel production of these countries mainly came from palm oil plantation as Indonesia was the biggest producer of palm oil followed by Malaysia. On the other hand, biodiesel feedstock for India basically comes from Jatropha which is one of the potential non-edible vegetable oils to be used as biodiesel feedstock.

As well known, the USA is one of the largest producers of biodiesel in the world by producing 135 million gallons in June 2016 as shown in Fig. 6 of US monthly biodiesel production since 2014 to 2016. It was 2 million gallons higher than production in May 2016. The production came from 95 biodiesel plants with the capacity of 2.1 billion gallons per year [25].

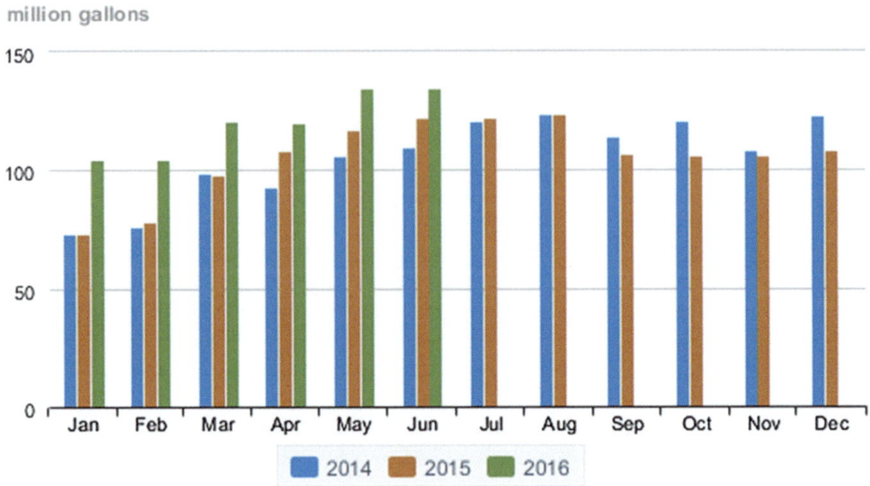

Fig. 6 US monthly biodiesel production 2014–2016 [14]

5 Development of Crop Production

5.1 Edible and Non-edible Oils

Recently, more than 350 vegetable oils have been investigated as a suitable biodiesel feedstock in the world. Currently, the high-quality feedstocks that are used for biodiesel production are edible vegetable oils such as soybean oil, rapeseed oil and palm oil from the USA, European and Malaysia, respectively. Table 3 shows more different feedstocks. Currently, Malaysia is the second largest palm oil producer in the world after Indonesia by producing over 17 million tonnes of crude palm oil (CPO) every year from 4.69 million hectares of palm oil plantations [26].

Utilization of the first-generation biodiesel creates economic, environmental and political concerns; as a result, additional arable agricultural lands are needed to increase production capacity [27]. However, lack of arable lands led to severe food shortage toward the group of developing countries such as Asia, Africa and South America. This is due to the high demand of the food by their large populations. Today, about 1% (approximately around 14 million hectares) of the available arable land in the world is used for the production of biofuels, which provides only 1% of global transport fuels. Therefore, the production of the first-generation biofuel crops is unmanageable and will affect the resources of food supply and the enormous production area land required [28].

Since the first-generation biodiesel is not suitable to replace petroleum-based fuel, the second-generation biodiesel appears as an alternative feedstock for biodiesel production. Meanwhile, the second-generation biodiesel primarily is a non-edible oil which includes Jatropha, grass, karanja, straw, rubber seed oil, cassava and many

Table 3 Main feedstocks of biodiesel [13, 21, 22]

Edible oils	Non-edible oils	Animal fats	Other sources
Soybeans (*Glycine max*)	*Jatropha curcas* Mahua (*Madhuca indica*)	Pork lard	Bacteria
Rapeseed (*Brassica napus L.*)	Pongamia (*Pongamia pinnata*)	Beef tallow	Algae (*Cyanobacteria*)
Safflower	Camelina (*Camelina Sativa*)	Poultry fat	Switchgrass
Rice bran oil (*Oryza sativum*)	Cotton seed (*Gossypium hirsutum*)	Fish oil	Terpenes Poplar
Barley	Karanja or honge (*Pongamia pinnata*)	Chicken fat	Microalgae (*Chlorella vulgaris*)
Groundnut	Cumaru		Miscanthus
Sorghum	*Cynara cardunculus*		Latexes
Wheat	*Abutilon muticum*		Fungi
Corn	Neem (*Azadirachta indica*)		
Coconut	Jojoba (*Simmondsia chinensis*)		
Sesame (*Sesamum indicum L.*)	Passion seed (*Passiflora edulis*)		
Peanut	Moringa (*Moringa oleifera*)		
Palm and palm kernel (*Elaeis guineensis*)	Tobacco seed		
Sunflower (*Helianthus annuus*)	Rubber seed tree (*Hevea brasiliensis*)		
Canola	Salmon oil		
	Tall (*Carnegiea gigantea*)		
	Coffee ground (*Coffea arabica*)		
	Nagchampa (*Calophyllum inophyllum*)		
	Croton megalocarpus		
	Pachira glabra		
	Aleurites moluccana		
	Terminalia belerica		

more. As shown in Table 3, there are more than 20 non-edible oils, which identified as potential biodiesel feedstock. Another example of second-generation feedstock is such the animal fats (e.g., beef tallow and pork lard).

5.2 Waste Materials

The major cost in the biodiesel production basically depends on the price of the feedstock. It was reported that the biodiesel costs in the range of 1.25–2.5 USD a gallon excluding the taxes, transportation distribution and the feedstock costs [29]. Because of that, the utilization of waste material edible oils or low-cost material as biodiesel feedstock are suggested to lower down the production cost. For example of waste materials are palm fatty acid distillate (PFAD), waste cooking oil, grease and soapstocks. In addition they also could avoid from the competition of food versus fuel supply, as well as reduce wastage disposal problem. Statistically, the EU was reported to produce about 0.7–1.0 million tonnes: Turkey 350,000 tonnes and Canada 120,000 tonnes of waste oils. Those uncollected oils normally go to waste through sinks and garbage and finally seep into the soil and water sources [30, 31].

For that reason, researchers focused on microalgae as the third-generation biofuels. Microalgae are considered to have potential as alternative renewable energy resources for the production of biofuel and substituting fossil fuels. Microalgae potential for biodiesel production is about 15–300 times more compared to conventional crops on an area basis (Alam et al. [27]). Microalgae are expected to be the most promising biodiesel feedstock due to its high oil lipid yield, easily cultivated and will not compete with food-based oil and arable land.

6 How to Maintain the Sustainability of Bioenergy?

6.1 Quantity of the Feedstock and Product

Nowadays, the biodiesel industries experienced a huge improvement in many aspects including technologies and management in production, thanks to raw material supplies that have assisted a production and created a line on margins for biodiesel producers. Because of the reduced dollar value, global biofuel production had demand for lower price of commodities and larger quantity feedstock to produce more product with good quality . As being discussed previously, there are numerous numbers of feedstocks that have quite a high potential such as microalgae oil, palm oil, corn oil, soybean oil, canola oil, Jatropha oil and more as shown in Table 4, with the expected amount of oil per acre of plantation. It is stated that the third generation of biodiesel from microalgae oil has very high potential to produce and supply more biodiesel with estimated production about 5000 gallons of oil per acre of land, equivalent to

Table 4 Types of potential plant (biodiesel's feedstock) and amount of product per acre [32]

Plant	Gal oil/acre	Plant	Gal oil/acre
Microalgae oil	5000*	Rice	85
Oil palm	610	Buffalo gourd	81
Macauba palm	461	Safflower	80
Pequi	383	Crambe	72
Buriti palm	335	Sesame	71
Oiticica	307	Camelina	60
Coconut	276	Mustard	59
Avocado	270	Coriander	55
Brazil nut	245	Pumpkinseed	55
Macadamia nut	230	Euphorbia	54
Jatropha	194	Hazelnut	49
Babassu palm	188	Linseed	49
Jojoba	186	Coffee	47
Pecan	183	Soybean	46
Bacuri	146	Hemp	37
Castor bean	145	Cotton	33
Gopher plant	137	Calendula	31
Piassava	136	Kenaf	28
Olive tree	124	Rubber seed	26
Rapeseed	122	Lupine	24
Opium poppy	119	Palm	23
Peanut	109	Oat	22
Coca	105	Cashew nut	18
Sunflower	98	Corn	18
Tung oil tree	96		

8.2 times higher than production of biodiesel from palm oil. Nowadays, numbers of research groups in Malaysia and other countries are focusing on the development of microalgae as the potential source for biodiesel production.

Basically, the selection on the starting material is depending on the region and climate of the country, for example, normal biodiesel has cloud point at around 4–16 °C, which depends on the amount of FFA consist in the product. If the feedstock contained high FFA component (e.g., palm fatty acid distillate and waste cooking oil), the biodiesel product should also have higher cloud point which will make them not suitable for the winter climate country. At cold temperature, the diesel fuel will form wax crystal that causes the product being clouded and also affects the performance of the engine. In 2009, the American Society for Testing and Materials (ASTM) suggested that for commercialized biodiesel, the cloud point must not exceed the

38 °F or 3.33 °C for winter grade of biodiesel oil. Otherwise, for the high cloud point biodiesel must be treated with the winterizing additives formulated such as blended with the 80% of petro-diesel to produce B20 of biodiesel with lowered cloud point [32].

6.2 The Compatibility of Economics and Policies

The commercialization of biodiesel to the market has been through a fluctuation and slow growth in the past few years. Since 2006–2007, biodiesel experienced a rapid growth for several years and caused the spikes in commodity price and collapse of the petroleum prices, which gave the hard lesson to the investor in biodiesel industries. As a result, some of the existing plants were taken down in production or closed in late 2008. Since then, few lessons and experienced had been extracted from the crisis which could be helped to assess the economics and benefiting larger-scale biodiesel production in various locations around the country.

The biodiesel industry plays an important role in order to make sure biodiesel always give an advantages to the global. One of the advantages of the biodiesel is to prolong the effectiveness of petroleum, durability and cleanliness of diesel engines. The biodiesel industry in the USA is supported by the Energy Policy Act (EPA) compliance strategy. In 2005, EPA was recognized as a number of energy management goals where EPA created Renewable Fuels Standard (RFS) to ensure a minimum volume of renewable fuel use in transportation. However, in 2007 RFS has been updated through the Energy Independence and Security Act (EISA) and renamed as RFS2. The updated standard had combined with diesel where previously it only includes gasoline. RFS2 also increased the overall usage of requirements for renewable fuel and added new categories of renewable fuel [33].

The production and use of biodiesel have increased rapidly in recent years due to the policies that aimed to enhance the energy security, reducing greenhouse gas emissions and accelerating agricultural development. However, biodiesel production recently has emerged the current issue in energy policy, rural development and natural resource management. Moreover, the conflicting views on the potential benefits of large-scale biodiesel production are also drawn attention to a range of environmental and socioeconomic issues that may arise from the large scale of biodiesel production. To replace the petroleum diesel with biodiesel, the food price and food versus fuel supply must be lowered and solved, respectively. Other than that, the impacts of emitted gas from burning of biodiesel towards the air and water qualities which are related to climate change should be studied and clarified.

According to Malaysia's National Biofuel Policy 2006 released on March 21, 2006, the five pillars of Strategic Trusts had been introduced and became the main guide for commercialization. They are:

1. Biofuel for transport,
2. Biofuel for industry,

3. Biofuel technologies,
4. Biofuel for export, and
5. Biofuel for cleaner environment.

Malaysia also introduced the Malaysian Biofuel Industry Acts 2007 (with the main objective to further regulate and facilitate the sector development of biodiesel that only limited to the palm olein and methyl ester. The act came into force on November 1, 2008, and provided for mandatory blending of biofuel with petroleum diesel and licensing of downstream activities, such as production, blending, storage, transportation, and export. This act forcing the manufacturer to apply two types of licenses: manufacturing license from the Malaysian Industrial Development Authority and license to use palm oil raw materials at the manufacturing plant from the Malaysia Palm Oil Board (MPOB). Furthermore, several initiatives had also been launched by the Government of Malaysia:

1. National Biomass Strategy (NBS) 2020:
2. To promote and intensify the utilization of biomass for high-value products,
3. To reduce the nation's CO_2 emissions,
4. To create more job opportunities,
5. To focus on oil palm biomass as a starting point,
6. To encourage the investor and plantation owners to create awareness on downstream opportunities,
7. Economic Transformation Program (ETP) 6.0:
8. Main objective to turn Malaysia into high-income country by 2020 by extending the production of palm bio-based products.
9. The ministry of energy, green technology and water:
10. To increase biomass power generation with the capacity of 800 MW by 2020 (500 MW of it is to be generated from oil palm biomass).

Figure 7 shows seven (7) benefits that could result when Malaysia Biodiesel Industry Acts 2017 is carried out.

The listed acts, initiatives and programs initiated by the Malaysia resulted positive trend of biodiesel production by producing high quantity of biodiesel in past several years and has known as one of the largest palm oil producers (producing 15 million tonnes a year) and largest exporters of palm oil (export 13.4 million tonnes a year) with the value of production increased up to 45 billion ringgit Malaysia or approximately around 11 billion USD. Perhaps, in the way to achieve the sustainability of bioenergy worldwide, these acts could be used and implemented to increase the bioenergy production for the other developing countries in the world who producing bioenergy from other crops such as corn, rapeseed, jatropha oils.

6.3 Politics and Acceptance of the Country

Bioethanol and biodiesel are the two most promising biofuel products as an alternative fuel in transportation. Therefore, the production of biofuel continues to be

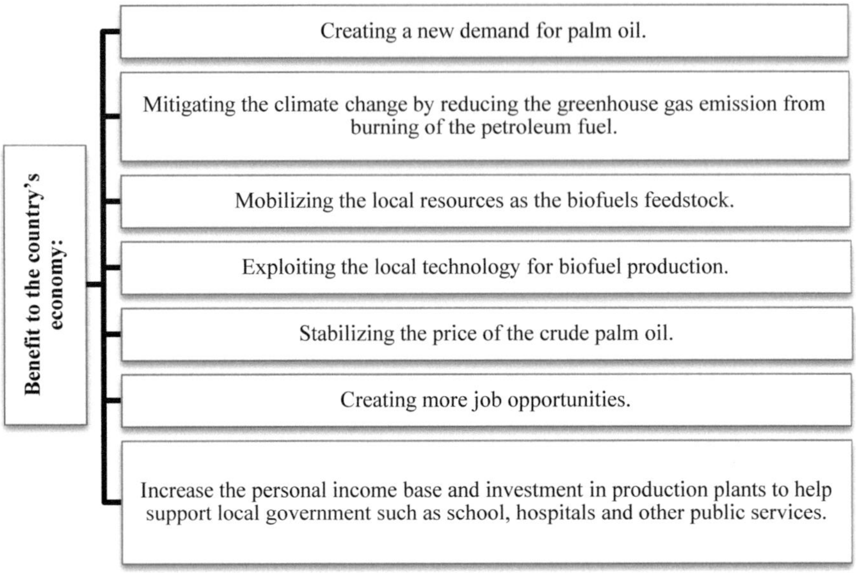

Fig. 7 List of expected benefits to the economy of the country

attractive in the global energy market. As increasing the use of biofuel as well as biodiesel, the governments are responsible to take initiatives to evaluate the potential of biofuel to the impacts on food security, especially to the poor countries. The environmental and ecological impacts also required further assessments at regional, national and international levels by monitoring of water resource quantity and quality including water pollution. The full life cycle assessments of land, forest clearing, and alternative feedstocks are also required to evaluate in terms of GHG emissions which resulting from biofuels.

The food price has increased over the past 5 years, and a number of reports suggest that factors associated with food price rise throughout the world [34–36]. Biodiesel feedstock can come from a variety of agricultural crops and many crops which produce oils can be used for biodiesel production. Since 2005, worldwide ethanol production has nearly doubled and biodiesel production has increased nearly three-fold. Brazil and the USA account for most of the world's ethanol production. Brazil uses sugarcane as a feedstock, while the USA uses nearly all corn. The European Union is the largest biodiesel producer, and rapeseed oil is its main feedstock [35]. The increment in US biofuels production in 2006–2008 was raised up nearly 14% of the escalation in corn prices while the soybean prices (soybeans and soybean oil) are also increased nearly 10% of the rise in soybean prices [34].

Over the last decade, very few studies provide a qualitative, quantitative and fully comprehensive description relative to the price of food commodity and the cause of price increased. One of the investigations conducted by Armah et al. [37] showed that the key reasons related to the increase of food prices are due to the increase of energy costs and the weakening of the US dollar. Besides, the conditions of global weather also affected the price of foods. For example, the severe weather events in 2007 such as floods during harvest in Northern Europe, drought in Southern Europe, Russia, Ukraine, Australia and Northwest Africa, late freeze followed by sudden drought in Argentina and hot dry growing conditions in Canada were affected yields worldwide [35].

Another concern on the impact of biofuel production is about carbon loadings and GHG emissions due to land use shifts resulting from biofuels. An evaluation of forest covered by Righelato and Spracklen (2007) reported that the substitution of ethanol and biodiesel for gasoline and diesel by 10% in the USA needs 43% of current US cropland for the biofuel feedstocks. However, a suggestion to use the cellulosic feedstocks as second-generation biofuels could be achieved and would give positive or neutral effects on GHG emissions.

Social acceptance of biodiesel. Social acceptance of biodiesel could be expressed in two forms which are active and passive acceptance. The clarification of these forms of acceptance has been studied by Sauter and Watson [38]; they reported that the social acceptance had been widely seen as passive consent rather than active by the public for a particular renewable energy project. The unawareness of social acceptance issue and wrong information may cause the misunderstanding of renewable energy technology.

In New Zealand, the acceptance of the biodiesel is reported by a survey on the various respondents from the engine manufacturers concerning the use of blends with higher biodiesel content. In summary, engine manufacturers of light vehicles generally accept the use of blends containing up to 5% biodiesel but often refer to the European standard for biodiesel (EN 14214). The BMW motor has moved to be consistent with a recent change in the European standard for diesel (EN 590:2009) and accepted the use of blends contained up to 7% of biodiesel. While, for the heavy vehicle engines, the Japanese-origin engine manufacturers have to limit the maximum blend level up to 5% of biodiesel only, but some other engine manufacturers accepted the use of blends with higher biodiesel content for some model of a specific engine [39].

Another survey was conducted on 571 face-to-face interviews with the public in the three main cities in Greece. The interviews were covered views on energy dependency and biofuels, climate change and collected socioeconomic data. The surveyed data has demonstrated that almost 90% of people believed that the climate change is related to fossil fuel consumption. Meanwhile, 50% of participants think that biofuels could be an effective solution and a quarter of them believed that priority should be given to biofuels compared to the other forms of renewable energy sources. The findings demonstrated that there is a lack of information about biofuels within the public where only 23.8% of participants knew the difference between bioethanol and biodiesel [40].

The social acceptance of biodiesel in Malaysia also has difficulty when the government started the biofuel program known as Envo Diesel program which encourages the people to use the processed palm olein instead of palm methyl ester due to its advantages in terms of production cost and process time. However, the used direct

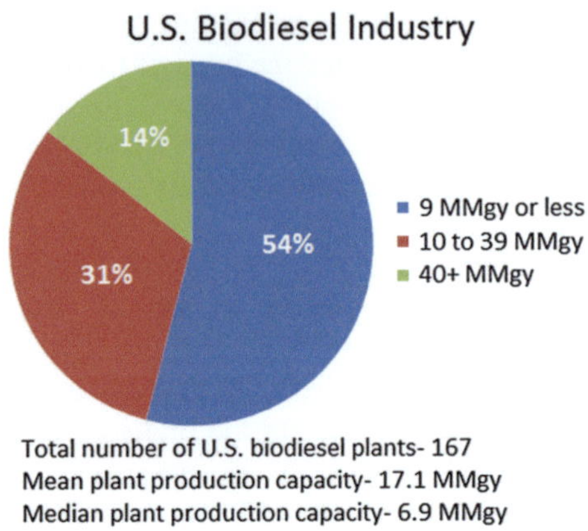

Fig. 8 Size of plant across US industries [43]

processed palm olein may cause clogging to the engine and this drawback discourages the engine manufacturers to use Envo Diesel [41]. However, due to lack of expertise in engine manufacturing, Malaysia's government had to change the plan from Envo Diesel to palm oil methyl ester biodiesel or known as B5 program, which was launched on June 2011 [42]. The acceptance issue is not only raised among engine manufactures but also stretch up from the company oil that concerns on their company profit, if the government implements the biodiesel engine instead of petrol diesel. This phenomenon will cause chaos to the petroleum industries, due to the competition with the biodiesel demand. Thus, social acceptance is going to look as a continuous issue and this issue can only be resolved when the public are fully understand on the particular situation.

7 Larger Scale

The biodiesel industry in the USA has significantly grown over the past decade through a significant increase in annual and capacity production and a transition from smaller-scale plants to larger-scale continuous producers. The pie chart in Fig. 8 highlighted the US biodiesel industry plant capacities in a year. From 167 biodiesel plants tracked by Biodiesel Magazine, it was observed that 54% of the total plants produce around 9 MMgy. Meanwhile, 31% of the plants produce around 10–39 MMgy of biodiesel. And the rest of the plants produce more than 40 MMgy yearly [43].

Moazami et al. [44] were investigated the large-scale biodiesel production using microalgae biomass in indoor open ponds. The study found that by varying the impeller speed will be obtained the higher lipid content and biomass where microalgae could access more CO_2, nutrients and illumination. At the optimum condition, microalgae would growth to maximum at 46 g $L^{-1}m^{-2}$ and lipid content of 52% after 14 days. The main components of hydrocarbon fraction from the extract of microalgae oil were oleic acid (45.4%), palmitic acid (23.4%), linolenic acid (12.2%) and stearic acid (7.1%). The high content of oleic acid in the fatty acid compositions makes it fully compliant for the manufacture of good-quality biodiesel. Thus, it can be estimated that microalgae strain PTCC 6016 has the capability to produce around 130 ton h^{-1} $year^{-1}$ biomass and 60,000 L of biodiesel.

However, number of impacts need to be considered when extrapolating to a larger scale of biodiesel production. Giampietro et al. [45] have given comprehensive assessment of the feasibility for large-scale biofuel production as an alternative to fossil energy depletion. Few obstacles have been addressed:

11. Biofuel systems seem appeared to be unable to match the demand for the energy sector of a developed society.
12. Food and environmental security should give more concern to society compared to the energy security for a world population.
13. Research into new processes of biofuel production should avoid repeating the mistake of bioethanol. Biomass is important for soil conservation, and the costs involved in their harvest, agricultural residues and by-products could contribute to be a more efficient and sustainable agricultural system.
14. The economic cost of biofuel (especially in the developed countries) mostly from the labor demand per unit of energy throughput delivered.

From the sustainability perspective, the use of biodiesel offers both advantages and disadvantages toward the environmental and socioeconomic impacts. The running out of fossil energy and its harmful emissions encourage countries to turn to renewable energy and to support innovative technology for the consumption of first- and second-generation biofuel.

8 Conclusion

In conclusion, the importance of the sustainability of bioenergy toward the survival of several sectors including transportation, import-export and others is well highlighted and discussed in this chapter, since biodiesel has benefits and hold great promises as one of the sustainable energy resources. To maintain the sustainability of the bioenergy is not an impossible task and maybe well organized if the entire government and peoples who involve in this sector realize the importance of the bioenergy supply and bioenergy demand in 100 years ahead. Furthermore, good management and investment on crop plantations could supply enough feedstock resources for the biodiesel production. Also, a good economic and government compliance with the advanced technology could assist to make reality become true which is to sustain the bioenergy as a second energy source after fossil fuel.

References

1. Amigun B, Sigamoney R, von Blottnitz H (2008) Commercialisation of biofuel industry in Africa: a review. Renew Sustain Energy Rev 12:690–711
2. Demirbas A (2008) Biofuels sources, biofuel policy, biofuel economy and global biofuel projections. Energy Convers Manag 49:2106–2116
3. Lam MK, Lee KT (2011) Production of biodiesel using palm oil. Biofuels (1st ed.), Elsevier Inc, pp 353–374
4. Aguilera RF, Eggert RG, Lagos G, Tilton JE (2009) Depletion and the future availability of petroleum resources. Energy J 30(1):141–174
5. Lim S, Lee KT (2012) Implementation of biofuels in Malaysian transportation sector towards sustainable development: a case study of international cooperation between Malaysia and Japan. Renew Sustain Energy Rev 16(4):1790–1800
6. Shu Q, Gao J, Nawaz Z, Liao Y, Wang D, Wang J (2010) Synthesis of biodiesel from waste vegetable oil with large amounts of free fatty acids using a carbon-based solid acid catalyst. Appl Energy 87:2589–2596
7. Rashid U, Anwar F, Moser BR, Ashraf S (2008) Production of sunflower oil methyl esters by optimized alkali-catalyzed methanolysis. Biomass Bioenerg 32:1202–1205
8. Fukuda H, Kondo A, Noda H (2001) Biodiesel fuel production by transesterification of oils. J Biosci Bioeng 92(5):405–416
9. Lim S, Lee KT (2012) Implementation of biofuels in Malaysian transportation sector towards sustainable development: a case study of international cooperation between Malaysia and Japan. Renew Sustain Energy Rev 16:1790–1800
10. Lim S, Teong LK (2010) Recent trends, opportunities and challenges of biodiesel in Malaysia: an overview. Renew Sustain Energy Rev 14(3):938–954
11. Pinzi S, Garcia IL, De Castro MDL, Dorado G, Dorado MP (2009) The ideal vegetable oil-based biodiesel composition: a review of social. Econ Tech Implications 3:2325–2341
12. AliO M, Mamat R, Abdullah NIKR, Abdullah AA (2008) Importance of palm biodiesel as a transportation fuel in Malaysia 2 Transportation Energy 119–128
13. Atabani AE, BadruddinI A, Mekhilef S, Silitonga AS (2011) A review on global fuel economy standards, labels and technologies in the transportation sector. Renew. Sustain Energy Rev. 15: 4586–4610
14. U.S Energy information administration, international energy outlook 2016. Available at: https://www.eia.gov/outlooks/ieo/pdf/0484(2016).pdf
15. Agency E (2010) IEA statistics. Statistics
16. John RP et al (2011) Bioresource technology micro and macroalgal biomass: a renewable source for bioethanol. Biores Technol 102(1):186–193
17. Ma F, Hanna MA (1999) Biodiesel production: a review. Biores Technol 70(1):1–15
18. Fu X, Li D, Chen J, Zhang Y, Huang W, Zhu YA (2013) Microalgae residue based carbon solid acid catalyst for biodiesel production. Bioresour Technol 146: 767–770
19. Shahabuddin M, Masjuki HH, Kalam V, Hazrat MA, Mofijur MA, Nazira M, Varman AM, Liaquat M (2012) Biofuel : potential energy source in road transportation sector in Malaysia, Innovation for Sustainable and Secure Energy
20. Atabani AE, Silitonga AS, Badruddin IA, Mahlia TMI, Masjuki HH, Mekhilef SA (2012) Comprehensive review on biodiesel as an alternative energy resource and its characteristics. Renew Sustain Energy Rev. 16: 2070–2093
21. Singh SP, Singh D (2010) Biodiesel production through the use of different sources and characterization of oils and their esters as the substitute of diesel: a review. Renew Sustain Energy Rev 14:200–216
22. Shahid EM, Jamal Y (2008) A review of biodiesel as vehicular fuel 12: 2484–2494
23. Eryilmaz T, Yesilyurt MK, Cesur C, Gokdogan O (2016) Biodiesel production potential from oil seeds in Turkey. Renew Sustain Energy Rev 58:842–851
24. Wetstone G, Thornton K, Hinrichs-rahlwes R, Sawyer S, Sander, Taylor MR (2016) REN21. Renewables 2016 Global Status Report

25. U.S. Energy information administration, Monthly biodiesel production report., (January). Available at: http://www.eia.gov/biofuels/biodiesel/production/(2015)
26. Fan X, Burton R (2009) Recent development of biodiesel feedstocks and the applications of glycerol: a review. Open Fuels Energy Sci J 2(1):100–109
27. Alam F, Date A, Rasjidin R, Mobin S, Moria H (2012) Biofuel from algae-is it a viable alternative? Procedia Eng 49:221–227
28. Nur Syazwani O, Rashid U, Taufiq Yap YH (2015) Low-cost solid catalyst derived from waste Cyrtopleura costata (Angel Wing Shell) for biodiesel production using microalgae oil. Energy Convers Manage 101: 749–756
29. Jaeger WK, Siegel R (2018) Economics of oilseed crops and their biodiesel potential in Oregons's Willamette Valley. Special Report 1081. Oregon State University
30. Nurfitri I, Maniam GP, Hindryawati N, Yusoff MM, Ganesan S (2013) Potential of feedstock and catalysts from waste in biodiesel preparation: a review. Energy Convers Manag 74:395–402
31. Chai M, Tu Q, Lu M, Yang YJ (2014) Esterification pretreatment of free fatty acid in biodiesel production, from laboratory to industry. Fuel Process Technol 125:106–113
32. Tickell J (2000) From the fryer to the fuel tank: The complete guide to using vegetable oil as an alternative fuel. 3rd Edition. Tickell Energy Consulting, Tallahassee, FL. 162
33. Schnepf R, Yacobucci BD (2013) Renewable fuel standard (RFS): overview and issues
34. Baier S, Clements M, Griffiths C, Ihrig J (2009) Biofuels impact on crop and food prices: using an interactive spreadsheet, 967
35. Trostle R (2008) Global agricultural supply and demand: factors contributing to the recent increase in food
36. HLPE. Biofuels and food security. A report by the high-level panel of experts on food security and nutrition of the committee on world food security, Rome 2013
37. Armah P, Archer A, Phillips GC (2009) Drivers leading to higher food prices: biofuels are not the main factor, 330–341
38. Sauter R, Watson J (2007) Strategies for the deployment of micro-generation: implications for social acceptance. Energy Policy 35:2770–2779
39. Campbell A (2010) Acceptance of the use of biodiesel fuels in vehicle and engine applications in New Zealand (September), 1–22
40. Savvanidou E, Zervas E, Tsagkarakis K (2010) Social acceptance of biofuels. Energy Policy 38(7):3482–3488
41. Jayed MH, Masjuki HH, Kalam MA, Mahlia TMI, Husnawan M, Liaquat AM (2011) Prospects of dedicated biodiesel engine vehicles in Malaysia and Indonesia. Renew Sustain Energy Rev 15(1):220–235
42. Chin HC, Choong WW, Alwi SRW, Hakim A (2013) Issues of social acceptance on sustainable biofuel development 2030: 25–27
43. Hoagland K (2013) Comparing plant capacities of U.S. ethanol & biodiesel industries. Biomass Magazine http://biomassmagazine.com/articles/9801/comparing-plant-capacities-of-u-s-ethanol-biodiesel-industries
44. Moazami N, Ashori A, Ranjbar R, Tangestani M (2012) Large-scale biodiesel production using microalgae biomass of Nannochloropsis. Biomass Bioenerg 39:449–453
45. Giampietro M, Ulgiati S, Pimentel D (2013) Feasibility of large-scale biofuel production. Bioscience 47(9):587–600

MIX
Papier aus verantwortungsvollen Quellen
Paper from responsible sources
FSC® C105338

If you have any concerns about our products,
you can contact us on
ProductSafety@springernature.com

In case Publisher is established outside the EU,
the EU authorized representative is:
Springer Nature Customer Service Center GmbH
Europaplatz 3, 69115 Heidelberg, Germany

Printed by Libri Plureos GmbH
in Hamburg, Germany